HISTORY OF THE COELACANTH FISHES

JOIN US ON THE INTERNET VIA WWW, GOPHER, FTP OR EMAIL:

WWW: http://www.thomson.com
GOPHER: gopher.thomson.com
FTP: ftp.thomson.com
EMAIL: findit@kiosk.thomson.com

A service of I(T)P®

HISTORY OF THE COELACANTH FISHES

PETER L. FOREY

Department of Palaeontology
The Natural History Museum
London, UK

CHAPMAN & HALL
London · Weinheim · New York · Tokyo · Melbourne · Madras

Published by Chapman & Hall, an imprint of Thomson Science, 2–6 Boundary Row, London SE1 8HN, UK

Thomson Science, 2–6 Boundary Row, London SE1 8HN, UK

Thomson Science, 115 Fifth Avenue, New York, NY 10003, USA

Thomson Science, Suite 750, 400 Market Street, Philadelphia, PA 19106, USA

Thomson Science, Pappelallee 3, 69469 Weinheim, Germany

First edition 1998

© 1998 The Natural History Museum

Thomson Science is a division of International Thomson Publishing I(T)P˙

Typeset in 10/12pt Palatino by Acorn Bookwork, Salisbury, Wiltshire

Printed in Great Britain by TJ International Ltd, Padstow, Cornwall

ISBN   0 412 78480 7

A catalogue record for this book is available from the British Library

Library of Congress Catalog Card Number: 97–68944

∞ Printed on acid-free text paper, manufactured in accordance with ANSI/NISO Z39.48-1992 (Permanence of Paper).

CONTENTS

PREFACE

Coelacanths are unusual amongst vertebrates in that they were known as fossils for over 100 years before the living representative, *Latimeria*, was discovered in 1938. During that century, many fossils had been discovered and described and considered to be an interesting diversion in fish design modified from Devonian rhipidistians – the ancestors of tetrapods or land-dwelling vertebrates. When the Recent coelacanth was discovered there was, naturally, great interest centred on what it might tell us about the biology of the tetrapod ancestor. To some extent that interest has evaporated with the current consensus placing *Latimeria* and its immediate fossil relatives more distantly related to tetrapods than was first thought. And *Latimeria*, as a deep-water marine fish, seems far from an ecological starting point for Devonian tetrapods, most of which are found in shallow freshwater deposits. Nevertheless, it holds some significance in this direction because, despite a more remote affiliation with tetrapods, there are still features of anatomy and physiology shared with fishes thought to be closer to tetrapods.

Throughout this concentration on *Latimeria*, little attention had been paid to coelacanth fishes as a group. The fact that nearly all coelacanths have very similar and distinctive body shapes has distracted our attention from other parts of the anatomy. Indeed, it is usual to consider all coelacanths as a single entity – 'We've seen one so we've seen them all!' In so doing, we tend to forget that coelacanths have a history of their own, a history which stretches back to, at least, the Middle Devonian and is mapped out by well over 100 described species (whether all can still be regarded as valid is addressed here). It is the major aim of this work to document that history and to plot changes that have taken place in order to evaluate how much we can rely on *Latimeria* to be representative of the group when discussing the role of coelacanths in the history of vertebrates.

My interest in coelacanth fishes began during conversations with the late P. Humphry Greenwood (The Natural History Museum, London), whose early life in the Republic of South Africa had brought him into close contact with coelacanth research. His enthusiastic stories about the living coelacanth encouraged me to look at the fossils. Throughout this work, I have gained much knowledge from working with Colin Patterson (The Natural History Museum, London) and Brian Gardiner (Kings College, London University), both of whom have freely shared their considerable knowledge of fish anatomy and classification. Their influence is as difficult to quantify as it is pleasurable to acknowledge. Particularly, I also wish to thank Richard Cloutier (Université des Sciences et Technologies de Lille) who shared with me his knowledge of Palaeozoic coelacanths and was kind enough to read the first draft, and more recently Per Ahlberg (The Natural History Museum, London), with whom I have discussed sarcopterygians in general. I express my thanks to Prof. Hans Fricke, Max Planck Institute, Seweissen, for discussions concerning the biology and conservation of

Latimeria. Hugh Owen and Richard Jefferies (The Natural History Museum, London) have given me general scientific encouragement. Conclusions and mistakes remain mine.

A great number of people have helped in many ways with technical support. Alison Longbottom and Sally Young (The Natural History Museum, London) have prepared latex casts, line drawings and graphs; Tim Parmenter, Phil Crabb, Harry Taylor and Pat Hart (Photographic Studio, The Natural History Museum, London) have all taken photographs of specimens and drawings during numerous stages of preparation of specimens which have been carried out by Lorraine Cornish, David Gray and Andrew Doyle of the Palaeontological Conservation Unit (The Natural History Museum, London). Chuck Hollingworth (University of Wales, Bangor), Martin Tribe and Ward Cooper (Chapman & Hall, London) have all spent time editing and steering this monograph through the publication process. Lynn Millhouse helped with figure labelling. To all I give my appreciation for their skill.

Over the years many people have made coelacanth specimens available to me and a number have offered hospitality at their respective institutions and I thank the following: S.M. Andrews (formerly of Department of Geology, Royal Museums of Scotland), T. Pettigrew (Sunderland Museum), J.A. Clack and K.A. Joysey (Museum of Zoology, Cambridge University), H.C. Bjerring (Naturhistoriska Riksmuseet, Stockholm), S.E. Bendix-Almgreen and N. Bonde (respectively Mineralogical Museum and Department of Historical Geology, University of Copenhagen), P. Wellnhofer (Bayerische Staatsammlung für Paläontologie und historische Geologie, Munich), G. Viohl (Jura Museum, Eichstätt), A. Ritchie (Australian Museum, Sydney), J.A. Long (Western Australian Museum, Perth), S. Wenz, P. Janvier, D. Goujet (Muséum National d'Histoire Naturelle, Paris), K.S. Thomson (formerly of Yale Peabody Museum, New Haven), D. Hamilla (Sandusky, Ohio), L. Grande (Field Museum of Natural History, Chicago), W.E. Bemis (University of Massachusetts at Amherst), J.G. Maisey and G. Nelson (American Museum of Natural History, New York), and B. Schaeffer (formerly of the American Museum of Natural History, New York). Financial support from the J. Dee Fellowship and the Robert O. Bass Scholarship, Field Museum, Chicago is gratefully acknowledged and my thanks go to the Committees who administer these grants.

Finally, the preparation of this monograph has spanned the duration of two Keepers of Palaeontology in whose department this work was done. My thanks are due to H.W. Ball (former Keeper) and L.R.M. Cocks for their patience and encouragement.

PETER L. FOREY
London, November 1997

GLOSSARY

In this book there are a number of terms used in connection with cladistic analysis which may be unfamiliar. Cladistic analysis is briefly explained in Chapter 1 but here some of the terms that crop up in later parts of the book are defined.

Apomorphy A derived character or character state. In evolutionary terms a derived character or character state is one that is a modification of an ancestral condition. Cf. plesiomorphy. See also autapomorphy, homology, synapomorphy.

Autapomorphy A derived character or character state (apomorphy) that is restricted to a single terminal taxon in a data set. An autapomorphy at a given hierarchical level may be a synapomorphy at a less-inclusive level: for instance, if the systematic problem was interrelationships of amniotes and one of the terminal taxa were birds then 'possession of feathers' would be an autapomorphy of birds. However, if the systematic problem was interrelationships of turtles, lizards, ravens and penguins then 'feathers' would be a synapomorphy of a group 'ravens+penguins'.

Binary character A character in which there are only two states (usually designated as 0 and 1). Cf. multistate character.

Bootstrap A statistical procedure for achieving a better estimate of the parametric variance of a distribution than the observed sample variance by averaging pseudoreplicate variances. The original data set is sampled with replacement to produce a pseudoreplicate of the same number of taxa and characters as the original. Each pseudoreplicate is analysed and the taxonomic groups identified are noted. This is carried out many times and percentages of the number of times a particular group is recovered is noted. This is presumed to represent the strength of evidence supporting that particular grouping.

Clade See monophyletic group.

Cladistics A method of classification that groups taxa hierarchically into nested sets and conventionally represents these relationships as a cladogram. See also phylogenetic systematics.

Cladogram A branching diagram specifying hierarchical relationships among taxa based upon homologies (synapomorphies). A cladogram includes no connotation of ancestry and has no implied time axis. Cf. phylogenetic tree.

Consensus method A method for combining the grouping information contained in a set of equally parsimonious cladograms for the same taxa into a single topology - the consensus tree. There are various ways in which a consensus tree may be constructed. In this work only strict consensus trees are used.

Consensus tree A branching diagram produced using a consensus method.

Consistency index (ci) A measure of the amount of homoplasy in a character relative to a given cladogram. The consistency index is calculated as the ratio of m, the minimum number of steps a character can exhibit on any cladogram, to s, the minimum number of steps the same character can exhibit on the cladogram in question. See also retention index. E.g. if a character appears only once on a cladogram/tree then the consistency index is '1': if it appears twice then the coonsistency index is 0.5. The consistency index of the entire cladogram/tree (ensemble consistency index) is the sum of all the separate consistency indices divided by the number of characters.

Convergent (characters) Convergent characters are those that pass the conjunction test of homology but fail both the similarity and congruence tests. In evolutionary terms convergent characters are those inferred to have arisen from different ancestral conditions. Also known as homoiologies. See also homology, parallelism.

Derived character See apomorphy.

Distance method A method of grouping taxa on overall similarity (this includes similarity in the absence as well as the presence of characters). In molecular systematics gene sequences are compared simply by the numbers of different nucleic acids and no account is taken of the position along the molecule.

Fully resolved cladogram A cladogram which shows relationships as strictly sister-group pairs: that is, there are no polytomies.

Ghost lineage A period of time which must have been occupied by fossils presently unknown.

Heuristic algorithm An algorithm for constructing cladograms that is not guaranteed to find the most-parsimonious solution but can deal with large numbers of taxa and characters.

Homology (homologue) (1) Two characters that pass the tests of similarity (they occupy the same topological position and are made of similar tissues), conjunction (they do not occur together in the same organism at the same time) and congruence (they specify the same group as another putative homology) are termed homologous. Also known as synapomorphy. (2) Character states that share modifications from another condition, e.g. wings of birds as modified forelimbs of other tetrapods. See also convergent, parallelism.

Homoplasy (adj. homoplastic) (1) A character that specifies a different group of taxa from another character. (2) Any character that is not a synapomorphy (homology).

Illogical coding See non-applicable coding.

Ingroup The group of taxa under investigation. Cf. outgroup.

Length (of cladogram or tree) The minimum number of character changes (steps) required on a cladogram to account for the data, or the summed fit of all characters to a cladogram.

Monophyletic group (monophyly) A group that includes a most recent common ancestor plus all and only all of its descendants. A monophyletic group is diagnosed by the discovery of homologies (synapomorphies). Also known as a clade. See also paraphyly, polyphyly.

Multistate character A character in which there are more than two states (e.g. 0, 1, 2 etc). Cf. Binary character.

Non-applicable coding A coding to denote character states that cannot logically be observed in some taxa (e.g. the condition of tooth characters in modern turtles). Computer programs evaluate these as missing data coded as question marks. Also known as illogical coding.

Ordered character A multistate character of which the order has been determined. Transformation between any two adjacent states costs the same number of steps (usually one), but transformation between two non-adjacent states costs the sum of the steps between their implied adjacent states. For example, in the ordered character, 0 r 1 r 2, the transformations 0 r 1 and 1 r 2 each cost the same number of steps but the transformation 0 r 2 costs twice as many (i.e. transformation proceeds as if via state 1).

Outgroup A taxon used in a cladistic analysis for comparative purposes, usually with respect to character polarity determination. Cf. ingroup.

Outgroup comparison An indirect method of character polarization that uses the information on character states in outgroup taxa to determine the relative apomorphy and plesiomorphy of character states found in the ingroup taxa.

Parallelism Two characters that pass both the similarity test and conjunction test of homology but fail the congruence test are termed parallelisms. See also convergent, homology.

Paraphyly (paraphyletic group) A group that remains when one or more components of a monophyletic group are not included. In evolutionary terms this is a group that includes a most-recent common ancestor plus only some of its descendants (e.g. Reptilia). See also monophyly, polyphyly.

Parsimony The general scientific criterion for choosing among competing hypotheses that states that we should accept the hypothesis that explains the data most simply and efficiently.

Phylogenetic tree An hypothesis of genealogical relationships among a group of taxa with specific connotations of ancestry and an implied time axis. Cf. cladogram.

Phylogenetic systematics A method of classification that utilizes hypotheses of character transformation to group taxa hierarchically into nested sets and then interprets these relationships as a phylogenetic tree. See also cladistics.

Plesiomorphy An apomorphy of a more inclusive hierarchical level than that being considered. In evolutionary terms a plesiomorphy is an ancestral or primitive character or character state. Cf. apomorphy.

Polarity A character or transformation series is said to be polarized when the direction of character change or evolution has been specified, thereby determining the relative plesiomorphy and apomorphy of the characters or character states.

Polymorphic character A character that can show two or more states within a single taxon, e.g. eye colour within the taxon *Homo sapiens*.

Polyphyly (polyphyletic group) A group based upon homoplastic characters assumed to have been absent in the most-recent common ancestor of the group. See also monophyly, paraphyly.

Polytomy A node on a cladogram/tree which leads to more than two collateral taxa.

Primitive character See plesiomorphy.

Rescaled consistency index The product of the consistency index and the retention index of a character.

Retention index (ri) A measure of the amount of similarity in a character that can be interpreted as synapomorphy on a given cladogram. The retention index is calculated as the ratio of (g-s) to (g-m), where g is the greatest number of steps a character can exhibit on any cladogram, m is the minimum number of steps a character can exhibit on any cladogram and s is the minimum number of steps the same character can exhibit on the cladogram in question. See also consistency index.

Root The starting point or base of a cladogram or phylogenetic tree.

Rooting The process of assigning a root to a cladogram whereby a taxon is chosen as the starting point for the cladogram or phylogenetic tree.

Sister-group(s) Two taxa that are more closely related to each other than either is to a third taxon.

Steps (on a cladogram or tree) The number of times that a character must be assumed to have changed when plotted onto a cladogram or tree. Cladogram or tree length is measured as the total number of steps. See also 'length'.

Strict consensus tree A consensus tree formed from only those groupings common to all members of a set of fundamental cladograms.

Successive (approximations character) weighting An iterative procedure for weighting characters a posteriori according to their cladistic consistency, which is usually measured by the consistency index or rescaled consistency index.

Symplesiomorphy A synapomorphy of a more inclusive hierarchical level than that being considered. For instance, jaws would be a symplesiomorphy of a group frog + lizard because the character jaws specifies a larger group - gnathostomes

Synapomorphy An apomorphy that unites two or more taxa into a monophyletic group.

Unordered character A multistate character of which the order has not been determined. In an unordered character, transformation between any two states, whether adjacent or non-adjacent, costs the same number of steps (usually one, see direction). For example, in the unordered character, 0 - 1 - 2, the transformations 0 r 1, 1 r 2 and 0 r 2 all cost the same number of steps.

1

INTRODUCTION

There can be few episodes in the history of ichthyology to rival the excitement following the announcement in the *East London Dispatch* of 20 February 1939, declaring that a coelacanth had been captured a few miles off the mouth of the River Chalumnae, South Africa. The story of its capture and its immediate fate have been well documented by both Ms Courtenay-Latimer (1979, 1989), who spotted the fish amongst the contents of an otherwise normal fish trawl, and J.L.B. Smith (1956), who identified and subsequently named it as *Latimeria chalumnae* (Smith, 1939a). That story relates that the fish arrived on a hot December day (23 December 1938) at the ill-equipped East London Natural History Museum and was quickly emptied of the soft tissue by a local taxidermist. But enough remained for Smith to publish a reasonably complete description (1939b) and bring *Latimeria* and other coelacanths to the attention of the scientific community and the public. But why should a single fish have created such excitement?

The first and most obvious answer centred on the fact that *Latimeria* was recognizably a member of a group of fishes – the coelacanths – well known from the fossil record but thought to have died out some 80 million years previously in the Chalk seas of southern England. Coelacanths had long been known as fossils since Agassiz (1839) described a single specimen of a tail discovered in a road cutting through the Permian Marl Slate of Durham (p. 308). Agassiz was impressed by the fact that the fin rays supporting the tail were hollow and he

coined the name *Coelacanthus* (Gr. κοῖλος hollow ἄκανθα spine) which has come to give the name to the group. During the following 100 years, some 80 species had been described ranging in time from the Middle Devonian (360 million years) to the Upper Cretaceous (80 million years). With one or two exceptions, the body form of coelacanths appeared to have remained the same, yet distinctively shaped, particularly with respect to the position and contours of the fins and the tail. It was this conservatism that helped Smith to recognize the East London fish as a coelacanth.

The conservatism of coelacanths heightened excitement in another, more significant direction. The theory of relationships prevalent in the 1930s suggested that coelacanths were descendants of rhipidistian fishes, the group popularly thought to be ancestral to tetrapods. The coelacanth therefore was considered as the nearest living relative of land-dwelling vertebrates. Given this theory it was to be expected that some trace of rhipidistian ancestry would be 'frozen' within the Recent coelacanth. In particular it was hoped that the detailed anatomy and functioning of the fins, respiratory, circulatory, reproductive and nervous systems would elucidate details of the transition of vertebrate life from water to land. One aim of this work is to assess how far this expectation has been realized. To do this, it is necessary to re-evaluate much of the knowledge of the soft anatomy which has been gained following the discovery of *Latimeria*. Additionally, and more importantly, it is necessary to re-evaluate the theory of relationship which led to that initial expectation. In recent years, there has been considerable debate over the systematic methods by which we infer and express our ideas of relationships, and by which we evaluate the fossil record. It is therefore pertinent to ask whether these differing methods lead to a different evaluation of the significance of *Latimeria* and other coelacanths.

The apparent conservatism of coelacanths alluded to earlier has been cited in connection with more general theories of evolutionary rates. Coelacanths have been seen as classic examples of a lineage of organisms which showed an initial rapid period of change or tachytelic evolution followed by long periods of slow, or bradytelic evolution (Simpson, 1944). In this they have been compared with lungfishes (Schaeffer, 1952a,b). Such assessments were originally based on rather scanty knowledge of superficial features and on vague theories of the interrelationships of the coelacanth species. Since those assessments were made, we have had the benefit of new material and new techniques for studying that material. And this, together with the opportunity to revise our reconstructions of fossil species in the light of the Recent model, considerably improves the 'facts' used in our discussions of evolutionary rates.

Another major aim of this work is to describe, or redescribe many of the fossil species of coelacanths and to provide diagnoses for all genera. This is intended to be the most durable part of this work. The description is done in comparative fashion with chapters devoted to regional anatomy (skull roof, neurocranium, cheek, lower jaw etc.) followed by a wider discussion of skeletal anatomy leading to ideas of the polarity of character transformation within coelacanths. Such conclusions are used to construct cladograms and phylogenetic trees.

This attempt to understand more about coelacanth evolution is timely because of the many excellent and wide-ranging studies of related fishes that have been carried out in recent years. We know considerably more about Devonian lungfishes through the discovery of superb three-dimensional skeletons from the Upper Devonian Gogo Formation of Western Australia (Miles, 1977) and the Lower Devonian of eastern Australia (Campbell and Barwick, 1982, 1983, 1984a, 1985, 1987). And from the same Western Australian deposit, we have equally well-preserved palaeoniscids which have allowed us the chance to be more precise about the primitive conditions of actinopterygians (Gardiner and Bartram, 1977; Gardiner, 1984a). These works complement earlier work by Patterson (1973, 1975, 1977a) which established a detailed phylogeny and character transformations for neopterygian actinopterygian fishes. With the now classic work on 'rhipidistians' carried out by Jarvik (1942, 1954, 1972, 1980) and recently supplemented by particular work on porolepiforms by Ahlberg (1989a,b, 1991), as well as the exciting new finds of unusual fishes from China (Chang, 1982; Chang and Yu, 1984), there is a rich source of comparative information potentially useful for our understanding of the interrelationships and history of the bony fishes (Osteichthyes).

1.1 *LATIMERIA*: 'LIVING FOSSIL' OR 'MISSING LINK'

Latimeria has often been popularly described both as a 'living fossil' and as a 'missing link'. Because much of the reputation of the coelacanth hangs on these terms, it is worth considering what exactly is meant. Such catch-phrases may be colourful but they are, in fact, difficult to define, given the numerous ways in which they have been used and the diversity of implied significances.

Because fossil coelacanths were known long before the discovery of the Recent representative, the term 'living fossil' as applied to *Latimeria* is literal. But this is not what is usually meant when this label is applied to such organisms as *Limulus, Platypus, Lingula, Tuatara, Metasequoia* or *Auracaria.*

Darwin (1859: 107) coined the term 'living fossil' and he considered them to be "remnants of a once preponderant order" and to be forms "which like fossils, connect to a certain extent orders at present widely sundered in the natural scale". Huxley (1908) wrote of such animals as intercalary types – forms that are morphologically but not temporally intermediate between two other living groups. They are therefore both relicts (Simpson, 1953, 1954b) and 'missing links'. However, a glance through essays contributing to a volume on living fossils (Eldredge and Stanley, 1984) shows a great diversity of candidates and reasons for their inclusion. Eldredge and Stanley (1984) described living fossils as living species anatomically very similar to fossil species occurring very early in the history of the lineage. In other words, they stand in place of extinct organisms. This view is shared by Greenwood (p. 163 in Eldredge and Stanley, 1984), who described the bichirs *Polypterus* and *Erpetoichthys* as "having retained a great number of primitive features lost by other extinct members of the Actinopterygii". A similar view is given by Tattersall (p. 32 in Eldredge and Stanley, 1984), who described *Tupaia* as a living model of the ancestral primate: i.e. it conforms with an idea of an ancestral morphotype. For Emry and Thorington (p. 23 in Eldredge and Stanley, 1984), the tree squirrel *Sciurus* can be described as a living fossil because it is the least derived of a very diverse group (Sciuridae). All of these descriptions imply relative plesiomorphy and morphological stability, and with this comes the implication of behavioural, molecular, physiological and perhaps ecological stability.

Other descriptions of living fossils incorporate time as a necessary parameter. Both Batten (p. 218 in Eldredge and Stanley, 1984) and Hessler (p. 181 in Eldredge and Stanley, 1984) justify the inclusion of *Neopalina* and

cephalocarids as living fossils because they are modern animals occupying a phylogenetic position that suggests a long, but unknown fossil record. Viewed in this light, many animals may be described as living fossils, the criterion being a lack of, or a large gap in the fossil record. *Latimeria chalumnae* is separated from its Cretaceous sister species by some 80 million years and certainly is unusual amongst fishes in this respect (there are some more notable examples such as lampreys and hagfishes, both of which are unknown as fossils since the Upper Carboniferous, a gap of some 300 million years). Yet other descriptions of living fossils, such as lungfishes, stress the restricted and often Gondwanic geographic distributions (Burton, 1954).

From these descriptions it is clear that there is a wide variety of ideas within the single phrase 'living fossil'. Significantly, the one point of agreement among the many commentaries on living fossils is that the 'living fossil' species need not have a long fossil record. In other words, there is no implication that the 'living fossil' species should have existed any longer than any other species.

We will return in Chapter 12, via the circularity of the bulk of this book, to consider *Latimeria* as a 'living fossil' by answering the following questions.

1. Is *Latimeria* a 'missing link' between fishes and tetrapods?
2. Is *Latimeria* anatomically close to the earliest coelacanths?
3. Does *Latimeria* signify a long, unrepresented fossil record?
4. Is it an organism with an unusually restricted geographic range?

1.2 HISTORY OF STUDIES ON FOSSIL COELACANTHS

Fishes which we now recognize as coelacanths were originally described as 'fossil fishes' without any clear statement as to their affinities, either to one another or to other fishes. If comparisons were made, they were made with actinopterygians. Prior to Agassiz's pioneering studies of fossil fishes the only fishes described that we now recognize as coelacanths were the Upper Cretaceans *Amia*? *lewesiensis* (Mantell, 1822), the Tithonian *Undina penicillata* (Münster, 1834) and an un-named fish from the Keuper of Coberg (Berger, 1832). Agassiz renamed the Chalk *Amia*? *lewesiensis* as *Macropoma mantelli* (Agassiz 1835) and at about the same time he described a specimen of a tail skeleton from the Marl Slate of Durham as *Coelacanthus granulatus* (Agassiz, 1839). Agassiz named his genus *Coelacanthus* and in 1844 he gave a fuller description of *Macropoma* and *Coelacanthus*, uniting these two genera, with *Undina*, in his family 'des Coelacanthes'. Unfortunately he included as coelacanths a number of other genera that we now assign to completely different fish groups. *Holoptychius* and *Glyptolepis* (porolepiforms), *Uronemus* Agassiz (a dipnoan), *Glyptosteus* and *Phyllolepis* (placoderms), *Gyrosteus* and *Ctenolepis* (actinopterygians), and *Psammolepis* (heterostracan).

Thiollière (1858), when describing a fossil fish fauna from the Upper Jurassic of Bugey, France, recognized the heterogeneity of Agassiz's 'Coelacanthes' and restricted it to *Coelacanthus*, *Undina* and *Macropoma*, naming this group 'Ortho-coelacanthes'; this was first time that coelacanth genera were correctly brought together.

Huxley (1861) published his essay on the classification of Devonian fishes and this proved to be one of the most influential in bringing together early information about lobe-finned fishes. He shared Thiollière's views on the composition of the coelacanths and grouped them as Coelacanthini and associated them with a variety of fishes we now recognize as osteolepiforms and porolepiforms in his Saurodipterini and Glyptodipterini, the actinopterygian *Polypterus* (Polypterini), and the lungfishes *Dipterus*

(Ctenodipterini) and *Phaneropleuron* (Phaneropleurini). He called this entire group the Crossopterygii. Members of the Crossopterygii could be recognized by the possession of two dorsal fins, lobed paired fins and, as Huxley thought, the absence of branchiostegal rays (considered to be replaced by jugular bones or gular plates). Within the Crossopterygii, the coelacanths could be distinguished by the shape of the median dorsal fins (the anterior dorsal fin is sail-like, the posterior dorsal fin is lobate), the symmetrical tail, the large paired gular (jugular) plates, the shape of the pelvic bones and the ossified air bladder. Huxley made his closest comparisons between coelacanths and ctenodipterines (lungfishes) pointing out that fishes of these two groups showed circular scales, a short lower jaw and a similar but unspecified arrangement of the teeth.

Cope (1871), in his classification of fishes established primarily on the basis of fin structure, recognized coelacanths as the sole representatives of his Actinistia; thus coelacanths became co-extensive with actinistians and the two words Coelacanthini (or Coelacanthiformes) and Actinistia are synonymous. Like Huxley, Cope associated coelacanths (actinistians) with crossopterygian fishes.

Therefore, by the 1870s coelacanths had been firmly associated with the lobe-finned fishes and amongst these the coelacanths were thought to be most closely similar to porolepiforms and lungfishes (Glyptodipterini and Ctenodipterini respectively) (Huxley, 1861; Lutken, 1868).

Throughout the 19th and beginning of the 20th century, many coelacanth species had been described from a variety of localities. The most complete of these were descriptions of Upper Jurassic coelacanths from the famous Solnhofen Limestone of Germany (Münster, 1842a; Quenstedt, 1858; Willemoes-Suhm, 1869; Reis, 1888); Upper Carboniferous, Jurassic and Upper Cretaceous forms from England (Egerton, 1861; Huxley, 1866;

Wellburn, 1902a,b; Woodward, 1909; Woodward, 1916), Upper Carboniferous and Lower Triassic forms from USA (Newberry, 1856, 1878) and the Lower Triassic of Madagascar (Woodward, 1910). However, it was Reis (1888, 1892) who summarized the 19th century knowledge of coelacanth anatomy and he did this through his work on the Solnhofen coelacanths, but he refrained from offering any views concerning the relationships of coelacanths to other fish groups.

Woodward (1891, 1898a,b) added considerably to knowledge of coelacanths by providing diagnoses of all known species and placing many in synonymy. Woodward was the most knowledgeable paleaoichthyologist at the turn of the century: he had seen many collections and corresponded with many scientists, such that his views were tempered with breadth of expertise across all fish groups. He accepted coelacanths as crossopterygians (*sensu* Huxley) but he was the first to suggest that they were 'degenerate'; that is, they had reduced or lost structures (infradentaries, opercular bones, radial supports in the median fins) formerly present in their ancestry, but the ancestral phantom was never caught.

The next major advance in coelacanth research came with the discovery of several new taxa from the upper part of the Lower Triassic of Spitzbergen and described by Stensiö (1921). Very little postcranial material was available but Stensiö conducted a detailed study of the head, including descriptions of the braincase in which he recognized the peculiarly-shaped prootic and basioccipital ossifications characteristic of Mesozoic coelacanths. He gave an overall summary of coelacanth organization and a point-by-point comparison with actinopterygians, dipnoans, rhipidistians and tetrapods. He was clearly impressed by the similarities in jaw structure between dipnoans and coelacanths but, on the basis of similarities in the pattern of dermal bones of the skull roof and the cheek,

he concluded that coelacanths should be classified with rhipidistians (osteolepiforms and porolepiforms).

In 1932 Stensiö described the fishes from the Lower Triassic of Greenland. This work included a description of *Laugia*, a genus which showed a well-preserved postcranial skeleton. Stensiö (1932) also provided summary diagnoses for many of the better-known genera and provided an extended account of the braincase of a Devonian coelacanth, *Diplocercides kayseri*, which he had earlier described (1922a,b) and which was based on a wax model reconstruction of a serial grinding series. This single specimen later (Stensiö, 1937) became known as *Nesides schmidti* Stensiö and has been used as a standard reference coelacanth taxon when comparing braincases of Devonian fishes. In hindsight, there is little justification for separating this single specimen as a new genus and species, and in this work it is placed in synonymy with *Diplocercides kayseri*. Stensiö (1937) completed his work on coelacanths by returning to work on Upper Devonian coelacanths from Germany, originally discovered by von Koenen (1895) and Jaekel (1927). This time he concentrated on the dermal bones and sensory canals of the skull. Through Stensiö's extensive writing on coelacanths, there is no clear statement of relationships amongst coelacanth genera/species but, in using rhipidistian bone terminology, it is clear that he thought coelacanths and rhipidistians were close relatives.

The discovery of the living coelacanth in 1938 provided the Recent anchor by which palaeontological studies were fixed. Schaeffer (1941, 1948, 1952a), in a series of papers, documented the subtle changes in the morphology of coelacanths leading to *Latimeria*. He studied large collections of the late Triassic/early Jurassic *Diplurus* and, as a result of comparisons with other coelacanths, he was able to propose the first, reasonably complete phylogeny of coelacanth genera (Fig. 9.2). Schaeffer (1952b) suggested that the early evolution of coelacanths was marked by rapid morphological change followed by stasis and, in this sense, the history of coelacanths paralleled the history of lungfishes (Westoll, 1949). However, the phylogenetic position of coelacanths within the phylogeny of fishes remained the same: they were regarded as descendants of rhipidistian fishes with *Latimeria* still regarded as the closest living relative of tetrapods. In fact, the knowledge gained from the study of *Latimeria* (Smith, 1939c; Millot & Anthony, 1958, 1965) had no immediate impact on ideas concerning the relationships of lobe-finned fishes. The change in the ideas of relationships of lobe-finned fishes came not with new data but with a different method of classification (Rosen *et al.*, 1981).

Further advances in the study of coelacanth anatomy came with the description of Carboniferous coelacanths from Great Britain (Moy-Thomas, 1937), the Lower Triassic coelacanths from Madagascar (*Whiteia*) by Lehman (1952) and several new species described from the Upper Carboniferous of Montana (Melton 1969; Lund and Lund, 1984, 1985). Together these studies increased the range of coelacanth diversity known to exist, but they did not contribute any significant information as to how coelacanth species were interrelated, and consequently little could be said about the evolution within the group.

Several classifications of coelacanth fishes were proposed that were cast in the die of evolutionary taxonomy (Chapter 9) and recognized variants on the theme that coelacanth evolution could be thought of in grade terms, passing from primitive Devonian forms such as *Diplocercides* which retain a solidly ossified braincase equipped with a basipterygoid process, to Mesozoic forms in which the braincase shows separate ossifications. *Latimeria* was sometimes separated as a separate family, and *Laugia* was placed in its own family (sometimes with *Coccoderma*) because it differed from all other coelacanths in the anterior position of the pelvic fins. These classifications, detailed in Chapter 9, mixed gradal classification with the occasional apomorphy-based clade.

More recent classifications (Forey, 1981, 1984, 1988, 1991; Cloutier, 1991a,b) are based on cladistic methodology and are continued here (Chapter 9). These classifications, which are in broad agreement with one another and reflect complementary studies (Cloutier on Palaeozoic forms, Forey on Mesozoic and Recent), recognize an essentially pectinate tree for Palaeozoic forms with two major monophyletic groups being recognized for the Mesozoic and Recent forms. Although these classifications raise problems of expressing results within the constraints of Linnaean taxonomy they, more importantly, demonstrate clear evolutionary trends throughout the group (Chapter 9). Many of our ideas of coelacanth evolution have been based on very patchy knowledge. To some extent this is a consequence of the fossil record. But to a greater degree this is lack of study of available material. The data matrix, which forms the basis for the classification and deduction of evolutionary patterns given here (Chapter 9), summarizes the major aim of this work.

A great deal of work has been directed towards describing the anatomy of *Latimeria*. This has been done chiefly by Smith (1939c), Millot and Anthony (1958, 1965) and Millot *et al.* (1978). There is, however, a vast amount of subsidiary literature on various aspects of anatomy, physiology, biochemistry and popular appeal, and this is usefully summarized by Bruton *et al.* (1991). Two major multiauthored volumes have been published about *Latimeria* (McCosker and Lagois, 1979; Musick *et al.*, 1991).

Chapter 2 attempts to summarize knowledge about the natural history of *Latimeria* and emphasizes behaviour, physiology and some of the conservation problems faced by this species. Chapters 3 to 8 describe the anatomy of fossil coelacanths and this infor-

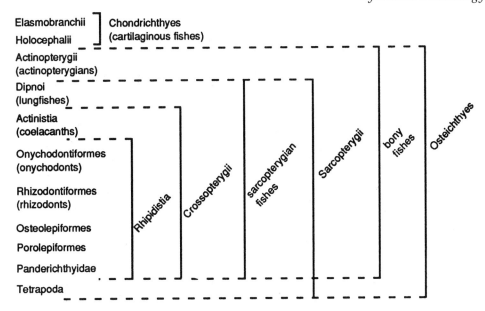

Fig. 1.1 Figure to show the relationship of group names used throughout the text.

mation is used in Chapter 9 to contruct a phylogenetic tree from which statements of evolutionary trends and rates can be made. The systematic position of coelacanths is discussed in Chapter 10. Chapter 11 gives diagnoses of coelacanth taxa. Throughout, a cladistic method of classification is followed (Hennig, 1966).

Throughout the text, a variety of fish group names are used, usually in the vernacular. Figure 1.1 shows a brief summary of the taxonomic meanings of these words. They are a mixture of paraphyletic and monophyletic groups.

1.3 SYSTEMATIC METHODOLOGY

As has been explained above, the reputation of *Latimeria* as a living fossil and as a missing link is based, largely, on an understanding of the evolutionary history of coelacanths and the relationships of coelacanths to other fishes and tetrapods. In order to reconstruct this history a phylogenetic

classification based on morphological observations is needed. At the ends of most of the chapters dealing with different aspects of coelacanth morphology there are lists of characters expressing the morphological variation. The characters are combined into a single data matrix and subsequently analysed in Chapter 9 to suggest hypotheses of relationships. This is extended in Chapter 10 to suggest a hypothesis of relationship between coelacanths and other groups of fishes and tetrapods based on further morphological and molecular data. The method used to erect these hypotheses is the cladistic method of classification which has been used widely during the last two decades. Here the principles of cladistic methodology are explained with the introduction of some terms which are used freely throughout the text.

Cladistics is a method of biological classification which groups taxa into a hierarchy of sets and subsets based on the most parsimonious distribution of characters. The results

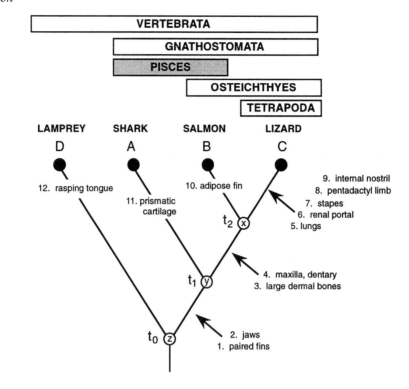

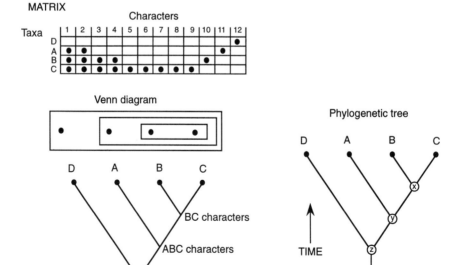

of these analyses are expressed in clado-grams which show the distribution of char-acters and represent a general pattern of relationships with which several evolutionary trees might be compatible.

Cladistic methods were originally formu-lated by Willi Hennig (1966) under the name 'Phylogenetic systematics'. Hennig's most significant contribution was to offer a precise definition of relationship and to outline how relationship might be detected. Hennig's concept of relationship is relative and is illu-strated in Figure 1.2A. In this figure the lizard (taxon C) is more closely related to the salmon (taxon B) than either is to the shark (taxon A) because the salmon and the lizard share a common ancestor, x which lived at time t_2 and which is not shared with any other taxon. Similarly the shark (taxon A) is more closely related to the salmon+lizard than to the lamprey (taxon D) because these taxa share a unique common ancestor, y which lived at time t_1. B and C are called sister-groups; A is the sister-group of the combined taxon B+C. The aim of cladistic analysis is to search for the sister-group hier-archy and express the results in branching diagrams called cladograms.

Sister-groups are discovered by finding shared derived characters (synapomorphies) inferred to have originated in the latest common ancestor. Synapomorphies can be thought of as evolutionary novelties or as homologies. From Figure 1.2A characters 3 and 4 are synapomorphies suggesting that the lizard and the salmon shared a unique common ancestor x. Shared primitive char-acters (symplesiomorphies) are characters inherited from more remote ancestry and are

irrelevant to the problem of relationship of the lizard and the salmon. For example, the shared possession of characters 1 and 2 in the salmon and lizard would not imply that they shared a unique common ancestor because these attributes are also found in the shark. Characters 1 and 2 are more universal to our understanding of the relationships between the lizard and the salmon. They may be useful at a higher hierarchical level to suggest common ancestry, y at t_1.

Synapomorphy and symplesiomorphy describe the status of characters relative to a particular systematic problem. Characters 3 and 4 are synapomorphies when we are interested in the relationships of the salmon or the lizard, but symplesiomorphies if the problem involves the relationships of differ-ent species of lizards or different species of salmon. Hennig recognized a third type of character: that is, those which are unique to one species or group, such as characters 5–9 in the lizard, 10 in the salmon, 11 in the shark and 12 in the lamprey. These he called autapomorphies which, in Figure 1.2A define the terminal taxa A – D.

Characters used to discover relationship are derived characters or character-states. Such character-states imply acceptance of transformation (absence $\rightarrow$ presence, or condition a $\rightarrow$ a'). Hennig suggested several criteria by which polarity of transformation may be recognized. The two most frequently used are ontogenetic transformation and outgroup analysis. The latter is most applic-able to palaeontological studies, and may be briefly stated: when a character exists in a variable state within the group under study (the ingroup), the condition which is also

Fig. 1.2 Cladistic classification. A. Phylogenetic tree to illustrate Hennig's concept of relationship, char-acter distribution and group membership. Solid circles (A–D) real taxa identified by autapomorphies (numbered 10–12). Open circles (x–z) hypothetical ancestral taxa existing at relative time $t_0 - t_2$. Char-acters 1,2 – synapomorphies of group ABC; characters 3,4 – synapomorphies of group BC. Named monophyletic groups are labelled as open boxes, paraphyletic group labelled as stippled box. See text for discussion.

found in taxa outside the group (the outgroup) is the plesiomorphic state. Cladistic analysis is concerned with choosing the particular hierarchical arrangement of ingroup taxa that is congruent with the greatest number of derived states. An alternative way of expressing this is that the most parsimonious solution is sought. There are a number of computer algorithms available to help in this task. In this work the PAUP program (Swofford, 1993) is used. It may be argued that it is inappropriate to apply the principle of parsimony to an exercise seeking to reconstruct phylogenetic relationship; after all, evolution may not have followed the most parsimonious course. In cladistic analysis the most parsimonious solution is sought because this is a universal criterion by which different hypotheses of relationship might be evaluated. We may accept a solution which is other than parsimonious but this requires additional assumptions which themselves require independent justification.

As a result of the relative definition of relationship we can identify three types of groups.

A monophyletic group contains the latest common ancestor plus all and only all descendants. In Figure 1.2A such groups would be BC(X), ABC(Y), DABC(Z). In the particular example using real taxa the monophyletic groups are named Osteichthyes, Gnathostomata and Vertebrata.

A paraphyletic group is one remaining after one or more parts of a monophyletic group have been removed. The group A+B (Pisces) is a paraphyletic group because one of the included members (B) is genealogically closer to C which is not part of the group Pisces.

A polyphyletic group is one defined on the basis of convergence, or on non-homologous (analogous) characters assumed to have been absent in the latest common ancestor.

Most systematists would agree with the desirability of recognizing monophyletic groups and have long respected the artificiality of polyphyletic groups. It is paraphyletic groups which are the source of debate, particularly amongst palaeontologists.

Cladistic classification insists that only monophyletic groups, which are recognized on the basis of synapomorphy be included. Paraphyletic groups obscure relationships because they are not real in the same sense, they do not have historical integrity because not all of the genealogical descendants are present. Furthermore, they cannot be recognized solely by synapomorphy but they have to be defined on a combination of presence and absence. Reptilia, for example, is a paraphyletic group recognized by having synapomorphies (amniotic membranes, cleidoic egg) of a larger group (Amniota) but lacking the synapomorphies of two contained amniote subgroups birds (feathers) and mammals (hair). Reptiles are distinctive amongst amniote animals only because they lack characters.

Paraphyletic groups have been popular in palaeontology because they are traditionally the ancestral groups (fishes ancestral to tetrapods, reptiles ancestral to birds and mammals). These paraphyletic groups can only be recognized on absence of characters of the presumed descendants. The problem is compounded in fossils because we cannot check conditions of soft anatomy used in the classification of Recent representatives and can never be certain that absence of features is real or a preservational artefact. In other words, nothing can be found to support an ancestral/paraphyletic group. In reality paraphyletic groups are usually delimited by authority.

A special case of ancestral/paraphyletic groups frequently occurs in palaeontology. This is the extinct, presumed ancestral group. Hennig called these stem-groups, and a well-known example relevant to the subject of this book is the Rhipidistia. Rhipidistians are paraphyletic, and are defined by convention as fishes with lobe-fins which lack the synapomorphies of tetrapods. When analysed using cladistic methods this group turns out to be composed of successively more derived taxa; some showing synapomorphies with tetrapods and to be genealogically closer than they are to other rhipidistians. By breaking up such a paraphyletic group into real historical entities (successive mono-phyletic taxa) we gain some insight into the sequence of character acquisition from which we may deduce something about evolution. Of course, for coelacanths in which there is only one living species the stem-group is particularly large.

When Hennig formulated his phyloge-netic system he did so in strictly evolu-tionary terms. His branching diagrams were phylogenetic trees with an implicit time axis in which hypothetical ancestors were located at the nodes, the nodes represented speciation or cladogenetic events with evolutionary transformation taking place along the branches. However, the construc-tion of a phylogenetic tree using cladistic methodology uses only the distribution of characters among the taxa under considera-tion. The result is the cladogram which has no time axis. The nodes denote a hierarchy of synapomorphies and the relationship can be represented as a Venn diagram of sets and subsets or may be written in paren-thetic notation in which there is no impli-cation of ancestry and descent (Figure 1.2B). A phylogenetic tree, which can only be drawn as a branching diagram, has the addition of a time axis which may be rela-tive or absolute. In this book the clado-grams constructed are translated to phylogenetic trees since the purpose is to understand something of the evolutionary history of coelacanths.

Institutional abbreviations

Throughout, there are references to coela-canth specimens in numerous institutions. The following is a list of the abbreviations used.

AM	Albany Museum, Grahamstown, Republic of South Africa
AMNH	American Museum of Natural History, New York, USA
AUB	American University Beirut, Lebanon
BGS	British Geological Survey
BMNH	The Natural History Museum, London, England
BSM	Bayerische Staatsammlung für Paläontologie und historische Geologie, Munich, Germany
CM	Carnegie Museum of Natural History, Pittsburgh, USA
FMNH	Field Museum of Natural History, Chicago, USA
GN	University Museum of Zoology, Cambridge University, England
Gö	Geological–Paleontological Institute and Museum, George-August University of Göttingen, Germany
GSM	Geological Museum, London, England; specimens now in the collection of the British Geological Survey, Keyworth, Nottingham, England
HM	Hancock Museum, University of Newcastle, England
ISI	Indian Statistical Institute, Calcutta, India
IVPP	Institute of Vertebrate Palaeontol-ogy and Paleoanthropology, Beijing, China
JME	Jura Museum, Eichstätt, Germany
KUVP	Division of Vertebrate Palaeontol-ogy, Museum of Natural History, University of Kansas, Lawrence, Kansas, USA
MCZ	Museum of Comparative Zoology, Harvard University, Cambridge, Massachusetts, USA
MGUH	Geological Museum (formerly Mineralogical Museum), University of Copenhagen, Denmark
MHNM	Muséum d'Histoire Naturelle, Miguasha, Quebec, Canada
MNA	Museum of Northen Arizona, Flag-staff, USA

MNHN Muséum National d'Histoire Naturelle, Paris, France

MRAC Muséum Royal l'Afrique Centrale, Tervuren, Belgium

MSB Museo de Geologia del Seminario de Barcelona, Spain

NMP National Museum, Geology Department, Prague, The Czech Republic

NRM Naturhistoriska Riksmuseet (Swedish Museum of Natural History), Stockholm, Sweden

PUGM Princeton University Geological Museum (this collection has now been relocated at Yale University, New Haven, USA, but the original numbers are retained)

RSM Royal Museum of Scotland, Geology Department, Edinburgh, Scotland

SAM South African Museum, Cape Town, Republic of South Africa

SM Sunderland Museum, Sunderland, England

SMC Sedgwick Museum, Cambridge University, England

SMNS Staatliches Museum für Naturkunde, Stuttgart, Germany

TM Tylers Museum, Haarlem, The Netherlands

TTCM Texas Technical University Museum, Lubbock, Texas, USA

UB Geological–Paleontological Institute, University of Berlin, Germany

UCL Palaeontological Collections, University College, London

UCMP University of California, Museum of Paleontology, Berkeley, USA

ULQ Musée de Géologie, Université Laval, Québec, Canada

UM Paleontological Museum, University of Michigan, Ann Arbor, USA

UMON University of Montana, vertebrate collections, Missoula, USA

UP Palaeontological Institute, University of Uppsala, Sweden

UT Institut und Museum für Geologie und Paläontologie, University of Tübingen, Germany

YPM Yale Peabody Museum, Yale University, New Haven, USA

ZM Paläontologisches Institut und Museum des Universität Zürich, Switzerland

2

NATURAL HISTORY OF *LATIMERIA*

2.1 INTRODUCTION

This chapter reviews some of the most distinctive features of the living coelacanth *Latimeria chalumnae*, and concentrates on details of the soft anatomy, physiology, feeding, locomotion and reproduction. Also included here are data related to geographical distribution, habitat and issues surrounding any conservation efforts which may be made.

Coelacanths are one group of lobed-finned fishes which include the lungfishes (Dipnoi) and a variety of extinct groups of predominantly Devonian fishes such as osteolepiforms, porolepiforms, rhizodonts and panderichthyids which have been frequently studied because of their possible evolutionary roles in the origin of tetrapods. It is commonly, if not universally accepted (Chang, 1991a) that the lobe-finned fishes are a paraphyletic group with some members genealogically more closely related to tetrapods than to other lobe-finned fishes. However, as explained in Chapter 9, there are many particular theories of interrelationships amongst these fishes.

2.2 ANATOMY

Latimeria, the only living coelacanth, is a heavily built, plump-bodied fish which amongst the Recent fish fauna shows a very distinctive profile (Fig. 2.1). It grows to 1.8 m and a maximum weight of 80 kg. The head is quite large and is covered with thick dermal bones (Fig. 2.5). The body is covered with large, deeply imbricating round scales which, in larger individuals, are ornamented with many small surface denticles (Fig. 11.9). The caudal peduncle is very deep and the tail is nearly symmetrical (the upper lobe is slightly longer than the lower lobe) and the body ends in a small terminal tuft of fin rays forming the supplementary caudal lobe. The tail is often described as 'trifid'. The anterior dorsal fin is sail-like and is supported by a few very stout fin rays. In the second specimen caught off the shores of Anjouan in the Comores Islands, the anterior dorsal fin and the supplementary lobe of the caudal fin appeared to be missing. The 'absence' of these fins led Smith (1953) to name a second species of living coelacanth, *Malania anjouanae*. It is now known that this second specimen was a mutilated individual, perhaps having suffered attack when young.

The second dorsal fin lies nearly opposite the anal fin and both are lobed with fin rays arranged around the lobes. The fins appear to be mirror images of each other. The fin rays appear to be arranged asymmetrically around the lobed bases but this is due to the pattern of scale cover. The fin rays, in fact, are arranged symmetrically around the endoskeleton (Figs 8.1, 8.3). Both pectoral and pelvic fins show prominent scale-covered lobes supporting a fringe of long fin rays. The pectoral fin is held high on the flank and, like the pelvic fin, is asymmetrical both externally and internally (Fig. 8.4). In many specimens, such as the one illustrated here (Fig. 2.1), the pectoral fin is held pointing ventrally with the longer fin rays

Fig. 2.1 *Latimeria chalumnae* Smith. A painting by Mr Gordon Howes of a small individual showing the typical coelacanth body form. This painting was made from a colour slide of a 'freshly dead' specimen caught in 1973 and displays the typical white blotch markings which show unique patterns enabling individuals to be recognized. The typical outline of a coelacanth is shown here, including the sail-like first dorsal fin, the lobed second dorsal and anal fins and the trilobed tail.

towards the 'leading' edge. When Millot and Anthony (1958) described this fin they did so under the assumption that this was the real life position. However, from an anatomical point of view it appears as if this is an orientation 180° out of its true position (Fig. 8.1), a fact reinforced by observations on the embryo. The pectoral fin is known to be capable of considerable rotational movement (see below) and this common and 'misleading' fin orientation may be due to unequal contraction of the fin muscles just before death.

Several characteristics of the head are better seen in the embryo than in the adult. The lateral profile of the head shows an agulation above and behind the eye (Fig. 2.2, i.j). This is the external manifestation of the intracranial joint which divides the braincase into anterior and posterior portions (Figs 2.5, 6.1(a)). *Latimeria* is the only living animal

with such a joint, although many extinct lobe-finned fishes had a similar joint. *Latimeria* is therefore our only model on which we can base ideas of the movement and function of this unusual feature (Chapter 7).

The eye is relatively large, more so in the embryo (Fig. 2.2) than in the adult (Fig. 2.1): this is an almost universal allometric difference in fishes. The relative size of the eye in the adult is more like that of actinopterygian fishes than that in other lobed-finned fishes. The mouth is relatively small and is surrounded along its upper edge by a very thick fold of muscularized skin – the pseudomaxillary fold (Fig. 2.2, psmax). This fold replaces the maxilla, which is absent (Fig. 2.5) in all coelacanths. The mouth is well equipped with teeth and some of these are fang-like. However, they are not borne by the marginal jaw bones but are instead found upon the coronoids in the lower jaw and the dermopa-

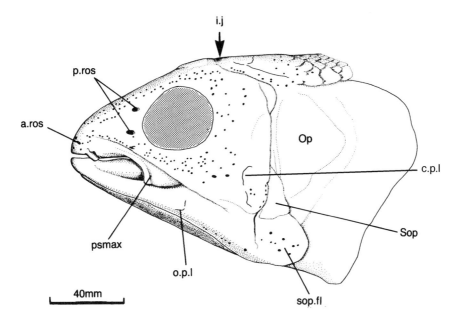

Fig. 2.2 *Latimeria chalumnae* Smith. Camera lucida drawing of the head of an embryo specimen (BMNH 1976:7:1:16) in left lateral view. The maxillary fold which replaces the upper jaw can be seen here, together with the large subopercular flap and the position of the external openings of the nostrils and the rostral organ. The characteristic dorsal profile of the head is due to the presence of an intracranial joint. Abbreviations: see Appendix.

latines in the palate (Fig. 2.5). On the ventral surface of the head (Fig. 2.3) there are paired gular plates between the jaw rami.

The cheek is also distinctive in *Latimeria*. The opercular bone is relatively small (Figs 2.2, 2.5) but this carries a very large soft-tissue opercular flap. Anterior to this the skin is deeply infolded (Fig. 2.2). Behind the eye this fold marks the position of the spiracular cleft and at the dorsal end of this fold is a small spiracular bone (Fig. 2.4, Sp). However, the spiracle is closed. At the ventral end of the fold of skin there is a prominent subopercular flap (Fig. 2.2, sop.fl), the dorsal end of which is supported by a subopercular bone (Fig. 2.5, Sop).

There are two external nostrils which superficially are placed far apart. The anterior (incurrent) nostril opens through a small soft-tissue papilla located at the margin of

the mouth at the anterior end of the pseudo-maxillary fold (Fig. 2.2). Within the skeleton the external opening is located between the premaxilla and the lateral rostral bones (Figs 2.5, 3.2). The nasal capsule is relatively small but the posterior (excurrent) nostril leaves the nasal cavity quite near the anterior nostril (Fig. 3.2) but runs posteriorly through a long soft-tissue subdermal nasal tube which opens to the surface immediately anteroventral to the eye. It also opens below the path of the infraorbital sensory canal, but this unusual topographic relationship is brought about by the course of the soft-tissue nasal tube.

The laterosensory system is very well developed in *Latimeria*. The individual canals as they pass through the bones are described later in Chapters 3–5, but the pores opening to the surface can easily be seen in the embryo. On the dorsal surface (Fig. 2.4) a

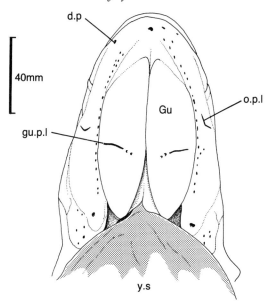

Fig. 2.3 *Latimeria chalumnae* Smith. Camera lucida drawing of the head of an embryo specimen (BMNH 1976:7:1:16) in ventral view. This shows the large paired gular plates, the ventrally-facing sensory pores in the lower jaw as well as the anterior end of the very large yolk sac still attached at the isthmus.

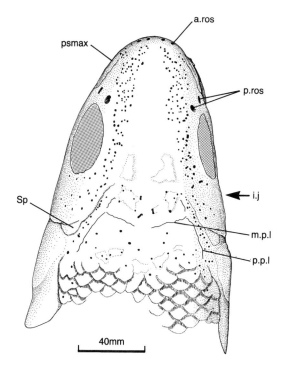

Fig. 2.4 *Latimeria chalumnae* Smith. Camera lucida drawing of the head of an embryo specimen (BMNH 1976:7:1:16) in dorsal view. The prolific development of the openings from the sensory canals are obvious, as are the pit lines in the temporal region. At this early growth stage the raised areas on the skull roof bones which are prominent in the adult are only faint indications beneath the skin.

dense aggregation of pores above the eye marks the path of the underlying supraorbital canal. This continues posteriorly as the otic canal. On the cheek (Fig. 2.2) openings from the infraorbital jugal and preopercular canal (Fig. 4.2(A)) can easily be seen, as can the openings from the subopercular branch. This last branch is not seen in other fishes and is a character of more derived coelacanths (Chapter 8). The sensory canal openings from the mandibular canal, including a large dentary pore (d.p). can only be seen in ventral view (Fig. 2.3). The skin also shows pit lines on the roof of the temporal region (Fig. 2.4), on the cheek and lower jaw (Fig. 2.2) and on the gular plate (Fig. 2.3). The snout is also marked by three large paired holes which open into the rostral organ (see below) which is unique to coela-

canths. At the tip of the snout there is a single anterior opening (Figs 2.2, 2.4, a.ros) and immediately in front of the eye there are two posterior openings placed one above the other (p.ros).

The skeleton is described in Chapters 3–8 but a few general aspects of the head skeleton can be pointed out here (Fig. 2.5). The cheek bones (Po, Lj, Sq, Pop, Sop) are robust but they are well separated from one another with a single large bone beneath the eye. The palate is triangular and the principal jaw joint between the quadrate and articular lies

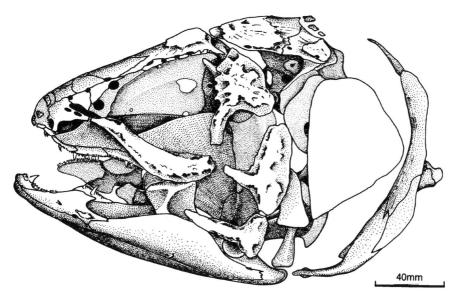

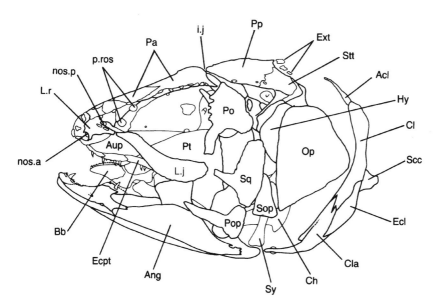

Fig. 2.5 *Latimeria chalumnae* Smith. Skeleton of the head seen in left lateral view. This is a compilation drawing based on Millot and Anthony (1965) and information from specimens. Some characteristic features seen here are the intracranial joint, the shape and orientation of the palate and the lower jaw, the rugose cheek plates which lie free from one another, and the shoulder girdle lying quite separate from the skull.

medial to the preoperculum (Pop) and vertically below the articulation between the palate and braincase and on a vertical level with the intracranial joint. Posterior to the main jaw joint, there is another joint between the posterior tip of the jaw and the symplectic (Sy) which is, in turn, attached to the hyoid arch (Chapter 7). This double, tandem jaw joint is a coelacanth character and the topological relationships between the palate, braincase and lower jaw determine a characteristic method of opening the mouth (Fig. 7.5). The shape of the lower jaw is also very distinctive in *Latimeria* and other coelacanths (Chapter 5): in particular there is a very large angular which forms much of the outer face of the jaw. One final feature that can be seen in Fig. 2.5 is the fact that the shoulder girdle lies free from the skull and contains an extracleithrum (Ecl) – a bone not seen in other vertebrates.

The eye

The eye has been studied on several occasions (Lenoble and Le Grand, 1954; Millot and Anthony, 1965; Millot & Carasso, 1965; Locket and Griffith, 1972; Locket, 1973, 1980). It is relatively large and appears to be adapted for vision in low light conditions. Thus, the retina, which is relatively simple and does not contain a fovea, consists mostly of rod cells and these contain a pigment which is particularly suited to absorption of low wavelenths: γ_{max} = 473 nm (Locket, 1980) which matches the maximum penetration of light in clear oceanic waters. There are cone cells contained in the retina which Locket (1980) ascribed to three types. The presence of cone cells may be significant because these are present in bony fishes and tetrapods but lacking in chondrichthyans. However, according to Locket there are relatively few and it is doubtful if there is any well-developed colour vision – it would not be expected at the depth at which *Latimeria* lives. Another adaptation of the eye of *Latimeria* to low light regimes is the presence of a tape-

tum. This layer consists of choroid cells containing guanine crystals and reflects light back through the retina such that the retinal cells are stimulated twice. The nature of the tapetum is very similar to that in sharks (Locket, 1980).

The brain and pituitary gland

The brain has been described by Millot and Anthony (1965), Nieuwenhuys *et al.* (1977), Northcutt *et al.* (1978) and, in part, by Northcutt and Bemis (1993). Many aspects of the brain still need to be described. Perhaps the most remarkable feature of the brain of *Latimeria* is the small size in the adult. It occupies less than 1.5% of the cranial cavity and is confined to the otico-occipital part of the skull. However, this is a result of remarkable negative allometric growth, because the young show a brain which occupies much of the cranial cavity with the telencephalon extending forward to reach the level of the eye (Anthony and Robinson, 1976: fig. 2). The brain is basically like that of a shark with very clear divisions into telencephalon, diencephalon, mesencephalon, cerebellum and medulla. As in sharks, the olfactory lobes are closely pressed against the olfactory sacs and there are very long olfactory tracts. The roof of the telencephalon is thickened as an evagination. The tectum of the diencephalon is also thickened and well developed, matching the possession of large eyes. The pineal is also well developed as two vesicles (pineal and parapineal) which are both in contact with the ventricle of the diencephalon. Locket (1980) reports that both pineal and parapineal are lined with sensory cells remarkably similar to the retinal cells and thus potentially photosensitive. However, the pineal complex is deeply embedded within the cranial cavity, precluding light stimulation.

Northcutt *et al.* (1978) compiled a quantitative analysis on parts of the brain and compared this with similar studies on other fishes. They concluded that the relative sizes of parts of the brain are similar to those of sturgeons and primitve sharks. In summarizing their work they conclude that the brain of *Latimeria* shows several features which by their widespread distribution are primitive gnathostome features. These include well-developed visual system, bilobed roof of the mesencephalon, small mesencephalic acustico-lateralis centre, moderate to well-developed cerebellum and an evaginated roof of the telencephalon. Northcutt (1987) has also summarized comparative information about specific parts of the brain. He points out that *Latimeria* is unique in showing a rostral body: a swelling which presses against the cerebral hemisphere at the base of the olfactory tract. Northcutt also described four points of similarity between the brain of *Latimeria* and *Neoceratodus* amongst lungfishes: (1) the thickened roof of the diencephalon (the dorsal thalamus) is laminated and protrudes into the third ventricle; (2) the development of olfactory tracts; (3) the cerebral hemispheres are evaginated and partly separated from each other by a non-neural septum ependymale; (4) the lateral wall of the mesencephalon contains a nucleus – the superficial isthmal nucleus receives neurones from the retina. These neural characters might support the theory that *Latimeria* and lungfishes are immediately related to one another. However, Northcutt (1987) emphasizes that not all of these similarities are shared with other lungfishes (lepidosirenids) and the polarity of other characters amongst vertebrates is not clearly established. Thus, we can sum up the present knowledge about the brain of *Latimeria* by suggesting that it is a basically primitive vertebrate brain which, in certain respects shows most similarities to that in lungfishes.

The pituitary gland of *Latimeria* has been described by Lagios (1975), who notes some remarkable similarities with the pituitary gland of elasmobranchs (Lagios, 1979). In both the pars distalis (pro- and mesoadenohypophysis) is extended ventrally so that it lies far from the brain and other parts of the pituitary. In *Latimeria* this may be explained by this part of the pituitary being 'left behind' during the allometric shrinking of the brain. The ventral lobe is lodged within the base of the basisphenoid (Fig. 6.12, pit.fos). This extension also receives a blood supply which is separate from that of the rest of the pituitary. The minute details of the ventral projection and the blood supply are not the same in both groups (Forey, 1980) and therefore this similarity may be of no phylogenetic significance.

Nasal sacs and rostral organ

The paired nasal sacs (Millot and Anthony, 1965) are relatively small and each contains a rosette of five papillae which serve to increase the surface area of the sensory epithelium. The anterior nostril opens into the nasal sac through a short papilla which may be seen externally (Fig. 2.2). The posterior opening from the nasal sac empties into a long nasal tube which runs subdermally to open near the eye (see above). Thus, except for the long posterior nasal tube, the nasal sacs are typically fish-like and there is no evidence of an internal nostril (choana). As mentioned above, the olfactory bulb lies directly in contact with the nasal sac.

Smith (1939c) described a large median space wholly contained within cartilage in the ethmoid region of the braincase. He thought that this was associated with the nasal sacs. Millot and Anthony (1956, 1965) recognized this ethmoidal cavity as containing a special rostral organ quite separate from the nasal sacs. The median rostral organ is surrounded by a layer of adipose tissue and is located dorsal to the paired nasal capsules. The organ opens to the outside through three paired tubes (one anterior and two posterior – see above) and there are no valves guarding any of these openings. The walls lining the rostral organ are infolded and the entire cavity, including the tubes leading to the exterior, is filled with a gelatinous substance.

The walls lining the tubes leading into the rostral organ are innervated by branches from the superficial ophthalmic branch of the anterodorsal lateral line nerve (Northcutt and Bemis, 1993). This suggests that the rostral organ is phylogenetically part of the laterosensory system. The cells lining the rostral organ are morphologically very similar to the cells lining the ampullae of Lorenzini which are embedded in the snout of elasmobranchs (Bemis and Hetherington, 1982). The latter are known to be capable of detecting weak electric potential differences such as those created by muscular activity of nearby organisms. Thus it is very likely that the rostral organ is concerned with electroreception (see also p. 25). The presence of an insulating layer of adipose tissue surrounding the rostral organ would be concordant with this view. Electroreception is widespread in fishes and is probably a primitive vertebrate character. Localization of electroreceptor cells into a rostral organ is a specialized feature.

Rosen *et al.* (1981) and Northcutt (1987) have suggested that the rostral organ of *Latimeria* may be homologous with the labial cavities of lungfishes. The labial cavity is a paired sac-like invagination in the skin which opens posteriorly near the angle of the mouth (Rosen *et al.*, 1981: fig. 19). The cells lining the labial cavity are innervated by branches of the anterodorsal lateral line nerve (Northcutt, 1987). To date the cells lining the cavity have not been examined and therefore a direct comparison with those of *Latimeria* cannot be made.

A further theory of homology between the labial cavity and the nasolachrymal duct of amphibians was suggested by Rosen *et al.* (1981), who point out similarities in ontogenetic development, adult topological relationships and the nature of the nerve innervation, which in both cases is via branches of the anterodorsal lateral line nerve.

Therefore, it is possible that the rostral organ is an extreme modification of a specialized part of the sensory system common to coelacanths, lungfishes and amphibians.

The ear

The inner ear has been described by Millot and Anthony (1965) and certain aspects of ultrastructure have been examined by Fritzsch (1987, 1989). The semicircular canals are developed in the normal vertebrate pattern and there is a short blindly ending endolymphatic duct arising from the utriculus. Beneath the semicircular canals there are the usual three labyrinth chambers: utricular, saccular and lagenar. The saccular is by far the largest and contains the only otolith, which is relatively featureless (Millot and Anthony, 1965: figs 47, 48; Nolf, 1985: fig. 80). Inasmuch as they are known, fossil coelacanth otoliths are similar (p. 185, Clack 1996). At the junction of the utriculur, saccular and lagenar chambers there is a medial opening, leading to a tube which connects left and right labyrinths, called by Millot and Anthony (1965) the canal communicante. The canal crosses the midline immediately beneath the level of the foramen magnum and therefore passes beneath the brain. An endocranial space which may have accommodated a similar canal is described in *Diplocercides* (Bjerring, 1972) as the endolymphatic occipital commissure. Thus, this connection between left and right inner ears is a primitive coelacanth feature. However, Fritzsch (1987) has argued that this duct, in *Latimeria*, which connects both inner ears should be compared with a perilymphatic space and is therefore like that of tetrapods.

Fritzsch (1987, 1989) has reported the presence of a basilar papilla within the inner ear of *Latimeria*. The basilar papilla is a thin membrane supporting a sensory epithelium and found lying between the sacculus and perilymphatic cavity (see above). The cells within the epithelium are supplied by a separate branch of the auditory (octaval) nerve. The basilar papilla is present in some amphibians and all amniotes, and in mammals it is much enlarged and lies along the cochlear. It is associated with hearing in air. Thus the occurrence of this structure in *Latimeria* is of considerable interest and suggests to Fritzsch (1989) that some of the prerequisites for hearing on land are already present in primary aquatic animals. The function of the basilar papilla in *Latimeria* remains unknown. Lungfishes apparently lack a basilar papilla, leading Fritzsch (1987) to suggest that *Latimeria* and tetrapods are sister-groups.

2.3 DIET AND FEEDING BEHAVIOUR

Various prey items have been recorded as stomach contents from captured coelacanths.

These prey are mostly deeper-water fishes which are found naturally living near the substrate. Additionally, there are a number of species which have been successfully used as bait. All evidence suggests that *Latimeria* is a selective carnivore. In the following list of prey found in the stomach of *Latimeria*, the natural depth occurrence of prey species, where known, is given in parentheses.

Feeding behaviour has rarely been observed. *Latimeria* is often described as a drift hunter. This scenario depicts *Latimeria* to passively drift with the current, using gentle movements of the fins to stabilize the body in the water column until it encounters prey, when it will suddenly lunge using the broad tail for rapid acceleration. Although *Latimeria* may passively drift for much of the time, such a chance method of finding food would seem insufficient for a large fish. We need

Lutjanidae	*Symphysanodon* sp., deepwater snapper (119–470 m)	McCosker (1979)
Polymixiidae	*Polymixia noblis*, Atlantic beardfish (183–640 m)	McCosker (1979)
Apogonidae	*Coranthus polyacanthus*, deepwater cardinalfish (150 m)	McCosker (1979)
Myctophidae	*Diaphus metopoclampus* (90–850 m)	Millot and Anthony (1958)
Synaphobranchidae	*Ilyophis brunneus* (1500 m)	Uyeno and Tsutsumi (1991)
Scyliorhinidae	*Cephaloscyllium sufflans* (40–440 m)	Uyeno and Tsutsumi (1991)
Berycidae	*Beryx decadactylus* (200–600 m)	Uyeno and Tsutsumi (1991)
Unidentified teleost vertebral column		
Entire cuttlefish	Species not recorded	McAllister (1971)
Cephalopod beak		Uyeno and Tsutsumi (1991)

Bait used in successful fishing efforts
Gempylidae	*Promeichthys prometheus*, roudi
Exocoetidae	*Cypselurus bahiensis*
Carangidae	*Decapterus* sp.
Scomberesocidae	*Tylosaurus choram*, crocodile needlefish
Lutjanidae	*Lutjanus* sp.
Labridae	*Thalassoma* sp.
'Tuna'	
Octopus	

more observations on feeding to confirm or refute this scenario.

Uyeno (1991) records a coelacanth taking a cuttlefish offered as bait by a diver. And Fricke observed a coelacanth snapping (unsuccessfully) at prey. In both cases the coelacanth lunged at the prey from short range and opened the mouth very rapidly. This, taken into consideration with the observation that most prey items have been swallowed whole (McCosker, 1979), suggests that the prey is sucked in and that the teeth are used to grip rather than to bite or mutilate prey. Such a mechanism is employed by many kinds of fishes and depends on the creation of a sudden, powerful negative pressure. Efficiency of suction will be a function of the speed of jaw opening, the size of the gape and the possibility and nature of jaw protrusion (as in many acanthopterygians). None of these aspects has been thoroughly examined in *Latimeria*, therefore it is not possible to evaluate the efficiency of the jaw mechanism. However, some theories of the jaw mechanism have been proposed and these are dealt with in Chapter 7.

From that which we understand about the jaw mechanism the presence of the intracranial joint, together with the unusual hyoid arch and tandem jaw articulation, would allow the jaws to be opened very rapidly to a wide gape in which the lower jaw is thrust forward. The powerful jaw adductors and the large basicranial muscles which span the intracranial joint beneath the braincase ensure a powerful bite.

2.4 LOCOMOTION

Shortly after *Latimeria* was discovered and the nature of the lobed fins described it was predicted that, when discovered in its natural environment, it would be seen to walk along the bottom. Smith (1956) entitled his autobiography *Old Fourlegs* with this expectation in mind. Walking would agree with the habits of a presumed relative of tetrapods. In

fact, this prediction has turned out to be both true and false. It is true in the sense that the coordination of the paired fins is like that of a tetrapod. It is false in the sense that *Latimeria* does not use the paired fins to propel itself over the substrate.

Initial observations of swimming were made by raising line-hooked coelacanths to the surface, releasing them and recording movements (Millot, 1955; Lockett and Griffith, 1972; Griffith, 1973; Thomson, 1973; Uyeno, 1991). To date, six coelacanths have been observed in this way. However, all of these authors acknowledge that the coelacanths were distressed and near to death and that the activities may not have been those of healthy individuals.

A new phase of 'coelacanth watching' began in 1986–7 when Hans Fricke observed six individuals from a submersible (Fricke *et al.*, 1987; Fricke and Schauer, 1987; Fricke and Plante, 1988). And in a later expedition (October–December 1989) another 38 were found (Fricke and Hissman, 1990; Fricke *et al.*, 1991). Extensive cine film was taken and subsequently analysed (Fricke and Hissman, 1992).

When first seen, the movement of *Latimeria* appears to be enacted in slow motion, with the paired fins and the second dorsal and anal fin all capable of extraordinarily complex movements which are seemingly uncoordinated. For most of the time, *Latimeria* appears to passively drift with the current, or swim slowly forward using principally the second dorsal and anal fins. Occasionally, it performs head-stands, holding the snout a few centimetres above the substrate, and it may even swim upside down for many minutes. During all these movements the head, body and tail is held rigid while the paired and median fins provide the thrust and stabilization. This type of movement corresponds most closely to the 'tetraodontiform' category of Breder (Lindsey, 1978). The caudal fin is used only when lunging at prey or when the animal escapes in a fright reaction. In these cases there is a rapid acceleration.

The pectoral fins have a narrow region of insertion to the body (Fig. 2.1) and are capable of flexing through $\pm$ 120° up and down and anterior–posterior and are capable of rotating $\pm$ 180° around the longitudinal axis. In lateral view the trace of a fixed point at the tip describes a figure-of-eight as the fin moves up and down (Fig. 2.6(A)). The angle of attack of the fin also changes so that the result is a movement like that of a bird's wing, which is capable of generating both lift and thrust.

The pelvic fins are attached to the body over a broader area and do not appear to be as mobile. Each fin can move through $\pm$ 120° up and down but does not appear to be capable of much antero–posterior movement; it appears to have some limited capability to rotate along the axis of the fin. Like the pectoral the fin tip describes a figure-of-eight but the trace is more asymmetrical (Fig. 2.6(A)). Fricke and Hissman (1992) suggest that the pelvic fins are used primarily as brakes for directional movement as well as stopping in a straight line. As such, the pelvics of *Latimeria* may function like the pelvics of many acanthopterygian teleosts.

The second dorsal and the anal are highly mobile. Each can swing through at least 90° each side of the vertical and is capable of rotating $\pm$ 180° around its own axis. Lockett and Griffith (1972) describe these fins as acting like a propeller with the leading edge dragging the rest of the fin which is convex anteriorly (Fig. 2.6(A)) and so creates thrust.

The sail-like first dorsal can be erected or depressed in the sagittal plane. Throughout most movements seen by Fricke and colleagues this fin was held in the erect position and probably serves as a stabilizer, to reduce the effects of roll and yaw. The caudal fin is capable of powerful flexing to create thrust but most of the observed time it was held rigid. The supplementary fin can move 90° either side of the midline.

After analysis of 6 hours of film, Fricke and Hissman (1992: table 1) described 12 types of movement. But for the majority of these, the coordination of fin movements was very similar. Most of the key points can be illustrated in Fig. 2.6(B,C). The left pectoral and right pelvic move in the same direction and at the same time and this is matched by coordinated but opposite movements of the right pectoral and left pelvic (Fig. 2.6(C)). This alternating movement is similar to the coordination seen in tetrapod limb movement as well as in the movement of lungfish paired fins. The second dorsal and anal move to the same side of the body at the same time (Fig. 2.6(C)) and the anal fin moves through a noticeably greater amplitude (Fig. 2.6(B)). These two fins will therefore almost touch one another on one side, then the other. Uyeno (1991) suggests that at this point, water will be squeezed backwards between the fins, so producing thrust. Thrust will also be produced by the propeller-like movement mentioned earlier. There also appears to be some coordination between the pectoral movement and the movement of the second dorsal and the anal in that the stroke-reversal point of the pectorals slightly precedes that of the second dorsal and anal (Fig. 2.6(C)). If these are the important fins for creating thrust, then this phase relationship is understandable.

The pelvic fins appear to act as stabilizing fins and as brakes rather than for providing locomotory thrust. They may also aid in turning by abbreviating a stroke while the body is flexed. The fast-start performance is good and has been calculated at $\sim$23 m s^{-1} (Fricke *et al.*, 1987), which compares to that of a trout (Webb, 1978). This rapid acceleration is produced by the body being thrown into an 'S' or 'U' curve with a rapid flexing of the tail. The deep lateral profile of the tail and caudal peduncle provides a large surface area of resistance against the water. During the fast start the paired fins, both dorsal fins and the anal fin reacted passively. The fast

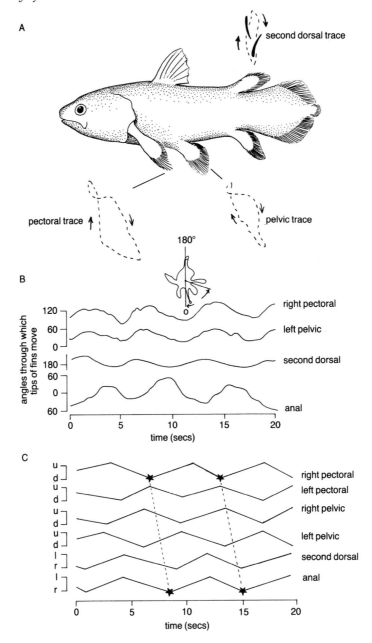

Fig. 2.6 Fin movements of *Latimeria*. (A) During slow forward movement the tips of both the pectoral and pelvic fins move through a figure-of-eight path. The pelvic trace is slightly more asymmetrical than the pectoral. (After Fricke and Hissmann, 1992.) The second dorsal and anal move from side to side with left and right sides of each fin alternating as the leading edge. A cross section of either fin would display a propeller-like hydrofoil. (After Locket and Griffith, 1972.) (B) Traces of angular movements of fins showing the relative amplitudes. The pectoral and opposite pelvic move in unison at roughly the same amplitude. The second dorsal and anal move in opposite directions and therefore beat to the same

Fig. 2.7 *Latimeria* frequently performs 'head-standing' movements, in which it rotates to a vertical position and holds the snout a few centimetres above the substrate. It is possible that it does this to hold the rostral organ close to the sea bed. The rostral organ is thought to be a specialized electroreceptive organ. Drawing based on specimens and photo cine-stills.

start may be used as a fright reaction or to lunge at prey.

A rather unusual movement of 'head-standing' is shown by *Latimeria*. During this movement, the caudal and supplementary caudal fins are flexed and the fish adopts an upright position with the snout hovering a few centimetres above the substrate (Fig. 2.7). It can maintain this station for up to a few minutes using both the paired and median fins as stabilizers. The supplementary caudal often moves 90° either side during head-standing. Although the function of this head-standing has not been physiologically determined, it is suspected that it may have something to do with electroreception by bringing the rostral organ (p. 21) close to the substrate to detect

side of the fish at the same time, with the anal moving through a greater angle than the second dorsal. (After Fricke *et al.* 1987.) (C) Schematic representation of fin coordination shown while the fish is drifting slowly forwards. Notice the alternating movements pectoral and pelvic (u, up; d, down) while the second dorsal and anal move to the same side at the same time (l, left; r, right). There appears to be some coordination between the pectoral and the anal (linked asterisks) with the pectoral preceding the anal by about 1.5 seconds. (After Fricke and Hissmann, 1992.)

minute potential differences created by the activity of prey. Fricke and Plante (1988) noted that the head-standing could be induced by experimentally creating a weak electric field near the fish. Uyeno (1991) reports that young ocean sunfish, *Mola mola* – which incidentally show the most extreme form of tetraodontiform locomotion – will feed in 'head-standing' position, even though these fishes do not have a specialized rostral organ.

Estimations of the swimming speed of *Latimeria*, apart from the acceleration speed (see above), are imprecise because, for most of the time, individuals have simply been observed to drift with the current. Constant forward speeds of 4.8 cm s^{-1} (Fricke and Hissmann, 1992) to 16 cm s^{-1} (Uyeno, 1991) have been estimated. This is very slow compared with speeds recorded for other fishes which are smaller and would otherwise be expected to be slower than *Latimeria*. These very slow speeds and the drifting behaviour emphasize the fact that *Latimeria* does not have a critical speed, necessary to maintain its level in the water, as do the negatively buoyant fishes such as scombroids and xiphioids. The high oil content of the liver and flesh of *Latimeria* suffice to make it neutrally buoyant.

In summary, the behaviour of *Latimeria* is most obviously compared to that of large groupers (e.g. species of the genus *Epinephalus*), which spend most of their time drifting with occasional bursts of swimming to capture prey or escape predators. The actual method of swimming is rather different. For most activities the body and tail is held rigid while the thrust is produced by the second dorsal and anal, supplemented by the pectorals. The pelvics and the first dorsal are primarily for stabilization and aid in turning and braking. The movements of the pectorals and pelvics are coordinated like the limbs of a tetrapod and the second dorsal and anal appear to be coordinated with movement of the pectorals.

2.5 OSMOREGULATION

The kidney of *Latimeria* is very unusual amongst vertebrates in being located ventrally within the abdominal cavity and posterior to the cloaca – a position which is attributed to the large fat-filled air bladder having displaced the kidney posteroventrally (Millot and Anthony, 1973; Millot *et al.*, 1978). The rectal gland (nodular or postanal gland) is a median gland, supplied by the posterior mesenteric artery and drained by the dorsal intestinal vein. It opens anteriorly into the posterior end of the hindgut just anterior to entry of the urinary ducts (male: in the female the urinary ducts open separately into the cloaca). The detailed structure of the gland has been described by Millot and Anthony (1958, 1972) and Lemire and Lagios (1979). The presence of mitochondria-rich cells and complex infolding of the cell membranes suggests active ion secretion and the contained high level of sodium-activated ATPase is concordant with salt secretion (Griffith and Burdick, 1976). This gland is very similar to the rectal gland of elasmobranchs, which is known to be concerned with excretion of excess sodium chloride (Burger and Hess, 1960). Holocephalans also have rectal glands, although the structure is slightly different (the gland is more diffuse). The African lungfish is also known to possess a cloacal gland of presumably similar function (Lagios and McCosker, 1977). This systematic distribution of a rectal/cloacal gland does not clearly indicate whether it is a primitive vertebrate character or whether it has arisen on several occasions. The function of excreting excess sodium and chloride ions is linked with a similar manner of osmoregulation in all these fishes.

Latimeria and chondrichthyans produce high levels of urea which they habitually retain in the plasma and body tissues, together with TMAO (trimethyl-amine-oxide), to use as an osmolyte and so maintain the osmolarity of the body equal or near to that of sea water. This similarity has been held by some (Løvtrup, 1977; Lagois, 1979) to be evidence that *Latimeria* and chondrichthyans are sister groups. This issue has been discussed on several occasions (Pang *et al.*, 1977; Lagios, 1979; Forey, 1980; Griffith, 1991) with the consensus that urea production is a general feature of vertebrates and that its retention may be derived and adaptive. It is unlikely to be a synapomorphy of *Latimeria* and chondrichthyans because congruence of other characters places the coelacanth as a bony fish.

Urea is produced via the ornithine–urea cycle, including the enzyme arginase, and is less toxic than more soluble ammonia which is the usual nitrogenous waste product in fishes. (As an interesting side issue, the urea is formed within mitochondria in most fishes but is formed outside the mitochondrion in lungfishes and tetrapods; Griffith, 1991.) However, urea is more energy expensive to produce. A wide variety of lower tetrapods produce urea for use in at least some stages of their life. Notably it is often produced by embryos such as in guppies (Griffith, 1991) and intermittently by adults (aestivating African and South American lungfishes, *Periophthalmus*). A few habitually retain urea for use as osmotic balance (e.g. *Rana crancrivora*, *Latimeria*, chondrichthyans) where there are concomitant specializations such as tolerance of the tissues and blood (Mangum, 1991) to high levels of urea (and chloride ions). *Latimeria* probably retains urea through showing a very low glomerular filtration rate, whereas both chondrichthyans and *R. crancrivora* show active urea resorption within the glomerulus. Strategies to avoid urea loss in both chondrichthyans and *Latimeria*, which in most fishes occurs across the gills, involve a reduced gill surface area/body weight ratio (Hughes and Morgan, 1973) as well as a thickened tissue barrier between the blood supply through the gills and the surface (Hughes 1976). Reduced permeability of the gill, which is usually the site of ion exchange,

may, in turn, explain the presence of the rectal/cloacal gland. Griffith (1991) argues that because urea synthesis is volume dependent while urea loss is surface area dependent, then osmotic balance via urea retention is more economical for large fishes than for small ones. Furthermore, the ability for the tissues to withstand high urea levels is obviously advantageous to livebearing fishes. It is of course very difficult to separate cause and effect here. However, Griffith (1991) proposes an evolutionary scenario for ureosmoregulation in fishes in which he proposes that urea synthesis is a primitive vertebrate character and that urea retention is a response to desiccatory environments. The early evolution of predatory and large gnathostomes included protracted embryonic development which favoured the retention of urea as a means of detoxifying ammonia produced as a metabolic breakdown of yolk. Lastly, the paedomorphic retention of ureosmoregulation into the adult stages may be a preadaptation of large fishes returning to the sea. Griffith proposes a number of consequences of this scenario which may be tested in the future by experimentation to see, for instance, when and where the gene controlling the production of arginase is turned on and off.

2.6 REPRODUCTION

The reproductive biology of coelacanths in general and *Latimeria* in particular has been the subject of many papers and there is ongoing controversy over the interpretation of prenatal growth of *Latimeria*. In *Latimeria* the sexes are separate, females tend to grow larger than males (Fig. 2.11), and there are some sexual diferences in the scales and folds of skin surrounding the cloaca (Millot and Anthony, 1960a,b; Millot *et al.*, 1978). However, there is no obvious intromittent organ in the male.

Watson (1927) described a specimen of *Undina penicillata* containing two small, incompletely ossified individuals set far back in the abdominal cavity with heads facing anteriorly and hence unlikely to have been stomach contents. This was taken as evidence that this species gave birth to live young.

Early observations of *Latimeria* showed no evidence of developing young, and those dissections that had been made (Millot and Anthony, 1974) revealed only eggs preserved in the right, functional oviduct (the left is non-functional; Millot *et al.*, 1978). At the same time Schultze (1972) described large eggs and young of the Carboniferous *Rhabdoderma exiguus*. Some of these young bear yolk sacs and are found in large numbers in the Upper Carboniferous Mazon Creek fauna. The young appear to range from small individuals carrying large yolk sacs to larger individuals with a reduced sac, to those without the yolk sac. These observations suggested that coelacanths were and are oviparous, laying large yolky eggs which develop free from the mother.

However, as Thomson (1991) recounts, doubts about this reproductive strategy began to surface. The eggs of *Latimeria* are very large (some are up to 9 cm in diameter and weigh 334 g and are the largest eggs recorded for fishes) and are protected only by a thin membrane. Furthermore, it had long been known that the strategy of retaining urea to regulate osmotic balance shown by *Latimeria*, as in cartilaginous fishes, can pose problems for developing young. In egg-laying sharks, which also produce large yolky eggs, the egg is surrounded by a tough impermeable chitinous egg case which provides the embryo with osmotic protection until urea synthesis creates a suitable internal environment. *Latimeria* has no shell gland capable of producing an egg case. This circumstantial evidence led Griffith and Thomson (1973) to predict that *Latimeria* may be ovoviviparous.

Such a prediction was confirmed when a pregnant female was dissected to show five embryos with yolk sacs attached contained

within the oviduct (Smith *et al.*, 1975; Atz, 1976). Inferences about the reproduction of coelacanths had come full circle, back to Watson (1927) who originally suggested that coelacanths were livebearers. The case of the Carboniferous *Rhabdoderma exiguum* remains open. Thomson (1991) suggests that the fact that young are found without the adults might suggest that they were born prematurely under some environmental stress (e.g. reduced salinity). However, the fact that so many hundreds of young specimens have been found alongside eggs implies that this species may have been oviparous, as Schultze (1972) suggested.

The fact that *Latimeria* gives birth to live young has now been established from a second specimen (Bruton *et al.*, 1992), but before this find there had been considerable speculation as to how the young grow within the oviduct. The initial discovery of young within the oviduct showed five embryos (popular literature often refers to them as 'pups'), each contained within an expanded portion of the oviduct and separated from its neighbour by a constriction within the oviduct wall. All were found facing anteriorly. The embryos ranged from 301 to 327 mm in a mother of 160 cm and were probably close to being born. The fact that each embryo carried a large yolk sac might suggest that they are entirely lecithotrophic, depending entirely on the yolk supply. However, Wourms *et al.* (1988, 1991) have constructed a complex scenario in which they suggest that *Latimeria* embryos supplement the yolk supply in at least the later stages of their prenatal development by eating supernumerary eggs and by utilizing a yolk-sac placenta whereby some nutrient may be passed from mother to embryo. The term 'placenta' is used but it should be pointed out that this refers to a simple contact between maternal oviduct wall and foetal yolk-sac wall.

The basis of this inference comes from calculations of changes in weight. Embryos that depend entirely on their own yolk supply show either no dry weight gain or slight dry weight loss throughout their development from the egg. Wourms *et al.* (1991) calculated a change in dry weight from egg to embryo from −7% to +30%. These calculations were necessarily vague because actual weights could not be measured but had to be estimated from wet weight of the egg and alcohol-preserved embryos and compared with known wet/dry weights of shark embryos. Those workers decided to accept the higher figures and concluded that there was indeed a dry weight gain, implying a source of food additional to the yolk contained in the yolk sac.

The idea that additional food supply comes from eating eggs, or the fragmented supernumerary eggs, is based on the observation that Suyehiro *et al.* (1982) found 30 eggs and Millot and Anthony (1974) found 19 eggs in separate females, yet only five young were found by Smith *et al.* (1975). Because each embryo lives within its own constricted part of the oviduct, then the egg material must be assumed to be passed down the oviduct as broken-up food material.

The presence of a yolk-sac placenta was inferred from several observations: an increased vascularization of the oviduct wall adjacent to the yolk sac of each of the embryos; the nature of the oviduct vascularization, which is similar to capillary networks associated with gaseous exchange and the passage of low-molecular-weight molecules in other vertebrates; the position of the embryos, which appear to remain constant; a large yolk sac, which maintains a large and constant size right up to the time of birth; and the apparent absence of a vitelline duct, which in lecithotrophic embryos connects the yolk sac to the embryo gut.

Heemstra and Compagno (1989) questioned the idea that there was intra-uterine oophagy in *Latimeria*, pointing out that there was little direct evidence. The discovery of a second pregnant female (Bruton *et al.*, 1992) has only increased this scepticism (Fricke and Frahm, 1992; Heemstra and Greenwood, 1992). This second female contained 26 late-term foetuses; thus, there was no need to explain the large numbers of supernumerary eggs. The yolk sac was variously developed among the different foetuses and in some it was reduced to an external yolk-sac scar, demonstrating clearly that an external yolk-sac did not survive the entire prenatal life and therefore could not function as a yolk-sac placenta, at least in late developmental stages.

Dissection of one of the foetuses (Heemstra and Greenwood 1992) revealed a small internal yolk sac which was in direct connection with the pyloric end of the stomach. Additional dissection of a foetus (BMNH 1976.7.1.16) from the first pregnant female also showed a direct connection between the yolk sac and the gut (Heemstra and Greenwood, 1992: fig. 7).

Fricke and Frahm (1992) also estimated the dry-freeze weight of the new foetuses to calculate that they were 15–31% below the dry weight of the eggs. Thus these figures are in disagreement with those of Wourms *et al.* (1991), but they are more likely to be correct because the foetus weight is measured directly and reflects a late-term individual. From this new evidence it appears as though hypotheses of complex and additional sources of food, other than from the yolk sac, are unnecessary to account for the development of *Latimeria* young. *Latimeria* appears to be an entirely lecithotrophic, ovoviviparous fish.

2.7 GEOGRAPHICAL DISTRIBUTION

The vast majority of the 180 or so recorded individuals (Bruton and Coutividis, 1991) have been caught around the islands of Grande Comore and Anjouan in the Comores Archipelago, Indian Ocean (Figs 2.8, 2.9 and see below). There have been three exceptions.

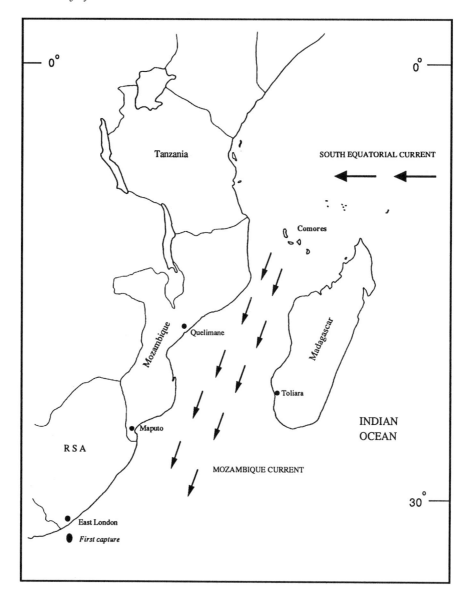

Fig. 2.8 Geographic distribution of *Latimeria*. The first coelacanth was captured in shallow water and landed at East London. Subsequent catches were located in deep water around the Comores Islands some 2000 km to the north. Thus, the first capture was explained as a stray which had got caught up in the strong Mozambique Current flowing down the east coast of southern Africa. However, there have been two further discoveries which suggest that the geographic range of *Latimeria* may be greater than originally thought. In 1992 a pregnant female was caught near Quelimane, Mozambique and landed at Maputo. In 1995 an individual was fished from deep water off Toliara, Madagascar.

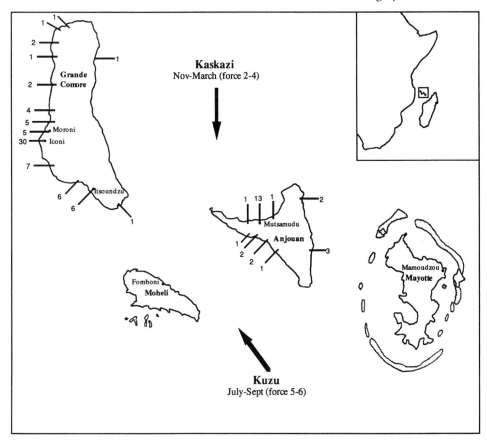

Fig. 2.9 Most coelacanths have been fished from deep waters off the west coast of Grande Comore and the north coast of Anjouan in the Comores Archipelago. This diagram shows the number of catches against the location of the fishing villages at which they were landed. Not too much can be read into the subtle variations in catch numbers per location because the sizes of the fishing villages vary and local fishers do not travel far from their home base. However, there is a clear bias towards the western side of Grande Comore. The strength, direction and times of the two major prevalent winds, which influence fishing effort, are shown (cf. Fig. 2.11).

The first specimen was caught off the mouth of the Chalumna River, near East London, South Africa. That individual was regarded as a stray (Smith, 1939a,b; Bruton, 1989) because most subsequent catches have been made around the Comores Islands, some 2000 km to the north, and up-current, of the site of first capture (Fig. 2.8). The second exception is the capture of a large pregnant female close to the Mozambique coast (Fig. 2.8) just north of Quelimane (Bruton *et al.* 1992). Once again this was thought to have been a stray, accidentally caught up in the strong southwardly flowing Mozambique Current. In the second instance, some tissue samples were taken from one of the embryos and analysed for DNA 'fingerprints' by comparing it to five individuals collected from Grande Comore (Schliewen *et al.*, 1993). There was no significant difference between the Mozambique specimen and the Grande Comore specimens and the conclusion was that they belonged to the same population. Late in 1995 a third 'non-Comoran' coelacanth was caught off Toliara (Tuléar) on the south-west coast of Madagascar some 1300 km south of Grande Comore (Fig. 2.8; Heemstra *et al.*, 1996). This capture is interesting because the fish was at depth (140–150 m) and is less likely to have been a stray caught up in surface currents. The local fishers at Toliara traditionally fish for sharks using hook and line but have recently switched technique to setting deep-sea shark nets (Heemstra *et al.*, 1996) and this less discriminatory method may capture more coelacanths. This capture off Madagascar is significant because there have been several claims by authorities in Malagasy that coelacanths have been caught around Madagascar. Until very recently, no specimens have been associated with any of these claims (Bruton 1989). Any single one of these catches may be explained as a stray but, together, they lend support to the idea that the endemic range of *Latimeria* includes the eastern coast of Africa and Madagascar as well as the Comores.

Within the Comores Islands, *Latimeria* has been recorded from the coasts of Grande Comore and Anjouan (a full inventory is given in Bruton and Coutividis, 1991: appendix 1); there are no records from Moheli or Mayotte, which are the two other islands in the Comores Island chain (Fig. 2.9). These are the only locations in the Comores at which *Latimeria* is definitely known, but before discussing possible reasons for such a limited distribution, some speculations about other populations of *Latimeria* must be mentioned.

Reports of *Latimeria* living in the Atlantic Ocean and the Mediterranean are more difficult to justify and are based on rumours of scales and silver models. Ley (1959) reports that, in 1949, a peculiar scale was sent by the owner of a gift shop near Tampa, Florida to The United States National Museum in Washington where it was identified as being a possible crossopterygian. The scale was alleged to have been given to the shop owner by local fishers working in the Gulf of Mexico. Unfortunately, the scale can no longer be traced and this claim of an Atlantic coelacanth must be dismissed as lacking any evidence.

Other records are based on the existence of two silver talismans which have been traced back to churches in Spain (de Sylva, 1966; Anthony, 1976). Of course, these amulets could have been crafted almost anywhere, but as one of them is reputedly over a century old there is the suggestion that these were crafted after the model of local fishes with the implication that there may be an eastern Atlantic or a Mediterranean population of coelacanths (de Sylva, 1966).

The silver talismans are slightly different from each other. The first (de Sylva, 1966) came from a church near Bilbao on the Atlantic coast of Spain. It was purchased by a chemist who believed the object to have been crafted at least a century ago. The amulet is 12 cm long and shows the typical coelacanth fin arrangement. The obvious difference from *Latimeria* is that the caudal peduncle is very narrow and is fashioned as being quite separate from the caudal fin. In all, this part of the amulet looks very like the shape of a tail of a teleost fish. The second dorsal and the anal fins are not clearly modelled but they appear to have been fashioned after a lobed fin of the coelacanth rather than the more usual broad-based ray fin, common in teleosts. The sail-like first dorsal fin, the large opercular flap and the obvious lobed paired fins are all features seen in the living coelacanth. It is very likely that a coelacanth was used as the model. What is particularly striking is the pose of the paired fins. The pectoral fins on the amulet are shown inserted high on the flanks, as in *Latimeria*, and they extend directly downwards. This is precisely the pose in which the first specimen was mounted in 1939 (Smith, 1956: pl. 2). It is possible that this amulet was crafted using a photograph of the holotype: such photographs were widely circulated in the press, including on the 'wanted poster' circulated by Smith (1956: pl. 3). This would mean that it could not have been modelled in the 19th century.

The second amulet (Anthony, 1976: pl. 9) was found in a Paris antique shop and is reputed to have come from Toledo, Spain. It is far more accurate than the first and is more finely sculptured. It is possible that the tail has been modelled separately from the rest of the body because the sculpture is far cruder and sharply delimited from the trunk. The dorsal fins are accurately crafted, although the first dorsal is placed further back than in the living fish. The paired fins are also well sculptured, although the pectoral is placed too low and too far back for an exact match with *Latimeria*. But other aspects, including the deep caudal peduncle and the shape of the tail, the modelling of a lip fold and the few but well-marked fangs within the mouth, are good likenesses to these features in *Latimeria*. And even the white blotches seen as markings in *Latimeria* have been reproduced along the flank of the amulet.

There is no doubt that this was modelled on a specimen of *Latimeria*. However, there is no reason to doubt that it was modelled after a Comores specimen. As Thomson (1991) points out, centuries of trading along the East African coast could easily have located a coelacanth caught by the local Comoran fishers and brought back to Europe.

So, these amulets do not prove that there are other populations of coelacanths living in the Atlantic or Mediterranean although they may well suggest that European knowledge of the living coelacanth outside of the Comores is considerably older than we think. In fact, the fishes of the coastal Atlantic and the Mediterranean are very well known and it is unlikely that a fish the size of *Latimeria* could go unnoticed, particularly after the 1938 find.

There have been extensive searches for *Latimeria* in the Western Indian Ocean (Forster *et al.*, 1970) in areas to the north of the Comores, including steep-profile islands superficially similar to the Comores (e.g. Aldabra, Assumption, Farquhar Islands). These have drawn a blank even though large fish species found living alongside *Latimeria* in the Comores, such as *Ruvettus pretiosus* (oilfish), *Prometichthys prometheus* (roudi) and the deep-water squaloid *Centrophorus granulosus*, are all found to be widely distributed. So, the present 'negative' evidence points to *Latimeria* having a very restricted distribution centred on the Comores and possibly including the south-east coast of Africa and west coast of Madagascar. So the question then becomes: is there something special about the Comores?

The geological history of the Comores suggests that the home of *Latimeria* is very young indeed. The Comores are situated on the Somali crustal plate where they rise out of the ocean on submarine plateaus which separate the ocean basins (Mart, 1988). It is thought that the Comores lie at the south-western end of a series of volcanic islands which form an arcuate chain which clips northern Madagascar before running north to the Farquhar Islands and the Seychelles (Fig. 2.10). Emeric and Duncan (1982) have plotted the age of volcanism on several islands and concluded that the most likely explanation for the existence of the Comores is that they were formed as the Somali plate moved north-east over a fixed hot spot within the mantle. Grande Comore is the youngest of the islands and here the oldest volcanism is dated at only 0.75–1.3 million years. Mount Kartala on Grande Comore remains an active volcano and last erupted in 1977. The oldest of the Comores islands is Mayotte, the most easterly and dated at 5.4 million years with Moheli (2.75 million years) and Anjouan (estimated at 3.5 million years) intermediate in age.

One of the theories put forward to explain

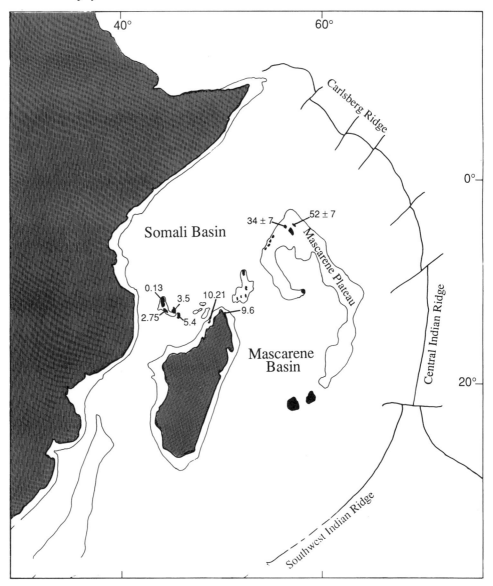

Fig. 2.10 Geology of the Comores Archipelago. The Comores Islands lie between two ocean basins on the Somali crustal plate. It is thought that the Comores came into existence as the plate moved over a crustal hot spot which periodically erupted, forming the volcanic islands which are youngest in the west and arc to the north-east to the Mascarene Plateau. The ages of the islands (million years) are shown after estimations by Emeric and Duncan (1982).

the limited distribution of *Latimeria* is derived from Darwin's concept of living fossils. He viewed them as competitively inferior. Thomson (1991) developed this idea to suggest that *Latimeria* avoids competition and retreats to selected and geographically restricted environments – in this case, relatively new environments which have not yet been colonized by serious competitors. This 'exodus' theory predicts both that *Latimeria* was formerly present around neighbouring islands but has now been usurped by competitively superior animals, and that the long-term survival of the coelacanth lineage is dependent upon the eruption of new volcanic islands.

This would appear to be a very chance-related survival strategy. Instead, if the distribution of *Latimeria* really is limited (but see above), this restriction may be due to specific habitat requirements such as steep sand-free slopes, crevices for hiding by day, ready access to deeper water and relatively cold water (see below). At present, the Comores may provide such habitats but these are equally likely to be found in neighbouring parts of the Indian Ocean. It may be that the particular distribution of such habitats is so widely-spaced that this precludes *Latimeria* populations from being more widely spread. However, considering that the first Madagascan individual was only found some 55 years after the first capture (which itself was 2000 km from the Comores), it would be unwise to say that we 'know' the distribution of *Latimeria*.

2.8 HABITAT

There have been predictions about the habitat preferences of *Latimeria*. When the first individual was trawled from a depth of 80 metres, White (1939) suggested that it may have been a stray from a natural home in deep oceanic waters. Smith (1939c) disagreed, pointing out that the thick scales and head bones, the tough flesh and the large eyes of *Latimeria* were unlike that expected in a bathypelagic or abyssal fish but more like that of a reef-dwelling fish. However, Smith also acknowledged that the shallow reefs along the East African coast were likely to have been thoroughly explored by fishers. He therefore predicted that populations of *Latimeria* would be found in rocky areas below

100 metres and thus below the reach of most line fishers. Smith's prediction turned out to be more accurate than White's.

After the second and subsequent coelacanths were caught in the Comores Islands, speculation about habitat preference became more precise. Catch records by Comoran fishers suggested that coelacanths lived at depths of 180–400 m around Grande Comore and Anjouan. Furthermore the catches were largely restricted to the west coast of Grande Comore and the north and east coasts of Anjouan (Fig. 2.9), both areas known to be steep and rocky. The catches around Grande Comore were thought to be associated with submarine freshwater aquifers (Forster, 1974; McCosker, 1979), formed by rain water percolating down through the porous lava flows. This gave rise to the idea that *Latimeria* was really a freshwater fish (as would be predicted in a tetrapod ancestor) trapped in a marine environment. Such aquifers do exist, at least temporarily in the rainy season, but they do not appear to occur much below 80 metres (Fricke *et al.*, 1991). Stobbs (1989) has further pointed out that freshwater aquifers do not occur around the coasts of Anjouan. Thus, if *Latimeria* is found around submarine aquifers, it is clearly not an obligatory association.

Thomson (1973) and Stobbs (1989) have both pointed out that the catch records throughout the islands may reflect the distribution and fishing habits of the local fishers (Stobbs and Bruton, 1991) rather than the true distribution of *Latimeria*. For instance, Stobbs (1989) emphasizes that fishers, using traditional boats and longlines, rarely fish more than 2 km from the shore and that they concentrate their fishing efforts in the calmer waters and at certain times of year. The catch records show clearly that coelacanths are usually hooked at depths greater than 100 metres (Fig. 2.11) and Stobbs suggests that the absence of catch records from the islands of Mohele and the south side of Anjouan may be due to the fact that here the 100

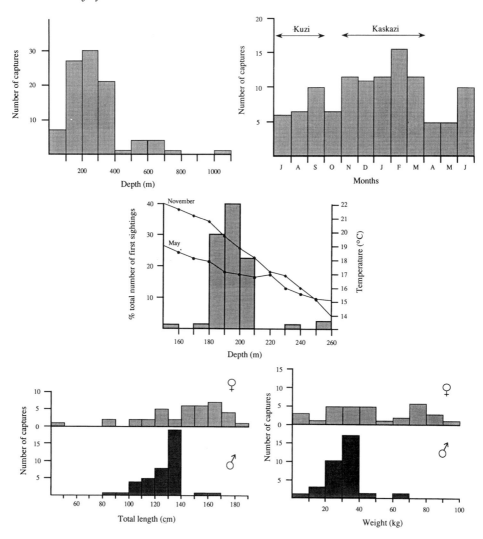

Fig. 2.11 Statistics of line-fishing catches for *Latimeria*. Top left: depth distribution of coelacanth catches, showing that most are caught between 100 and 400 metres. These measurements are made by knowing how much line is played out, and as Stobbs (1989) points out, the effect of the strong current may exaggerate the estimated depth. Centre: a more valid assessment of depth occurrence is that by Fricke *et al.* (1991), who recorded the depth at which a coelacanth was first observed during a dive. Most were encountered between 180 and 210 metres. There is a difference in depth of first encounter between May and November reflecting the seasons and depth of the 18 °C isotherm. Top right: catch records through the year, showing that most coelacanths are caught during November to March at the time of the lesser of the two dominant winds (Fig. 2.9) and probably reflect fishing effort. Bottom left and right: there is a decidedly skewed distribution in length and weight between captured males and females, the latter being by far the heavier (from data summarized in Bruton & Coutouvidis, 1991).

metre isobath occurs well beyond 2 km from shore, beyond the range of traditional fishing methods. Presumably increased fishing effort further from the coast could provide a test of this hypothesis.

Given the fact that our present knowledge of coelacanth catches is largely restricted to certain coasts within the Comores Islands (Fig. 2.9), then some associations may be made. The west coast of Grande Comore consists of steep submarine profiles (Fricke and Plante, 1988) and there are numerous lava caves developed down to about 220 metres where sand cover obliterates most deeper caves (Fricke et al., 1991). On the east coast, where catches are very rare, sand cover is far more extensive in shallower water. The north and east coasts of Anjouan have submarine profiles which are also steep with eroded caves and limited sand cover.

The inference from this is that *Latimeria* prefers steep, rocky and deeply dissected submarine profiles with little sand cover between 100 and 400 metres.

Direct observations from submersibles have been carried out by Fricke and his colleagues (Fricke et al., 1987, 1991; Fricke, 1988; Fricke and Hissman, 1990) who, in the course of four expeditions, have made some 204 dives. Most of these dives have been made on the south-west coast of Grand Comore. Coelacanths have been observed at 150–250 metres with the vast majority encountered at 180–210 metres (Fig. 2.11). And during the first expedition a single individual was followed through a nocturnal migration from 198 to 117 metres. Another, radio-tagged individual was followed through a migration from 200 to 500 m (Fricke and Hissman, 1994).

It is generally believed that the minimum depth distribution of *Latimeria* is governed by temperature because the maximum oxygen saturation of coelacanth haemoglobin occurs at 15–18 °C. *Latimeria* is unlikely to be able to tolerate temperatures exceeding 22 °C for any length of time. The depth of observed coela-

canths does correlate with the depth of the 18 °C thermocline. This varies from 228 metres in November – the warm, stable summer season – to 174 metres in May – the cold season (Fricke et al., 1991: table 1). And Fricke further noticed that the depths of the first encounters while diving were shallower in the cold season (Fig. 2.11). The implications of these observations are that the upper limit of the natural distribution of *Latimeria* coincides with the 18 °C thermocline while the lower depth may be controlled by the availability of barren rocky substrate.

During the third and fourth expeditions undertaken by Fricke (Fricke et al., 1991; Fricke and Hissman, 1992), coelacanths were seen at 180–253 metres along the west coast of Grande Comore in water temperatures of 15–19 °C and variable water current speeds of 5–16 cm s^{-1}. Fricke found that during the day, coelacanths rested up in caves 2–3 metres deep within the lava cliffs. Within these caves the water temperature was generally higher and in some cases reached 23 °C (Fricke et al., 1991: table 1). This observation, and the fact that the water temperature of the Mozambique find (Bruton et al., 1992) was 28–30 °C, calls into question the preferred temperature tolerance (Hughes and Itasawa, 1972).

Fricke observed that the coelacanths left the caves after sunset to drift close to the substrate and, presumably, to feed. Some left the caves to migrate to deeper water. Just before dawn, coelacanths returned to the nearest cave. This excursion behaviour is quite common for reef fishes. Coelacanths have white blotch markings irregularly scattered over the body. This provides a camouflage against a cave wall encrusted with sponges and oysters (Fricke et al. 1991: fig. 12). The white markings also allow coelacanths to be individualy recognized – at least when analysing film taken during the dives. From this Fricke was able to show that, over a period of days, an individual may use at least five caves spread along 8 km of lava cliff (Fricke *et al.*, 1991: fig. 1). Fricke also found that three of the individuals that he filmed in 1987 were found in the same place in 1989, suggesting that coelacanths are home-loving.

A number of coelacanths have been hooked and brought to the surface alive where they have survived for a few hours. But they do not seem able to survive for very long and this denies any attempts to keep coelacanths in aquaria. There have been a number of suggestions for their vulnerability. It is possible that the coelacanth is poor at getting rid of excess lactic acid accumulation within the muscle tissue as a result of struggling at the end of a line. Another suggestion centres on the observations of Hughes (1976, 1980) that the gill area/body mass ratio is very low, meaning that any increased metabolic rate, which may be induced either by struggling on a line or by an increase in water temperature, would lead to increased oxygen demands which the gill surface area could not meet. The temperature difference between the surface and the natural depth at which coelacanths live is at least 10 °C and this could induce hypoxic stress (Hughes and Itasawa, 1972). The haemoglobin of the coelacanth shows that maximum efficiency at picking up oxygen and releasing carbon dioxide is reached at 15 °C. A 10 °C rise in temperature, as well as an increase in acidity due to muscular activity, would significantly move the dissociation curve such that coelacanth haemoglobin would be unable to release sufficient oxygen to the tissues.

2.9 GENETIC DIVERSITY, POPULATION SIZE AND CONSERVATION

One phenomenon of relict species with restricted distributions is a reduced genetic diversity within small populations. If true, this may have some implication for survival. Little is known of genetic variability in *Latimeria* but there has been one preliminary study conducted to try and evaluate this.

Setter and Brown (1991) carried out electrophoretic studies for 14 enzymes taken from a variety of tissues of two coelacanth specimens. They found that few of the enzymes had more than two isoloci and that there was no variation between the two individuals studied. Given that this sample was small it remains true that both of these facts suggest that there is very little genetic variability.

Evaluation of the need and requirements for the conservation of the coelacanth demand some knowledge of population size and structure. This has been very difficult to acquire. Catch records, which may give some indication of availability of individuals, have been compiled from time to time (Millot *et al.*, 1972; McAllister and Smith, 1978; McCosker, 1979; Thys van den Audernaerde, 1984) and that given by Bruton and Coutividis (1991) is by far the most complete and includes sex, length and weight data where known. To date, 172 captures have been recorded at an average rate of four per year since 1954 (there have been as many as 10 caught in some years – Bruton and Coutividis, 1991: fig. 4). Although local fishers do not specifically target the coelacanth, because it is usually caught as a bycatch (see below), this is a very low number and suggests either low population size or that most coelacanths live outside the fishing areas. About half of the captured individuals have been sexed and there appear to be approximately equal numbers of males and females, with both the modal length (♀ 160 cm, ♂ 130 cm) and modal weight (♀ 75 kg; ♂ 35 kg) substantially larger for females (Fig. 2.11). Females also account for the largest individuals (Bruton and Armstrong, 1991). Amongst the catches and direct underwater observations (Fricke and Hissman, 1990; Fricke *et al.*, 1991), fewer than 10% of individuals are less than 100 cm. This may mean that the young do not live within the fishing area or that they stay close to the submarine caves.

Fricke *et al.* (1991) estimate a population

size of between 210 and 400 individuals for Grande Comore and Bruton and Armstrong (1991) estimate a total of 500 for Grande Comore and Anjouan. This is based on extrapolation from repeated observations of cave inhabitants along an unspecified 8 km stretch of coastline (this represents approximately 10% of the west coast of Grande Comore). Bruton and Armstrong (1991) extrapolate this figure to a total population of 500 for Grande Comore and Anjouan. There are many unquantifiable variables in this calculation but the figure remains alarmingly low.

Age calculations for *Latimeria* are similarly difficult to make. Based on measurements of the growth rings of 12 specimens from a full-term foetus to an adult female (180 cm), Hureau and Ozouf (1977) calulated that coelacanths live at least 12 years. This calculation is based on the assumption that two growth rings are laid down each year, coincident with rainy seasons. Balon *et al.* (1988) disagreed, suggesting instead that, like many tropical fishes, an internal annual growth rhythm results in a single ring per year. Thus, under this assumption coelacanths would reach at least 20 years. Estimated growth rates range from 6.5 to 13 $cm\,year^{-1}$ (Hureau and Ozouf, 1977; Uyeno, 1984; Balon *et al.*, 1988). The age of maturity is unknown. The female dissected to reveal five full-term foetuses (AMNH 32949) is 160 cm long and, assuming the young to have been developing inside the mother for a year, this means that maturity is reached at least after 9 years (two growth rings per year) or 17 years (one growth ring per year).

Estimates of fecundity can only be based on the two records of females bearing young. The first pregnant female contained five young (Smith *et al.*, 1975) and the second (Bruton *et al.*, 1992) contained 26. Whether these are extremes is unknown, but these disparate figures clearly make considerable differences to our ideas of recruitment to the population (as well as to ideas of reproductive biology – see above). To date, only one very small coelacanth measuring 42 cm has been caught (most other 'small' coelacanths known are between 80 and 100 cm). This skewed size distribution may mean that we have not yet found where the young live, but it may also mean that few survive to reach maturity.

With so little known, it is difficult to make informed judgements about conservation status. But the need for some 'endangered' recognition is emphasized by recent observations by Fricke *et al.* (1995) and Browne (1995). During his series of submersible dives (1988–1994), Fricke had observed some 60 individuals on each occasion within a short stretch of coastline. The latest dive observed only 40, a substantial drop. This may be a natural population fluctuation, a fact that could be established with further dives. But equally it could be due to a change in fishing habits (Fricke *et al.*, 1995). Coelacanths are usually caught as bycatch when fishing for the oilfish, *Ruvettus pretiosus*. The oil within the flesh of *Ruvettus* is used both as a laxative and as a mosquito repellent. Local fishers traditionally fished at night, when the fishes bite, using baited longlines from small outrigger canoes (Stobbs, 1989; Stobbs and Bruton, 1991). The coelacanth was not favoured and, until scientists showed interest, hooked coelacanths were probably thrown back and no doubt some survived, although attempts to keep coelacanths alive at the surface have so far failed (Thomson, 1991). The prize of a coelacanth which may be traded to individual scientists, museums, aquaria or to official buildings certainly encouraged coelacanths to be landed. As long as this demand was low, the few coelacanths caught probably did little harm to the population. Fishing effort was confined to very near shore because the fishers, using the rather crude canoes, could not venture far from shore. In the late 1980s there was a change in local fishing habits which took the fishers further out into more productive waters. Instead of canoes, international aid provided fibreglass boats powered with outboard motors and effectively took the fishers away from coelacanth territory. Now those motors have broken, forcing the fishers back into shallow waters (Fricke *et al.*, 1995), threatening the coelacanth once again.

There are a number of actions that can be undertaken to reduce the numbers of coelacanth catches.

- Fishers can be encouraged to fish away from nearshore or at much shallower depths than are frequented by coelacanths. This can be done by either repairing the outboard motors or employing fish attracting devices (Fricke *et al.*, 1995).
- Alternative laxatives and malarial salves can be encouraged to dimish the need for the oilfish.
- The scientific demand can be reduced through the use of high-quality models and cine film of live coelacanths.
- Awareness of the coelacanth can be increased through official international conservation organizations.

A considerable amount has been done to satisfy the last. In 1987 the Coelacanth Conservation Council (CCC) was set up (Bruton, 1988). The aims of the council are various.

1. To upgrade the coelacanth in the IUCN *Red Data Book* of endangered species from 'K' category ("Insufficient known to justify action") to 'V' category ("Vulnerable to extinction if nothing changes".

2. To reduce the incentive in illegal trade. At present the coelacanth is listed in Appendix I of CITES (Convention in International Trade in Endangered Species). Many countries are signatories to this convention, unfortunately the Comores is not one of them.

3. To attempt to establish a coelacanth national park (Bruton and Stobbs, 1991) in which fishing is restricted and in which an educational centre may be built.

4. To coordinate research on the coelacanth and to raise funds to help carry out that research.

5. To investigate ways in which local fishing can be undertaken without undue pressure on the local coelacanth population.

6. To publish newsletters, updates of research activities and other literature making scientists and media aware of the current status of the coelacanth.

3

DERMAL BONES OF THE SKULL ROOF

3.1 INTRODUCTION

The dermal bones covering the dorsal surface of the neurocranium are divided into those anterior and those posterior to the intracranial joint. The anterior bones include the parietals (frontals), nasals, supraorbitals, tectals and the rostrals. The premaxilla and preorbital may also be considered here because they are invariably closely associated with the rostrals and tectals. Together these bones cover the dorsal surface of the ethmosphenoid and they may be considered to form a parietonasal shield. In previous literature this portion of the dermal roof has been called the fronto-ethmoid shield but it is decided here to adopt tetrapod terminology to bring coelacanths into line with other sarcopterygians.

The bones behind the level of the intracranial joint are the postparietals (parietals), supratemporals (tabulars) and intertemporals (supratemporals). The extrascapulars may also be considered here: although they are not primitively associated with the underlying neurocranium they are often preserved with the skull roof in fossils. The bones posterior to the joint may be considered as forming the postparietal shield (parietal shield of previous literature).

Among coelacanths there is little difficulty in recognizing topographic homologues within the parietonasal and postparietal shields. Nevertheless, a variety of names have been used for the roofing bones (see for instance Schaeffer, 1952a: table 2), reflecting different authors' ideas on coelacanth rela-tionships and different ideas on the history of the osteichthyan dermal skeleton (e.g. whether it is considered legitimate to assume phylogenetic loss, fusion or fragmentation). The theoretical problems accompanying these different assumptions and the practical consequences to which they lead have been discussed by several authors (Jardine, 1970: 345; Miles, 1977: 220–225; Patterson, 1977b). In each of the individual descriptions given below, the bone synonyms used for that particular genus are noted to facilitate direct comparison between this and earlier literature.

The parietonasal shield of coelacanths has remained relatively conservative, permitting a few general remarks to be made. Above the orbit and the posterior half of the ethmoid region, the dermal bones are arranged in two paired series related to the supraorbital sensory canal. A mesial series, the parietonasal series, consists of parietals posteriorly and nasals anteriorly; the lateral series, the supraorbito-tectal series, consists of supraorbitals posteriorly and tectals anteriorly. The supraorbital sensory canal runs between these two series following a sutural course. Anteriorly to these paired series there are a variable number of rostrals, some of which are occasionally wedged between the nasal series of either side and distinguished as internasals. On either side of the anterior tectals there is always a large, robust rostral element – the lateral rostral, which carries the infraorbital sensory canal and is associated with both anterior and posterior nostrils.

The chief variations that occur in the

dermal shield affect the number of rostrals and internasals, parietal number, development of the pit lines upon the postparietal shield, extrascapular morphology, the relative superficiality of the more posterior elements and the relative proportions of the parietonasal/postparietal shields.

3.2 DESCRIPTIONS OF GENERA

Latimeria

The skull roof of adult *Latimeria* has been described in detail by Smith (1939c), Jarvik (1942) and Millot and Anthony (1958). The dermal bones of an embryo specimen (BMNH 1976. 7.1.16) are illustrated in Figs 3.1 and 3.2. A comparison of these figures with that provided by Smith (1939c: fig. 3) or Millot and Anthony (1958: pl. 19) shows that even at this relatively early growth stage, most of the bones are well developed. In the embryo specimen there is a considerable amount of cartilage of the neurocranial roof exposed between the opposing parietonasal series and immediately in front of the lateral rostral. Even in adult specimens the dermal bones never completely cover this ethmosphenoid cartilage. The bones are unornamented, although there may be ridges present which radiate from the growth centres. The sensory canals send primary tubules through the bones or, more usually, through large pores between bones (Fig. 3.2). The branching of the primary tubules into the many surface pores (Figs 2.2, 2.4, 4.3) occurs in the skin.

The parietonasal series is here taken to include that row of bones which carry the supraorbital sensory canals along their lateral edges, these being grooved for their reception. The holotype and BMNH 1976: 7:1:16 show five members in this series, each successive member being larger than its anterior neighbour. Anteriorly there are three nasals (Fig. 3.2, Na), (Jarvik's nasals 3, 4 and 5; Smith's rostral, postrostral and parietonasal). Posteriorly there are two parietals (Fig. 3.1, Pa), anterior and posterior (respectively Jarvik's nasal 6 and frontal; Smith's frontals). The specimen illustrated by Millot and Anthony shows only four bones in this series, termed nasals 1, 2 and 3, followed by the parietal. The anteriormost nasal occupies the territory claimed by the first and second nasals in the holotype and BMNH 1976: 7:1:16. The nasals and the anterior parietal are unornamented, showing only radiating lines of growth from the ossification centres. The sutural relations between these bones are simple, one bone directly abutting against another. The posterior parietal is a more complex bone which has been described in some detail by Smith (1939c: 30–32). The raised area upon the dorsal surface of the adult posterior parietal is not nearly so well developed in the embryo (it can be seen as a faint irregular outline beneath the skin – Fig. 2.4) but its lateral edge is marked by a pronounced undulating ridge. A very thick layer of connective tissue (Millot and Anthony, 1958: pl. 15) is attached to this ridge, which spreads ventrolaterally over a thin lateral lamina of the parietal and is attached to the dorsal and anterior margins of the postorbital within an excavation (Fig. 2.5, Fig. 4.1, exc.Po). The anterior and posterior parietals meet in a complicated interdigitating suture in which the anterior parietal partially overlaps the posterior parietal. This sutural relationship is found nowhere else on the parietonasal shield.

Smith thought that the posterior parietal ossified from two, possibly three centres: one of which lies towards the posterior edge of the raised portion and which forms most of the horizontal lamina; a second which forms a posteroventral process arising from the undersurface and which becomes attached to the basisphenoid; and possibly a third which forms a posterolateral prong of the posterior parietal. With the material available, however, only one centre of ossification can be identified from which grows the horizontal lamina forming the roof and the

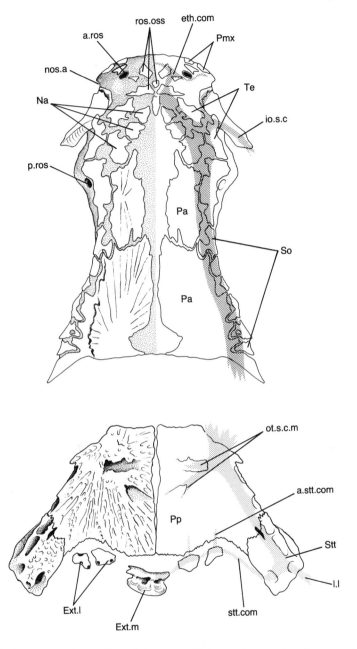

Fig. 3.1 *Latimeria chalumnae* Smith. Skull roof drawn as if the parietonasal and postparietal shields have been pulled apart. Path of sensory canals shown by stipple on right side. Cartilage areas of ethmo-sphenoid cartilage are shown on the left side as light stipple. Based on a camera lucida drawing of the left side of an embryo specimen (BMNH 1976.7.1.16) in which the raised areas characteristic of the adults of this species have not yet developed. Not all extrascapulars or rostral ossicles as seen in the adult have ossified.

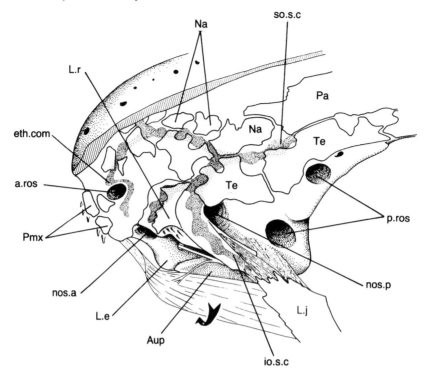

Fig. 3.2 *Latimeria chalumnae* Smith. Camera lucida drawing of the snout of an embryo specimen (BMNH 1976.7.1.16) in dorsolateral view to show positions of the nostrils, openings of the rostral organ, the paths of the sensory canals (fine stipple) and the insertion of the pseudomaxillary fold (reflexed – arrow). Note the interruption of the infraorbital sensory canal at the level of the anterior nostril.

posteroventral process. This latter process deserves further comment. It has usually been interpreted as a separate bone, the pleurosphenoid (Millot and Anthony, 1958: 7, pl. 24) or alisphenoid* (Smith, 1939c: 31), which has become fused to the frontal. These authors do not make it clear as to whether they think this fusion has taken place ontogenetically or phylogenetically. In either case, this fusion would have to have taken place

---

*The term alisphenoid was introduced into coelacanth literature by Stensiö (1921), who thought that it corresponded to the alisphenoid (= pterosphenoid) of higher actinopterygians such as *Amia* and teleosts. Stensiö's later remarks (1932: 21), however, made it clear that he did not consider the so-called alisphenoid of coelacanths to be the phylogenetic homologue of the halecostome pterosphenoid. Stensiö was working within a framework of relationship which recognized coelacanths as rhipidistian derivatives. At first (1921: 123) he considered the coelacanth alisphenoid as the product of phylogenetic fusion; thus "...the following bones of the Rhipidistia may in coelacanths be fused: ... the frontal [parietal] with the alisphenoid and the dermosphenotic,..." However, no separate alisphenoid has ever been recorded in rhipidistians. Stensiö later (Holmgren and Stensiö, 1936: 348 0.5pt>–349) changed his views by considering the coelacanth alisphenoid as representing an ossification in the alisphenoid thickening of the rhipidistian neurocranium. This embodies the idea that the ethmosphenoid of rhipidistians fragmented and that an endochondral alisphenoid portion became attached to the frontal in coelacanths. Once again no evidence is offered in support of this hypothesis. It still depends on the assumption that an endochondral bone has become fused with (or perhaps replaced by) a dermal bone. It is argued in the main text that such theories of fusion are unnecessary. It is more easily interpreted as a downgrowth of the frontal. A separate alisphenoid does not occur in coelacanths.

between a presumed cartilage bone, the alisphenoid or pleurosphenoid, and a dermal bone. It should be possible to detect ontogenetic fusion when a more complete growth series of *Latimeria* becomes available. The embryo *Latimeria* used in this work, however, shows the adult condition, with a single centre of ossification. Furthermore, the bone of this posteroventral process is laminate, like that of the main body of the frontal, and differs markedly from the spongy bone of the basisphenoid. There is thus no evidence of ontogenetic fusion.

A hypothesis of phylogenetic fusion between a separate alisphenoid and the parietal is difficult to substantiate. Patterson (1977b: 95) details four kinds of evidence which may be required before such a postu-

late can be accepted; the first of these is that in the primitive condition, two bones existed where now there is one. But there is no coelacanth known in which the posteroventral process is a separate ossification. Because neither this, nor any of the other requirements are satisfied in coelacanths, I conclude that the posteroventral process cannot be considered phylogenetically as a separate ossification.

It is more reasonable to interpret this posteroventral process as a hypertrophied lamina which descends from the undersurface of the skull roof. Descending laminae are seen in placoderms (Stensiö, 1963: fig. 42c; Miles and Westoll, 1968: 398; Young, 1978), primitive teleosts (Patterson, 1975: figs 146, 147) and *Eusthenopteron* (Jarvik, 1944). The

---

In other vertebrate groups, an ossification of the braincase in the posterior wall of the orbit, above the basisphenoid and in front of the otic capsule, has been given a variety of names. The bone was initially identified as an alisphenoid in mammals. Gregory and Noble (1924) argued convincingly that it is not a primary braincase element and cannot be considered as the homologue of the topographically comparable bone in non-mammalian vertebrates. They suggested the term laterosphenoid for the primary braincase element. Further study, chiefly by de Beer (1937) and Goodrich (1930), led to the creation of further terms: pterosphenoid for the bone in halecostomes and pleurosphenoid (or laterosphenoid). The bone in crocodiles and birds. Rieppel (1976) reserved the term laterosphenoid for the bone in snakes. The implication of these different terms is that the bones are not homologous in halecostomes, snakes, birds and crocodiles.

The arguments for non-homology are centred on embryological data. Thus, the pleurosphenoid of crocodiles and birds is an ossification within the embryonic pila antotica. In snake embryos there is no pila antotica. Therefore, Rieppel argues, the adult bone in the same position is not homologous with the pleurosphenoid of crocodiles (a sentiment already hinted at by de Beer 1926: 315). Rieppel therefore restricts the term laterosphenoid to that bone in snakes. Thus, for the authors mentioned so far, the essence of a pleurosphenoid is the cartilage precursor, the pila antotica. The essence of the snake laterosphenoid is the absence of a cartilage precursor in the chondrocranium, that is, it is a membrane bone (*sensu* Patterson 1977b: 79).

When we consider actinopterygians, the procedure of extrapolating from embryological to adult conditions to determine homology leads to difficulties. *Acipenser*, *Lepisosteus*, *Amia* and teleosts have a bone in the position of a pleurosphenoid (or laterosphenoid). Considering the adult condition only, no-one has ever doubted their homology. In *Acipenser* and *Lepisosteus* the pleurosphenoid develops as an ossification within the pila antotica. *Amia* lacks a pila antotica. Instead, a cartilage bar grows down lateral to the head vein rather than medial to it. This lateral bar, the pila lateralis, ossifies and the resulting ossification is called a pterosphenoid (de Beer 1926: 296). In teleosts such as *Salmo* there is neither a pila antotica nor a pila lateralis. The adult ossification developing in membrane, should therefore be called a laterosphenoid. But de Beer, quite rightly, was content to call it a pterosphenoid because it is obviously the phylogenetic homologue of the *Amia* pterosphenoid. It is also clear that the topographic homologue in adult *Lepisosteus* and *Acipenser* must also be considered the phylogenetic homologue of the pterosphenoid of *Amia*, irrespective of its embryological precursor. To rely on embryological criteria would result in the acceptance of three non-homologous structures in the actinopterygian lineage: pleurosphenoid, pterosphenoid and laterosphenoid.

The difficulties mentioned above stem from a belief that embryology is an all-powerful tool in recognizing adult homologies. This is, however, a misunderstanding which is best summed up by de Beer (1937: 39): "... the centres of ossification need not bear any relation to the positions of the original centres of chondrification". In searching for homology the only valid criteria are correspondence in position, shape and congruence with other character distribution; and in reconstructing phylogeny we use comparable life stages. I can see no reason why the pterosphenoid, pleurosphenoid and laterosphenoid cannot be considered phylogenetic homologues and suggest the use of the word pleurosphenoid as it is the one most widely used in the literature.

process in *Latimeria* and most other coelacanths is particularly complex; it buttresses the skull roof against the antotic process of the basisphenoid and will be referred to as the parietal descending process. This parietal descending process also forms the anterior half of the dermal part of the intracranial joint and has been satisfactorily described by Millot and Anthony (1958).

The tectal/supraorbital series (Figs 3.1, 3.2, Te, So) flanks the nasals and the parietals and forms a continuous row of bones from the snout to the posterolateral corner of the posterior parietal. Three large anterior members of the series are designated as tectals (Jarvik, 1942: from anterior to posterior, the rostro-nasal, naso-antorbital and the parafronto-antorbital of Smith) which, by definition, roof the ethmoid region. There may be a small tectal anteriorly (Smith, 1939c, fig. 3, 'rostral') but this is absent from the embryo specimen used here. The posteriormost tectal is always the largest of the series and it lies above the posterior openings from the rostral organ (Fig. 3.2, p.ros). Behind this there are six supraorbitals which lie deeply embedded in connective tissue.

The extreme tip of the snout is covered in a mosaic of loosely connected bones, the pattern of which is shown in Figs 3.1 and 3.2. Between the nasals, there are two small median internasals (inter-rostral and meso-rostral of Smith; postrostrals of Jarvik) which lie one behind the other. A union between the supraorbital sensory canals perforates the posterior of these (Fig. 3.1) and because this lies partly between the nasals it can be called an internasal. Anterior to the nasals, internasal and median rostral, there is a chain of three paired rostrals which overlie the ethmoid commissure. These bones have been misleadingly called median (Smith) or medial (Jarvik) rostrals. The topographic homologue of the actinopterygian antorbital lies within this chain. The anterior opening of the rostral organ lies either within the middle rostral or between the middle and lateral members of the series (Fig. 3.2, a.ros).

The premaxilla in the embryo is represented as two or three small, tooth-bearing splints on either side. Two are shown in Fig. 3.2 (Pmx). A fragmented premaxilla persists into the adult (Smith, 1939c). However, Schultze and Cloutier (1991) suggest that there may be some specimens which show a single paired premaxilla. Despite this variation, the premaxilla is never associated with the ethmoid commissure or the anterior opening of the rostral organ.

The lateral rostral (latero-rostro-nasal of Smith) is a very large bone (Figs 2.5, 3.1, 3.2, L.r) and contacts the two anterior tectals along their lateral margins. In coelacanths the lateral rostral has a very characteristic shape which has been described for *Latimeria* by Smith (1939c: 39) and Millot and Anthony (1958: 14). The infraorbital sensory canal (Fig. 3.2, io.s.c) runs through the posterior tubular portion before looping dorsally through the main body of the bone. The canal follows the anterior margin of the lateral rostral before skirting the posterior rim of the anterior nostril (nos.a). A ventral process of the lateral rostral descends to contact the lateral ethmoid and on the posterior edge of this process the surface of the bone is roughened. It is to this point that part of the tough connective tissue of the pseudomaxillary fold (Fig. 2.2, psmax) is attached.

The postparietal shield of the adult has been described in detail by Smith (1939c) and Millot and Anthony (1958) and there is little to add to their descriptions. The postparietal shield of the embryo specimen is relatively broader than in the adult (compare Fig. 3.1 with Millot and Anthony, 1958: pl. 19) and this reflects slightly shorter postparietals (intertemporals of Smith and Millot and Anthony). The prominent raised areas on the postparietal (Fig. 3.1, Pp) of the adult are not clearly marked in the embryo. The supratemporal (Stt) has an exposed face which is densely pitted and pierced by large pores along the otic canal and the supratemporal commissure. The lateral and posterior edges are strongly downturned and smooth where they clasp the neurocranium. Millot and Anthony describe (1958, 1965) a levator operculi which is said to originate from the posterolateral corner of the supratemporal. Such a muscle was not found in the dissection of the embryo. There is, however, a small ligament connecting the posterolateral edge of the supratemporal to the anterodorsal tip of the operculum. The descending process of the supratemporal (Fig. 6.1(A), v.pr.Stt) is very large in *Latimeria*; it reaches anteroventrally to encircle the posterior edge of the facial foramen and it also forms the hind wall of the temporal excavation (fossa temporalis of Smith). Cloutier (1991a,b) claims that a descending process of the supratemporal (his tabular) is absent from *Latimeria*; this is clearly an error.

The postparietal also has a large ventral process (Fig. 6.1(A), v.pr.Pp) (descending apophysis of the intertemporal of Millot and Anthony) which descends to the level of the facial foramen and which forms much of the median wall of the temporal excavation. The lateral edge of the postparietal is very thick in transverse section (Millot and Anthony, 1965: fig. 24) and this depth is almost entirely occupied by the enlarged otic sensory canal, the bone itself being quite thin. The anterior edge of the postparietal is thickened to form articulatory surfaces of the intracranial joint.

The extrascapulars form a chain of small ossicles, each barely larger than the contained sensory canal, and which lie within a posterior embayment of the skull roof between the supratemporals (Fig. 3.1; Millot and Anthony, 1958: pl. X). In the embryo specimen examined only the median (Ext.m) and two lateral extrascapulars (Ext.l) were ossified. In the adult there are four lateral extrascapulars plus a larger median extrascapular (Millot and Anthony, 1958; Smith, 1939c: fig. 2). Because there are gaps within the chain of extrascapulars of the embryo it is probable

that more will ossify during future growth. The 'post-temporal' described by Millot and Anthony is the first lateral line scale (see also Smith, 1939c).

The paths of the sensory canals can be seen in Figs 2.2–2.4, 3.1, 3.2, 4.3). They have been most completely described by Hensel (1986) and the innervation described by Northcutt and Bemis (1993). The supraorbital canal (so.s.c) runs between the suprorbital/tectal and parietonasal series and turns medially above the lateral rostral to meet its partner in the midline. There is also a connection between the supraorbital and infraorbital canal at the level where the infraorbital canal loops dorsally through the lateral rostral, in addition to a junction between the supraorbital canal and the ethmoid commissure (Fig. 3.2). The supraorbital canal is innervated by the superficial ophthalmic branch of the anterodorsal lateral line nerve (Northcutt and Bemis, 1993). This nerve also innervates the anterior and posterior tubes leading into and out of the rostral organ. On the snout (Fig. 3.2) the infraorbital canal (io.s.c) runs forward within the lateral rostral and ends just above the anterior nostril (nos.a). The ethmoid commissure (eth.com) is found free in the skin between the rostral ossicles as it loops over the anterior opening to the rostral organ (a.ros). There is therefore an interruption between the infraorbital canal and ethmoid commissure at the level of the anterior or incurrent nostril.

The supraorbital canal (so.s.c) continues back to join the otic and infraorbital canals just behind the level of the intracranial joint. This contrasts with the observations of Millot and Anthony (1958: fig. 10), who suggest that the supraorbital and otic canals do not join. The anteriormost neuromast organ within the otic canal is almost twice as large as others on the head but the significance of this is unclear. At the level of this junction the otic canal sends off a median branch which, in turn, immediately divides into two branches (ot.s.c.m) opening to the roof through two or more large fenestrae. The otic canal continues posteriorly along the edge of the postparietal and supratemporal and the junction with the supratemporal commissure and the lateral line occurs within the supratemporal (Fig. 3.1). The supratemporal commissure (stt.com) runs through the extrascapular series and sends off anterior branches (a.stt.com) which lie in the skin above the postparietals. The neuromasts within the postparietal are innervated by the middle lateral line nerve (formerly called the otic ramus of VII, Northcutt and Bemis, 1993), while the neuromasts in the supratemporal and those forming the supratemporal commissure are innervated by branches of the supratemporal lateral line nerve (Northcutt and Bemis, 1993).

The pit lines lie superficially in the skin and do not mark the underlying postparietals. A middle and posterior pit line can easily be recognized (Fig. 2.3, m.p.l, p.p.l). There is also a short pit line which lies anterior to the middle pit line: this is left unlabelled in Fig. 2.3. It may correspond to the line labelled by Hensel (1986: fig. 6) as an anterior pit line but the patterns are not precisely the same. It is doubtful that a true anterior pit line is present in *Latimeria* because all of the pit lines associated with the postparietals are innervated by either the middle or the supratemporal lateral line nerves (these two nerves join and hence the separate fibres cannot be traced – Northcutt and Bemis, 1993). There is no innervation by the anterodorsal lateral line nerve as would be expected for the anterior pit line.

Miguashaia

The skull roof has recently been described by Cloutier (1996). His description is used for coding characters in Table 9.1 and his reconstruction is reproduced here (Figs 3.3(A), 4.4). The skull roof is long and narrow and markedly constricted at the level of the intracranial joint. The joint margin is nearly straight and it appears that there is little or no gap between the two moieties. This is a condition like that in other joint-bearing sarcopterygians rather than that in more derived coelacanths.

The parietonasal shield is approximately as long as the postparietal shield and the small orbit is located well forward. Details of the snout are poorly known. The premaxilla bears 6–9 small pointed teeth and carries a dorsal lamina which, according to Cloutier, is pierced by the anterior opening to the rostral organ. Cloutier (1996) restores the ethmoid commissure as passing beneath the opening to the rostral organ: if confirmed, this is a unique condition in coelacanths for in most it passes above the opening.

The naso-parietals are represented by at least four elements which form an anteriorly diverging series (Fig. 3.3(A)) which leaves a large rostral area unknown, although isolated rostrals can be seen in some specimens. It is possible that the snout was covered by a mosaic of small rostrals and that some of these may have been median internasals. The distinction between nasals and parietals is not clear in *Miguashaia* because there is no difference in the nature of the sutures between successive elements. However, the orbit is roofed entirely by a single pair of parietals (Pa) occupying territory usually filled by two pairs in most other coelacanths. Schultze (1973) describes anterior and posterior parietals in a young individual so it is possible that *Miguashaia* is polymorphic for this character. The supraorbital–tectal series contains at least five elements of which four can be called supraorbitals because they lie above and behind the eye. Because the anterior end of the skull roof is poorly known, the number of tectals is uncertain. The supraorbital sensory canal passes through the centre of ossification of the supraorbital series (Fig. 4.4) and this appears to be a uniquely primitive condition, seen in no other coelacanth.

The postparietal shield is long and narrow

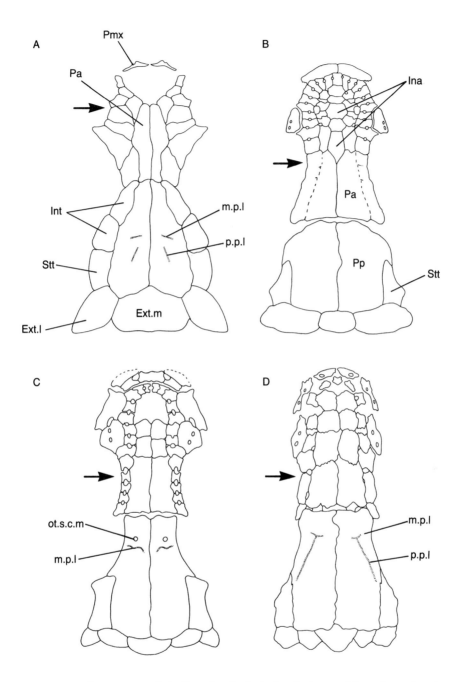

Fig. 3.3 Diagrammatic illustrations of skull roofs of relatively plesiomorphic coelacanths. Arrow marks position of the centre of the orbit. (A) *Miguashaia bureaui* Schultze from Cloutier (1996) with permission. (B) *Hadronector donbairdi* Lund and Lund, modified after Lund and Lund (1985). (C) *Caridosuctor populosum* Lund and Lund, modified after Lund and Lund (1985). (D) *Rhabdoderma elegans* (Newberry).

and consists of postparietals, dermosphenotic (intertemporal), intertemporals (supratemporal) and supratemporal (tabular). The postparietal is described as 'L'-shaped by Cloutier, by which he means that the anterior portion is narrower than the posterior portion and the middle and posterior pit lines are located at roughly mid-level within the bone (Fig. 3.3(A)). As in all coelacanths, except possibly *Diplocercides* (see below), there is no anterior pit line.

Miguashaia is the only coelacanth in which the postparietals are flanked by three bones which carry the otic sensory canal through the centres of ossification. The extrascapulars are represented by a very large median and paired laterals; the latter carry the triple junction of otic, lateral line and supratemporal commissure (Fig. 4.4). The ornament consists of small, closely spaced tubercles present on all roofing bones.

Diplocercides (including *Nesides*)

The pattern of roofing bones has been described by Stensiö (1937) who described *Nesides* and *Diplocercides* (see Chapter 11). The parietonasal and postparietal shields are known in *Diplocercides kayseri* (*Nesides schmidti*) and, very imperfectly, in *D. heiligenstockiensis* (Jessen, 1966, 1973). By far the best specimens are 'specimen a' of *D. kayseri* and the holotype of *Nesides schmidti*. A latex cast of the former specimen and illustrated in Fig. 3.4 forms the basis for the following remarks.

The most obvious feature of the parietonasal shield of *Diplocercides* is the row of at least four large, median internasals (Ina, postrostrals of Stensiö). These are flanked by four nasals and these, in turn, by four tectals. Stensiö described a specimen (his 'specimen d', 1937: pl. 7, fig. 1) of *D. kayseri* as having a single series of bones on either side of the internasals. This was said to be like *Nesides* which was similarly restored (Stensiö, 1937: fig. 2). 'Specimen d' shows an impression of

the visceral surface of the anterior roofing bones, in which the lateral edges are clearly incomplete so it is impossible to say whether there was one or two series either side of the internasals. Similarly, in *Nesides* the roofing bones of the snout are not well preserved (Stensiö, 1937: pl. 9) and as Stensiö (1937: 44) remarks, there 'are no certain traces of sutures between' supraorbitals or between supraorbitals and parietals; this may also apply to the tectals and nasals. The restoration of a single series of bones flanking the internasals (Stensiö, 1937: fig. 2) is therefore questionable.

The two pairs of parietals are seen in 'specimen a' (Fig. 3.4), the right anterior parietal is considerably larger than the left. The labelling given here differs from the restoration given by Stensiö, who illustrates a paired postrostral in the place of the right anterior parietal recognized here. As usual in coelacanths, the anterior parietal partly overlaps the posterior parietal and the mutual suture line is irregular and oblique. Elsewhere on the parietonasal shield the sutures are of the simple abutting type. One of the supposed differences between *Nesides* and *Diplocercides* is the length of the parietals (allegedly longer in *Nesides*) but there appears to be no difference between the specimens illustrated by Stensiö (1937, pl. 5, fig. 2 – *D. kayseri* and pl. 9 – *Nesides*).

There are approximately six supraorbitals and these are perforated by a row of small pores leading from the supraorbital canal (Stensiö 1937: pls 5, 9). The canal is wide (so.s.c) and straddles the sutures between parietals and supraorbitals (Fig. 3.4). In specimen 'd' the floor of the supraorbital sensory canal above the orbit is pierced by three foramina for branches of the superficial ophthalmic ramus of the anterodorsal lateral line nerve (f.br.s.opth). In the wax model of the braincase (*Nesides*) the canal can be traced as a trough as far as the middle of the ethmoid region (Jarvik, 1954: fig. 4B) but no further forward.

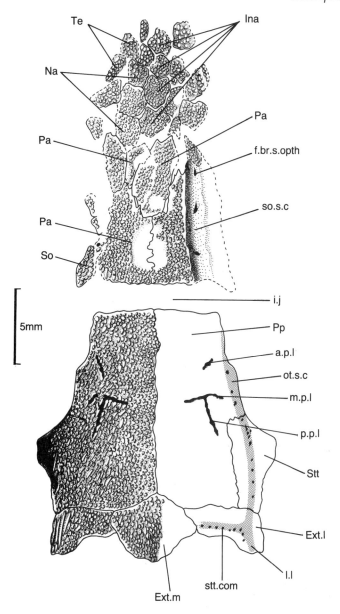

Fig. 3.4 *Diplocercides kayseri* (von Koenen). Skull roof in dorsal view; the two halves of the skull have been drawn as if slightly apart. The parietonasal shield has been drawn as a camera lucida drawing. The postparietal shield is slightly restored. Specimen 'a' of Stensiö (1937).

The lateral rostral is described by Stensiö and, more completely, by Jarvik (1942). It is simpler than that of most coelacanths, consisting of a simple tube around the infra-orbital canal which, as usual, turns down anteriorly. Several foramina for branches of the buccal nerve pierce the medial wall of the lateral rostral (Stensiö, 1937: pl. 10, fig. 1;

Jarvik, 1942: pl. 16, fig. 2). The preorbital (antorbital of Stensiö, posterior tectal of Jarvik) is known only in the holotype of *Nesides*. Even here it is incomplete, but it can be seen that the bone is perforated by two holes (Jarvik, 1942: pl. 16, fig. 4). Both Stensiö and Jarvik interpret these two openings as posterior nostrils but comparison with *Latimeria* suggests that, at least one and probably both represent the posterior opening of the rostral organ. In many coelacanths (e.g. *Allenypterus* – Fig. 3.5), the preorbital is pierced by two, equally sized holes which lie one behind the other. These are interpreted as the two posterior openings of the rostral organ. But in *Diplocercides* the anterior opening is considerably smaller than the posterior and it is therefore not obvious that both openings are associated with the rostral organ. In *Latimeria* the posterior nasal tube leaves the nasal capsule above the lachrymal and runs medial to the lachrymal to open to the surface at the lip margin. It is impossible to decide if this also happened in *Diplocercides* or if the anterior pore within the preorbital is the posterior nostril. This latter interpretation would leave the posterior opening to represent a double posterior opening of the rostral organ, conditions closely approached in *Spermatodus* (Fig. 3.12). The anterior nostril lies along the anterior edge of the lachrymal (Jarvik, 1942, fig. 77).

The postparietal shield has been described by Stensiö (1937: 9–10) and little needs to be added to the description given there. The postparietal (Pp, parieto-intertemporal of Stensiö, 1921: 'parieto'-dermopterotic of Stensiö, 1947) is relatively flat compared with the postparietal of most other coelacanths, where some transverse curvature is always evident. The joint margin is straight. There is no postparietal descending process developed. The supratemporal (Stt) ends at the same level as the postparietal such that the posterior margin of the postparietal shield is straight and not embayed. There is no supratemporal descending process. The extrascapular series consists of a large median extrascapular (Ext.m) which shows a narrow overlap with a lateral extrascapular (Ext.l). The path of the sensory canal is traced as a series of small pores and this shows that the junction of the otic canal, supratemporal commissure and the lateral line is located within the lateral extrascapular.

There is no evidence for a median branch of the otic canal. The pit lines are well developed as deep grooves. There is a short, anteriorly placed pit line which lies anterior to the centre of growth of the postparietal. Behind this lies the transversely orientated middle pit line, which is continuous with the longitudinal posterior pit line. The postparietal pit lines of *Diplocercides* (*Nesides*) were discussed at length by Stensiö (1937). The anteriorly placed pit line has here been labelled as an anterior pit line (a.p.l) but this identification may be incorrect because in *Latimeria* (Fig. 2.3) there is also an anterior line but it is not innervated by the nerve usually associated with that pit line in other fishes (see above). The bones of the skull roof are ornamented with many closely spaced tubercles which form a dense covering.

Euporosteus

The unique specimen of *Euporosteus* has been described and figured by Jaekel (1927), Stensiö (1937) and Jarvik (1942). The description of the parietonasal shield given by Stensiö is the most complete. Only fragments of the posterior parietals and the right postorbital are preserved; information about the remainder of the skull roof must be inferred from the pattern of ridges and grooves which mark the dorsal surface of the endocranium (Fig. 6.3). The parietonasal shield must have been relatively broad. On either side of the midline there is a row of matrix infillings occupying the openings of the supraorbital sensory canal (p.so.s.c). There are 11 such openings on the left side. The canal therefore opened through the bones in a series of large, regular pores as in *Libys* and *Diplurus*. Medial to these, there is a prominent groove which runs parallel to this row of pores. This groove marks the path of the inner wall of the supraorbital canal, and the broken area on the right side of the specimen shows the canal to be wide and to be pierced from below by foramina for branches of the superficial ophthalmic branch of the anterodorsal lateral line nerve (f.br.s.opth).

A series of ridges runs over the dorsal surface of the nasal capsule; some represent paths of blood vessels, others are straight and are impressions of sutures. But it is not always clear as to which represents which and there must obviously be some doubt about the restoration given by Stensiö (1937: fig. 9). Stensiö restores a series of hexagonal internasals (postrostrals). The specimen shows evidence of the two most posterior of these (Ina). Stensiö further suggests that the internasals are flanked by a single series of bones representing the nasals and tectals of other coelacanths (see also remarks on *Diplocercides*). He did this to bring Devonian coelacanths into line with rhipidistians (1937: 6). However, the specimen shows longitudinal ridges running between successive pore infillings (Fig. 6.3; Jaekel, 1927: fig. 49; Stensiö, 1937: fig. 10, pl. 12, fig. 1; Jarvik, 1942: pls 14, 15). I suggest that these are evidence of a suture separating tectals laterally from nasals medially. Above the orbit there is similar evidence of this suture separating parietals from supraorbitals (So). There is also some evidence of transverse ridges running medially from several of the pore infillings above the nasal capsule, suggesting a one-to-one correspondence between pores and nasals/tectals. Under this assumption, there would be approximately six tectals followed by about four supraorbitals.

The posterior parietal is long and there is a small and simple parietal descending process (Fig. 6.3) developed posteriorly. The centre of ossification lies approximately midway along the bone and on the inner edge of the supraorbital sensory canal. An anterior parietal is probably represented by that element called naso-postrostral by Stensiö. Fragments of the preorbital (antorbital of Stensiö, 1937, posterior tectal of Jarvik, 1942) can be seen on the right side of the specimen. There are two pores (unlabelled in previous figures and omitted in Jarvik's restorations) which pierce the wall of the ethmoid region. These are the posterior openings of the rostral organ (p.ros): they lie one behind the other and are rather high up on the side wall of the ethmoid region.

The supraorbital sensory canal is wide and runs parallel to its antimere throughout most of the parietonasal shield. Anteriorly it curves inwards towards its partner in the midline. The canal leaves the parietal posteriorly immediately lateral to the parietal antotic process and above the foramen for the superficial ophthalmic branch (Fig. 6.3, f.s.opth).

Allenypterus

The skull roof has been described by Lund and Lund (1985). I have examined several specimens and find several important differences from their description. A restoration is provided as Fig. 3.5 and is shown in lateral view in Fig. 4.6. I am unable to confirm the details of the snout given by Lund and Lund.

In some respects *Allenypterus* resembles very derived coelacanths (splint-like premaxilla and the proportions of the parietonasal and postparietal shields). The parietonasal shield is very long compared with the postparietal shield, the ratio being nearly 3:1. As restored by Lund and Lund (1985: figs 60, 61) the premaxillae are splint-like without a dorsal flange and bear very small pointed teeth. Cloutier (1991a: data matrix) records that the premaxillae are fragmented but Lund and Lund show a single paired bone which is separated in the midline by a median rostral (this latter configuration of the premaxillae and rostral is doubted by

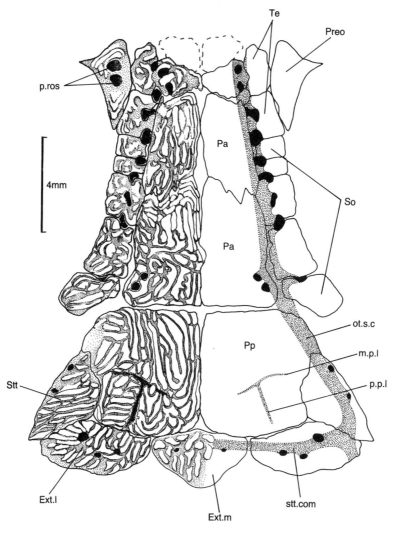

Fig. 3.5 *Allenypterus montanus* Melton. Restoration of the skull roof based on FMNH 10939, PF 10940, PF 10942. The paths of the sensory canals are shown as stipple on the right side.

Cloutier, 1991a: 398). Thus, there does seem to be some ambiguity about the condition of the premaxillae and associated rostral ossicles and I have coded the condition relative to this part of the skull as uncertain data. The lateral rostral (Fig. 4.6, L.r) is relatively small and, unlike other skull roof bones, is unornamented and shows a small ventral process.

Behind the snout the parietonasal shield contains at least two nasals and anterior and posterior parietals which are flanked by at least seven elements within a tectal/supraorbital series. Cloutier (1991a) agrees with the relatively low number of bones within the naso-parietal series but Lund and Lund (1985: fig. 61) record many more. Lund and Lund (1985: fig. 61) also suggest that there is an additional longitudinal series called 'medial supraorbitals' wedged between the naso/parietal and tectal/supraorbital series. Like Cloutier, I have failed to find this series in well-preserved specimens and the roof is like that of other coelacanths. The preorbital (Preo) is a small triangular bone wedged between the lachrymojugal and tectal. It is pierced by two closely spaced posterior openings from the rostral organ.

There are two pairs of parietals above the orbit, with a digitate suture between the smaller anterior and larger posterior pair. The intracranial joint margin of the posterior parietal is straight with no indentations. The presence or absence of a descending process of the parietal is unknown. The supraorbital canal runs its usual sutural course and opens by medium-sized pores. Most of the pores are located close to the medial margin of the supraorbital/tectals and open between successive elements, but there may be additional pores lying wholly within the posterior parietal (Fig. 3.5).

The postparietal shield is very short and broad and the postparietals (Pp) are almost equidimensional (an unusual feature in coelacanths). The supratemporal extends laterally. It is clear from the nature of preservation that the skull roof is strongly vaulted (Fig. 4.6) so that the full width of the postparietal shield is usually preserved in lateral view. Lund and Lund (1985) restore a separate intertemporal but such a separate element was absent in specimens examined here. I have been unable to determine if descending processes were present on either the supratemporal or the postparietal but Cloutier (1991a: data matrix) suggests that both were absent.

There are three extrascapulars, a median (Ext.m) plus paired laterals (Ext.l), and the latter carry the triple junction of otic, supratemporal and lateral line sensory canals. The extrascapulars are very scale-like and are only loosely associated with the skull roof which shows a straight posterior margin.

Hadronector

The skull roof (Fig. 3.3(B)) has been described by Lund and Lund (1985) and by Cloutier (1991a). Three latex casts showing the skull roof were examined here (CM 30712A, CM 27308A, CM 30711A). A small portion of the skull roof is shown in Fig. 4.7(C).

The bones are heavily ornamented with coarse tubercles which are closely spaced and tend to run into ridges posteriorly. This coarse ornament makes interpretation of bone patterns very difficult. The parietonasal shield is only slightly longer than the postparietal shield. The intracranial joint margin is straight. The premaxilla is small but there is a dorsal flange perforated by the anterior opening for the rostral organ (Cloutier, 1991a: fig. 44) as in *Diplocercides*. The parietonasal series consists of small nasals, at least three, which lie anteriorly and two pairs of parietals which lie above the orbit. Lund and Lund (1985: fig. 44) restore a double series of bones lying lateral to the parietals. I find no evidence of this. The preorbital (antorbital of Lund and Lund) is large and pierced by well-spaced posterior openings from the rostral organ. The lateral rostral has a remarkably well-developed ventral process (Lund and Lund, 1985: fig. 37).

The postparietal shield consists of the postparietals and supratemporals (Figs 3.3(B), 4.7, Pp, Stt). I find no evidence of separate intertemporals. The middle and posterior pit lines are very short and located well anteriorly within the postparietals. There are three extrascapulars (a median plus paired laterals). The median extrascapular is considerably smaller than the laterals (cf. *Miguashaia* and *Diplocercides*).

Caridosuctor

The skull roof is very similar to that of *Rhabdoderma elegans* in proportions, composition and ornamentation. It is shown as a simple outline diagram in Fig. 3.3(C), taken from

Lund and Lund (1985). In the parietonasal shield there are two pairs of parietals above the orbit. The posterior parietal is much the larger and covers most of the orbit. The tectals (supraorbitals of Lund and Lund, 1985) and supraorbitals are numerous, at least 12 can be seen. The premaxilla is relatively large and bears three or four robust teeth. Lund and Lund (1985) mention that the premaxillae of either side are separated by a median rostral and this is shown in the figure. If this observation is confirmed this is the only coelacanth in which this condition is known. The large preorbital (antorbital of Lund and Lund, 1985) is pierced by two well-spaced posterior openings from the rostral organ. The intracranial joint margin is simple and not markedly digitate and it is not known if a parietal descending process was present.

The postparietal shield is relatively narrow and the posterior margin is straight with no embayment and implying that the lateral extrascapular remains free from the supratemporal. Lund and Lund (1985) describe a separate intertemporal (their supratemporal), but in the specimen I have examined showing this region, no such bone is present. There is a median plus two lateral extrascapulars and the junction between the otic sensory canal and the supratemporal commissure lies within the lateral extrascapular. A middle pit line (lateral pit line of Lund and Lund) is well developed and there is a large pore near the centre of ossification, suggesting that a median branch of the otic canal was present (Fig. 3.3(C), ot.s.c.m). A posterior pit-line has not been seen.

Rhabdoderma

The skull roof is well known only in *R. elegans* and has been briefly described by Moy-Thomas (1937) and Forey (1981). A camera lucida drawing of the parietonasal shield is given in Fig. 3.6 and an outline restoration of the entire roof is shown in Fig.

3.3(D). The parietonasal shield is about the same length as the postparietal shield. The parieto-nasal series consists of anterior and posterior parietals which roof the orbit and at least two pairs of nasals (postrostrals of Moy-Thomas). Successive elements of the series become larger posteriorly. As with the coelacanths in which many individuals are available for study (*Macropoma lewesiensis*, *Whiteia woodwardi*, *Diplurus newarki*), there is considerable variation in sutural patterns but, as usual, the anterior and posterior parietals contact one another by a complex interdigitating suture. The parietal descending process (Fig. 3.6, v.pr.Pa) of *Rhabdoderma* is relatively small and slender, and does not bear any complex facets such as are seen in many other coelacanths (e.g. *Latimeria*, *Laugia*, *Spermatodus*, *Macropoma*). The simplicity of this process resembles the (even smaller) process in *Euporosteus* (Fig. 6.3(A)).

The snout is covered with paired premaxillae (Pmx), the anteriormost nasal (Na) and a median internasal (Ina). These bones are largely in contact with one another, separated only by pores. The position of these pores suggests that the ethmoid commissure (eth.com) ran along the dorsal edge of the premaxillae (it was incorrectly restored by Forey (1981: fig. 1) as running through the body of the premaxilla). There is a small median opening within the mutual suture of the premaxillae, as in *Macropoma* and *Whiteia*. This pore is presumably related to the ethmoid commissure. The posterior edge of the premaxilla is notched and this notch marks the position of the anterior nostril (nos.a). The posterior border of the anterior nostril is indicated as a notch along the anterior border of the ventral process of the lateral rostral (v.pr.L.r). The premaxilla (Pmx) is relatively large, bears three or four large conical teeth and the dorsal lamina is pierced centrally by a large opening, interpreted here as the anterior opening of the rostral organ (a.ros).

The supraorbito-tectal series consists of at

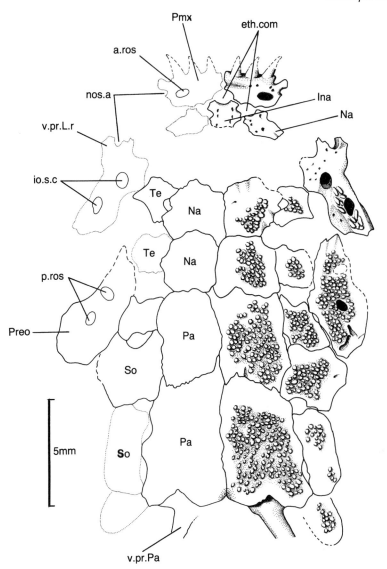

Fig. 3.6 *Rhabdoderma elegans* (Newberry). Parietonasal shield in dorsal view. Camera lucida drawing of cast of Museum of Zoology, Cambridge GN.238. For full restoration see Fig. 3.3(D).

least two tectals which lie adjacent to the nasals and four supraorbitals which lie alongside the parietals. The second and third supraorbitals are characteristically large.

The lateral rostral (rostral of Moy-Thomas, 1937) is relatively small and the ventral process is poorly developed. The tubular portion is pierced by two large openings (io.s.c) belonging to the infraorbital sensory canal. In contrast, the preorbital (Preo) is relatively large, it forms part of the orbital margin and is pierced by two large openings for the posterior openings of the rostral organ.

The postparietal shield is longer than broad (Fig. 3.3(D)), with the supratemporals forming a relatively small percentage of the total area. As in *Diplocercides*, the posterior margins of the supratemporals and postparietals (intertemporals of Moy-Thomas) are straight: that is, the posterior margin of the skull roof is not embayed. In large specimens of *R. elegans* the suture between the postparietal and the supratemporal is obliterated (e.g. BMNH P.6613a), as in large specimens of *Macropoma*. The joint margin of the postparietal is indented above the facet on the underside of the postparietal which articulates with the parietal antotic process. The adjacent contours of the parietal and postparietal thus form a gap immediately above the dorsal articulation between the braincase moieties.

The otic sensory canal is carried within a large tube developed, as usual, on the undersurface of the postparietal. There is no descending process on the postparietal, but the descending process of the supratemporal is well developed (Fig. 6.5). The extrascapular series consists of a larger median plus two smaller lateral extrascapulars. All are scale-like and the overlap is laterad.

The sensory canals open to the surface through small pores. These pores are noticeably smaller than those that pierce the parietonasal shield, but they are comparable in size with those on the cheek. Both middle and posterior pit lines are developed (Fig. 3.3(D), m.p.l, p.p.l). In some specimens there may be a sensory pore, sometimes double, present immediately anterior to the middle pit line (e.g. CM 44094B, 44033, 43959) and this implies that a medial branch of the otic sensory canal was developed. In some specimens the pores opening through the supratemporal are slightly enlarged (BMNH 30572). Lund and Lund (1985) suggest that the possession of large pores on the supratemporal is a feature that distinguishes the North American specimens of *R. elegans* from the British specimens, but I consider it

to be no more than an individual variation of, at least, the latter. Both middle (termed 'lateral pit line' by Lund and Lund, 1985) and posterior pit lines are developed and, as in *Diplocercides* and *Whiteia*, these together form a 'T'-shaped groove on the postparietal. The posterior pit line seems to vary in length from specimen to specimen. It is shown at its maximum development in Fig. 3.3(D).

In *R. elegans* most of the bones of the skull roof are heavily ornamented; the exceptions are the snout bones, which are quite smooth. The ornament consists of irregularly sized tubercles which, in small individuals, are confined to the centre of each bone (Fig. 3.6). The form of ornament pattern varies from species to species (absent in *R. huxleyi*) and has been commented on by Forey (1981); it is mentioned again in the diagnoses of *Rhabdoderma* species (p. 333).

A postscript to the above description is necessary in view of remarks made by Lund and Lund (1985). Those authors consider that the material of *R. elegans* from Linton (the type locality) differs consistently from the British material in a number of features of which the skull roof features may be mentioned here. According to Lund and Lund, the specimens from Linton show a separate intertemporal (their supratemporal) lateral to the postparietal and, in one specimen, an additional non-canal-bearing bone was wedged between an intertemporal and the postparietal. I have not examined the same material as Lund and Lund but the specimens of Linton *Rhabdoderma* to which I have had access (p. 334) show no such multiplicity of bones. Lund and Lund also remark that the supratemporal reaches posteriorly well beyond the parietals, and that occasionally there may be a duplication of the postparietals which they liken to that recorded for *Mawsonia* by Wenz (1975). I could not confirm either of these observations. But it is very unlikely that the supernumerary bones could be similar to those in *Mawsonia*

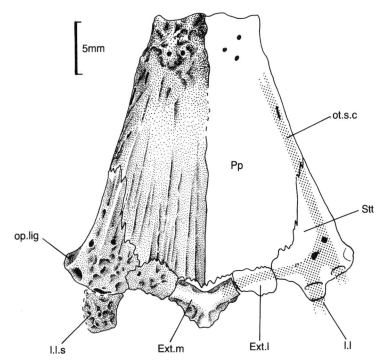

Fig. 3.7 *Coelacanthus granulatus* Agassiz. Restoration of the postparietal shield based on BMNH P.3340. The path of the sensory canal is shown on the right-hand side as mechanical stipple.

because in that genus, these bones are canal-bearing extrascapulars incorporated to the skull roof (p. 75). In sum, I see no reason for believing that the British specimens of *R. elegans* differ significantly from the specimens from the type locality in any aspect of the skull roof morphology.

Coelacanthus granulatus

The skull roof of this species is incompletely known, particularly in aspects of the parieto-nasal shield. Overall the skull roof is narrow, especially at the level of the intracranial joint. The postparietal shield, which is shown restored in Fig. 3.7, is relatively long, equal to about 80% the length of the parietonasal shield.

The most complete description of the parietonasal shield is given by Schaumberg

(1978), who used material from the Kupfer-schiefer of Reichelsdorf and to which little can be added. The material studied here comes from the Marl Slate of Durham and forms the basis of the remarks below.

The anterior and posterior parietals show the typical complex suture seen in many coelacanths (cf. Schaumberg, 1978: fig. 5, where the anterior parietals are labelled as na$_1$). A ventral view of the parietals (MGUH 2344b) shows that the anterior parietal must have been largely underlain by the posterior parietal, as in *Whiteia* (p. 72). The overlapped portion of the posterior parietal, figured by Schaumberg (1978, fig. 2) has been restored by that author as an elongate and strut-like tectal (Schaumberg, 1978: fig. 6). A similar strut is seen in HM HMG.26.52 where it is obviously part of the parietal. The tectal series is, instead, formed by the more usual series of small elements (Schaumberg, 1978: figs 3 and 9).

There are at least two pairs of nasals. Little else of the snout covering can be seen in detail. BMNH P.3335, a small specimen (115 mm SL (standard length)), and HM HMG.26.52 show a few tiny ossicles in the snout region. The edges of each of these ossicles are emarginated for sensory pores. The shape and relative size of these ossicles recall those in young *Latimeria* (Fig. 3.1). I cannot confirm the restoration of the premaxillae given by Schaumberg. That author shows the premaxilla to be represented as three small tooth-bearing plates, similar to those in *Latimeria*. Given the highly fragmented nature of the anterior part of the roof of *Coelacanthus granulatus*, this is very likely. There is no preorbital.

The postparietal shield is described as being vaulted (Schaumberg, 1978 – as parietal shield). While this is probably the case, it is likely that the vaulting was not marked because nearly all specimens are preserved in dorsal view. The postparietals are very large (78% of the total area of the shield) and, as noted by Schaumberg (1978), are very thick throughout. The undersurface of the postparietals is damaged in all specimens examined but there does not appear to be any evidence of a descending process. The supratemporal is small but there is a prominent descending process as in *Latimeria* and most other coelacanths.

The restoration of the extrascapular series (Fig. 3.7) differs from that given by Moy-Thomas and Westoll (1935: fig. 2) and Schaumberg (1978: fig. 5). Previous authors were uncertain about the condition of the extrascapulars. I agree with Schaumberg in recognizing a prominent median extrascapular (Ext.m). This lies free from the postparietals. On either side, there is a single lateral extrascapular (BMNH P.3340) wedged in between the supratemporal and median extrascapular. This is tightly sutured with both the supratemporal and postparietal. The

otic sensory canal appears to join the supratemporal commissure within the supratemporal (Fig. 3.7). A stout lateral line scale is present posterior to the supratemporal (l.l.s).

Schaumberg (1978) also notes that the extrascapular series is closely associated with the parietals. However, he records three or four lateral extrascapulars and identifies the junction of the otic sensory canal and the supratemporal commissure as lying within a separate lateral extrascapular. My Kupferschiefer material is not sufficiently well preserved to allow me to comment on this discrepancy.

Most of the bones of the parietonasal shield are without ornament, although Schaumberg does indicate (1978: fig. 9) granular ornament on the lateral rostral. Ornament on the postparietal shield appears to be confined to rugose bone above the centre of ossification on the postparietals and over the entire surface of the supratemporal and extrascapulars. Schaumberg (1978: 172) mentions the presence of tubercles, implying surface enamel tubercles as found in many coelacanths. Such tubercles are not seen on the Marl Slate specimens, nor on Kupferschiefer material examined in this work. There is no trace of pit lines, and this may simply mean that the bones were sunk deeply beneath the skin, a fact that would agree with a lack of ornament. There is, however, a group of pores clustered near the centre of ossification of the parietal, implying that there was a medial branch of the otic canal.

Laugia

The skull roof of *Laugia* shows a narrow parietonasal shield and a relatively long postparietal shield (Fig. 4.10). Despite the large number of specimens by which *Laugia* is known, very few show a well-preserved parietonasal shield and this part of the roof remains incompletely known. Some emenda-

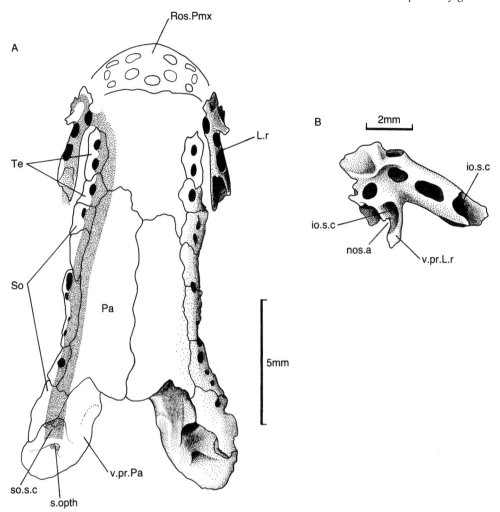

Fig. 3.8 *Laugia groenlandica* Stensiö. (A) Partial restoration of parietonasal shield in dorsal view. Stippling on left-hand side denotes path of supraorbital sensory canal. (B) Restoration of lateral rostral in left lateral view. Based on MGUH VP2310, VP2311, VP2316.

tions and additional remarks can be given to Stensiö's (1932: 53) original description. The chief area of uncertainty concerns the roof of the nasal capsules (Fig. 3.8). Only one pair of parietals (Pa) can be identified and they cover, at least, the entire orbit. Their anterior limit is poorly known. The intracranial joint margin is indented and the descending process of the parietal (v.pr.Pa) is broad and inclined at a very shallow angle from the

plane of the parietonasal shield. Figure 3.8 shows that the dermal joint is a far more complicated structure than that illustrated by Stensiö (1932: fig. 18). In the specimen on which the restoration shown in Fig. 4.10 is based, the joint is in the closed position, yet a considerable gap remains between the parietal and the postparietal. The parietal is narrow and without ornament. MGUH VP2315A shows a lateral view of the descending process of the parietal in which can be seen a canal for the superficial ophthalmic branch of the anterodorsal lateral line nerve lying immediately ventral to the entry of the supraorbital sensory canal (Fig. 3.8).

There are approximately five supraorbitals. Each is narrow and without ornament and is perforated by several small pores leading from the supraorbital canal. At least two tectals can be identified (MGUH VP2311 and VP2309) and, like the lateral rostral and snout elements, these are perforated by very large pores, unlike those within the supraorbitals.

The dorsal covering of the snout has been figured by Stensiö (1932: pl.v) as a solid bony hemisphere, bearing teeth along the oral margin and perforated by several pores showing a symmetrical arrangement around the midline. Stensiö called this large element a 'rostralo-premaxillary' and compared it to that in *Macropoma*. At this point the bone is very thin. It is possible that this large element is, in fact, a series of small rostral elements between which there are pores for the openings of sensory canals, similar to the condition in Nielsen's 'undetermined coelacanth' (Nielsen, 1936: fig. 9). MGUH VP2310 shows a crushed snout in which there are indications of a (MGUH VP2315b) narrow band of villiform teeth, the innermost of which are slightly larger than the marginal teeth. These may be the vomerine teeth.

Behind the 'rostralo-premaxillary' Stensiö restored a large nasalo-antorbital (1932, fig. 19). This is clearly in error. There is, instead, a large lateral rostral (Fig. 3.8(B)) of the usual coelacanth type, consisting of a tubular portion and a more robust anterior portion. The openings of the infraorbital sensory canal are very large and there is one pore directed dorsally at the point where the canal turns downwards suggesting that a connection probably existed between the infraorbital and supraorbital canals at this point, as in *Latimeria*. The ventral process (v.pr.L.r) of the anterior portion is grooved along the inner face and these grooves interlock with ridges upon the lateral ethmoid forming a very tight union between the nasal capsule and the dermal covering at this point. Similar conditions are seen in *Macropoma* and *Whiteia* and it is probable that, were more coelacanths known by well-preserved material, this would be a general condition.

The postparietal shield (Figs 3.9, 4.10) is relatively long compared with the parietal shield. The bones are smooth, with only a few tubercles of ornament confined to the supratemporal and the extreme lateral edge of the postparietal. There are no pit lines marking the bones.

The undersurface of the postparietal (Fig. 3.9(B)) is marked by a well-developed longitudinal ridge which follows the path of the otic sensory canal (ot.s.c). There is no descending process on the postparietal. Instead, the prootic rises up to meet the postparietal ridge. Anteriorly, the ridge is expanded to form a facet (fa.i.j) for articulation with the descending process of the parietal. The lateral edge of the postparietal is bevelled immediately behind the point where the infraorbital canal enters the bone. The bevelled edge is overlapped by the postorbital. The suture separating left and right postparietals is sinusoidal anteriorly and, in one specimen examined (MGUH VP2308a), there is a small separate median element. Similar individual variants have been described in the 'osteolepid' *Gyroptychius* by Jarvik (1948: 64) who regarded such variants as evidence of an ancestral median series of roofing bones predicted by Allis (1935). Here, such a bone is regarded as an individual variant with no phylogenetic significance.

The supratemporal shows a well developed descending process (Fig. 3.9(B), v.pr.Stt). Posterior to this the dorsolateral edge of the supratemporal and lateral extrascapular is excavated as a shallow pit into which the opercular ligament was inserted (Fig. 3.9(A), op.lig). As Stensiö (1932: 53)

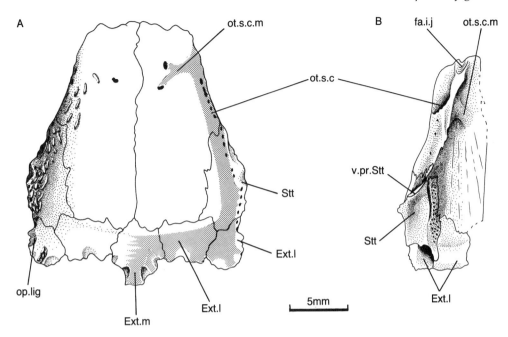

Fig. 3.9 Postparietal shield of *Laugia groenlandica* Stensiö. (A) Restoration in dorsal view, based on MGUH VP2308a, VP2310, VP2311, VP2315a, VP2316. Path of sensory canal shown by stipple on right-hand side. (B) Undersurface of right postparietal and supratemporal as preserved in MGUH VP2315a. Camera lucida drawing.

noted, the suture between the postparietal and supratemporal is long, at least equal to half the length of the postparietal shield. The undersurface of the supratemporal is inflated as the floor of the otic canal and two foramina for supratemporal lateral line nerve pierce the floor of the canal as in *Macropoma* (Fig. 3.21(B)). Medial to the otic sensory canal the undersurface of the supratemporal is pitted. The posterior end of the supratemporal shows a complex sutural connection with the lateralmost extrascapular. In dorsal view (Figs 3.9, 4.10), the supratemporal ends on a level with the postparietal. But, in ventral view (Fig. 3.9(B)), the supratemporal reaches back beneath most of the lateral extrascapular. Thus the supratemporal forms a small shelf receiving the lateral extrascapular. A supratemporal of similar shape is seen in *Coccoderma* (Fig. 3.10(A)) and *Spermatodus* (Fig. 3.14).

The remainder of the extrascapular series, a median plus lateral extrascapular, are closely united to one another and to the parietal and supratemporal. There is a narrow overlap where the median overlaps the adjacent lateral and this, in turn, overlaps the lateralmost.

The sensory canals are very large, particularly posteriorly, and open to the surface through small pores. There are 15–25 pores along the main otic canal. Stensiö (1932: figs 18, 19) restores two groups of pores within the postparietal but the specimens examined here do not show such clustering. The floor of the otic canal within the postparietal is pierced by three foramina (Fig. 3.9(B)) for branches of the middle lateral line nerve.

There is a short medial branch given (ot.s.cm) off from the main otic canal which runs towards the centre of ossification. The medial branch opens to the surface by two small pores. The junction of the otic canal, the main lateral line and the supratemporal commissure lies at the suture of the supratemporal and the lateralmost extrascapular. The supratemporal commissure opens posteriorly through five pores within the extrascapulars (cf. Stensiö, 1932: 57).

Coccoderma

The skull roof of *Coccoderma* is in many respects similar to that of *Laugia*. The postparietal shield is restored in Fig. 3.10(B) and the skull roof is seen in lateral views in Figs 4.11 and 5.7. The parietonasal shield remains poorly known and it is possible that, as in *Laugia*, the central part of the roof anterior to the parietals was poorly ossified. At least one pair of long parietals is seen, although there may be a smaller anterior parietal present (Fig. 5.7). The snout consists of star-shaped ossicles. In BMNH P.8356 they have clearly been disturbed during fossilization but it is possible that they fit together to form a relatively consolidated snout pierced by large pores. The lateral rostral is very similar in shape to that of *Laugia* in that the ventral process is slender and there are large sensory openings. A preorbital may be absent: a bone in the appropriate position is seen in BMNH P.8356 but it is not pierced by foramina and may be a tectal. There are about seven supraorbitals. The openings of the supraorbital sensory canal are small and numerous above the orbit but anteriorly they become substantially larger as in *Laugia*.

The postparietal shield is better known (Fig. 3.10). The joint margin is rather narrow, with the articulation facets upon the postparietals situated very close together near the midline. The postparietal (Pp) is long and relatively narrow and there is a long sutural connection with the supratemporal (Stt). The latter bone shows a posterior shelf which received the lateral extrascapular (Ext.l) simi-

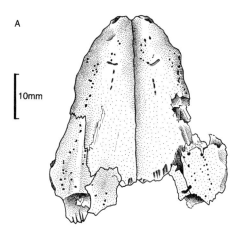

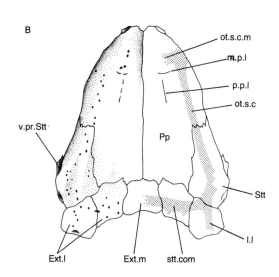

Fig. 3.10 Postparietal shield of *Coccoderma suevicum* Quenstedt. (A) Camera lucida drawing of postparietal shield as preserved in BSM 1870.XIV.506. (B) Restoration including details of extrascapular series from BMNH P.8356. Path of sensory canals shown by stipple on right-hand side.

lar to that in *Laugia* and *Spermatodus*. The posterior margin of the skull roof is slightly embayed but the lateral extrascapular remains free and the union of the otic canal (ot.s.c) with the lateral line (l.l) and the supratemporal commissure (stt.com) occurs within the suture between the extrascapular and the supratemporal. Middle and posterior pit lines are present and the otic canal sends off a medial branch to open through a few small pores near the anterior tip of the post-parietal. The sensory canals open through many very small pores.

Sassenia

The parietonasal shield is only known imperfectly, from *S. groenlandica* (p. 337) where it can be seen in two specimens, MGUH VP 2327a (the holotype) and MGUH VP2326 (Fig. 3.11). That which is known is very similar to the parietonasal shield of *Spermatodus*, particular points of resemblance being the shapes of the lateral rostral and preorbital. The latter is large, ovoid and is pierced by two posterior openings of the rostral organ. These openings (p.ros) lie very close together, separated only by a narrow bar, as in *Spermatodus*. The lateral rostral has a superficial ornamented face which covers the tubular portion and a deeper-lying unornamented lamina and, as in *Spermatodus*, the openings from the sensory canal lie immediately ventral to the ornamented portion.

The posterior end of the parietonasal shield is markedly convex in the transverse plane. The posterior parietal is short and broad and the posterior edge which forms the joint margin is digitate. The supraorbital which lies opposite the junction of anterior and posterior parietals is long and similar to that of *Spermatodus* (cf. Figs 3.11 and 3.12). The openings for the supraorbital sensory canal cannot be seen clearly, but it is probable that they are represented as minute pores scattered on either side of the sutures separating the parietals from the supraorbitals. They have been restored in this way in Fig. 4.13. The main supraorbital sensory canal is large. All of the exposed surfaces of

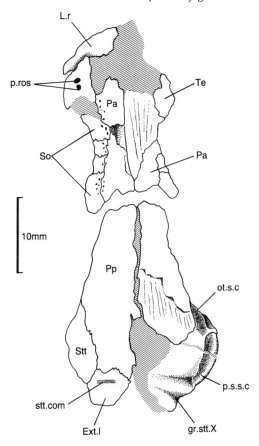

Fig. 3.11 Skull roof of *Sassenia groenlandica* n. sp. as preserved in MGUH VP2326. The dorsal surface of the neurocranium is exposed posteriorly on the right-hand side. Stipple indicates matrix.

the bones are covered with an ornament of regular, close-set rounded tubercles. The postparietal shield (Figs 3.11, 4.13) is relatively long and narrow and as in *Spermatodus*, the supratemporal (Stt) is very small. There is at least one lateral extrascapular, which is tightly sutured to the supratemporal and postparietal, and one smaller extrascapular more loosely associated with the supratemporal. Little detail of sensory canals or pit lines can be seen on any specimens examined here.

Spermatodus

Previous comments on the skull roof (Cope, 1894; Hussakof, 1908; Westoll, 1939) have been very brief due to the nature of the specimens studied. A newly discovered specimen (AMNH 4612, Fig. 3.12) allows a more complete description. No single specimen shows both halves of the skull intact so it is impossible to be certain of precise measurements. However, by comparing measurements of individual bones from specimen to specimen, it can be estimated that the postparietal shield is equal to 80% of the length of the parietonasal shield.

The parietonasal shield (Fig. 3.12) is narrow throughout and markedly vaulted in transverse section at the level of the posterior parietal. The constituent bones are uniformly covered with minute, round and closely spaced tubercles. The spaces between the smaller bones which lie anteriorly are occupied by small, irregularly shaped tesserae, the larger of which also bear the characteristic ornament.

There are four nasals as in *Diplocercides* (Fig. 3.4), each approximately as long as wide, and there is a gradual increase in absolute size from the anteriormost to the posteriormost. The anterior and posterior parietals (Pa) are very unequal in size; the former is less than half the length of the latter and, in this, *Spermatodus* is similar to *Macropoma* and *Diplocercides*. The intracranial joint margin of the posterior parietal is deeply notched and these contours are mirrored in the anterior margin of the postparietal (UM 10310). The descending process of the parietal (v.pr.Pa) is massive and an oblique ridge probably provided a point of attachment for joint ligaments.

The supraorbito-tectal series shows four supraorbitals and at least four tectals. The posteriormost supraorbital is not distinct from the posterior parietal in the specimen figured (Fig. 3.12) but this is probably an age-related phenomenon. The second

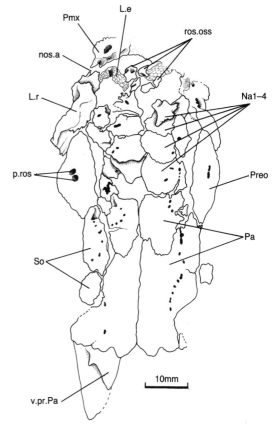

Fig. 3.12 Parietonasal shield of *Spermatodus pustulosus* Cope. Camera lucida drawing of AMNH 4612. Note the juxtaposition of the two posterior openings from the rostral organ and the enlarged descending process of the parietal.

supraorbital which lies opposite the suture between anterior and posterior parietals is characteristically large (cf. *Sassenia*). There is a large space between the premaxilla and the anteriormost tectals and nasals. In AMNH 4612, the only specimen showing this area, there are several rostrals but the disturbed preservation does not permit more precise restoration. No obviously symmetrical bones can be seen so it is possible that internasals are absent. The spaces immediately around the rostrals were filled with small tesserae and in life, the surface of the

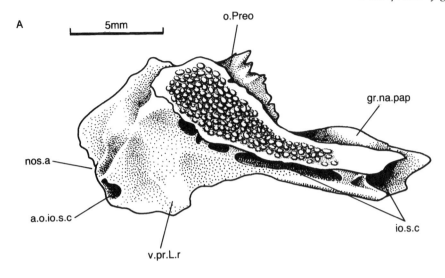

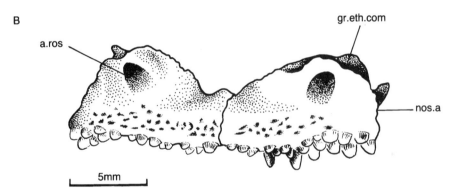

Fig. 3.13 *Spermatodus pustulosus* Cope. (A) Lateral rostral in left lateral view. (B) Premaxillae in anterior view. Camera lucida drawings of AMNH 4612.

snout must have been covered with a tessellated armour.

The lateral rostral (Figs 3.12, 3.13(A)), as usual, is robust with a relatively small ornamented area covering the posterior tubular portion. The thick anterior half lies deep and it is probable that overlying tesserae lay in the covering skin. The dorsalmost angle of the lateral rostral fits closely against two tectals and the anteroventral angle of the preorbital (Fig. 3.13(A), o.Preo). A notch is left between the preorbital and the tubular portion of the lachrymal; this space represents the point of exit of the posterior nasal tube from the nasal capsule. The superficial course of the nasal tube is not obvious but there is a shallow groove (Fig. 3.13(A), gr.na.pap) on the lachrymal which is similar to a groove seen in *Macropoma lewesiensis* implying that a soft tissue nasal tube crossed the bone.

The premaxilla (Figs 3.12, 3.13(B)) is large, unornamented and perforated by the anterior opening of the rostral organ (a.ros). The premaxilla therefore includes a rostral element (cf. *Latimeria*). The ventral surface of the premaxilla bears granular teeth which can be distinguished from ornament because of their larger size and the fine striations near to the base of each tooth. The ventrolateral angle of the premaxilla is notched, marking the position of the anterior nostril (nos.a). The preorbital (Fig. 3.12, Preo) is large, ovoid and perforated by two foramina for the posterior openings of the rostral organ. As in *Sassenia*, these openings lie in tandem and are very close together, separated only by a very thin bar of bone.

The path of the sensory canals, in so far as they can be traced, is similar to that in *Latimeria*. A broad band of small pores, straddling the sutures between the parieto-nasal and supraorbito-tectal series, marks the position of the supraorbital sensory canal. The infraorbital canal runs forward from the jugal into the tubular portion of the lateral rostral where it loops dorsally before descending along the anterior margin to open through a terminal pore (Fig. 3.13(A), a.o.io.s.c) just behind the anterior nostril. As the canal runs through the tubular portion it opens through small pores located just below the ornamented area. The ethmoid commissure (Fig. 3.13(B), gr.eth.com) appears to run along the dorsal rim of the premaxilla where there is a faint groove.

The postparietal shield is relatively narrow (Fig. 3.14), particularly at the level of the intracranial joint. The supratemporals contribute a small fraction (15%) to the total area of the parietal shield and the overall shape and proportions are very similar to those of *Sassenia groenlandica*.

In the postparietal shield (Fig. 3.14) the supratemporal bears a pronounced descending process and, as Westoll (1939) remarks, the pit for the insertion of the opercular ligament (op.lig) is well developed. Isolated supratemporals (UM 16165) show that there was a pronounced posterior process which must have lain deep beneath the skin and which provided the site of epaxial muscle

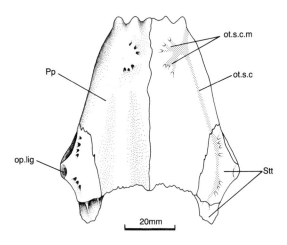

Fig. 3.14 Restoration of postparietal shield of *Spermatodus pustulosus* Cope, based on UM 10310 and 16165. Path taken by otic sensory canal shown by stipple: the canal is very narrow in this species, unlike that of most coelacanths.

insertion as well as supporting extrascapulars. The undersurface of the postparietal shows that a descending process was absent (cf. Westoll, 1939: 12).

The surface of the postparietal and supratemporal is covered, for the most part, with tiny closely set tubercles. UM 10310 suggests the covering may be absent anterolaterally and posteriorly near the midline where tiny tesserae replace tubercles.

The medial branch of the otic sensory canal (Fig. 3.14, ot.s.c.m) opens through a double cluster of medially directed pores within the parietal and a double cluster of laterally facing pores within the supratemporal. The pattern of pores suggests a conventional coelacanth pattern of sensory canals and such is restored in Fig. 3.14. The main otic sensory canal is quite narrow as can be seen in specimens broken through the postparietal. The otic canal leaves the supratemporal through a single posterior opening. No pit lines were seen.

The extrascapular series remains virtually unknown. UM 10310 shows that there was probably a large, irregularly shaped median extrascapular. On either side the extrascapular series appears to be represented by many small ossicles. This is almost certainly a specialized feature of *Spermatodus*.

Wimania and Axelia

The dermal bone pattern of the parietonasal shield of these two genera is so similar that they can be treated together. The descriptions given by Stensiö (1921) form the basis of the following remarks because I have not examined original material of either genus. According to Stensiö's description these two genera are very similar to each other and the restoration of *Wimania sinuosa* (Stensiö, 1921: fig. 20), particularly the restored path of the sensory canals, has been influenced by interpretation of the more complete material of *Axelia robusta*. The restoration of the parietonasal shield of *Wimania* is based upon impressions of the undersurface (Stensiö, 1921: 66).

The parietonasal shield of both genera is broad; that of *Axelia* is short and blunt while the anterior part of the snout is unknown in *Wimania*. The parieto-nasal series consists of two large bones. Posteriorly, there is a large parietal (fronto-dermosphenotic of Stensiö) which completely covers the orbit and, in *Wimania*, the posterior half of the nasal capsule as well. Such a large parietal is seen also in *Laugia* (Fig. 3.8) and *Euporosteus* (Fig. 6.3), but unlike the latter genus the radiation centre of the parietal lies close to the intracranial joint margin. The descending process of the parietal is large (Stensiö, 1921: pl. 5, fig. 1, pl. 12). Anteriorly there is a large nasal (postrostral of Stensiö) and in *Axelia* there may also be a small internasal (Stensiö, 1921: pl. 14, fig. 4). There are several supraorbitals along the lateral edge of the frontals and these are preceded by a larger element called, by Stensiö, 'nasalo-antorbital'.

Stensiö justified the term 'nasalo-antorbital' on the basis of his interpretation of *Axelia*

robusta where he restored (1921: fig. 43) the supraorbital canal, specifying the nasal component, and two 'nasal' openings specifying the antorbital component. If we translate Stensiö's terms for equivalent terms used here, this compound bone would be a tectal-preorbital; the 'nasal' openings would represent posterior openings of the rostral organ. Irrespective of the names for the openings we might apply, such a presumed compound bone is otherwise unknown in coelacanths (although Stensiö, 1932, did suggest that a similar compound bone is present in *Laugia*), and it therefore might be of relevance for ideas of immediate relationship between the two genera *Axelia* and *Wimania*. However, it needs to be pointed out that the interpretation of this bone with a sensory canal and two openings is not clear cut. The presumed path of the sensory canal is not clear in the plates given by Stensiö (1921: pls 11–14) and it is restored in an unusual position, i.e. neither running along a suture as is normal in coelacanths, nor through the centre of ossification, as in most other bony fishes. The openings are shown in Stensiö's earlier restoration but omitted from a slightly altered restoration (1932: fig. 15B). Furthermore, the condition of the bone in the same specimen showing the openings is differently shown in plates 13 and 14, fig. 1 of Stensiö's work. Some recently prepared photographs of the relevant specimen of *Axelia robusta* (UP P195), together with a plaster cast, kindly supplied by Ms S. Stuenes of Uppsala University, show almost no detail in this area of the snout. In sum, confirmation of the rather unusual restoration of the snout in *Wimania* and *Axelia* must remain until more material is forthcoming.

The supraorbital series of *Axelia* consists of four large quadrangular bones separated from the parietals by large sensory canal pores. The supraorbitals of *Wimania* are not well known, but they are restored by Stensiö (1921: fig. 21) as small ossicles. The supraorbital sensory canal is restored by Stensiö

(1921: fig. 43) for *Axelia*. The restoration is unusual for two reasons. First, the anterior part of the canal is shown sweeping across the dorsal edge of the presumed compound nasalo-antorbital (see above). Second, the posterior end is shown as joining with the infraorbital canal within the posterolateral angle of the parietal. Normally, in coelacanths the canals join within, or immediately lateral to the postparietal. The photographs provided by Stensiö show little evidence of the supraorbital canal other than the primary pores, and it has not been possible to check this restoration.

There is little to add to the descriptions given of the postparietal shield by Stensiö (1921). The supratemporal bears a well-developed descending process. There are several free extrascapulars in *Wimania*.

Chinlea

The skull roof of *Chinlea* has been described and figured by Schaeffer (1967) and Elliott (1987). The parietonasal shield of *Chinlea* is very long compared with the postparietal shield and at least half of this length lies in front of the orbits, which emphasizes the long-snoutedness of this form. The lateral profile of the roof is decidely concave, which perhaps explains why three of five specimens showing the roof are broken transversely above the front of the orbit.

There are two pairs of elongate parietals which together extend most of the length of the parietonasal shield. The anterior parietals are distinctly longer than the posterior parietals, an unusual feature amongst coelacanths. Three, small and approximately rectangular nasals lie anteriorly. The tectal/supraorbital series is composed of approximately 12 rectangular bones of which five (occasionally six) lie above the eye and may be considered as supraorbitals. Schaeffer (1967: fig. 14) restores a supraorbital lying lateral to the postparietal, an observation based on AMNH 5653 (Schaeffer, 1967: pl. 28, fig. 1). However,

it should be noted that two other specimens (AMNH 5652 and that illustrated by Elliott, 1987) do not show this posterior supraorbital, the space being occupied by the dorsal extension of the postorbital.

The lateral rostral is elongate, in keeping with the long snout. There is no preorbital: the lachrymojugal contacts the tectal/supraorbital series dorsally. There is a small gap left between the lateral rostral and neighbouring bones and this represents the position of the posterior nostril. Small rostral bones were certainly present (Schaeffer, 1967: pl. 28, fig. 2) but their number and arrangement is not clearly seen on any specimen. The premaxillae are little more than splints each bearing four or five small teeth.

The sensory canal openings are not obvious except for two elongate pores on the lateral rostral. Most of the roofing bones fit tightly together and it is therefore likely that the openings were small and scattered within the rugose ornament characteristic of this genus.

The postparietal shield is rather square in dorsal view, the supratemporal is relatively large and the extrascapulars are tightly sutured to the postparietals. Elliott (1987) restores four extrascapulars spanning the embayment between the supratemporals. Thus, there is no median extrascapular shown and this might be a resemblance to *Mawsonia*. However, Schaeffer (1967: fig. 14) indicates that a median element is present and the condition of the extrascapulars needs to be checked against more material. All of the bones of the skull roof are ornamented by coarse rugae.

Whiteia

The skull roof has been described by Moy-Thomas (1935b), Lehman (1952) – *W. woodwardi* and *W. tuberculata*, and Nielsen (1936) – *W. nielseni* n. sp. A figure, showing the skull roof of the species from western Canada was given by Schaeffer and Mangus (1976, fig. 19B). Beltan (1968, pl. 49C) figures a specimen of *W. woodwardi* (MNHN IP.230) interpreted as the anterior part of the skull roof of *W. woodwardi* but this is really the basibranchial dentition (Fig. 7.6(D)). The parietonasal shield is proportionately longer and narrower than that of many other coelacanths except *Chinlea*, *Axelrodichthys* and *Mawsonia*. Most of the bones in all species are smooth; the posterior parietal, the last two supraorbitals and postparietals may, in some specimens, show a few tubercles. The skull roof of *W. woodwardi* is described here with differences between the species noted. A restoration of the skull roof is shown in Figs 3.15 and 4.15.

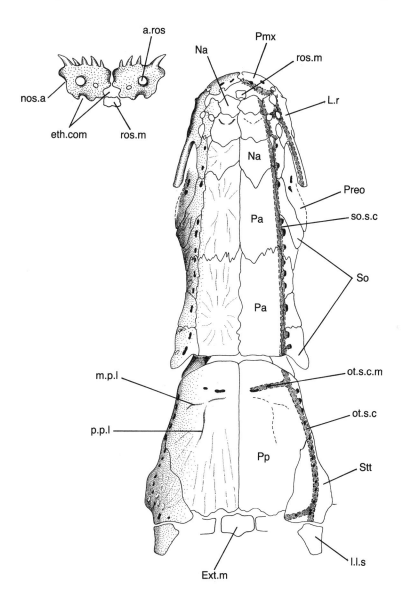

Fig. 3.15 *Whiteia woodwardi* Moy-Thomas. Restoration of the skull roof with slightly enlarged inset of premaxillae seen as if flattened out. Based on BMNH P.16636, P.17167, P.17169, P.17201.

The parietonasal series consists of the usual elements, three nasals (post-rostrals of Nielsen; posterior rostrals of Moy-Thomas) and two parietals. The exposed portions of the anterior and posterior parietals are, in most individuals, of almost equal size (Fig. 3.15) but in young individuals (e.g. BMNH P.17204 – which is the specimen used by Moy-Thomas 1935b for his fig. 2) the anterior parietals are considerably shorter. There is a very large overlap surface on the posterior parietal for the anterior parietal. Indeed, in the East Greenland species, the posterior parietal extends beneath almost the entire length of the anterior parietal so that, in the many specimens (e.g. MGUH VP2328, VP2330, VP2332) showing impressions of the undersurface of the skull roof, there appears to be a single parietal.

There are three tectals (nasals of Nielsen, antorbitals of Moy-Thomas) followed by four or five supraorbitals which flank the frontals. The anterior supraorbital is the largest of the series and is produced ventrally to lie near the lachrymojugal and so nearly occlude the preorbital from the orbital margin. The posterior supraorbital was called the dermosphenotic by Moy-Thomas (1935b) in the belief that it contained the union of the infraorbital with the supraorbital canals. Nielsen (1936: fig. 11) demonstrated that the infraorbital canal does not penetrate the supraorbital but joins the main lateral line in the edge of the post-parietal, as in *Latimeria*.

The lateral rostral is of the usual elongated shape. The ventral process is very poorly developed (Fig. 4.15) in contrast to that in *Latimeria* and Macropoma. The posterior tubular portion is slender and particularly long, reflecting slight elongation of the snout. An elongate, narrow preorbital (antorbital of Nielsen, tectal of Lehman) (Figs 3.15, 4.15, Preo) lies wedged between the tubular portion of the lachrymal below and the posterior tectal and anterior supraorbital above. It is perforated by two holes (poster-ior nares of Nielsen, 1936; Lehman, 1952), representing the posterior openings of the rostral organ (Fig. 4.15, p.ros). These openings lie in tandem in *Whiteia*, not one above the other as in *Latimeria* (Fig. 3.2).

The premaxillary teeth are borne upon a single premaxilla (Fig. 3.15 inset) which is also perforated by several pores, some of which presumably represent openings from the ethmoid commissure. This bone is there-fore a premaxilla plus rostral. In *Latimeria* the ethmoid commissure loops above the ante-rior opening of the rostral organ. Interpreting *Whiteia* on the *Latimeria* model suggests that one of the two pores which are completely enclosed by bone represents the anterior opening to the rostral organ (a.ros). The dorsal pore (notch) and the pore in the midline are openings along the ethmoid commissure (eth.com). The hole formed by the posterior (lateral) edge of the premaxilla and the anterior edge of the lateral rostral is the position of the anterior nostril. There were probably some additional rostral and internasal elements lying between the premaxilla and the nasals. One specimen figured by Lehman (1952, pl. 4D) shows several star-shaped bones but it is difficult to restore these to their mutual positions.

The paths of the sensory canals (Fig. 3.15), in so far as they can be traced, are similar to those in *Latimeria*. The supraorbital canal is carried between the parietonasal and the supraorbito-tectal series but mostly within the former. The tectals and supraorbitals are perforated by several small secondary tubes in *W. woodwardi* (Fig. 3.15) and *W. nielseni* and by a few large primary pores in *W. tuberculata* (Lehman, 1952, fig. 14, pl. 4B). In *W. woodwardi* there are large pores between the tectals and the nasals and between the middle tectal and the lateral rostral (Fig. 4.15) where the supraorbital canal turns ventrally to join the infraorbital canal. There appears to be only a single pore from the infraorbital canal as it passes through the tubular portion of the lateral rostral (Fig. 4.15).

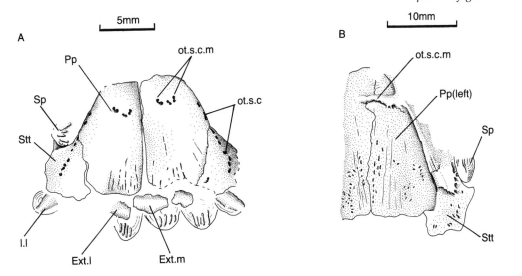

Fig. 3.16 Postparietal shield of *Whiteia*. (A) *W. woodwardi* Moy-Thomas. Camera lucida drawing of cast of BMNH P.17204. (B) *Whiteia nielseni* n. sp. Camera lucida drawing of undersurface as preserved in MGUH VP.2331c; note the medial branch of the otic canal meets its partner in the midline.

The postparietal shield is known in all species except *Whiteia* sp. from the Lower Triassic of western Canada (Schaeffer and Mangus 1976). A single description will suffice for all species because, aside from slight variation in ornament, no significant difference is known to exist between the species in postparietal shield morphology.

The postparietal shield (Figs 3.15, 3.16) is wider than long and the supratemporals (supratemporo-extrascapular of Nielsen, posterior dermopterotic of Lehman) form a relatively small fraction of the total area of the shield. The posterior margin of the shield is embayed (cf. Moy-Thomas, 1935b: fig. 2). The anterior margin of the postparietal (fronto-parieto-intertemporal of Nielsen, 'parieto'-dermopterotic of Lehman) is straight without any indentations along the joint margin. The supratemporal bears a stout descending process. The ventral surface of the postparietal is marked with the usual longitudinal ridge suggesting the path of the otic sensory canal; but unlike *Macropoma* and *Latimeria*, there is no descending process.

Instead, the ventral ridge is slightly inflated beneath the centre of ossification of the postparietal.

The path of the sensory canal has been described and figured by Nielsen (1936: figs 11, 12) for *W. nielseni* and it appears to be similar in *W. woodwardi* where, however, only the pores are evident. Nielsen (1936: 29) remarked that the medially directed branch of the otic sensory canal within the postparietal was well developed in *Whiteia* and that the canals of either side 'nearly meet in the mid-line of the parietal shield'. It is possible that the canals did meet, because in one specimen (MGUH VP 2331c) of *W. nielseni* the matrix infilling of the sensory canal is perfectly continuous across the mutual postparietal suture (Fig. 3.16(B)). The medial branch of the otic sensory canal opened to the surface through a series of pores but the number and distribution of these pores is variable, although in nearly all specimens they are arranged in two groups. The main otic canal opens along the lateral edge of the postparietal by two to four pores.

The supratemporal contains the triple junction of the otic, supratemporal commissure and the main lateral line. Within the supratemporal the otic canal opens by 6–10 pores. Lehman shows these pores as opening in two groups, corresponding to the two neuromast organs of *Latimeria*. Such a distribution may also be seen in BMNH P.17204 (*W. woodwardi*). However, in other specimens a clustered distribution is not obvious.

The extrascapular series (Figs 3.15, 3.16) consists of a median plus, at least, two lateral extrascapulars. It is likely that there was some individual variation in the number of extrascapulars because Lehman found three lateral extrascapulars on the left side of one specimen of *W. woodwardi* (Lehman, 1952: fig. 7).

The postparietal shield is generally without ornament although some larger specimens of *W. woodwardi* show a few tubercles restricted to the supratemporal (Fig. 3.15), and in some specimens of *W. nielseni* there are a few tubercles on the posterior half of the postparietal. Lehman (1952: fig. 5) records middle and posterior pit-line grooves on the postparietal but it should be noted that among the specimens used in this work these were only seen in BMNH P.16636 (Fig. 3.15) as very shallow grooves. However, enough remains to suggest that the pit lines were developed in normal coelacanth fashion, with a transverse middle pit line and longitudinal and slightly oblique posterior pit line.

Diplurus

The skull roof has been most completely described by Schaeffer (1952a) and only a few points need be emphasized and emended here. I use a redrawing of Schaeffer's restoration (Fig. 4.16(A)) to illustrate the following remarks. There is no significant difference in the bone pattern between the two species of *Diplurus*.

The parietonasal shield is relatively broad in keeping with the general broadness of the head. The proportions of the shield are similar to those of *Axelia*. There are three principal bones in the parietonasal series: a nasal (postrostral of Schaeffer), which covers most of the nasal capsules, and two parietals, anterior and posterior, which roof the orbit. Although there is considerable variation in the pattern formed by these bones (Schaeffer, 1952a: fig. 5), the posterior parietal is always separate and is the largest of the series. Many specimens show sutures which run only part way between bones and I interpret this to mean that some suture obliteration took place during growth (cf. *Macropoma*). The joint margin of the posterior parietal is straight and there is a slender descending process of the parietal (Schaeffer, 1952a: pl. 7).

In front of the nasal there are many tiny ossicles but no specimen examined in this study was well enough preserved to allow a restoration of either the number or arrangement. Together they must have formed a loose mosaic.

I agree with Schaeffer's restoration of the supraorbito-tectal series except in one small particular. The largest member of the series, here identified as the posterior tectal, has a distinct notch at the ventrolateral angle (Fig. 4.16). This notch lies adjacent to the pronounced groove on the lachrymojugal (L.j). Schaeffer suggested that this groove might represent the location of the posterior nostril. I interpret this as the posterior opening(s) of the rostral organ. Schaeffer (1952a: fig. 2) described the rather unusual lateral ethmoid of *Diplurus* in which the posterior margin is produced as three finger-like processes which delimit two notches, one above the other. In specimen PUGM 1493a, the ventral notch lies beneath the notch in the margin of the posterior tectal and therefore probably represents an endoskeletal opening from the rostral organ. Schaeffer mentions another possible location for the posterior nostril as being represented by the deep notch in the dorsal margin of the lachrymal, immediately behind the anterodorsal process. This notch is certainly in the same position as the opening of the posterior nostril from the nasal capsule in *Latimeria*. There is, however, no distinct notch in the lachrymal of *Latimeria* and a nasal tube leads back along the dorsal edge of the lachrymal. The position of the posterior nostril of *Diplurus* must remain uncertain.

The paths of the sensory canals in the parietonasal shield, in as much as they can be traced, are similar to those in other coelacanths. The supraorbital canal opens to the surface by way of large pores between adjacent supraorbitals/tectals, as in *Libys* and *Euporosteus*.

The postparietal shield is, in many respects, similar to that of *Chinlea*, *Mawsonia* and *Axelrodichthys*. Particular points of resemblance are the proportions (in which the shield is about half the length of the parietal shield and the supratemporals are relatively large) and the overall shape (in which the lateral margin of the postparietal is flared away from the intracranial joint and then is angled at the suture with the supratemporal to give a lateral edge running parallel with the midline – Schaeffer, 1952a: fig. 4); there is no descending process on the supratemporal. In *Diplurus newarki* there are about seven extrascapulars, which are very small ossicles that lay free from the postparietals (extrascapulars are unknown for *D. longicaudatus*).

Mawsonia

The parietonasal shield is incompletely known in all five species of *Mawsonia*. *Mawsonia tegamensis* and the *Mawsonia* sp. from the Santana Formation (Maisey, 1986b) are the most completely known in this respect, but even here the rostrals are unknown. The middle portion of the parietonasal shield is known in *M. lavocati* (Wenz, 1981: fig. 3, pl. 7) and the posterior end is known in *M. gigas* and *M. libyca*.

There are at least four bones in the parietonasal series of *M. tegamensis*: two parietals preceded by, at least, two nasals. The anterior and posterior parietals are of equal size (as in *Whiteia*) and each is longer than wide. The nasals of *M. tegamensis* are quadrangular and as wide as the anterior frontal, whereas the nasal of *M. lavocati* is elongate and considerably narrower than the anterior frontal. These differences are considered by Wenz (1981: 9) to distinguish these two species, and she also cites the more elongate form of the anterior parietal of *M. lavocati*. The intracranial joint margin of the frontal is complex (e.g. *M. gigas* – Woodward, 1907: pl. 7, fig. 2) with a pronounced bevelled edge and a very large descending process of the parietal (processus ventral, Wenz, 1981; alisphenoid, Weiler, 1935). The canal for the superficial ophthalmic (labelled as the supraorbital sensory canal by Weiler, 1935: fig. 20) runs through the descending process of the parietal to open into the posterodorsal corner of the orbit (Wenz, 1975: fig. 2).

The supraorbito-tectal series of *M. tegamensis* and *M. lavocati* consists of large bones which are nearly as wide as those of the parietonasal series and this is an unusual feature among coelacanths. Weiler (1935) identified some isolated bones as supraorbitals in *M. libyca*. These are equidimensional, with an irregular outline, and are covered with coarse stellate ornament. As illustrated by Weiler (1935: pl. 1, figs 19–21), these bones are more similar to rostral ossicles than to supraorbitals. The ornament of *Mawsonia* consists of very coarse bony ridges which either radiate from the centres of growth or, as in the posterior parietal of *M. gigas* (Woodward, 1907: pl. 7, fig. 2) and *M. libyca* (Weiler, 1935: pl. 1, fig. 19), run longitudinally. This type of ornament appears to be characteristic of *Mawsonia* and *Axelrodichthys* among coelacanths. The path of the supraorbital sensory canal can be traced only as a groove running along the lateral edge of the posterior parietal and pierced medially by nerve foramina (*M. tegamensis* – Wenz, 1975: fig. 2b). Wenz (1975, 1981) did not find any evidence of sensory canal pores within the parietonasal shield of either *M. tegamensis* or *M. lavocati* and this led her to regard this absence as a characteristic of *Mawsonia* (Wenz, 1981: 14). The condition of the sensory canal cannot be checked in other species.

The postparietal shield is known with varying degrees of completeness in the five species of *Mawsonia* (*M. minor* is regarded as a juvenile *M. gigas* – p. 327). The shield is best known in *M. gigas*, and particularly from the *Mawsonia* sp. from the Santana Formation (Maisey, 1986b, 1991) and in *M. tegamensis* (Wenz, 1975).

Mawsonia is unusual among coelacanths in showing a second pair of 'postparietals' which are here interpreted as extrascapulars that have become incorporated into the postparietal shield. The postparietal shield is short and broad. In *M. gigas* and *M. tegamensis*, the two species in which complete skull roofs are known, the length of the postparietal shield is equal to approximately half the length of the parietonasal shield. These proportions are similar to those of *Diplurus*, *Libys*, *Macropoma*, *Latimeria* and, probably, *Holophagus*. As in *Latimeria*, the anterior border of the postparietal meets the lateral edge through a high angle, so that the lateral margin shows a marked divergence posteriorly. However, unlike *Latimeria*, the lateral margin of the supratemporal in *Mawsonia* lies parallel to the midline. This produces a characteristically shaped skull roof, best seen in *M. tegamensis* (Wenz, 1975: pl. 1, fig. 1), and most closely approached by the shape of the postparietal shield in *Diplurus* and *Chinlea*. The specimen of a postparietal shield illustrated by Weiler (1935: pl. 3, figs 11, 13) as *M. libyca* is exceptional in being relatively much narrower. The outline of the posterior margin of the postparietal shield (where known) shows a shallow embayment between the supratemporals.

The postparietal shield is markedly convex in the transverse plane, with a noticeable depression along the midline immediately behind the intracranial joint. The postparietal shows a very well-developed anterior process which articulates with the posterior edge of the descending process of the parietal. The interlocking is particularly complex and has been described by Casier (1961). The postparietal shows a descending process but this is very short (Wenz, 1975: pl. 4, fig. 2c). Wenz (1981: 9) distinguished *M. lavocati* from *M. tegamensis* on the basis that, in the former, there is no bridge of bone joining the anterior and descending processes of the postparietal. I find this difference to be of doubtful significance: a bridge seems to be developed in one specimen of *M. gigas* (BMNH P.10356) but not in another (BMNH P.33370). The supratemporal is equidimensional and relatively large. The lateral edge is thickened and developed as a bevelled, unornamented surface which marks the site of insertion of the opercular ligament.

Wenz (1975: 183) noted that she could not find a descending process of the supratemporal in *M. tegamensis*, and suggested its absence as a character of *Mawsonia*. This suggestion is probably correct because well-preserved specimens of *M. gigas* (BMNH P.3370 and Maisey, 1991) show no indication of such a process.

Posterior 'postparietals' have been recorded in *M. gigas* (including *M. minor*) (Woodward, 1907: 137; Wenz, 1975: 183; Campos and Wenz, 1982; Maisey 1986b, 1991) and *M. tegamensis* (Wenz, 1975: pl. 1, fig. 1). Posterior postparietals were probably also present in *M. libyca* because the specimen illustrated by Weiler (1935: pl. 1, figs 6, 7) – the left half of a postparietal shield – shows three different areas of ornamentation on the dorsal surface and three different orientations of radiating lines on the undersurface. In all these instances these posterior postparietals are as large, or nearly as large, as the neighbouring supratemporals, they are closely sutured to both supratemporals and the normal coelacanth postparietal lying anteriorly. The supratemporal commissure penetrates these posterior postparietals (Maisey 1986b: fig. 3) or runs within a groove along the posterior edge. It therefore seems reasonable to regard these postparietals as extrascapulars that have become incorporated with their associated sensory canal into the skull roof. This is a view which is reinforced by the fact that separate extrascapulars are otherwise unknown in *Mawsonia* and that a closely similar species, *Axelrodichthys araripensis*, shows a median plus lateral canal bearing bones incorporated to the skull roof. In any event, the presence of these extrascapulars as an integral part of the skull roof is a derived character, and to show a bilateral arrangement without a median element is a further derived condition within coelacanths which may also be seen in *Chinlea* (p. 70).

The paths of the sensory canals are generally difficult to trace because the surface pores are very small and largely obscured by the very coarse ornamentation. However, broken surfaces and swellings on the undersurfaces suggest that the arrangement is closely similar to that of *Latimeria*. Casier (1961) noted that in *Latimeria* there are two medial branches given off from the otic canal at the level of the postparietal ossification centre (Fig. 3.1). One of these branches passes transversely to open onto the skull roof, the other passes posteromedially and these are called the transverse and postparietal branches respectively by Casier (1961: fig. 10). He suggested that in *M. ubangiensis*, only the postparietal branch is present and that this is hypertrophied to run parallel to the main otic canal. He restored the canals of *M. gigas* in similar fashion (Casier, 1961: fig. 9A). On the other hand, Wenz (1975: fig. 1) describes only a transverse branch in *M. tegamensis* which runs directly to the midline suture. One specimen (BMNH P.33370) of *M. gigas* shows clearly that there is a medially directed branch, comparable to the transverse branch

described by Wenz (1975). The holotype of *M. gigas* shows the beginnings of a canal running parallel to the main otic canal but its true extent is unknown. It is possible therefore that *M. gigas* had both branches and was similar to *Latimeria* in this respect.

The ornamentation consists of prominent ridges. The pattern of ornament varies considerably between individuals of the same species. For instance, in small individuals of *M. tegamensis* the pattern is of shallow, irregular grooves (Wenz 1975: pl. 1). Large individuals show deep pits (Wenz 1975: pl. 5, fig. 6a), and between the two there are varying developments of grooves and pits. In some of the larger species (*M. gigas, M. libyca, M. ubangiensis*) the ornament can be almost crocodile-like.

To summarize the comparative information on the postparietal shield of *Mawsonia*, there are five distinctive aspects: the great width of the supraorbitals, the incorporation of extrascapulars as an integral part of the postparietal shield with the loss of the median extrascapular; the lack of a descending process upon the supratemporal; the development of thick bone throughout the entire postparietal shield, and the nature of the very coarse ornamentation of pits and grooves.

Axelrodichthys

The skull roof has been described by Maisey (1986b, 1991) and in many respects it is very similar to that of *Mawsonia*. The postparietal shield is less than half the length of the parietonasal shield (Fig. 4.17). The parietonasal shield is very narrow and is formed by two small pairs of parietals in front of which lie at least two pairs of elongate nasals. The supraorbitals are very large and there are only four present and at least two tectals. The snout is very poorly known but there appear to be small, star-shaped ossicles, including one median rostral. The premaxillae are small and splint-like.

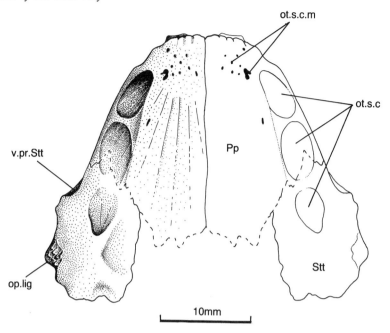

Fig. 3.17 *Libys polypterus* Münster. Restoration of the postparietal shield based on BMNH P.3337 and BSM 1870.xiv.502. Note the extremely large openings of the sensory canals characteristic of this species.

The postparietal shield is virtually identical to that of *Mawsonia*, the chief difference being that there is a median extrascapular fused into the skull roof. As in *Mawsonia* the ornament is of coarse rugae and pits, and this makes it difficult to trace the paths of the sensory canals which are, however, thought to be like those of other coelacanths.

Libys

The postparietal shield of *Libys* has been noted by Reis (1888). Three specimens showing the postparietal shield were examined in this study: BMNH P.3337 and BSM 1870.xiv.502 (Reis 1888: pl. 3, fig. 1) – *L. polypterus*, BSM AS.i.801 (Reis, 1888: pl. 2, figs 1, 2) – the holotype of *L. superbus*. The postparietal shield is known only in external view in *L. polypterus* and from the internal aspect in *L. superbus*. There is every reason to suspect that there is only one species.

The shield is broader than long (Fig. 3.17) and was probably strongly vaulted because all three specimens show the postparietal shield crushed laterally. The supratemporal reaches considerably beyond the posterior limit of the postparietal and it bears a lateral pitted area for the opercular ligament (op.lig). The postparietal (Pp) is thick anteriorly but posteriorly the bone is thin and broken, making it difficult to be certain of the shape of the posterior margin. Distinctively the otic sensory canal (ot.s.c) is very large and opens to the surface by very large vacuities, two in the postparietal and one within the supratemporal. The undersurface of the postparietal and supratemporal is inflated as the floor of the sensory canal. The holotype of *L. superbus* shows the presence of a narrow descending process of the postparietal. Other specimens show the proximal part of a descending process of the supratemporal.

The path of the sensory canal within the supratemporal is not clear but slight swellings on the dorsal surface suggest that otic, lateral line and supratemporal commissure canals met near the posterior edge of the supratemporal. There is at least one extrascapular medial to the supratemporal. The size of the extrascapular suggests that there were comparatively few bones in the extrascapular series; that is, *Libys* was similar to the majority of coelacanths in this respect but unlike *Macropoma* and *Latimeria* (p. 47).

The dorsal surface of the postparietal shield is smooth although there are a few pores, presumably related to the medial branch of the otic sensory canal (ot.s.c.m). These pores lie above the centre of ossification which, in *Libys*, lies very close to the joint margin. Reis (1888: pl. 2, figs 1, 2) illustrates patches of ornament at the anterior end of the postparietal and the posterior end of the frontal of *L. superbus*. However, the specimen (BSM AS.I.801) shows these to be sutural surfaces with the postparietal and parietal antimeres.

Holophagus

The parietonasal shield has been described by Gardiner (1960), who drew attention to the poor preservation of this part of the skeleton in all known specimens. The most completely preserved parietonasal shield is seen in BMNH P.7795 which shows that the anterior roofing bones are very thin, crushed and broken, making it difficult to restore the pattern. There are two pairs of parietals, anterior and posterior; each is elongate and the posterior is slightly longer and considerably more robust than the anterior. Together, they roof the orbit. There are at least two pairs of nasals which have irregular outlines and their outer edges are perforated with many small pores. BMNH P.7795 and P.2022 show that the tip of the snout is covered with several small, perforated bones. The

premaxilla (BMNH P.2022) is small, it lacks a dorsal lamina and bears 3–4 pointed teeth.

The supraorbital/tectal series consists of at least six elements (Gardiner, 1960: fig. 55). The 12 elements shown by Woodward (1891: fig. 53) are certainly too many. Each element is perforated by many small pores and in this, and their regular shape, they resemble those of *Macropoma*. The tectal which contacts the lachrymal (Gardiner, 1960: fig. 55) is noticeably larger than others in the series. The lateral rostral (rostral of Gardiner) is large and of complex shape, very similar to that of *Macropoma*. Ornament is confined to a few rounded tubercles on the posterior frontal and to a more dense covering of tubercles on the supraorbitals.

All five specimens of *H. gulo* examined show parts of the postparietal shield such that it is possible to provide a partial resoration (Fig. 3.18). In many respects the shield is similar to that of *Macropoma* and *Latimeria*: the shield is strongly vaulted; both the postparietal and the supratemporal bear a well-developed descending process (BMNH P.3344), the supratemporal is relatively large and the posterior margin of the shield is markedly embayed. The embayment presumably contained extrascapulars as judged by comparison with near relatives, but these are not preserved in the material examined.

The postparietal shield is relatively short compared with the pareitonasal shield. Precise measurements are not possible but the midline length of the postparietals is approximately half that of the parietonasal shield. The postparietals are composed of thick bone anteriorly but posteriorly the bone becomes very thin, as in *Macropoma* and *Libys*, so that the posterior edge is broken in all specimens. There are no raised areas upon the postparietals and ornament is confined to a few tubercles present near the lateral edge of the postparietal.

The sensory canals opened through many pores, which are particularly abundant on the supratemporal and above the ossification

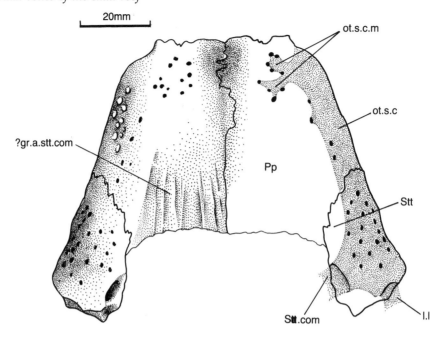

Fig. 3.18 *Holophagus gulo* Egerton. Restoration of the postparietal shield based on BMNH P.2022, P.2022a, P.7795. The path of the sensory canal is shown on the right-hand side by stipple.

centre of the postparietal. The distribution of pores follows a prominent medial branch of the otic canal within the postparietal and indicates that the supratemporal commissure joined the otic canal within the supratemporal (Fig. 3.18).

Macropoma

The skull roof of *Macropoma lewesiensis* and *M. precursor* shows two main differences from that of *Latimeria*. First, the snout ossifications are highly consolidated in *M. lewesiensis* (the snout of *M. precursor* remains unknown). Second, the bones lay more superficially as inferred from three observations: the parietals and nasals and the postparietals bear ornament; the posterior parietals and the postparietals do not have raised portions; and the bones are pierced by many tiny sensory canal pores, implying that the primary tubules of the supraorbital

sensory canal had branched before they perforated the covering bone. There are several minor differences between the species of *Macropoma* discussed here and these will be referred to at the relevant points in the following description. Restorations of the parietonasal shields and the postparietal shields of *M. lewesiensis* and *M. precursor* are given in Figs 3.19, 3.20 and 3.21. The skull roof of *M. lewesiensis* is also shown in lateral view in Fig. 4.19 and the parietonasal shield of *M. precursor* in Fig. 6.11.

The parietonasal shield is relatively longer and narrower than that of *Latimeria* and is more convex in cross section. In lateral view the profile is slightly concave (Fig. 4.19), particularly in larger specimens. The parietonasal series consists of five elements, three nasals (Na) and two parietals (Pa). The posterior parietal is, as usual, the largest and most complicated of the series and it meets and slightly underlaps the anterior parietal in

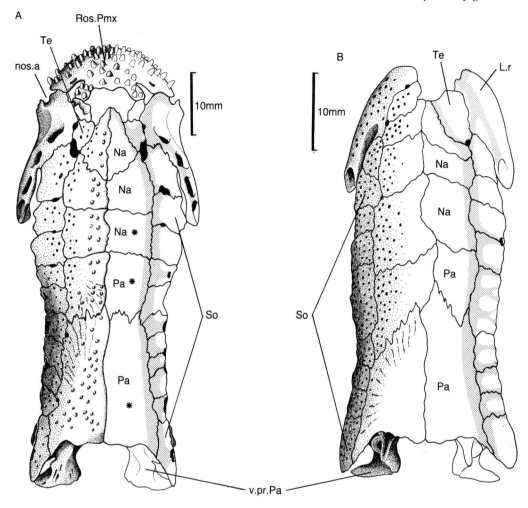

Fig. 3.19 Restorations of the parietonasal shields of *Macropoma*. (A) *M. lewesiensis* (Mantell), based on BMNH 4207, P.742, P.33540, P.42028. Asterisk denotes the centre of ossification. (B) *M. precursor* Wood-ward, based on BMNH 35700, P.10916, P.10917, P.11002. Path of sensory canals shown by stipple. Note the enlarged openings from the infraorbital sensory canal within the lateral rostral of *M. lewesiensis*.

a digitate suture. In *M. lewesiensis* this suture runs transversely (Fig. 3.19(A)) but in *M. precursor* (Fig. 3.19(B)) it is oblique. The descending process of the parietal (alisphe-noid, Watson, 1921: 323) is best described with the help of Fig. 3.19. The pedicel of the process (v.pr.Pa) is produced as an elbowed ridge which slid in the groove on the post-

parietal contribution to the intracranial joint (Fig. 3.21(B), fa.i.j). At the dorsal end of the pedicel there is a crescentic groove which received a matching ridge on the postparietal shield. The ventral end of the parietal descending process is expanded to a thin fan which spreads over the dorsal surface of the basisphenoid (Fig. 6.12, o.v.pr.Pa). The process and the basisphenoid are very closely associated and in most disarticulated specimens they are found joined but separate from the roofing portion of the posterior parietal. This state of preservation probably led Watson (1921) to consider the process as an ossification (the alisphenoid) separate from the parietal. A shallow groove runs across the lateral face of the descending process (Fig. 3.19(B)), immediately below the elbow. This groove contained the superficial ophthalmic branch of the anterior lateral line nerve which ran forward to innervate the supraorbital sensory canal. In *Latimeria* there is a similar groove (Millot and Anthony, 1958: pl. 23) or, as in the holotype of *Latimeria*, this nerve may pass through a foramen in the parietal descending process (Smith, 1939c: pl. 18, J). The course of the superficial ophthalmic above the orbit may have differed slightly between the two species of *Macropoma*. In specimens of *M. precursor* (e.g. BMNH P.11002) the supraorbitals and parietals do not meet along their entire lengths of the undersurface. The floor of the sensory canal is therefore incomplete. A similar condition exists in *Latimeria*, where the canal is underlain by the ethmosphenoid cartilage, and branches of the superficial ophthalmic, supplying the supraorbital neuromasts, pass through the fenestrae within the incomplete floor (Smith, 1939: 38). The main trunk of the superficial ophthalmic of *Latimeria* runs through the ethmosphenoid cartilage (Millot and Anthony, 1958: pl. 27). In *M. lewesiensis* the floor of the supraorbital sensory canal is complete (BMNH P.33540), the ventral flange of the parietal entirely meets the matching flange of the supraorbitals. For this reason it is probable that the superficial ophthalmic, or at least those branches which supplied the neuromasts, must have entered the sensory canal at the level of the intracranial joint and ran forwards within the floor of the canal.

The anterior parietals and the nasals (ethmoid plates, Woodward, 1909) are considerably smaller than the posterior parietals, and they meet their antimeres leaving no gaps in the midline (cf. *Latimeria*, Fig. 3.1). There are variations in the shape and relative sizes of these bones and this variation extends to the right and left sides of the same individual. The anteriormost nasal is always triangular and is generally the smallest bone of the series. In large specimens (BMNH P.33540, SMC 9123), sutures between the first two pairs of nasals are not visible.

The tectal/supraorbital series (marginal ossicles, Huxley, 1866; parafrontals, Woodward, 1891; Watson, 1921) consists of 11–12 elements in *M. lewesiensis* and 10–11 in *M. precursor*. The three anterior ones may be designated as tectals, as in *Latimeria*. In *M. precursor* the tectals are of about equal size but in *M. lewesiensis* the posteriormost tectal is always considerably larger, as in *Latimeria*. The supraorbitals form a chain which reaches back to the posterior edge of the parietonasal shield. Most of the supraorbitals are as long as broad but the posteriormost, which forms the posterolateral corner of the parietonasal shield, is longer than broad. In *M. precursor* the supraorbital which lies above the centre of the orbit is long and rectangular, approximately twice as long as its neighbours. There is often variation in the number of supraorbitals between the right and left sides of the same individual. The lateral margins of the two posterior supraorbitals are produced ventrally as a small lip which is larger in *M. precursor* than in *M. lewesiensis*. This lip lies adjacent to the anterior margin of the postorbital and it is probable that tough connective tissue, which stretched between the skull roof and the postorbital as in *Latimeria*, was attached to this lip. The undersurfaces of the posterior three or four supraorbitals of *M. precursor* are grooved and this was also a site for the attachment of connective tissue.

The lateral rostral (prefrontal, Huxley 1866) is a large bone sutured with the two anterior tectals and occasionally also with a nasal (Fig. 3.19(B), left side). Its shape is very similar to the lateral rostral of *Latimeria*, with the main body produced ventrally as a process which rests against a shallow pit on the lateral ethmoid. The anterior margin of the ventral process (Fig. 6.11, v.pr.L.r) is contoured, marking the position of the anterior nostril (nos.a). The posterior end of the lateral rostral is developed as a tube which encloses the infraorbital sensory canal and has a groove on the dorsal surface as in *Latimeria* (Fig. 3.2) for the attachment of connective tissue joining the lachrymal with the jugal. The lateral rostral separates the anterior from the posterior nostrils. Watson (1921: 334) thought that the posterior nostril opened to the surface behind the ventral process, i.e. below the infraorbital canal. Comparison with *Latimeria* suggests, however, that it opened from the nasal cavity within the notch bordered by the posterior tectal and the lateral rostral. It therefore lay above the sensory canal, in the hole described by Watson (1921: 330) and attributed to an opening of the infraorbital sensory canal. In *Latimeria* there is a long nasal tube, which runs posteriorly from the opening of the nasal capsule, across the inside of the tubular portion of the lachrymal, and opens to the surface below the infraorbital sensory canal at the anterior tip of the jugal. There was probably a similar long nasal tube in *Macropoma* because the anterior end of the lachrymojugal is marked by a groove which runs in the same direction (Fig. 4.18, gr.na.pap). The tip of the snout in *M. lewesiensis* is developed as a heavily ossified hemisphere which is loosely attached to the lateral rostral, anteriormost nasal and the lateral ethmoid. It is usually missing in specimens of *M. lewesiensis* and is unknown in *M. precursor* but its absence from specimens of the latter species is assumed to be an artefact of preservation. This ossification has been interpreted either as fused premaxillaries (Huxley, 1866; Watson, 1921) or as coalesced vomers (Huxley, 1866; Woodward, 1909). This apparently single ossification is, in fact, represented by, at least, a pair of bones on either

side of the midline (Fig. 3.20). There is a premaxilla and a rostral but there is no associated vomer. The vomer is found in its more usual position, beneath the lateral ethmoid. The premaxilla is the larger of the two components and is identified by the presence of replacement teeth. It is partially separated from its antimere by a large pore (Fig. 3.20), corresponding to the median sensory pore in *Latimeria* (Millot and Anthony, 1958, fig. 8, p.s.m.e.m.). In three of the five specimens that show the premaxillae, the separation of the two premaxillae cannot be seen. But in two specimens the suture may be inferred; in BMNH P.10918 the premaxillae do not meet below the sensory canal pore and in BMNH P.33373, an isolated portion of one premaxilla of a small individual, the median margin of the premaxilla is entire (i.e. not broken). The outer surface of the premaxilla bears many blunt tubercles which are largest ventrolaterally, immediately anterior to the premaxillary teeth. The teeth are found towards the lateral edge of the ventral margin and are distinguished from the tubercles in being larger and in having pointed, recurved tips. There are one or two functional teeth and these are associated with resorption sockets.

The rostral is only seen on the left side of one specimen (BMNH 4270, Fig. 3.20) although several specimens show a recess on the premaxilla in which it is located. The rostral is a posteriorly incomplete ring and carries small tubercles on a raised portion anterodorsally. A foramen represents the anterior opening to the rostral organ. The curved anterior margin of the rostral rests on a matching ledge on the inner surface of the premaxilla. The total area occupied by the premaxilla and the rostral in *Macropoma* is equivalent to that area claimed by three premaxillae, three rostrals and the anterior internasal in *Latimeria*. Whether there were ossification centres in young *Macropoma* comparable to those in *Latimeria* cannot be decided with available material.

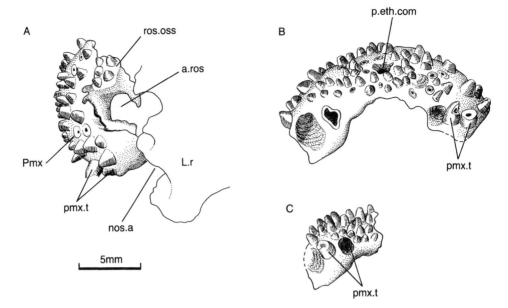

Fig. 3.20 Consolidated snout of *Macropoma lewesiensis* (Mantell). (A) Snout in left lateral veiw as preserved in BMNH 4270. This is an individual in which there appears to be a separate rostral ossicle containing the anterior opening to the rostral organ. (B) Snout in ventral view as preserved in BMNH 4270. (C) Ventral view of isolated right premaxilla as preserved in BMNH P.33373 showing a tooth and accompanying replacement socket. All camera lucida drawings.

The path of the sensory canal (Fig. 3.19) runs between the parietonasal and the supraorbital/tectal series. A few pores open to the lateral edges of the supraorbitals posteriorly and the dorsolateral faces of the anterior supraorbitals. Most of the large pores lie between the supraorbitals. Additionally, there are many tiny pores opening from the sensory canal to the surface over both the nasals and parietals and the supraorbitals and tectals. Between the anterior and middle tectals there is a pore (large in *M. lewesiensis*, small in *M. precursor*) which led to the junction of the supraorbital with the infraorbital canal (the commissural canal of Jarvik, 1942). Anterior to this junction, the supraorbital canal continues in the lateral edge of the anteriormost nasal, but beyond this level it cannot be followed.

The infraorbital canal within the lateral rostral follows a course identical to that in *Latimeria*; that is, it runs close to the dorsal margin and is represented as a gutter along the anterior margin. In *M. lewesiensis* (Figs 3.19(A), 4.19) there are three or four large pores opening laterally, a condition similar to that in *Latimeria*. In *M. precursor* many small pores perforate the lateral wall of the canal (Figs 3.19(B), 6.11). The medial surface of the lachrymal is perforated by four foramina which transmitted branches of the ramus buccalis lateralis VII.

The pattern of sensory canals through the ethmoid region is largely unknown. It is possible that, if an ethmoid commissure were developed, it ran beneath the bones as it does in *Latimeria*. The sensory canal pore separating the premaxillae suggests that at least a portion of the commissure was present. Elsewhere there is no indication of sensory pores on the snout, but it should be pointed out that the roofing bones do not

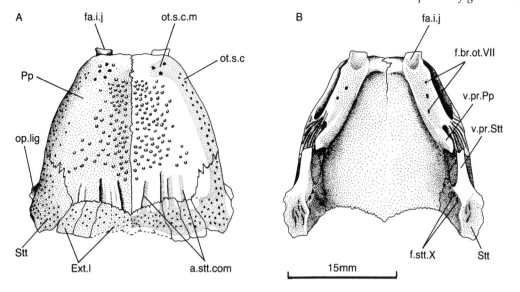

Fig. 3.21 Restoration of the postparietal shield of *Macropoma*. (A) *M. precursor* Woodward in dorsal view. (B) *M. lewesiensis* (Mantell) in ventral view. The extrascapular series is incompletely known for *M. lewesiensis*.

completely meet here (Fig. 3.19) so it is possible that the ethmoid commissure ran in soft tissue within this space.

The postparietal shield has been described by Huxley (1866), Woodward (1891, 1909) and Watson (1921) and *M. lewesiensis* and *M. precursor* are very similar to each other except in minor details of ornament pattern. Small specimens of *M. lewesiensis* (BMNH 49110, 49837) do not show the tubercular ornament seen in the adult. *Macropoma* is similar to *Latimeria* in having well-developed ventral processes on both the supratemporal (Fig. 3.21, v.pr.Stt) and the postparietal (v.pr.Pp) and in showing anterior branches of the supratemporal commissure which penetrate the postparietals (Fig. 3.21(A), a.stt.com). *Macropoma* differs from *Latimeria* in lacking raised areas on the postparietals, in the number and form of the extrascapulars and in the form of the openings of the sensory canals.

The outlines of the postparietal shield and

constituent bones can be seen in Fig. 3.21. The shield is strongly vaulted. As in *Latimeria*, small specimens of *M. lewesiensis* show a relatively broader shield than the larger specimens (this could not be checked for *M. precursor*).

Watson (1921: fig. 5) described two bones lying lateral to and sutured with the postparietal along the line of the otic sensory canal. Watson considered these bones to be posterior supraorbitals (his parafrontals). Stensiö accepted Watson's reconstruction and claimed that these elements represent the intertemporals, thus bringing the coelacanth postparietal shield into line with the osteolepiform postparietal shield and justifying the term parieto-intertemporal for the single bone in the vast majority of coelacanths. I have examined many specimens of *Macropoma*, including most of those used by Watson, but have never seen separate intertemporals nor any evidence to suggest that there is more than one centre of ossification within the postparietal. I do not believe that *Macropoma* has separate intertemporals and suggest two alternative explanations for Watson's observations: either the 'intertemporals' are displaced supraorbitals or the lateral edge of the postparietal shield was fractured (see also Westoll 1939: 19) along the line of the otic sensory canal in a specimen examined by Watson. It is of significance to note that the greatest transverse curvature occurs above the otic sensory canal where the bone is very thin and might therefore be expected to be a line of weakness.

The undersurface of the postparietal (Fig. 3.21(B)) is raised as a prominent longitudinal ridge forming the floor of the otic sensory canal. The floor is pierced by three foramina for the middle lateral line nerve (f.br.ot.VII) which, by comparison with *Latimeria*, would have supplied the three neuromast organs within the postparietal (Millot and Anthony, 1965: fig. 2). Much of the lateral wall of the otic canal within the postparietal is open. Posterolaterally, the undersurface of the postparietal lateral to the sensory canal is pitted in *M. precursor* and ridged in *M. lewesiensis* (Fig. 3.21). The undersurface of the postparietal in this area forms the roof of the temporal excavation and it is probable that these roughened areas provided origin for the palatal levator muscle. The anterior end of the ventral thickening of the postparietal bears a concave facet and ridge (Fig. 3.21, fa.i.j) which match the contours of the parietal descending process. The descending process of the postparietal is developed as in *Latimeria* and arises from the floor of the sensory canal immediately in front of the postparietal/supratemporal suture.

The supratemporal is a separate element, although in large specimens of *M. lewesiensis* the suture between it and the postparietal is often obscured. It is, however, distinct on the undersurface and the suture is always obvious in small specimens (Watson, 1921: 328). The lateral edge of the supratemporal is produced as a small process which lay adja-cent to a process on the operculum and provided a point of attachment for an opercular ligament as in *Latimeria*. The undersurface of the supratemporal is inflated due to the passage of the otic sensory canal.

The descending process of the supratemporal (Fig. 3.21, v.pr.Stt) arises from the floor of the canal and passes anteroventrally as a thin lamina. At the base of the process there are two foramina, accompanied by a groove in *M. precursor*, which lead into the lumen of the sensory canal. These foramina mark the course of the supratemporal nerve to innervate the two otic sensory neuromasts within the supratemporal (Millot and Anthony, 1965: fig. 52). In *M. precursor* the undersurface of the supratemporal and postparietal is marked with ridges (BMNH P.12884) or pits for the insertion of epaxial musculature into the post-temporal fossa.

A complete extrascapular series is seen in *M. precursor* where there are nine including a median extrascapular. In *M. lewesiensis*, only the two most lateral extrascapulars are known, although several specimens do show fragments of the median extrascapular. It is probable that the extrascapular chain was similar in the two species. The extrascapulars are larger than those of *Latimeria*; they abut one another and form a complete covering to the nape. They are longer than broad, with the supratemporal commissure contained in the posterior half and the anterior ends overlapping an unornamented area on the postparietals.

The path of the otic sensory canal and the supratemporal commissure is similar to that in *Latimeria*. The short, medially directed branch of the otic canal opens to the surface through scattered pores close to the ossification centre of the postparietal. These pores are larger than those lying along the main canal. Elsewhere the sensory canals, including the anterior branches of the supratemporal commissure open through many small pores.

3.3 DISCUSSION OF SKULL ROOF OF COELACANTHS

From the descriptions given above it may be seen that the bone pattern of the skull roof of coelacanths is relatively conservative, with the chief variations concerning the number and complexity of the snout ossicles, the presence or absence of a preorbital, numbers of parietals and supraorbitals, development of processes connecting the dermal bones to the underlying neurocranium and several features of the pit lines and extrascapular series.

The dermal roofing bones of coelacanths are always clearly divided by an intracranial joint into an anterior parietonasal shield and a posterior postparietal shield. The joint lies between the parietals and postparietals. In some coelacanths this joint margin is straight and simple (e.g. *Latimeria*) while in others it is strongly interdigitate (e.g. *Spermatodus*). While this variation may imply some functional differences, there appears to be no correlation between this and other shape or proportional variation in head morphology. Comparison with the joint margins in osteolepiforms and porolepiforms suggests that the straight margin is plesiomorphic (character 1).

In all coelacanths the postparietal shield is always equal in length to, or shorter than, the parietonasal shield and this contrasts with some porolepiforms (e.g. *Holoptychius*) or osteolepiforms (e.g. *Barameda, Marsdenichthys*). Nevertheless within coelacanths the proportions of the lengths of the parietonasal and postparietal shields vary considerably. In general, those cladistically more derived coelacanths (Chapter 8) show short postparietal shields and long parietonasal shields (Forey, 1991) but the match between relative proportions and cladistic rank is not perfect. For instance, *Allenypterus* – a cladistically plesiomorphic genus – has a very short postparietal shield, comparable to that in cladistically derived forms such as *Holophagus* and *Macropoma*. The variations in shield proportions cannot easily be translated to discrete codes to be used in phylogenetic analysis but they can be plotted onto existing phylogenetic trees (Forey, 1991).

The parietonasal shield consists of relatively few bones showing a condition somewhat intermediate between the macromeric pattern seen in actinopterygians and the mosaic patterns seen in porolepiforms and lungfishes. *Miguashaia* was said by Cloutier (1991b: appendix) to have a snout covered by a mosaic of ossicles but, in reality, the snout remains too imperfectly known to be certain (Cloutier, 1996). Some coelacanths, such as *Coelacanthus granulatus*, show a more highly fragmented snout than others but it is difficult to quantify and code these conditions. On the other hand, some coelacanths show a highly consolidated snout (character 2) where the rostral bones are tightly sutured to one another and to the premaxillae.

Despite the minor variations in the number of rostral bones, there is some regularity in the pattern. Coelacanths may show one or several median rostrals (postrostrals) and where these lie between the nasals they may be called internasals (character 3). Comparison with other sarcopterygians and actinopterygians suggests that the presence of a single median rostral is plesiomorphic. In those few instances where the snout is known in sufficient detail, the ethmoid commissure runs along the anterior margin of the median rostral (or most anterior median rostral/internasal). And in this respect coelacanths are similar to *Eusthenopteron* (Jarvik, 1944: fig. 16). However, this is unlike the pattern in most other sarcopterygians where the ethmoid commissure pierces the median rostral (e.g. *Panderichthys* – Vorobjeva and Schultze, 1991 or *Holoptychius* – Jarvik 1972) because in those fishes the canal runs through the centre of ossification. The situation in coelacanths may be a reflection of sutural paths taken by the sensory canal system over much of the roof.

The premaxillae show some variation. Comparison with actinopterygians and with other sarcopterygians suggests that paired premaxillae showing a toothed margin and a dorsal lamina (character 5) carrying all or part of the ethmoid commissure is the primitive osteichthyan condition. Compound bone names (rostropremaxillary, nasorostropremaxillary – Jarvik, 1972) are sometimes used to imply the inclusion of canal-bearing bones.

The anterior opening of the rostral organ, which is a feature unique to coelacanths, may or may not be included within the premaxilla. This depends on whether the dorsal lamina of the premaxilla is developed or not (character 5). In those forms in which the premaxilla is represented by only the tooth-bearing portion, the anterior opening of the rostral organ lies completely above the premaxilla and it fails to mark it in any way. Additionally, the tooth-bearing portion may be divided into several small paired premaxillae (character 4) which is also regarded as a derived condition (see also Cloutier, 1991a).

In those coelacanths in which the dorsal lamina of the premaxilla is developed, the ethmoid commissure runs near to or along the dorsal margin, taking a sutural course and the lamina is pierced by the anterior opening to the rostral organ. Cloutier (1991a) recognized two characters in relation to the dorsal lamina and the path of the ethmoid commissure. The first (his character 5) is the presence/absence of a dorsal lamina (with a third state – emarginate – which refers to the presence of a median opening of the ethmoid commissure). The second (his character 6) records whether the ethmoid commissure perforates the premaxilla. The only taxon mentioned by Cloutier as lacking a dorsal lamina, yet showing an ethmoid commissure perforating the premaxilla, is *Polyosteorhynchus*. But the photograph given by Lund and Lund (1985: fig. 51) shows clearly a large emargination which is almost certainly the anterior opening of the rostral organ, suggesting the presence of a dorsal

lamina. In this work I prefer to code conservatively and record the presence/absence of a dorsal lamina (character 5) which may be recognized if the anterior opening of the rostral organ is included either as a foramen or as a notch.

The roofing bones of the parietonasal shield behind the snout are represented as the mesial parietonasal (fronto-nasal) and the lateral tectal–supraorbital series. The supraorbital canal runs between these series in most coelacanths. The definition of these series is given in the description of *Latimeria* (p. 42). Lund and Lund (1985) have described the presence of three longitudinal series on either side of the midline in *Allenypterus*, *Hadronector* and *Polyosteorhynchus*, with the third row being designated as medial supraorbitals. *Allenypterus*, however, conforms to the usual coelacanth pattern and Cloutier (1991a: 399) reports that *Polyosteorhynchus* also has just two rows. This leaves *Hadronector* as the only coelacanth reported with three rows. My observations, admittedly on poor material, do not support those of Lund and Lund (p. 55). But even if there were three rows in this genus, this would be an autapomorphic character and would lead to no systematic conclusions.

The number of elements present in the parietonasal series and the tectal–supraorbital series varies between coelacanths, and there is some individual variation and even between sides of one individual (e.g. *Macropoma*, *Diplurus* – Schaeffer 1952a). Within the parietonasal series, those elements present above the orbit are distinguished as parietals and there may be one or two (anterior and posterior) pairs. If two are present, the sutural contact between successive pairs is always more complicated, with deeper overlap, than between the parietals and the nasals. Sometimes the parietal pairs are nearly equal in size, sometimes markedly dissimilar (character 8). Comparison with 'osteolepiforms' and porolepiforms suggest that a single pair of parietals (recognized as bones lying above the orbit and carrying the anterior pit line – absent in most coelacanths) might suggest that this is the plesiomorphic condition. However, panderichthyids and tetrapods show two pairs of bones in the same position: the posterior is recognized as parietal (it lacks a pit line but is pierced by the pineal) and the anterior as frontal. Thus it is difficult to be certain of the plesiomorphic condition. Here the presence of one or two pairs of parietals is recorded (character 7).

Anteriorly there are a variable number of nasals but the exact condition is rarely known, making it difficult to use this variation as a character. Cloutier (1991a: character 8) used as a character the number for bones lying anterior to the parietals (or the anterior parietal if two were present).

The number of bones within the tectal/ supraorbital series is also highly variable, including recognition of variability between individuals and from left to right within the same individual. Furthermore, it is sometimes difficult to be certain of the exact number, particularly near the snout. However, coelacanths do seem to fall into two categories: those which have few (< 7) and those which have many (> 10) and this has been coded as a character (9).

Lateral to the tectals the coelacanth skull roof shows a complex-shaped lateral rostral, which is usually a large bone bearing the anterior limit of the infraorbital within a tubular portion and a ventrally directed limb which contacts the underlying lateral ethmoid. The description given for this bone in *Latimeria* (p. 46) is sufficient for most coelacanths although there is some slight variation in the shape and proportions of the ventral process.

In some coelacanths there is a preorbital (tectal of Cloutier, 1991a) wedged between the tubular portion of the lateral rostral and the tectal series (character 10). If the preorbital is present, it is always pierced by the posterior openings of the rostral organ.

The posterior parietal (or single parietal) of

most coelacanths is developed posteroventrally as a descending process (see p. 44 for discussion of this process) which is sutured to the basisphenoid (character 11). It appears to be absent in *Diplocercides* and it may be in other plesiomorphic coelacanths, but this region of the skull is poorly known in many genera.

The postparietal shield of most coelacanths is superficially like that of porolepiforms – a similarity which has been used by Andrews (1973) to suggest relationship between coelacanths and porolepiforms and the naming of this group as 'Binostia'. The similarity extends to the facts that the otic canal passes through the postparietal which most authors take as evidence of the loss or the fusion (Andrews, 1973; Vorobjeva and Schultze, 1991) of the postparietal with lateral canal-bearing bones (variously called dermosphenotic, intertemporal or supratemporal). Most other authors consider it to be homoplasious. Evidence for the latter view might come from the fact that some coelacanths have been described as having additional canal-bearing bones lying lateral to the postparietal. Lund and Lund (1985) reported that *Allenypterus*, *Caridosuctor*, *Hadronector*, *Lochmocercus*, *Polyosteorhynchus* and *Rhabdoderma elegans* all show such bones. For most of these taxa I can find no evidence of additional bones, but I cannot confirm presence/absence in *Lochmocercus* or *Polyosteorhynchus*. Cloutier (1996) describes two canal-bearing bones (labelled as intertemporal and supratemporal) in *Miguashaia*. This is possibly the only coelacanth with separate canal-bearing bones lying alongside the postparietals but it would be sufficient to suggest non-homology between the postparietal shields of porolepiforms and coelacanths. As with the parietal, both the postparietal and the supratemporal may bear descending processes (characters 13 and 14). These descending processes reach ventrally to brace the prootic.

An extrascapular series is always present

in coelacanths, although the number of bones may vary. The series consists of a median and lateral elements and where there is overlap it is lateral. In most coelacanths the extrascapulars lie free from the skull roof and behind the level of the neurocranial roof, and this is regarded as the primitive condition (character 15). In some coelacanths, extrascapulars have been incorporated into the skull roof proper and this is regarded as a derived condition. That is, direct association with underlying neurocranial structures is secondary (character 16) and has probably happened on several occasions within osteichthyans and by at least two processes. Either the extrascapular series, in whole or in part, has 'moved' forward to come to lie above the neurocranium, or there has been fusion between extrascapulars and original braincase bones.

Examples of the first mechanism may be found in the actinopterygian *Acipenser* and in the coelacanth *Mawsonia*. In both instances, the median extrascapular(s) lie above the posterior part of the neurocranium and are firmly sutured to bones readily identifiable as parietals by virtue of the presence of pit-lines and/or sensory canals and by their own particular relationship to underlying neurocranial structures. The extrascapulars retain the supratemporal commissure and, therefore, these canal-bearing bones are additional elements to the skull roof. This reinforces the idea that it is the extrascapulars with the enclosed sensory canal which have 'moved' and not the canal alone. Separate extrascapulars behind the skull roof are absent from *Mawsonia*, while in *Acipenser* there is a large median anamestic 'nuchal' judged as a synapomorphy of *Paleosephurus* + Acipenseridae (Gardiner, 1984b: 150). Immediate relatives of both *Mawsonia* and *Acipenser* each show normal extrascapulars lying free from the skull roof; and this implies that the incorporation has taken place independently. Incorporation of extrascapulars into the skull roof as distinct elements is therefore scored

as a derived feature within coelacanths (see also discussion p. 266).

A second possible process whereby extrascapulars can be incorporated to the skull roof is by fusion between skull roofing bones and the canal-bearing extrascapulars. Ontogenetic fusions of this kind are known to take place between the extrascapulars and parietals of some cyprinids (Lekander 1949), and similar fusions are suspected in the characoid *Hydrocynus lineatus* (Weitzmann, 1962: 22).

Phylogenetic fusion of extrascapulars with the parietal is considered to have taken place in clupeoids (Patterson, 1970) and presumably also mastacembelids (Travers, 1984). In both of these instances the parietal carries the supratemporal commissure, and in clupeoids this is primitively contained within a bone-enclosed tube. The territory occupied by the parietal plus commissure of clupeoids is occupied solely by the parietal in outgroup teleosts such as elopomorphs and osteoglossomorphs in which separate canal-bearing extrascapulars occur. Patterson (1977b: 177) favoured a theory of fusion over alternative theories such as loss of parietals and invasion by extrascapulars, or shift of canal with loss of the extrascapular. This is because the clupeoid parietal retains usual topographic relationships to adjacent bones and underlying endocranium, and because the association of the endoskeletal supraoccipital with a sensory canal (dermal skeleton) could only have come about by fusion. This reasoning is convincing, but it means that phylogenetic fusion – the phylogenetic replacement of two bones by one – need not be accompanied by one bone occupying the territory of two 'ancestral bones'. Thus a theory of phylogenetic fusion may be independent of the territory occupied.

There may be a variant of the fusion process in teleosts whereby there is phylogenetic fusion between canal-bearing extrascapulars and an underlying endochondral bone – the supraoccipital. Such is presumed to have happened in clupeomorphs (Patterson, 1970), kneriids and mastacembelids (Patterson, 1977b). However, ontogenetic fusion has rarely been seen. Taverne (1973) records such a fusion for one species of *Mastacembelus* but his observations could not be confirmed by Travers (1984: 72). In coelacanths there is no suggestion that extrascapulars have fused with an endochondral element, and in this work such teleostean examples are only relevant to the more general notion that extrascapulars may become incorporated to the skull roof by fusion.

In some coelacanths the lateral extrascapular lies behind the skull roof and contains the tripartite junction between the otic canal, the lateral line and the supratemporal commissure (character 21). This is here judged to be a primitive osteichthyan condition because it is widely distributed in primitive actinopterygians (Gardiner, 1984a), 'osteolepiforms' and porolepiforms (Jarvik, 1972), rhizodontiforms (Andrews, 1985), onychodontiforms (Jessen, 1966; Andrews, 1973) and primitive lungfishes (Miles, 1977; White 1965) where the 'Z' ('H') bone is considered as the homologue of the lateral extrascapular.

However, in most coelacanths the position occupied by the lateral extrascapular is occupied by an extension of the supratemporal which also contains the triple junction of the sensory canals. The observation suggests a case of phylogenetic 'fusion', although it is accepted here that it is impossible to distinguish between true fusion (a phyletic sequence in which formerly separate centres of ossification become coincident) or loss and subsequent invasion and canal capture by the remaining element. Thus, for the condition in most coelacanths I agree with the ideas of Stensiö (1921: 63) who called this bone a supratemporo-extrascapular even though I am not confident that we can identify the process. For this

reason I choose to refer to this as the supratemporal, this being the original skull roof component.

The fusion of the lateral extrascapular with the supratemporal results in the posterior profile of the skull roof being embayed with the remaining extrascapulars lying in between. In *Coelacanthus granulatus*, according to the restoration given here, there are only three extrascapulars. But these are topographically comparable to those lying within the embayment in those coelacanths showing the derived supratemporal. Thus, in terms of numbers of extrascapulars I would argue that *Coelacanthus granulatus* is more comparable with *Laugia* (showing five) or *Rhabdoderma* (five) than with *Diplocercides* (three). The most derived coelacanths such as *Macropoma* and *Latimeria* show seven to nine extrascapulars, which is a derived condition based on outgroup comparison.

Therefore, three features in this region of the skull roof may be associated: fusion of supratemporal and lateral extrascapular, supratemporal carrying the triple junction of otic canal, lateral line and supratemporal commissure; and the posterior embayed profile to the skull roof. Association between the triple joint of the sensory canals and the posterior embayment was pointed out by Forey (1991) as his character 36 and it is used below. Cloutier (1991b) also used this character but expressed it as the posterior limit of the tabular (supratemporal of this work) level with or reaching behind the level of the postparietals. According to Cloutier, the tabular is that bone which supports a segment of the otic canal and the occipital commissure (supratemporal commissure). Thus many primitive coelacanths (e.g. *Diplocercides*, *Rhabdoderma*, *Allenypterus*), like nearly all other sarcopterygians, would not on Cloutier's definition have a tabular, because that bone does not contain the occipital commissure.

Another variation noted within the extrascapular series is the relationship between the extrascapulars and parietals (Forey 1991; character 19). Sometimes (e.g. *Laugia, Coelacanthus granulatus, Chinlea*) there is a very close fit, perhaps with a complex suture. In other instances (e.g. *Whiteia, Macropoma, Latimeria*) the extrascapulars lie free and are usually much smaller ossifications immediately around the sensory canal. The former condition is judged plesiomorphic here because this appears to be the condition in primitive members of other osteichthyan groups. The special sutural association between extrascapulars and parietals of *Mawsonia* is regarded as a different derived condition because the extrascapulars come to lie above the neurocranium.

The supraorbital sensory canal usually takes a sutural course between the tectal/supraorbital series and the naso-parietal series. An exception to this appears to be that in *Miguashaia* the canal runs through the centre of ossification of the supraorbitals (its course anterior to the orbit remains unknown). The latter condition is plesiomorphic.

In most coelacanths the otic canal runs through the postparietals to pass into the supratemporal, where it sometimes sends off a medial branch within the anterior third of the postparietal. In some coelacanths it runs through the supratemporal to join the lateral line and the supratemporal commissure within the lateral extrascapular: in many coelacanths the triple junction between the three sensory canals lies within the supratemporal (character 21).

The supratemporal commissure passes through the extrascapulars and in some coelacanths it sends off anterior branches which run through or pass over the postparietals (character 22).

Pit lines mark the postparietals of most coelacanths (character 26). Most coelacanths show only the middle and posterior pit lines (character 24) but there are differences in the position of these lines. They may lie within the posterior half of the postparietals and

comparison with other sarcopterygians suggests this to be the plesiomorphic condition. Alternatively many coelacanths show the pit lines lying within the anterior third (character 25) of the postparietals, although there is often a backward extension of the posterior pit line.

Cloutier (1991a: character 9) has remarked that the pattern of branching of the supraorbital canal within the supraorbitals varies. He notes that in some coelacanths the canal opens through the dermal bone via single large pores, usually located between successive supraorbitals, whereas in others there are pairs of openings, suggesting that the branches from the main canal bifurcated. There is also a third condition where the canals open through many small openings. It needs to be stressed that the nature of the openings through the bone need not match the opening within the skin (see description of *Latimeria*).

Ornamentation of the skull roofing bones may consist of enamel-capped tubercles and ridges or the bones may be devoid of ornamentation or be marked by coarse rugosities in the bone surface (character 27).

Skull roof characters

Based on the above survey the following characters relating to the skull roof are listed and coded in Table 9.1.

1. Intracranial joint margin straight (0), strongly interdigitate (1).
2. Snout bones lying free from one another (0), snout bones consolidated (1).
3. Single median rostral (0), several median rostrals (internasals) (1).
4. Paired premaxillae (0), fragmented premaxillae (1). For those coelacanths with a consolidated snout it is asssumed that paired premaxillae are present. The coding for *Coelacanthus granulatus* is based on the descriptions given by Schaumberg (1978).

5. Premaxilla with dorsal lamina (0), without dorsal lamina (1). For those species with consolidated snout the condition of this character cannot be decided. As with the previous coding, the condition in *Coelacanthus granulatus* is taken from Schaumberg (1978).

6. Anterior opening of the rostral organ contained within premaxilla (0), within separate rostral ossicle (1). Those outgroup taxa which do not have a rostral organ must be scored as illogical coding for this character.

7. One pair of parietals (1), two pairs (2).

8. Anterior and posterior pairs of parietals of similar size (0), dissimilar size (1). Illogical coding must be used for those taxa with only a single pair.

9. Number of supraorbitals/tectals; fewer than eight (0), more than 10 (1).

10. Preorbital absent (0), present (1).

11. Parietal descending process absent (0), present (1).

12. Intertemporal absent (0), present (1).

13. Postparietal descending process absent (0), present (1).

14. Supratemporal descending process absent (0), present (1).

15. Extrascapulars sutured with postparietals (0), free (1).

16. Extrascapulars behind level of neurocranium (0), forming part of skull roof (1).

17. Number of extrascapulars: three (0), five (1), more than seven (2). In most coelacanths it is assumed that the lateralmost extrascapular has 'fused' with the supratemporal (see character 21). In those cases the supposedly fused bone is counted as an extrascapular.

18. Posterior margin of the skull roof straight (0), embayed (1).

19. Supraorbital sensory canal running through centres of ossification (0), following a sutural course (1).

20. Medial branch of otic canal absent (0), present (1).

21. Otic canal joining supratemporal canal within lateral extrascapular (0), in supratemporal (1).

22. Anterior branches of supratemporal commissure absent (0), present (1).

23. Supraorbital sensory canals opening through bones as single large pores (0), bifurcating pores (1), many tiny pores (2).

24. Anterior pit line absent (0), present (1).

25. Middle and posterior pit lines within posterior half of postparietals (0), within anterior third (1).

26. Pit lines marking postparietals (0), not marking postparietals (1).

27. Parietals and postparietals ornamented with enamel-capped ridges/tubercles (0), bones unornamented (1), bones marked by coarse rugosities (2).

28. Parietals and postparietals without raised areas (0), with raised areas (1).

4

CHEEK BONES AND SENSORY CANALS

4.1 INTRODUCTION

The bones covering the cheek and opercular region of all coelacanths are developed in a characteristic pattern which has fewer elements than the cheek of other sarcopterygian fishes except perhaps onychodonts. Most of the bones carry the sensory canals. Beneath the eye, there is a single lachrymojugal occupying the territory of the lachrymal and jugal of other sarcopterygians and carrying the infraorbital sensory canal. Behind the eye lies a single postorbital in which the infraorbital sensory canal continues before turning posteriorly as the otic canal within the postparietal. The jugal sensory canal leaves the infraorbital canal at the posteroventral angle of the orbit and the canal is carried by a single squamosal which is usually a large bone. The jugal canal turns ventrally within the squamosal, the change of direction usually being located at the centre of ossification, and runs within a preoperculum before entering the angular within the lower jaw. The cheek pit line is present in some coelacanths; it may mark the centre of ossification of the squamosal and also mark the preoperculum. The ventral end of the cheek pit line is sometimes distinguished as the quadratojugal pit line as it marks a small separate bone in actinopterygians and most sarcopterygians. In most coelacanths, the quadratojugal pit line marks the same bone that carries the preopercular canal and therefore this is

often called a preoperculo-quadratojugal. I have called this bone a preoperculum because there is no evidence of fusion; I have considered the presence of the canal, which is always there, to be more important than the pit line.

In some coelacanths, separate quadratojugal and preoperculum bones have been described. But in all instances (see below) there is considerable doubt about the presence of separate bones and it is very likely that coelacanths lack a quadratojugal.

Two anamestic bones are commonly found in the coelacanth cheek. Dorsally, many coelacanths show a spiracular (sometimes called postspiracular), which lies close to the skull roof posterior to the postorbital and overlies the dorsal end of the spiracular groove. It corresponds in position to the prespiracular of porolepiforms. But unlike the porolepiform prespiracular, the coelacanth spiracular is never sutured with neighbouring bones.

Ventrally there is a suboperculum, but the overlap relations of the coelacanth suboperculum are rather different from that of actinopterygians and other sarcopterygians. In these latter fishes the suboperculum is overlapped by the operculum dorsally and the preoperculum anteriorly. However, while the coelacanth suboperculum is overlapped by the preoperculum, it overlaps the operculum (where it is large enough to do so). I have retained the name suboperculum but it may represent a posterodorsally positioned submandibular, which is otherwise absent from coelacanths.

4.2 DESCRIPTIONS OF GENERA

Latimeria

The cheek bones of adult specimens have been described by Millot and Anthony (1958) and most thoroughly by Smith (1939c). The associated sensory canals and pit lines have been described by Millot and Anthony (1965) and Hensel (1986) and the innervation by

Millot and Anthony (1965) and Northcutt Bemis (1993).

The account given here concentrates on a description of the cheek bones in an embryo specimen (BMNH 1976.7.16); it contrasts this with the adult, and notes variation in cheek pattern which may have relevance to theories of cheek bone homology. This can be used as the model by which to reconstruct and describe conditions in fossil coelacanths.

The bones of an embryo specimen are shown in Figs 4.1 and 4.2(B). A lachrymoju-gal (L.j), postorbital (Po), spiracular (Sp), two

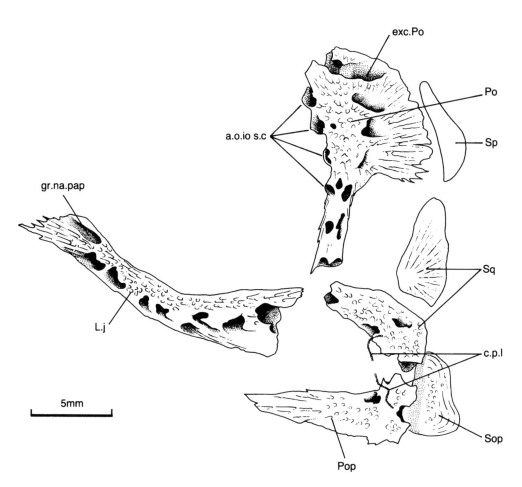

Fig. 4.1 *Latimeria chalumnae* Smith. Cheek bones of left side, camera lucida drawing of an embryo specimen (BMNH. 1976.7.1.16). The position of the cheek pit line (c.p.l) is drawn in although it does not mark the bone but lies free in the skin.

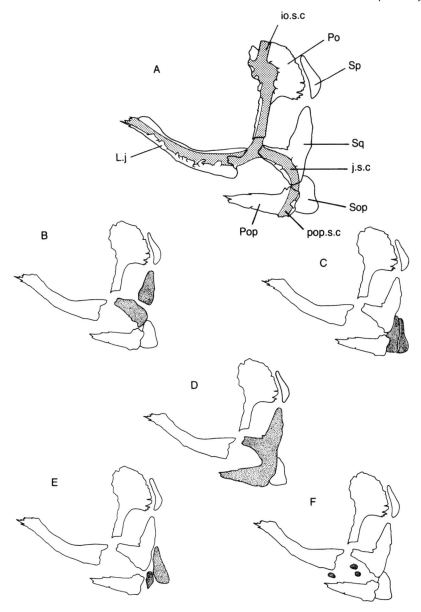

Fig. 4.2 *Latimeria chalumnae* Smith. Diagram to show variation in the pattern of cheek bones. All diagrams drawn from left side. (A) From Millot and Anthony (1958, fig. 10). (B) From BMNH 1976.7.1.16 showing a double squamosal. (C) From Millot and Anthony (1958, pl. 33), showing a double suboperculum. (D) Based on Millot and Anthony (1958, fig. 12), showing a fused squamosal and preoperculum. (E) From Smith (1939c, pl. 14), showing two unequal-size bones in place of usual suboperculum; the anterior of which was called 'interoperculum' by Smith. F. From Smith (1939b, pl. 15) showing small ossicles lying above preoperculum and termed quadratojugal ossicles by Smith.

squamosals (Sq), preoperculum (Pop) and suboperculum (Sop) are present. The position of the pit line in the overlying dermis is shown in Fig. 4.1 (c.p.l) and the path of the sensory canal is shown in Fig. 4.2(A). Most of the bones are sculptured with pits and grooves but there are no ornament tubercles such as are present in the adult.

The lachrymojugal is a curved bone, much narrower in the embryo than in the adult. At the anterior tip there is a smooth trough close to the dorsal margin and this contained the soft tissue surrounding the posterior nasal tube (Fig. 4.1, gr.na.pap). The main sensory canal is contained within the bone, entirely ventral to this groove. In the adult (Millot and Anthony, 1958: pl. 32), the groove containing the nasal tube crosses obliquely and superficial to the sensory canal. Therefore, the development of a groove containing the posterior nasal papilla varies. The external opening of the posterior nostril within the soft tissue of both embryo and adult is similar and lies ventral to the path of the main infraorbital canal immediately above the maxillary fold. It needs to be stressed that the posterior nostril opens from the nasal capsule dorsal to the infraorbital canal.

Posteriorly, the lachrymojugal meets the ventral limb of the postorbital and the anterior limb of the squamosal in a tripartite junction. The postorbital is similarly shaped in both embryo and adult. The anterodorsal angle of the postorbital is excavated (Fig. 4.1, exc.Po) and receives a tough ligamentous connection with the posteriormost supraorbital.

The squamosal is formed as two elements in BMNH 1976.7.1.16 (Fig. 4.1). Ventrally, there is a tubular canal-bearing bone which surrounds the sensory canal. Dorsally, there is a thinner, more scale-like bone with the centre of ossification located anteroventrally, adjacent to the centre of ossification of the canal-bearing squamosal. The presence of two squamosals, one canal-bearing and one anamestic element, is seen in at least two

embryo specimens (Fig. 4.2(B) and AMNH 32949SW) and also on the right side of the holotype (Smith, 1939c: pl. 16), which is an adult specimen. There is therefore the possibility that two squamosals represents the primitive embryonic condition in coelacanths.

In other adult specimens so far investigated, and on the left side of the holotype, the squamosal is a single element with a ventral canal-bearing limb and a thinner dorsal limb. The overall shape matches perfectly the combined shapes of the two elements in the embryo.

Ventral to the squamosal there is a strap-shaped preoperculum which carries the preopercular canal vertically through its posterior portion. The canal-bearing portion is distinctly marked off from the anterior blade-like portion. In one specimen (Fig. 4.2(D)), the preoperculum has fused dorsally with the squamosal.

A small anamestic suboperculum is present and is found lying behind and slightly overlapped by both the ventral squamosal and the preoperculum (Fig. 4.1, Sop). It is scale-like and has a clear exposed portion and is overlapped by the preoperculum. It is found within the preopercular flap of skin (Fig. 4.3) and this flap overlaps the operculum. The relationships between the operculum and suboperculum of *Latimeria* are therefore different from those same-named bones in actinopterygians, where the operculum overlaps the suboperculum, and these two are clearly part of a continuous opercular gular series (Gardiner, 1984a).

There is some variation in the development of the suboperculum: it may be small, as in the holotype, or very large (Millot and Anthony, 1958: pl. 14). It may be represented as a double element (Fig. 4.2(C)). In the holotype Smith identified a smaller element lying anteroventral to the suboperculum as an interoperculum (Fig. 4.2(E)) but it is not recorded in other specimens and is doubtfully homologous with the actinopterygian interoperculum. Smith also noted the presence of three small ossicles lying within the space between the squamosal, jugal and preoperculum (Fig. 4.2(F)). He suggested that these ossicles may represent the quadratojugal. These ossicles have not been recorded in other specimens and their identification as remnants of a quadratojugal must remain very doubtful.

The smallest element within the cheek series is the spiracular (postspiracular), which lies posterior to the postorbital. The size varies considerably. In the embryo (Fig. 4.1) it is elongate and relatively large, whereas in the adult it is more equidimensional and relatively small (Millot and Anthony, 1958: pl. 33).

The lateral line canals follow the characteristic sarcopterygian pattern. The infraorbital canal (io.sc.c) leaves the otic canal within the lateral edge of the postparietal and descends through the postorbital to the lachrymojugal. Within the postorbital, the canal runs along the orbital margin and sends off both anterior and posterior branches, whereas within the lachrymojugal, only ventral branches are seen. At the junction of postorbital and lachrymojugal the infraorbital canal gives off the jugal canal (Fig. 4.2(A), j.s.c). This canal runs horizontally along the ventral edge of the squamosal before turning ventrally as the preopercular canal. Branches are given off from both sides of the main canal within the squamosal and the preoperculum. Ventrally the preopercular canal (Fig. 4.2, pop.s.c) turns anteriorly as the mandibular canal to run within the lower jaw. At the point where the canal enters the angular, the posteriorly directed subopercular canal (Fig. 4.3) is given off to pass into the preopercular flap. The path of this canal across the hyoidean fold appears to be interrupted (Hensel 1986). This subopercular branch is probably a characteristic of some more derived coelacanths (Chapter 5).

The ramifications of the branches of the main canals and the pore openings through the skin can be shown to increase markedly throughout the life of an individual (Hensel, 1986; Fig. 4.3). This increased complexity of pore openings in the skin is reflected, to a lesser degree, within the canal bones by an

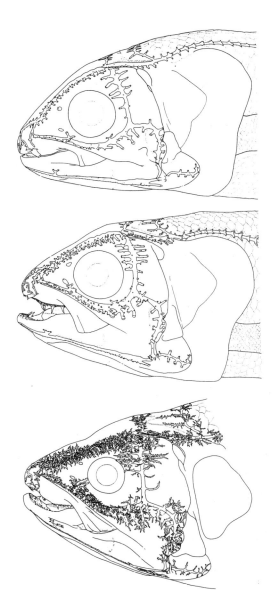

Fig. 4.3 *Latimeria chalumnae* Smith. Figure to show the increasing elaboration of the sensory canal system during life: (A) embryo specimen; (B) young; (C) adult. After Hensel (1986).

increase in the number of pores. This has probably come about by the ontogenetic subdivision of a single large pore in the embryo (cf. Fig. 4.1 with the pattern of pore openings in the adult lachrymojugal illustrated by Millot and Anthony, 1958: pl. 32).

A single, sometimes interrupted cheek pit line (vertical pit-line of Millot and Anthony, 1958) is present and this runs in a sinuous course between the ossification centres of the squamosal and the preoperculum. In *Latimeria* the pit line (Fig. 4.1, c.p.l) is entirely superficial within the skin, failing to mark the underlying bones. This suggests that we can never be certain that a cheek pit line is truly absent in fossil coelacanths, although it would be expected to mark the bones of those coelacanths carrying well-developed ornamentation.

Miguashaia

The cheek is poorly preserved in all specimens showing this region where the cheek bones are crushed against the underlying palate. It has been described by Schultze (1973) and Cloutier (1996) and these descriptions are based on five specimens (BMNH P.58693, P.62794; MHNM 06-264, 06-494; ULQ 120). This poor preservation is unfortunate because several obvious aspects of the cheek are superfically primitive and resemble other sarcopterygians rather than coelacanths. I have used the restoration provided by Cloutier (1996) and this forms the basis for the comments below.

The proportions of the cheek are very different from those of other coelacanths but are very similar to those in porolepiforms, onychodonts and osteolepiforms in that the orbit is small, anteriorly placed and there is a long postorbital cheek region. The overall proportions reflect primitive sarcopterygian conditions. Although the preservation is poor it can be seen that the cheek bones cover most of the cheek and they are sutured closely to one another and there are probably small overlap surfaces.

The jugal portion of the lachrymojugal may be partly seen in BMNH P.62794 but it is broken anteriorly. Cloutier (1996) suggests that the jugal was separate from the lachrymal which he identifies as represented by small bone fragments in this specimen. However, in my view, the poor preservation makes it impossible to be certain whether this interpretation is correct. A separate lachrymal and jugal is a primitive sarcopterygian condition. Thus, dependent on interpretation, the fused lachrymojugal is either a synapomorphy of coelacanths (Schaeffer, 1952a) or coelacanths minus *Miguashaia*. The jugal extends far posteriorly and there are long contact surfaces with the postorbital and, particularly, with the squamosal.

The postorbital, as restored by Cloutier (1996), shows a very short orbital margin which is like that in other sarcopterygians (except lungfishes) and unlike that in more derived coelacanths. However, this interpretation may be questioned because, in both specimens showing a postorbital, the dorsal margins are clearly broken, and in the restoration reproduced here (Fig. 4.4), there is a very large gap between the postorbital and the skull roof. If more complete specimens should show this to be a real gap, then this would be a highly unusual condition, and one perhaps unique to *Miguashaia*. Despite the difficulties in interpreting the true margin of the postorbital, it is a large bone.

The squamosal (squamosal + quadratojugal of Cloutier) is also large and reaches far dorsally to (probably) contact the skull roof. There is no evidence that a separate quadratojugal is present and the labelling used by Cloutier is based on the observation that the bone covers the areas usually occupied by the squamosal and quadratojugal in porolepiforms and osteolepiforms.

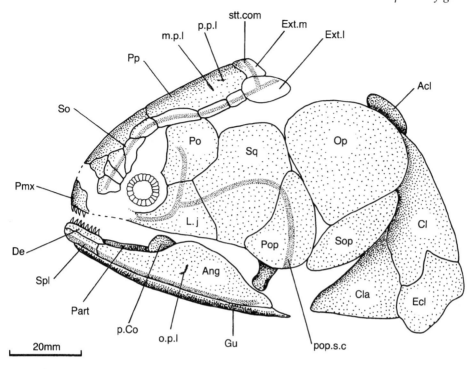

Fig. 4.4 *Miguashaia bureaui* Schultze. Restoration of head in left lateral view, from Cloutier (1996) with permission. In this form, the cheek is much longer and the eye much smaller than in most other coelacanths; both conditions are regarded as plesiomorphic for sarcopterygians.

The preoperculum is also large, with a rounded posterior profile. It is sutured with the squamosal along a sinuous margin and lies posteroventrally to that bone. A large suboperculum is present and although its true relationships to the surrounding bones cannot be seen, it is much more like that of osteolepiforms and porolepiforms than other coelacanths.

The sensory canals are developed as narrow tubes which appear to pass through the centres of ossification and send off many narrow branches which open to the surface through tiny pores. Within the postorbital the main infraorbital canal runs at a distance from the anterior margin.

In summary, the cheek bones of *Miguashaia* are primitive in forming a solid covering to the cheek which is long and this reflects

an oblique jaw suspension and the small eye.

Diplocercides

The cheek plates and associated sensory canals and pit lines have been described for *D. kayseri* (including *Nesides schmidti*) and *D. jaekeli* by Stensiö (1937, 1947) and for *D. heiligenstockiensis* by Jessen (1973) and Cloutier (1991a). The cheek of *D. kayseri* is restored in Fig. 4.5 and the notes below supplement and comment upon previous descriptions. Remarks on *D. heiligenstockiensis* are included separately (p. 103) because this is one species recorded as having a quadratojugal (Cloutier, 1991a).

The cheek of *D. kayseri* consists of six bones: lachrymojugal, postorbital, squamosal (squamoso-preopercular of Stensiö), preoperculum (preopercular-quadratojugal of Stensiö), spiracular (suprasquamosal of Stensiö) and the suboperculum. These bones completely cover the cheek and fit tightly together. Stensiö (1937) suggested that the cheek plates did not fit tightly together but there is a clear overlap area developed along the anterior edge of the squamosal for overlap by the postorbital in specimen 'b' of Stensiö (1937: pl. 11, fig. 2).

The postorbital is wider dorsally and shows the infraorbital canal (Fig. 4.5, io.s.c) running through the centre rather than at the anterior margin as in most coelacanths. It is completely ornamented. The squamosal is also triangular with the apex reaching the skull roof at the point where the supratemporal expands laterally. The jugal sensory canal (j.s.c.) enters the bone along the anterior margin and runs posterodorsally before turning acutely ventrally as the preopercular canal (pop.s.c). The centre of ossification occurs at the angle. This very sharp angulation of the sensory canal is characteristic for all species of *Diplocercides*. The spiracular is a tiny oval scale-like element lodged between the postorbital and the squamosal.

The preoperculum is the largest element in the cheek and fits tightly against the squamosal although there appears to be no overlap between them. The preopercular sensory canal probably ran vertically, close to the posterior margin (see below). The centre of ossification lies anterior to the canal at the point where the cheek pit line marks the bone. Both the squamosal and the preoperculum are completely ornamented with small

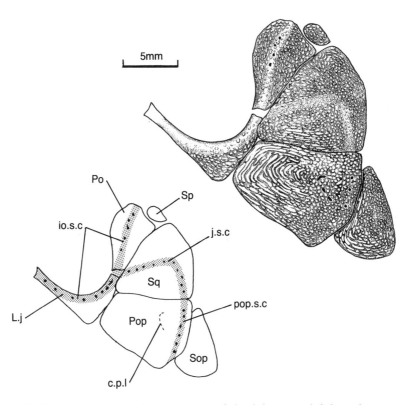

Fig. 4.5 *Diplocercides kayseri* (von Koenen). Restoration of cheek bones in left lateral view; sensory canals stippled. Based on a latex cast of specimen 'b' figured by Stensiö (1937, pl. 11, fig. 1).

tubercles and ridges. The suboperculum is present as a large triangular scale-like element with a centrally positioned ossification centre. It is preserved in the presumed near-natural position in Gö 470-1 (holotype of *Nesides schmidti*, Stensiö 1937: pl. 9) and found isolated in specimen 'a' of Stensiö (1937). The anterior edge is devoid of ornament and this implies that it is overlapped by the preoperculum and it has been reconstructed in this way (Fig. 4.5).

The lachrymojugal (Fig. 4.5, L.j) is known only from Gö 470-1 (holotype of *Nesides schmidti*) and is illustrated by Stensiö (1937: pl. 10, fig. 2). The bone is present beneath the orbit and it does not appear to extend anteriorly even though it has been reconstructed in this way by Stensiö (1937: fig. 3). The anterior part of the bone is narrow, with the infraorbital canal occupying the entire width. Posteroventrally there is an ornamented expansion, best seen in the holotype of *D. jaekeli* (Stensiö, 1937: pl. 8, fig. 1), imparting a characteristic shape seen in these two species but not in *D. heiligenstockiensis* (Cloutier, 1991a: fig. 3).

The path of the main sensory canals through the cheek bones is shown in Fig. 4.5 but the nature of the canal openings is not clearly seen in either *D. kayseri* or *D. jaekeli*. A few tiny openings are seen overlying the canal in the postorbital and the preoperculum in Gö 470-1. Several specimens of *D. kayseri* show calcite infilling of the sensory canals and it is apparent that there are no obvious secondary branches. The canal openings are better seen in *D. heiligenstockiensis* (Jessen, 1973: pl. 24), where there are tiny pores opening directly from the main sensory canals within the lachrymojugal, squamosal and the preoperculum. It is very likely that other species of *Diplocercides* were similar in this respect. Specimen 'b' shows four small nerve foramina piercing the medial surface of the postorbital and Gö 470-1 shows six nerve foramina piercing the lachrymojugal (Stensiö 1937: pl. 11, fig. 1 and pl. 10, fig. 2). This is

comparable with the innervation seen in *Latimeria*. The details of pit lines are poorly known because the surfaces of the squamosal in all specimens of *Diplocercides* are very imperfectly preserved. A small portion of the cheek pit line is present on the preoperculum as described by Stensiö (1937).

It is necessary to add a few notes here about the structure of the cheek of *D. heiligenstockiensis* because it has been interpreted as having a structure different from that in the other species and to have significance in the identification of a separate quadratojugal in species of this genus.

In one specimen of *D. heiligenstockiensis* (NRM P.7775) there are three elements occupying the lower cheek area: a large anterior bone bearing the preopercular canal; an equally large triangular element which lies in tandem with the canal-bearing bone, and a slightly smaller triangular bone which lies ventral to the other cheek plates and behind the lower jaw (Jessen, 1973: pl. 24, fig. 1). Jessen reconstructed the cheek as a text figure (Jessen, 1973: fig. 3A) but indicated only two bones: an anterior canal-bearing bone, identified as the combined quadratojugal and preoperculum, and the suboperculum, which is represented in the specimen by either of the non-canal-bearing bones. Cloutier (1991a: fig. 3) reinterpreted this specimen to suggest that the anterior canal-bearing bone is the quadratojugal (the preopercular canal having been captured by that bone), the middle bone is the preoperculum (anamestic) and the posterior bone is the suboperculum.

For the purpose of interpreting *D. kayseri*, in which only two elements are present, Cloutier opts for the anterior as the anamestic quadratojugal and the posterior as the anamestic preoperculum. The preopercular canal is presumed to pass between these two cheek elements because as he claims (1991a: p. 404) it does so "...In many actinistian taxa, [where] the trajectory of the canal lies between the quadratojugal and the preopercular". In reality, there are very few, perhaps no, coelacanths in which there is a distinct quadratojugal and preoperculum with the canal lodged between them. He assumes (1991a: appendix 2) that the subopenculum is present in *D. kayseri* (and *Nesides schmidti*) even though a third element has not been found in that taxon. As he admits, the path of the preopercular canal is difficult to trace in specimens of *D. kayseri*. However, in *D. jaekeli* the element which clearly corresponds in shape to the bone beneath the squamosal of *D. kayseri* has a prominent canal running vertically along the hind margin (Stensiö, 1937: pl. 8, fig. 1). Also, in specimen 'b' of *D. kayseri* (Stensiö, 1937: pl. 11, fig. 2), the preopercular canal leaves the ventral margin of the squamosal in such a position that its trajectory would clearly pass through the posterior half of the element beneath. Thus, I would maintain that this canal-bearing element represents at least a preoperculum because, as Cloutier says (1991a: p. 404) "...the preopercular, by definition, carries the preopercular canal." It may also contain the topographic homologue of the quad-ratojugal because it carries the cheek pit line.

D. heiligenstockiensis would appear to be similar to other species with a canal-bearing preoperculum and a subopenculum lying immediately behind. The main difference between the cheek of *D. heiligenstockiensis* and that of *D. kayseri* is the relatively large size of the subopenculum in the former species. This leaves the third element of the cheek (subopenculum of Jessen, 1973: pl. 24 and Cloutier, 1991a: fig. 3) unexplained. It may be an extra subopenculum lying within the operculogular series (see, for instance, the variation in *Latimeria*, Fig. 4.2). It differs from a subopenculum in three respects: the extreme ventral position; the lack of any obvious overlap area; the contour of the bone which bears an inturned and orna-mented anterior lamina. This last feature might suggest that it represents the clavicle or interclavicle of the shoulder girdle. It is

possible that further material will resolve this issue.

I conclude from this discussion that, for *Diplocercides* species, there is a normal coela-canth complement of cheek bones that does not include separate quadratojugal and preo-percular elements. The shapes and relative sizes of the cheek bones do vary slightly between the species, and there is some varia-tion in ornament patterns (see species diag-noses, Chapter 11).

Allenypterus

The cheek has been briefly described by Lund and Lund (1985) and certain characters of the separate bones and canals have been coded by Cloutier (1991a,b). The cheek is shown in the restoration of the head (Fig. 4.6). This restoration differs substantially from the illustrations given by Lund and Lund (1985: figs 57, 60, 61).

The cheek contains a lachrymojugal, squa-mosal, postorbital, preoperculum and subo-perculum. Lund and Lund (1985) illustrate a large spiracular which is shown in unchar-acteristic fashion to overlap the operculum. A spiracular is also recorded by Cloutier (1991a). I have not been able to discover a spiracular but there is space between the postorbital and the operculum, so it is possi-ble that a small element was present. The cheek bones abut one another and comple-tely cover the cheek: however, I do not find them deeply overlapping as recorded by Lund and Lund (1985), but it is noticeable that the squamosal, preoperculum and subo-perculum do overlap the ventral end of the operculum.

The lachrymojugal is tubular and angled slightly anteriorly. The postorbital is reduced to a narrow tube surrounding the sensory canal. Posterior to the ventral half of the postorbital lies the squamosal which is a relatively small triangular bone which fails to meet the skull roof. The preopercu-lum is the largest bone of the cheek, cover-

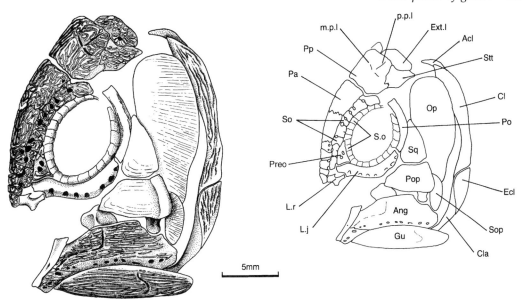

Fig. 4.6 *Allenypterus montanus* Melton. Restoration of the head, based mainly on FMNH PF10939. The head is shown in natural attitude in which the skull roof faces almost directly anteriorly.

ing the entire ventral half, and is considerably larger than the squamosal. The preopercular canal lies along the posterior edge. Lund and Lund (1985) illustrate a separate preoperculum (canal bearing) and an anamestic quadratojugal; I have been unable to confirm this observation. The suboperculum is small and scale-like and is overlapped by the preoperculum and, in turn, overlaps the operculum. As such it satisfies two of the three states of character 30 of Cloutier (1991a).

The infraorbital canal opens within the lachrymojugal through large pores. Openings within the postorbital have not been seen. The trajectory of the jugal canal through the squamosal could not be seen on specimens examined here. Lund and Lund (1985) record this canal as passing approximately through the centre of the bone and, judging from the positional relationships of adjacent bones with contained canals, this interpretation seems reasonable. None of the cheek bones is

ornamented and pit lines do not mark the bones.

Hadronector

The cheek has been described by Lund and Lund (1985) and figured by Cloutier (1991a: fig. 4).

The cheek is completely covered by a postorbital, spiracular, lachrymojugal, squamosal, preoperculum and suboperculum. All of these bones fit tightly together and there are overlaps between at least the postorbital, squamosal and preopercular. The lachrymojugal is relatively narrow with a slight expansion posteriorly. The postorbital is similarly narrow, although it is not restricted to the sensory canal tube. The squamosal is distinctively shaped: it is narrow and deep; it reaches far dorsally and is prevented from meeting the skull roof only by a small spiracular which may be present and represented in specimen CM30712A as a small patch of discrete ornament wedged between the lateral extrascapular, postparietal and postorbital (Fig. 4.7(A)). There is a sigmoid overlapping suture between the squamosal and the preoperculum (Fig. 4.7(B)). In one specimen (Fig. 4.7(A)) these two bones appear to have fused, a feature which has also been observed in individuals of *Latimeria* (Fig. 4.2(D)). All of the cheek bones are ornamented with flat-topped ridges which tend to abut one another and run longitudinally.

The interpretation of the bones in the lower cheek varies between authors. Both Lund and Lund (1985) and Cloutier (1991a) identify three bones lying in an anterior–posterior series: these are the quadratojugal, preoperculum and suboperculum. I can find two only, which do, however, occupy the total area covered by the three bones identified by other authors. The anterior bone recognized in this work – the preoperculum – is equivalent to the anteriormost of the three bones recognized by Lund and Lund and by Cloutier. Cloutier and I agree that this bone carries the pit line and carries the preopercular canal along the posterior margin. I call this preoperculum after the contained sensory canal; Cloutier calls it quadratojugal after the pit line.

The suboperculum that I recognize here is equivalent to Cloutier's middle bone of the series of three that he recognizes and which he calls preoperculum; we agree that this bone is anamestic. I identify it as a suboperculum because it is anamestic and it has the usual overlap relationships seen in other coelacanths; it is overlapped by the bone in front and overlaps the operculum behind. From specimens examined in this work I have been unable to recognize the third element labelled by Cloutier as suboperculum and which is shown overlying the ventral end of the shoulder girdle. This 'suboperculum' is illustrated as two elements lying one above the other and it possible that they may be scales which have become anteriorly displaced. Alternatively, it is possible

that the bone pattern of *Hadronector* shows individual variation in this region with, as one of the variations, a double suboperculum as in some specimens of *Latimeria* (Fig. 4.2(C)).

The restoration provided by Lund and Lund (1985: fig. 43) is more difficult to relate to the specimens. Their restoration shows an anterior quadratojugal carrying only the pit line, a middle preoperculum bearing the sensory canal along the posterior edge, and a narrow and deep suboperculum. This restoration is based on a photographed specimen (Lund and Lund, 1985: fig. 38) from which it differs in the relative proportions and shapes of the bones. In fact, the photographed specimen agrees much more closely with the camera lucida drawing provided by Cloutier (1991a: fig. 4), which is not surprising because that figured by Cloutier and that photographed by Lund and Lund are part and counterpart of the same specimen (CM 27307).

I conclude that the illustration given by Cloutier is the more accurate and, together with information from specimens examined here, suggest that the preopercular canal runs along the posterior edge of the anterior(most) element which also carries the pit line. I call this bone the preoperculum. A suboperculum is present and an extra suboperculum may exist as a variant.

The sensory canal opens to the surface through tiny pores between the ornament tubercles and, as Lund and Lund (1985) note, the more anterior pores within the lachrymojugal are larger than those posteriorly or within the jugal and preoperculum. Such a gradient in the size of sensory pores is seen in other coelacanths such as *Laugia*.

Caridosuctor, Polyosteorhynchus and *Lochmocercus*

The cheek of these three monotypic genera from the Bear Gulch Limestone are all very poorly known, yet are potentially interesting

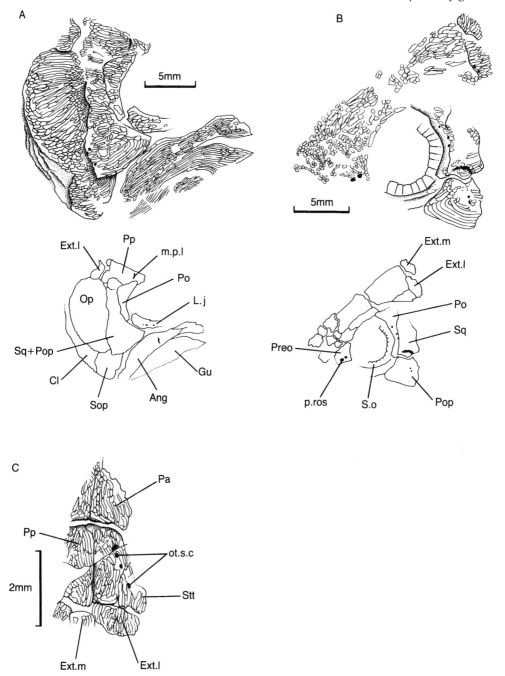

Fig. 4.7 *Hadronector donbairdi* Lund and Lund. Camera lucida drawings of specimens showing parts of the head. (A) Cast of head preserved in left lateral view as preserved in Carnegie Museum CM30712A. (B) Cast of head preserved in left lateral view as preserved in Carnegie Museum CM27308A. (C) Cast of posterior part of skull roof as preserved in Carnegie Museum CM30711A.

because all three have been described as having a separate preoperculum and quadratojugal (Lund and Lund, 1985) and this has been accepted by Cloutier (1991a,b). I have been unable to check the conditions described by Lund and Lund. But where I have seen specimens of other taxa described as having separated a quadratojugal and preopercular (*Allenypterus*, *Hadronector*), I have found difficulty in accepting the restorations provided by Lund and Lund (1985). Therefore I have scored the presence of a quadratojugal as questionable in these taxa.

Similarly, I can offer little comment on the other features of the cheek, other than a few observation of gross appearance.

In all three taxa, the cheek appears to be completely covered with tightly sutured bones: lachrymojugal, postorbital, squamosal, spiracular and preoperculum are present; the lachrymojugal is not expanded or angled anteriorly. All bones are heavily ornamented and pit lines mark the squamosal and preoperculum (quadratojugal of Lund and Lund).

Rhabdoderma

The cheek bones are known in detail in *Rhabdoderma elegans* where they have been described and illustrated by Moy-Thomas (1937), Forey (1981) and Lund and Lund (1985). The following description therefore is restricted to this species, although a few notes are added about other *Rhabdoderma* species. In the literature a distinction has been made between specimens of *Rhabdoderma elegans* occurring in the type locality of Linton, Ohio, USA and those occurring in Europe. Lund and Lund (1985) regard these specimens as belonging to different species and perhaps separate genera. Part of their evidence relates to the anatomy of the cheek bones. Cloutier (1991a,b), in his cladistic analysis of the Palaeozoic coelacanths, accepts the taxonomic distinction and codes these taxa separately with different coding for elements of the cheek. In the compilation of the following account I have examined European specimens which show the cheek particularly well (BMNH 48055, P.6286, P.6613a, P.6663a, P.10473, P.10474) and specimens from the type locality (CM 43938, 43955, 43990, 43033, GN. 238) and include some notes relating to the apparent distinctiveness.

The cheek of *Rhabdoderma elegans* is made up by six bones: lachrymojugal, postorbital, spiracular (prespiracular of Lund and Lund, 1985), squamosal, preopercular (quadratojugal of Lund and Lund, 1985) and subopercular (preopercular of Lund and Lund, 1985). The bones abut or overlap one another and together form a complete covering to the cheek. They are shown in Figs 4.8 and 4.9.

The lachrymojugal is a narrow bone, restricted to a tube surrounding the infraorbital canal beneath the orbit. It is not expanded or angled anteriorly. The postorbital is triangular and carries the sensory canal along the anterior border.

The squamosal is the largest element in the cheek and has an overlapped edge (Fig. 4.8, o.Po) for contact with the postorbital. The jugal canal crosses the centre of the bone and turns ventrally as the preopercular canal near to the posterior edge. The angle of the canal marks the centre of ossification which is therefore placed far posteriorly in this species. The upper portion of the cheek pit line also crosses the centre of ossification.

There are two alleged differences between the squamosal of European and American specimens. Cloutier (1991a) records the squamosal of the American form as being triangular and distinguishes this from the pentagonal shape as seen in the European form. Cloutier used the restoration based on European specimens and figured by Forey (1981: fig. 6) for this assessment. The American specimens I have examined all show a triangular squamosal. The shape of the squamosal in the European form varies (cf. Figs 4.8 and 4.9), being pentagonal in

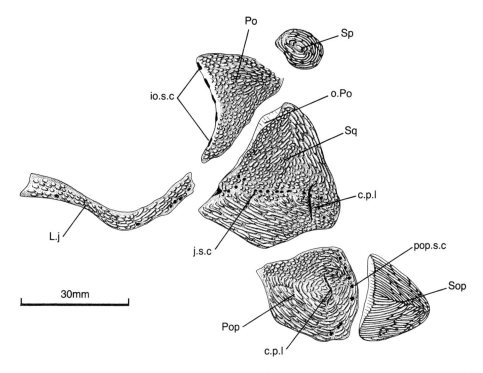

Fig. 4.8 *Rhabdoderma elegans* (Newberry). Restoration of the cheek bones in left lateral view. The bones have been drawn as if pulled apart slightly to show narrow overlap area between the squamosal and postorbital. Based on BMNH P.5379, P.6663a, P.10473.

some specimens while in others it is triangular. Therefore, this distinction may not be clear cut. Another difference concerns the openings of the sensory canal within the squamosal. The main canal opens through a series of tiny pores along the main canal, as well as through short branches. In the American specimens there are additionally two large pores located above the angle of the canal which must represent a prominent dorsal branch. These pores are usually absent from the European specimens but are present in some (e.g. Fig. 4.9). The presence or absence of these pores may therefore be of little significance in distinguishing American and European specimens. The presence of these large pores in at least some individuals leads me to record the presence of prominent

branches within the jugal canal in *Rhabdoderma elegans* (character 46).

The lower part of the cheek is occupied by two bones set in tandem. These are the preoperculum and suboperculum in Figs 4.8 and 4.9. Lund and Lund (1985) refer to the anterior bone as the quadratojugal because it bears the lower part of the cheek pit line at the centre of ossification (Figs 4.8 and 4.9, c.p.l). However, they refer to the posterior bone as the preoperculum because they restore the sensory canal passing vertically through the centre of the bone (Lund and Lund, 1985: figs 4, 5). Lund and Lund's restoration led Cloutier to record the presence of an independent quadratojugal in the American forms and the absence of such a bone in the European form. I have been

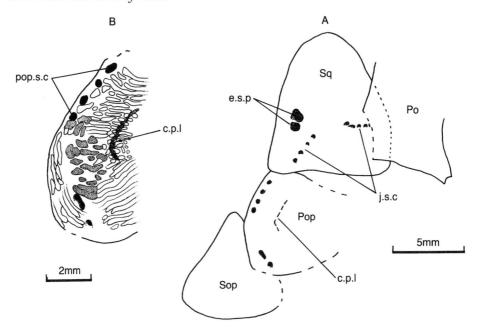

Fig. 4.9 *Rhabdoderma elegans* (Newberry). (A) Camera lucida outline of part of the cheek as preserved in BMNH P.6663a to show positions of the sensory canals and the posterior openings of two large pores from the jugal canal. (B) An enlargement of the preoperculum to show the development of second-generation tubercles overlying and replacing those lying posterior to the cheek pit line.

unable to check the path of the sensory canal in the American specimens but it clearly passes along the posterior edge of the anterior bone in the European specimens, justifying the identification as a preoperculum. The suboperculum is anamestic and scale-like (Fig. 4.8, Sop). It is overlapped by the preoperculum.

The spiracular is a small oval bone with scale-like ornament. In one specimen (BMNH P.6663a) it carries a small pit line. All cheek bones are heavily ornamented with tubercles and ridges. In some specimens (Fig. 4.9(B)) there is evidence of one generation of tubercles overgrowing an earlier generation.

As mentioned above, the cheek is very poorly known in other species of *Rhabdoderma*. The cheek is partly known in cf. *Rhabdoderma* (Forey and Young, 1985: fig. 3a). In that form, the same complement of bones

is present but the postorbital is quadrangular and the preoperculum is relatively much smaller.

Coelacanthus granulatus

The cheek series of *C. granulatus* is very poorly known, chiefly because it is represented by very thin bones which are barely larger than the canals they carry. Schaumberg (1978) has described the cheek, based on two specimens (his specimens B and C) from the German Kupferschiefer. I have examined casts of this material and three specimens from the English Marl Slate which show elements of the cheek (SM D.435, HM 926.52 and BMNH P.3340). I agree, in most respects, with the restoration provided by Schaumberg (1978: fig. 6).

The lachrymojugal, postorbital and squamosal as preserved in SM D.435 are illustrated in Fig. 5.4(A) and agree closely with specimen C illustrated by Schaumberg (1978: fig. 9).

The lachrymojugal is the largest element and is about the same relative size as in many other coelacanths. The surface of the bone is very heavily sculptured, and in HM 926.52 appears to be fragmented into many small mosaic elements. The anterior end is not expanded and there is no obvious angle anteriorly, despite the restoration provided by Schaumberg.

The postorbital is relatively tiny and tube-like, although it has the same ornament pattern as the lachrymojugal. The squamosal is also represented as a narrow tube devoid of ornament. In SM D.435 it is very small and confined to a narrow tube surrounding the jugal canal (Fig. 5.4(A), Sq). In Schaumberg's specimen C it appears to be much longer and curves posteroventrally to enclose part of the preopercular canal as well.

I have been unable to observe a separate preoperculum but one is illustrated by Schaumberg (1978: fig. 9) where it is a large rectangular bone orientated vertically. This bone shows no ornamentation and is therefore very different from the lachrymojugal and the postorbital. No other specimen appears to show this element and, in consequence, I have scored the presence of a separate preoperculum as questionable in this species.

Laugia

The cheek bones of *Laugia* have been described previously by Stensiö (1932: fig. 19, pl. 5, fig. 3, pl. 7, figs 2, 3) on the basis of several specimens. I have examined many more and confirm Stensiö's observations. The description below is intended as a supplement.

The cheek is restored in Figs 4.10 and 4.12(A). The only bones present are the lachrymojugal, postorbital and squamosal and these lay free from one another. The lachrymojugal is a very narrow tube-like bone, totally occupied by the infraorbital canal. The anterior end is neither expanded nor angled anteriorly. The infraorbital canal opens along the ventral edge by 15–20 small pores which in MGUH VP2308b are arranged in groups each of three or four. On the inner surface there are five or six nerve foramina.

The postorbital is broad dorsally and narrows to a tube ventrally. The infraorbital canal describes a curve through the bone curving posteriorly at the dorsal edge (Fig. 4.12(A), io.s.c). Therefore the canal does not lie at the anterior edge throughout its course and in this respect it bears some resemblance to the pattern in *Diplocercides* (p. 103). It is, however, more derived in showing prominent anterior branches (MGUH VP2309) as well as small posterior openings.

The squamosal is triangular and contains the jugal canal (Fig. 4.12, j.s.c), which runs close to the ventral margin of the bone before turning ventrally to leave the squamosal at the posteroventral corner. There are only pores opening from the ventral margin of the canal.

All of the cheek bones are ornamented with rounded or elongate tubercles (Fig. 4.10) and there is no pit line marking the squamosal.

Coccoderma

The cheek plates are known for *Coccoderma suevicum* from a single specimen studied here (BMNH P.8536) and illustrated in Fig. 5.7 and in Lambers (1992: fig. 2.3). The cheek is also partly preserved in the holotype of *C. nudum* (Fig. 4.11) which is probably a juvenile *C. suevicum* (p. 306).

In many respects the cheek is very similar to that in *Laugia*. The two are compared in Fig. 4.12. A lachrymojugal, postorbital, squamosal and preoperculum is present (the preoperculum is absent in *Laugia*). None of the bones contact each other.

The lachrymojugal is a narrow tube,

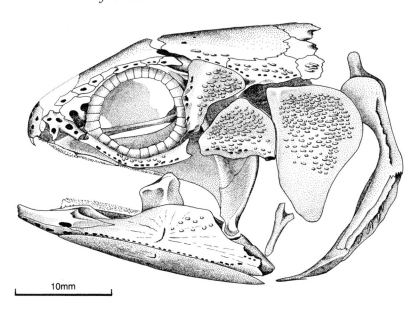

10mm

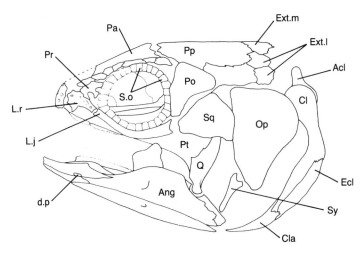

Fig. 4.10 *Laugia groenlandica* Stensiö. Restoration of the head in left lateral view, based on MGUH VP.2308a–b, VP.2311, VP.2316a–b.

wholly occupied by the infraorbital canal and which opens through many small pores which show some grouping (Fig. 5.7) similar to the *Laugia* pattern. There is no anterior expansion and the bone ends at the anterior margin of the orbit. The postorbital is rela-

tively smaller than that in *Laugia* but is of similar shape and, as in *Laugia*, the contained canal curves away from the anterior edge dorsally. As it does so the canal gives off two anterior branches (Fig. 4.12, a.o.io.s.c).

The squamosal is longer than deep; the

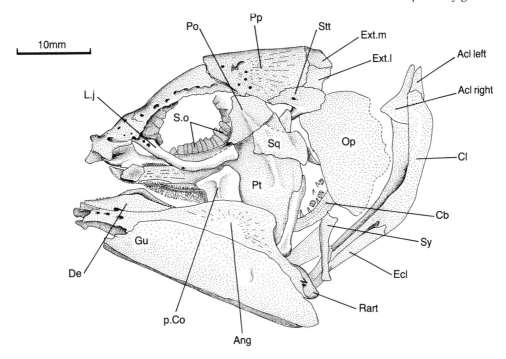

Fig. 4.11 *Coccoderma suevicum* Quenstedt. Camera lucida drawing of the head in left lateral view of holotype of *Coccoderma nudum*. Compare this figure with Fig. 5.7 and note especially the shapes of the squamosal, angular and principal coronoid.

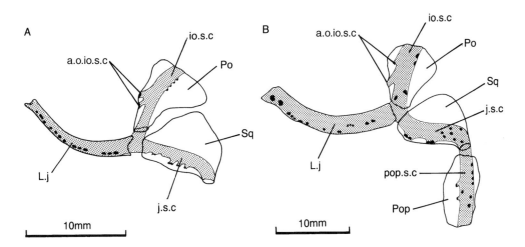

Fig. 4.12 Restoration of the cheek bones in (A) *Laugia groenlandica* Stensiö. (B) *Coccoderma suevicum* Quenstedt. Restored paths of sensory canals stippled.

exact margin is not entirely clear from available specimens. It is certain, however, that the jugal canal runs along the ventral margin as in *Laugia*. There are lateral pores from the canal as well as ventral pores.

There is an upright and rectangular preoperculum (Fig. 5.7, Pop) carrying the preopercular sensory canal, which runs closer to the posterior margin and opens through several small pores. The preoperculum is not seen in the specimen of *C. nudum*, but this may be due to poor preservation.

The main sensory canals on the cheek are very wide and this is also similar to the conditions in *Laugia*. The cheek bones of *Coccoderma* are perfectly smooth.

Sassenia

The cheek of *Sassenia tuberculata* was described by Stensiö (1921: 86, pl. 10, figs 1, 3, 6) but the material on which this was based was highly fragmentary. The cheek is more completely known in *S. groenlandica* and this is described here. Identifiable differences between the species concern only the shape of the preoperculum.

The cheek consists of the usual four bones which abut one another closely but which do not overlap, and a small spiracular which is loosely associated. Three specimens examined here (MGUH VP2326, VP2327a,b, VP3258) show cheek elements and information is combined in the restoration (Fig. 4.13).

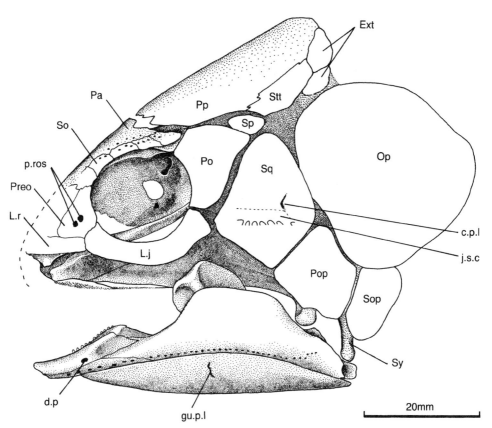

Fig. 4.13 *Sassenia groenlandica* sp. nov. Restoration of the head in left lateral view. Based on MGUH VP.2326, VP.3258, VP.2327a–b.

The lachrymojugal is relatively deep, parallel-sided and it curves beneath the orbit to become sutured to both the pre-orbital and the lateral rostral. It is neither expanded nor angled anteriorly. The post-orbital is broad and, characteristically, the anterodorsal corner is truncated (Fig. 4.13, Po). The infraorbital sensory canal appears to lie close to the anterior margin, although the exact course is difficult to trace and there are no prominent anterior and posterior branches.

The squamosal is very large and forms the major element in the cheek; it reaches far dorsally but just fails to contact the skull roof. The jugal canal passes through the centre of the bone and sends off many short, tiny branches along the ventral and posterior edge of the main canal. Two specimens (MGUH VP2327a and 131) show a very small pit line located near the angle of the jugal canal. The preoperculum is of similar shape but much smaller than the squamosal and lies well behind the level of the principal coronoid. This reflects the relatively oblique suspensorium in species of this genus. The path of the sensory canal within the preoperculum cannot be followed in the specimens examined. The preoperculum of *S. tuberculata* (Stensiö, 1921: pl. 10, fig. 6) is triangular rather than pentagonal as in *S. groenlandica*. The suboperculum is relatively large. Ornament is present as a continuous cover of small, flat-topped and closely spaced tubercles on all cheek plates.

Wimania

The cheek bones are known only in *Wimania sinuosa*, and very incompletely (Stensiö, 1921: fig. 25). Evidence of a lachrymojugal. postorbital, squamosal (squamoso-preopercular) and preoperculum (preoperculo-quadrato-jugal) is present. A small bone labelled 'x' by Stensiö probably belongs to the ventral limb of the postorbital. The lachrymojugal shows an anterior angle. The rather unusual and complex margins of the postorbital and squamosal restored by Stensiö are probably artefacts of preservation.

Axelia

Stensiö (1921) described the cheek bones as being reduced to a series of tiny ossicles, but because the material used by Stensiö is very poor this needs to be carefully checked.

Garnbergia

The cheek is very poorly known (Martin and Wenz, 1984: fig. 1) but appears to be covered by closely fitting lachrymojugal (infraorbital), postorbital and probably a squamosal (dorsal preoperculum) and preoperculum, but this area of the cheek is very badly preserved. One feature that may be recorded here is that the lachrymojugal turns forward anteriorly and has a broad contact area with a tectal.

Chinlea

The cheek of *Chinlea sorenseni* has been described by Schaeffer (1967: 324, fig. 14, pl. 28) and Elliott (1987: figs 1–3) and only a few notes of commentary need be added. There are differences between the cheek restorations provided by these two authors.

The cheek bones are robust, generally broad and they abut one another, forming a cover to most of the cheek. The lachrymojugal is relatively deep and as it extends anteriorly beyond the orbit it is angled and contacts two tectal bones. The anterodorsal corner of the lachrymojugal is also excavated and this excavation probably contained the soft tissue surrounding the tubes leading from the rostral organ. The postorbital is large, although it differs in shape between the two published restorations, being ovoid in one (Elliott 1987) and pentagonal in the other (Schaeffer, 1967). In both restorations, however, the postorbital is located in a position straddling the intracranial joint, which was probably a true life position (e.g. AMNH 5652 and AMNH 5653). This means that the infraorbital canal certainly took a diagonal course through the centre of the postorbital to join the otic canal at the anterior end of the postparietal.

The squamosal is triangular and the preoperculum is polygonal but the restorations differ in the relative sizes. Schaeffer suggests that the squamosal is the larger and that it is separated from the skull roof by a small spiracular (prespiracular of Schaeffer). Elliott (1987: fig. 2) suggests that the preoperculum is the larger and that the squamosal extends dorsally to contact the skull roof. I have been unable to confirm which of the two restorations is correct; indeed, there may be individual variation (which would imply polymorphic codings for character 34 below). The path of the sensory canal through the squamosal and preoperculum could not be followed.

The ornament of the cheek bones consists of a coarse rugose bone surface and the sensory canal must have opened through minute pores.

Whiteia

The cheek bones of *Whiteia* species are well known. Those of *W. woodwardi* (Fig. 4.14) and

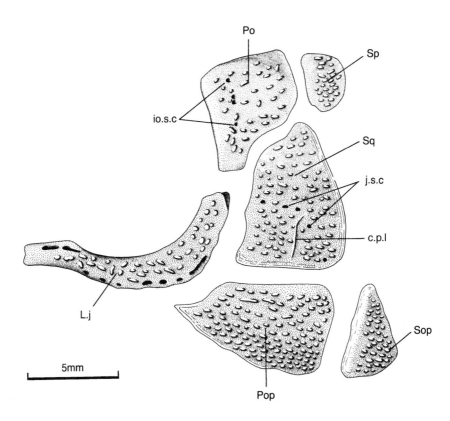

Fig. 4.14 *Whiteia woodwardi* Moy-Thomas. Restoration of cheek bones in left lateral view. The bones have been drawn as if pulled apart slightly. Based on BMNH P.16636–7. Note the two dorsal openings from the infraorbital canal at the anterior end of the lachrymojugal. These are characteristic of *Whitea* spp.

W. tuberculata have been described by Moy-Thomas (1935b) and Lehman (1952) and those of *W. nielseni* by Nielsen (1936). In this study nine specimens of *W. woodwardi* (BMNH P.17200-1, P.17303, P.17204-5, P.17206-7, P.16636-7, P.17210-1, P.17212-3, P.17167, P.17169, P.19500-1), one specimen of *W. tuberculata* (BMNH P.17203) and nine of *W. nielseni* (MGUH VP2323a, VP2328, VP2329, VP2330, VP2331a, VP2332, VP2333, VP2334, VP2335a) were examined. The species of *Whiteia* are very similar to one another in the pattern of cheek bones, the chief differences being ornament patterns.

The cheek is completely covered by a full complement of bones which fit tightly together with matching contours (Fig. 4.15). There is, however, no overlap between most of the bones, the exception being the usual overlap of the suboperculum by the preoperculum.

The lachrymojugal (Fig. 4.14, Lj) is a long curved element which is angled anteriorly to run some considerable way in front of the eye and beneath the preorbital. This reflects the slightly elongated snout of *Whiteia*. The lachrymojugal is relatively deep beneath the eye, particularly so in large specimens of *W. nielseni*. The infraorbital canal opens through four or five ventral pores and, unusually, it also opens through two large dorsal pores at the anterior angle.

The postorbital is expanded posteriorly in its dorsal half while ventrally it is tube-like. The infraorbital sensory canal runs along the anterior margin and opens directly through small anterior and posterior pores. Thus extensive secondary branches are not developed in *Whiteia*. There are considerable differences between the shape of the postorbital in individuals of *W. woodwardi*. Some, such as BMNH P.17210-1, P.17204 and P.17167, show a narrow postorbital in which the dorsal half is not much expanded. In these specimens the squamosal is noticeably deeper than wide and the entire cheek has a 'narrow' aspect. In the remainder of the specimens examined here, best exemplified by P.17212, the postorbital and the squamosal are relatively broad and these individuals may be described as 'broad-cheeked'. Because there is no clear-cut division between these two, and because other individual cheek variations (ornament details, absolute size of individual) are not congruent with cheek width, this is not a basis for dividing the species.

The squamosal (squamoso-preopercular of Nielsen, 1936) is triangular but the relative height and width do vary, at least within individuals of *W. woodwardi*. In *W. tuberculata* the squamosal is taller than wide. The jugal canal passes through the centre of the bone and is angled at the centre of ossification. The path of this canal, as well as other cheek canals is very well seen in the specimen of *W. nielseni* n.sp figured by Nielsen (1936: fig. 12). The canal sends off both ventral (anterior) and dorsal (posterior) short branches. A vertical cheek pit line is usually developed on the squamosal (Fig. 4.14, c.p.l) where it originates at the centre of ossification and courses ventrally. There is also some evidence (BMNH P.17169) of a horizontal cheek line as seen in palaeoniscids (Gardiner, 1984a), *Amia* (Pehrson, 1944), *Polypterus* and *Eusthenopteron* (Jarvik, 1944). This is the only case in coelacanths where this – presumably primitive – character has been observed.

The preoperculum is also a large bone meeting both the squamosal and, in *W. woodwardi* and *W. nielseni*, the lachrymojugal. In *W. tuberculata* it is somewhat smaller and fails to meet the lachrymojugal (Lehman, 1952: pl. 2, figs a–c). The preopercular canal runs along the posterior margin. The cheek pit line may also be present upon the preoperculum (BMNH P.19501) but, amongst the specimens examined here, this is the exception rather than the rule. Lehman (1952: pl. 4, figs A,B) suggested that the bone called preoperculum here was separated to a quadratojugal and a preoperculum in one speci-

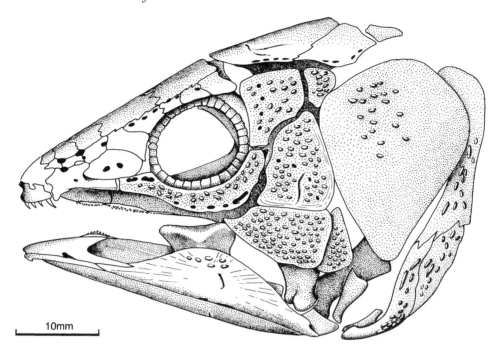

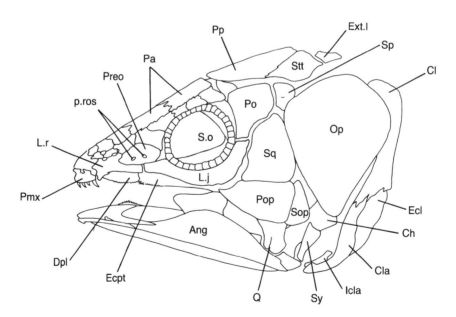

Fig. 4.15 *Whiteia woodwardi* Moy-Thomas. Restoration of the head in left lateral view. Composite from several specimens in BMNH collection.

men of *W. tuberculata*. This apparently occurs on the right side of the specimen only and the 'suture' may be a crack. Since this is the only specimen of *Whiteia* species in which a separate element has been recorded, I have scored a quadratojugal as being absent from the genus.

The suboperculum (preoperculum of Moy-Thomas, 1935b) is a small scale-like element with a marked overlap area anteriorly. Likewise the spiracular (suprasquamosal of Moy-Thomas, 1935b and Nielsen, 1936) is a small scale-like element which may be missing in some of the small specimens with a 'narrow' cheek.

As mentioned above, the ornament pattern varies between the species, although in all species, every cheek element is ornamented. In *W. woodwardi* there are irregularly sized tubercles sparsely scattered. In *W. groenlandica* the ornament is of closely crowded but irregularly sized tubercles. In *W. tuberculata* there are few tubercles of regular size and spacing.

Diplurus

The cheek bones of *D. longicaudatus* have been described most thoroughly by Schaeffer (1952a) and are figured here (Fig. 4.16(A)). The cheek bones are separated from one another. Postorbital, lachrymojugal, squamosal and preoperculum are present. The lachrymojugal is expanded anteriorly and is excavated for the posterior openings of the rostral organ. The excavation lies opposite a matching excavation in the adjacent tectal. The postorbital is relatively narrow with a long abutting surface with the squamosal, and in many respects this contact is very similar to the pattern seen in *Mawsonia* and *Axelrodichthys* (Fig. 4.17). Both the squamosal and preoperculum are small and triangular. The sensory canals open by a large pore within the postorbital, squamosal and preoperculum and by 4–5 pores within the lachrymojugal.

Ticinepomis

The cheek bones have been figured by Rieppel (1980) and although the restoration, reproduced here (Fig. 4.16(B)), is tentative, there are some unusal aspects. A lachrymojugal (margins unclear), postorbital, squamosal (margins unclear) and a preoperculum are present. The cheek bones must have lain free from one another. The lachrymojugal has a high postorbital limb not seen in other coelacanths. The postorbital is correspondingly very small. That which can be seen of the squamosal suggests that it was very small and bar-like. The preoperculum is substantially larger than the squamosal and this is an unusual relational proportion. Nevertheless, there is some similarity in the cheek proportions and pattern between *Diplurus* and *Ticinepomis* and this may reflect the relatively anterior position of the jaw articulation in these two genera.

Mawsonia

The cheek has been described for *M. gigas* by Woodward (1907) and Maisey (1986b) and for *M. tegamensis* by Wenz (1975). It is highly distinctive. The cheek is composed of a lachrymojugal, postorbital, squamosal and, in *M. tegamensis*, a preoperculum is present. This part of the cheek is poorly known for *M. gigas* and the presence/absence of a preoperculum is uncertain. All of the cheek bones are free from each other and all carry the rough rugose ornament characteristic of *Mawsonia*.

The lachrymojugal is very long reaching well forward beyond the level of the orbit. At the anterior end the bone turns abruptly dorsally, where it is excavated for the path of the posterior tubes opening from the rostral organ. The postorbital is large and carries a prominent anterior process which lies beneath the eye and partly overlies the lachrymojugal. The sensory canal runs through the centre of the postorbital. The

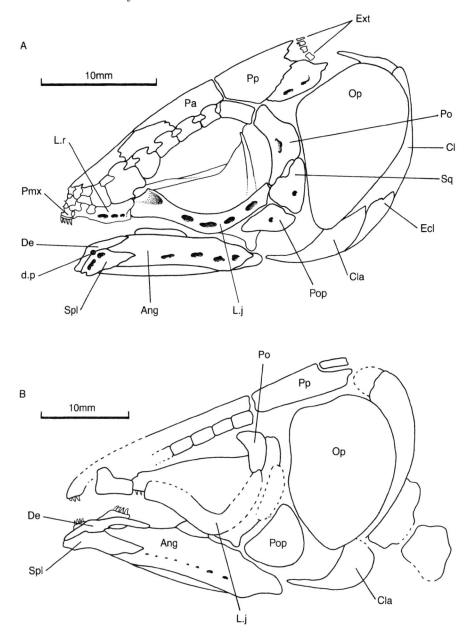

Fig. 4.16 Restoration of heads in lateral view. (A) *Diplurus newarki* (Bryant), from Schaeffer (1952a). (B) *Ticinepomis peyeri* Rieppel, from Rieppel (1980). Note similarities in the shapes of the lower jaws, clavicles and postorbitals.

squamosal (dorsal preoperculum of Wenz, 1975) is relatively small and triangular, whereas the preoperculum (ventral preoperculum) is polygonal.

Axelrodichthys

Maisey (1986b, 1991) described the cheek, which consists of postorbital, lachrymojugal, squamosal and preoperculum. It is briefly described below and illustrated (Fig. 4.17). The preoperculum lies free but the other bones are sutured at their mutual contacts.

The postorbital (Po) is distinctively shaped, with a prominent curved orbital margin and a small anteroventral process which partially overlies the posterior end of the lachrymojugal. The postorbital of *Axelrodichthys* may

therefore be considered to have an anterior process although it is much more poorly developed than in *Mawsonia*.

The lachrymojugal is deep, strongly curved and expanded anteriorly where it is excavated for the posterior openings of the rostral organ (gr.p.ros). There is no preorbital and the anterior end of the lachrymojugal reaches to the tectal series but there is no indication that the cheek and skull roof were in sutural contact at this point.

The squamosal (Sq) and preoperculum (Pop) are both relatively small, the preoperculum particularly so. These bones show both a canal-bearing tube and a lamina portion so that the canals do not wholly occupy the bones.

Ornament upon all cheek bones consists of

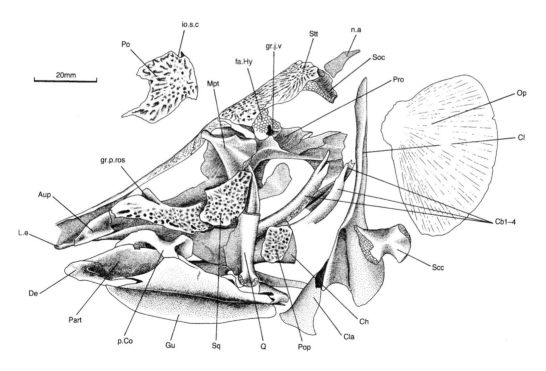

Fig. 4.17 *Axelrodichthys araripensis* Maisey. Camera lucida drawing of head in left lateral view as preserved in FMNH G.8567A. The postorbital has been drawn displaced dorsally to show underlying metapterygoid: the long sutural connection between the ventral edge of the postorbital and the dorsal edge of the squamosal can be visualized. Note also the strong medial expansion of the cleithrum, an unusual feature of coelacanths but also seen in *Diplurus*.

a rugose bone surface. The postorbital is more coarsely ornamented than the other bones. The contained sensory canals are very large and they open to the surface through many tiny pores which are difficult to see amongst the coarse ornament. As in *Mawsonia* and *Chinlea*, the postorbital spans the intracranial joint and this means that the infraorbital sensory canal (io.s.c) tips into the postorbital towards the posterior end of the dorsal margin. From here it runs anteroventrally and the junction between the infraorbital and jugal canals is developed as a wide lacuna between the lachrymojugal, postorbital and the squamosal.

Libys

The cheek bones of *Libys* are best known from specimens of *L. polypterus*, particularly BSM 1870.XIV.502, 503 and BMNH P.3337. The cheek bones are very thin, unornamented, and in all specimens they are preserved crushed over the underlying palatal bones such that the precise shapes are difficult to see.

Postorbital, lachrymojugal, squamosal and preopercular bones are present and these lie free from one another. The postorbital is rectangular and the sensory canal runs obliquely from the posterodorsal angle to the anteroventral corner. There are three or four large anteriorly directed pores. The lachrymojugal is deep and curves beneath the orbit to contact the tectal series anterior to the orbit. The anterior end of the lachrymojugal is greatly expanded and is very similar to that in *Macropoma*. The squamosal is also large but the true margins cannot be seen in any specimen studied here. However, BMNH P.3337 appears to show an intact ventral margin and the jugal sensory canal runs immediately behind this level much as in *Laugia, Coccoderma, Macropoma* and *Latimeria*. The preoperculum is clearly divided to a posterior tubular portion in which the sensory canal opens through a single large

pore, and a longitudinally elongate blade-like portion which lay along the dorsal edge of the angular. The preoperculum is very much like that of *Macropoma, Undina* and *Latimeria*.

Holophagus and *Undina*

The cheek bones are well known in *Holophagus gulo*, less well known in *Undina penicillata* and *U. cirinensis*. Cheek bone patterns in these taxa are sufficiently similar to be described together and they are shown in the whole fish restoration of *Holophagus gulo* (Fig. 11.8). The cheek bones of *Undina cirinensis* are shown by Saint-Seine (1949).

The cheek is completely covered by a postorbital, lachrymojugal, squamosal, preoperculum and suboperculum. All cheek bones are ornamented with small rounded tubercles, and the sensory canals, where they can be seen, open through tiny pores. Although the cheek bones do not overlap each other they closely abut one another with matching margins: the exceptions being the suboperculum (see below) and the preoperculum of *U. cirinensis* as shown by Saint-Seine (1949: fig. 34), which appears to lie free.

The postorbital is rectangular in the dorsal half and drawn out to a ventrally directed tube and of very similar shape to that in *Macropoma* and *Latimeria*, but there is no anterodorsal excavation as in the postorbital of these two genera. The sensory canal runs close to the anterior margin but I have been unable to confirm if there are anteriorly directed branches of the canal.

The lachrymojugal is relatively narrow and only slightly expanded anteriorly, although it does curve to reach the tectal series. There is only a narrow excavation for the posterior tubes from the rostral organ. The squamosal is large and expanded dorsally to fit the posterior contour of the postorbital. I have been unable to identify the path of the jugal canal in *Undina penicillata* or *Holophagus gulo* but it is significant that Saint-Seine (1949) shows the sensory canal of *U. cirinensis* running at the ventral margin. This is a derived condition similar to that in *Macropoma* and *Latimeria*.

The preoperculum in *Holophagus* and *Undina penicillata* is a relatively deep bone which extends forward to lie along the dorsal edge of the angular and cover the principal coronoid. It therefore matches the shape of the preoperculum of *Macropoma* and *Latimeria*. However, there is no clear separation between the posterior tubular and canal-bearing portion and the anterior blade-like portion. The suboperculum is very small and vertically elongated. In *Holophagus gulo* the suboperculum is uniformly ornamented over its entire surface and this suggests that it lay completely free from the preoperculum instead of showing the normal coelacanth overlap. Conditions in species of *Undina* could not be checked.

Macropoma

Woodward (1909) and Watson (1921: fig. 5) have provided descriptions of the cheek bones of *M. lewesiensis*, which is the best-known species. *M. precursor* is also well known and differs from the type species only in the form of ornament, and *M. speciosum* agrees with the other species in all known details.

Lachrymojugal, postorbital, squamosal and preoperculum are present and a suboperculum has been described (see below). The cheek bones are separated from one another and the contours do not match. The peripheral areas of the postorbital, squamosal and preoperculum show smooth areas (Fig. 4.18) suggesting they lay embedded within skin. They are shown in restored position in Fig. 4.19.

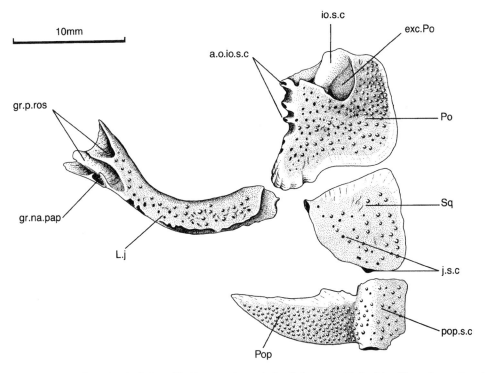

Fig. 4.18 *Macropoma lewesiensis* (Mantell). Restoration of cheek bones of left side. Note the anterodorsal excavation in the postorbital and the blade-shaped preoperculum. Based on several specimens in BMNH collection.

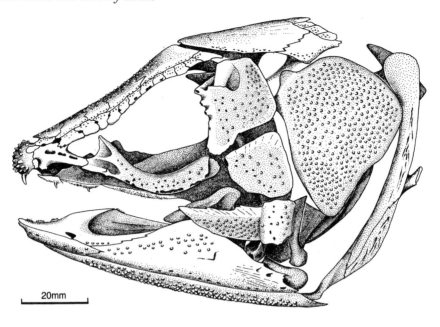

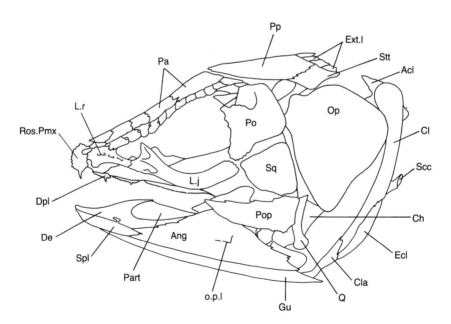

Fig. 4.19 *Macropoma lewesiensis* (Mantell). Restoration of the head and shoulder girdle in left lateral view. Based on several specimens in BMNH collection.

The lachrymojugal (suborbital of Watson, 1921) is expanded anteriorly where it meets the tectals. Two deep and parallel excavations received the posterior tubes leading from the rostral organ (Fig. 4.18, gr.p.ros), and a small ventrally placed groove, partially bridged with bone, carried the posterior nasal papilla which must have opened at the mouth margin as in *Latimeria*.

The postorbital is large and completely covers the prootic region of the braincase and denies the presence of a spiracular. The postorbital is deeply excavated anterodorsally (Fig. 4.18, exc.Po) and the floor of this excavation is rough. The excavation lies opposite a ledge upon the posterior supraorbital and these areas were probably linked by tough connective tissue as in *Latimeria*.

The squamosal (Sq) is small and triangular while the preoperculum is of an unusual shape but one that is similar to that of *Latimeria* and, perhaps, *Libys* (the cheek of this genus is poorly known). The preoperculum shows a swollen tubular portion posteriorly and a thin blade-like division anteriorly. Watson's restoration (1921: fig. 5) shows the blade-like portion lying closely against the lachrymojugal but in my restoration it lies against the dorsal edge of the angular where there is a matching smooth unornamented area (Fig. 5.13, o.Pop). The ornament on all bones of *M. lewesiensis* and *M. speciosum* is developed as small tubercles. The bones of *M. precursor* are pitted and without tubercles.

A suboperculum has been described and illustrated for *M. lewesiensis* by Watson (1921: fig. 5). The identity of this element must remain doubtful. Watson restores it lying ventral to the canal-bearing portion of the preoperculum, a position which is unlike the suboperculum (which lies posterior to and is overlapped by the preoperculum). I have seen a small bone lying ventral to the preoperculum in only two of many specimens examined here: one in BMNH 49834 was mentioned by Watson and another in GN

256. In both cases the bone is very thin, has ill-defined edges, and is ornamented with spine-like denticles like the scale ornament rather than head ornament. I have coded the presence of a suboperculum in *Macropoma* as questionable.

The sensory canals pass through the bones within very large canals which open to the surface through many tiny pores. Within the postorbital, the infraorbital canal runs at the anterior margin and sends off four prominent anterior branches and at least two smaller posterior branches. The infraorbital canal occupies most of the depth of the lachrymojugal and the medial surface is pierced by four foramina for the ramus buccalis VII, suggesting the presence of four neuromasts as in *Latimeria* (Millot and Anthony, 1965). Within the squamosal, the jugal canal runs at the ventral margin. The distribution of pores within the bone suggests that a small posterior branch was given off as the jugal canal turns ventrally. No pit line grooves are present.

4.3 DISCUSSION OF CHARACTERS RELATING TO COELACANTH CHEEK BONES

From the descriptions of the cheek bones and sensory canals, the following characters are recognized here.

29. Cheek bones sutured to one another (0), separated from one another (1). This character is scored '1' if there are no overlap areas upon the principal bones of the cheek (postorbital, squamosal and preoperculum), or if there are definite spaces between the bones (e.g. as in *Latimeria* or *Macropoma*). Where a suboperculum is present, then this is always overlapped by the preoperculum. This particular overlap is ignored when coding this character. The nature of the overlap between cheek bones has previously been used by Forey (1991). Cloutier (1991a, his character 23)

describes the cheek bone overlap relation as three states (complete suture, loose articulation of postspiracular and loose articulation of postspiracular and postorbital). The polarity decisions adopted by these authors remain the same, but I find that in all coelacanths in which a separate postspiracular can be identified, the postspiracular is always loosely associated with the postorbital, making one of these character states redundant.

30. Spiracular (postspiracular) absent (0), present (1). The presence or absence of this bone is recorded here. Some taxa such as *Whiteia woodwardi* may show individuals lacking the spiracular, but when coding this, such individual variation is ignored. Cloutier (1991a, his character 24) has used the size of the postspiracular as a character, preferring to link small size and absence together as a single state. In view of the great individual variability in shape and size in several coelacanths, a simple reference to presence/absence is preferred here.

31. Preoperculum absent (0), present (1). See also Forey (1991).

32. Suboperculum absent (0), present (1). See also Forey (1991) and Cloutier (1991a, his characters 30, 31, 32), who included three further parameters of the suboperculum: position, size and shape. With respect to size and shape, then these two character states probably convey the same taxonomic information. Those coelacanth taxa showing a suboperculum longer than deep also show a quadrilateral suboperculum (*Miguashaia*, *Polyosteorhynchus* – Cloutier, 1991a, Appendix 1). Another coextensive combination is the presence of a triangular suboperculum which is also deeper than long, present in many coelacanths. The three exceptional taxa recorded by Cloutier are *Latimeria*, *Macropoma* and *Lochmocercus*, which are each recorded as having a suboperculum deeper

than long and of quadrilateral shape. For *Latimeria* the coding of the shape as quadrilateral is surely an error; the suboperculum is basically triangular. For *Macropoma* the suboperculum is unknown (p. 125). This leaves the condition in *Hadronector*, which I have been unable to check. Cloutier has illustrated the suboperculum (1991a: fig. 4, his *Prop*) where, indeed, it is deeper than long. In specimens examined in this work the relative proportions could not be seen, although I confirm that the element is quadrilateral rather than triangular. Thus, among coelacanths, *Hadronector* would be autapomorphic in this combination: suboperculum deeper than long and of quadrilateral shape. With respect to the position of the suboperculum, Cloutier (1991a) distinguishes three states (articulation or suture with either the operculum or preoperculum or neither). I find that there is a consistent relationship between the suboperculum and the surrounding bones whereby the preoperculum overlaps the suboperculum and this in turn overlaps the operculum.

The precise form of the suboperculum is variable amongst coelacanths: it may be large or small; it may be double or single as individual variants within a species (p. 98 – *Latimeria*). The relative size may have some significance but I have found it difficult to adopt a discrete coding for this. As a result a simple presence/absence is coded for.

33. Quadratojugal absent (0), present (1). The presence of a quadratojugal, separate and distinct from the preoperculum in coelacanths has been discussed above. The individual cases are dealt with in the descriptive sections above.

34. Squamosal limited to mid-level of cheek (0), extending behind the postorbital to reach skull roof (1). Cloutier (1991a, his character 25) categorizes different shapes

of the squamosal into a multistate character. This is not accepted here, chiefly because I have found difficulty in ascribing discrete descriptors to the subtle differences in shape. In osteolepiforms and porolepiforms the squamosal also reaches the skull roof. This, therefore, may be a generalized character for sarcopterygians. Actinopterygians, which lack a squamosal, cannot be compared.

35. Lachrymojugal not expanded anteriorly (0), expanded anteriorly (1). This condition is found only in those taxa in which the preorbital is absent. However, there are taxa that lack the preorbital and which do not have an expanded lachrymojugal. The anterior expansion often bears a prominent groove which in life contained soft-tissue tubes opening posteriorly from the rostral organ.

36. Lachrymojugal ending without anterior angle (0), angled anteriorly (1). In some coelacanths the lachrymojugal reaches anteriorly well beyond the orbit and shows a distinct angle as it does so (best illustrated in *Whiteia*, Fig. 4.15). In most cases this angulation is associated with slight snout elongation. This character corresponds to that used by Cloutier (1991b, his character 64).

37. Squamosal large (0), reduced to a narrow tube surrounding the jugal sensory canal only (1). This corresponds to state '3' of character 25 (squamosal shape bar-like) as used by Cloutier (1991a).

38. Preoperculum large (0), reduced to a narrow tube surrounding the preopercular canal only (1). Species in which the preoperculum is absent are scored as 'not applicable'.

39. Preoperculum undifferentiated (0), developed as a posterior tube-like canal-bearing portion and an anterior blade-like portion (1). A polygonal undifferentiated preoperculum is regarded as plesiomorphic.

40. Postorbital simple, without anterodorsal excavation (0), anterodorsal excavation in the postorbital (1). See p. 98 for the significance of this character.

41. Postorbital without anterior process (0), with anterior process (1). This character corresponds to that used by Cloutier (1991a, his character 62).

42. Postorbital large (0), reduced to a narrow tube surrounding the sensory canal only (1). This character corresponds to that used by Cloutier (1991a, his character 63).

43. Postorbital entirely behind level of intracranial joint (0), spanning the intracranial joint (1).

44. Infraorbital canal within postorbital, with simple pores opening directly from the main canal (0), anterior and posterior branches within the postorbital (1). This character reflects the fact that, in some more derived coelacanths, there is an elaboration of the sensory canals.

45. Infraorbital sensory canal running through centre of postorbital (0), running at the anterior margin of the postorbital (1). Taxa coded '1' for character 42 will automatically have a coding of '1' here.

46. Jugal sensory canal simple (0), with prominent branches (1). As with character 44, this reflects elaboration of the sensory canal system. In some coelacanths this is shown by two large pores lying separate from the main trajectory of the jugal canal as it passes through the squamosal (e.g. *Rhabdoderma*).

47. Jugal canal running through centre of bone (0), running along the ventral margin of the squamosal (1). This character was previously used by Forey (1981, 1991) and Cloutier (1991a). Taxa coded '1' for character 9 will automatically have a coding of '1' here.

48. Pit lines marking cheek bones (0), failing to mark cheek bones (1).

49. Ornament upon cheek bones tubercular (1) or absent or represented as coarse superficial rugosity (2).
50. Infraorbital, jugal and preopercular sensory canals opening through many tiny pores (0), opening through a few large pores (1).
51. Lachrymojugal sutured to preorbital and lateral rostral (0), lying in sutural contact with the tectal–supraorbital series (1).
52. Sclerotic ossicles absent (0), present (1).

The states for these cheek bone characters for coelacanth taxa are tabulated in Table 9.1. Trends within the cheek that may be deduced from the coelacanth classification are discussed on p. 238.

5

LOWER JAW

5.1 INTRODUCTION

The lower jaw of coelacanths has long been recognized as being highly apomorphic (Stensiö, 1921; Schaeffer, 1952a; Andrews, 1973; Forey, 1981; Lund and Lund, 1985; Maisey, 1986a; Schultze, 1987; Panchen and Smithson, 1987; Cloutier, 1991a,b) with respect to the condition in other teleostomes. Particular points are the short dentary, only two infradentaries (angular and splenial), large angular which is very large and expanded dorsally, tandem jaw articulation, long prearticular, coronoid series in which the posteriormost is enlarged and projects well above the dorsal margin of the angular (sometimes expressed as a vertically orientated coronoid and separated from anterior coronoid), absence of submandibulars.

The significance of these characters is discussed and elaborated in the context of the relationships of coelacanths within the sarcopterygians (Chapter 10). In this chapter the lower jaws of coelacanth species are described as far as known and a list of characters that might be used to erect a classification of coelacanths is drawn up and discussed.

5.2 DESCRIPTIONS OF GENERA

Latimeria

The lower jaw has been described by Smith (1939c) and Millot and Anthony (1958, 1965). Together, these accounts are comprehensive and it is only necessary to mention salient points to provide a reference point to interpret the jaws of fossil coelacanths, as well as to make comparisons with the lower jaws of other fishes.

Meckel's cartilage is ossified in three pieces. Anteriorly, there is a mentomeckelian (Fig. 5.1, Mm) which is little more than a shell of perichondral bone forming the symphysis and sheathing Meckel's cartilage as far as one-third the length of the dentary. Posteriorly there are articular (Art) and retroarticular (Rart) ossifications which share in the formation of the articular glenoid fossa. The retroarticular has a posterior cap of cartilage and this provides a point of articulation for the ventral end of the symplectic. The lateral surface of the retroarticular is roughened for the insertion of a powerful anterior mandibulohyoid ligament (Chapter 7, p. 199).

The articular fossa is formed as a cartilage-lined double concavity. The medial concavity is slightly larger than the lateral and, with respect to the long axis of the lower jaw, this is displaced anteriorly. However, when the jaw is orientated in life position, the two concavities line up transversely.

The lateral face of meckel's cartilage is covered by a very large angular (Ang) and much smaller dentary (De) and splenial (Spl) bones. In the adult, each of these bones has a raised area which lay at the skin surface and a deeper-lying peripheral area. In the embryo specimen examined, there is no distinction between raised and sunken areas, and the bones appear to lie entirely superficially.

The dentary is excavated posteriorly into a

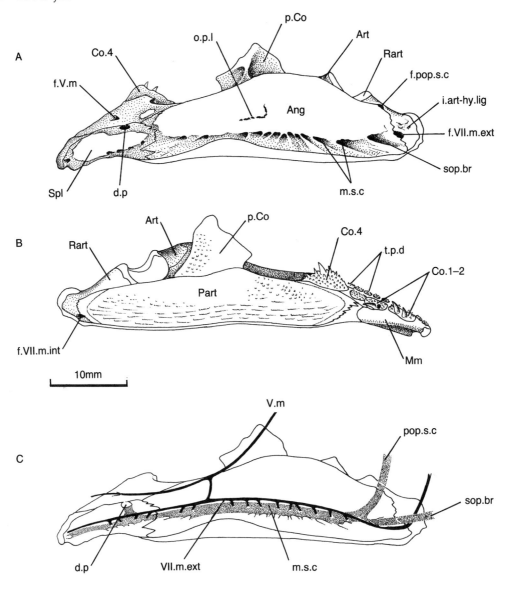

Fig. 5.1 *Latimeria chalumnae* Smith. Lower jaw of left side: (A) lateral view; (B) medial view; (C) view as if transparent, showing paths of mandibular nerves and sensory canals. All based on figures in Millot and Anthony (1958) and BMNH 1976.7.1.16.

hooked-shaped process and the inner contour of the hook receives the maxillary fold of skin and a small labial muscle. The dentary is overlain by a series of small dentary tooth plates (t.p.d) each bearing a shagreen of tiny teeth. There are eight such plates in the embryo specimen and at least 14 in the adult described by Millot and Anthony (1958). These tooth plates are separate from the dentary and lie free in the skin.

The inner side of Meckel's cartilage is covered by a very long prearticular (Part) surmounted by a series of coronoids. All of the bones bear teeth. The coronoids have a characteristic arrangement. Anteriorly there are four small coronoids which form a series lying medial to the dentary. The largest of this anterior series is coronoid 4 (Co.4) (precoronoid of Smith, 1939c) which lies opposite the hook-shaped process of the dentary. Coronoids 1, 2 and 4 each carry larger fang-like teeth, accompanied by replacement sockets.

The fifth coronoid is much larger: it lies opposite the deepest part of the angular and is separated from those in the anterior coronoid series by a diastema. This coronoid will be referred to in this work as the principal coronoid (P.Co).

The inner surface of the principal coronoid is covered with a shagreen of tiny teeth. The outer surface is marked by a strong ridge which follows the contours of the adductor muscle. In *Latimeria* there is no contact between the principal coronoid and the angular except through a powerful ligament (Millot and Anthony 1958: fig. 19A).

The path of the mandibular sensory canal has been described by Millot and Anthony (1958, 1965) and Hensel (1986). Here, a few points are made about the influence of the canal on the bones through which it passes. The main mandibular sensory canal runs down from the preopercular to penetrate the angular at a level just lateral and posterior to the articular glenoid. From here the canal (m.s.c) runs anteroventrally within the angular and turns anteriorly to run close to the ventral margin, continues into the splenial to join with its antimere across the symphysis.

The main canal is large and as it passes through the angular it opens ventrally by way of several large pores through the bone (Fig. 5.1(A)). There are eight such pores in the embryo specimen and 12 in the adult illustrated by Millot and Anthony (1958, 1965). This suggests that throughout growth

there is an increase in the number of pores, perhaps arising as the result of the subdivision of original pores. This probably accompanies the great proliferation of secondary canal branches and pores within the skin that develop throughout the life of *Latimeria* (Fig. 4.3; Hensel, 1986). The posteriormost pore in the angular is directed backwards and this marks the position where a subopercular branch (sop.br) of the preopercular canal leaves the bone to ramify through a subopercular flap of skin.

The paths taken by the nerves through the jaw have been described by Millot and Anthony (1958, 1965) and Bemis and Northcutt (1993) and the pattern is very similar to that described for nerves running within the jaw of *Amia* (Allis, 1897). The external mandibular ramus VII (antero-ventral lateral line nerve of Bemis and Northcutt, 1993) enters the lower jaw through a foramen (f.VII.m.ext) between the retroarticular and the angular, immediately below the point of insertion for the powerful hyomandibular/ retroarticular ligament (i.art-hy.lig). This branch runs along the inside of the angular and sends lateral branches to supply the neuromasts of the mandibular sensory canal.

The internal mandibular ramus VII (f.VII.m.int) enters the mandible on the medial side of the jaw articulation between the posterior margin of the prearticular and Meckel's cartilage. It runs forward on the inside of the cartilage to innervate the intermandibularis muscles. This is a much smaller nerve than the external mandibular VII.

The mandibular branch of V (V.m) runs along the dorsal edge of Meckel's cartilage anterior to the principal coronoid. It supplies branches to the labial muscle before piercing the dentary through a small foramen immediately above the dentary pore to send branches to the skin of the lower lip. There is an anastomosis between V and VII within the meckelian fossa (Fig. 5.1(C)).

Miguashaia

The lower jaw (Fig. 4.4) has been described by Cloutier (1996), who makes some emendations to the previous description by Schultze (1973). It remains very poorly known but it is possible to recognize the typical coelacanth aspect: that is, a short dentary and splenial and a large angular and principal coronoid.

The angular is very similar in shape to that in *Rhabdoderma* and the oral pit line (Fig. 4.4, o.p.l) is located at the centre of ossification. It is restored as being short and not extending to the dentary. Both the dentary and the splenial are small and shallow bones. There is no hook-shaped process on the dentary and the approximately eleven dentary teeth are firmly ankylosed to the bone. The principal coronoid is triangular and inclined forwards.

Diplocercides (Nesides)

The lower jaw has been described by Stensiö (1937, *Diplocercides kayseri*, *Nesides schmidti*), Jessen (1966, *D. heilegenstockiensis*) and by Janvier (1974) and Janvier and Martin (1979, *Diplocercides* sp.). There are some reported differences between the species and these are noted below. There are also many similarities and these are best emphasized through a common description.

Meckel's cartilage (Fig. 5.2, Mm) is ossified anteriorly as a mentomeckelian and posteriorly as an articulo-retroarticular. The mentomeckelian is relatively long compared with that in other coelacanths. The angular appears to vary slightly in shape and this variation has been used to distinguish between species. In *D. kayseri* (Fig. 5.2, Stensiö, 1937: fig. 7) the angular is very low throughout its length and there is no steeply inclined border anterior to its maximum depth. In *N. schmidti* (Stensiö, 1937: fig. 3) the angular is deeper and more steeply inclined. Stensiö (1937) used this feature as part of his

reason for distinguishing *Nesides* from *Diplocercides*. *D. ? heilegenstockiensis* (Jessen 1966: fig. 15F 'third infradentary') and *Diplocercides* sp. (Janvier and Martin, 1979: fig. 1(B)) are intermediate in shape and do question the validity of using subtle differences in shape of the angular to distinguish species. Despite these differences in shape, the angulars are similar in showing a well-developed oral pit line (unknown in *Diplocercides* sp.) which begins anterior to the maximum depth of the angular and reaches well forward. In *D. kayseri* the pit line is known to continue across both the dentary and the splenial (Fig. 5.2(A)) as an interrupted groove. In specimens of *N. schmidti* and *D. ? heilegenstockiensis* the dentary and splenial are too poorly preserved to check this feature, but it is probably significant that the pit line continues to the anterior end of the angular in both. It is highly likely, therefore, that conditions were as in *D. kayseri* and this has been assumed in the data matrix (Table 9.1).

The dentary (unknown in *Diplocercides* sp.) is relatively longer than in many other coelacanths, the oral border constituting nearly one-third the length of the jaw. It bears a row of tiny teeth (det) which are fused with the supporting bone, and this condition is regarded here as primitive. In *Diplocercides kayseri* there is a narrow band of tiny teeth. In *D. ? heilegenstockiensis* Jessen (1966: fig. 15I) illustrates a single row of larger teeth.

The splenial is noticeably shorter than the dentary, more comparable in relative size to the splenial in other coelacanths. The coronoid series shows a large principal coronoid lying above the prearticular, at the level of the maximum depth of the angular. Anteriorly there are at least two coronoids in *Diplocercides kayseri* and three in *D. ? heilegenstockiensis*. Other species are unknown in this respect. In *D. kayseri* the two anterior coronoids that are known are of very unequal size as in more derived coelacanths. However, they do bear a shagreen of teeth which are of uniform size and similar to

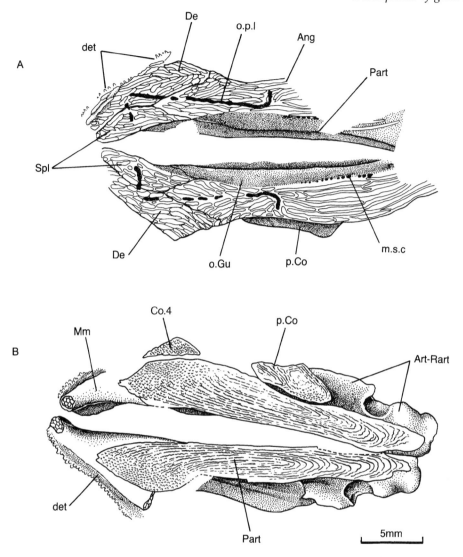

Fig. 5.2 *Diplocercides kayseri* (von Koenen). Camera lucida drawing of casts of lower jaws as preserved in specimen 'a' of Stensiö (1937). (A) External surfaces of jaw rami. Note the long oral pit line reaching forwards to the dentary and splenial. (B) Internal surfaces of jaw rami. Note the linear arrangement of teeth on the prearticular, a pattern often seen in more plesiomorphic coelacanths.

those of the prearticular. In *D. ? heilegenstock-iensis* the anterior coronoids are all of about the same size; but each bears a dentition which consists of both large and small teeth. The principal coronoid does vary in shape between the species from triangular in

Nesides ? heilegenstockiensis to long and low in *Diplocercides kayseri*. The dentition upon the prearticular and the principal coronoid tends to be arranged as tiny teeth coalesced at their bases and arranged into ridges running parallel to the bone margins.

The mandibular sensory canal runs along the ventral margin of both the angular and splenial and opens to the surface through many small and regularly spaced pores (m.s.c). Some variation in the disposition of the pores at the posterior end of the angular has been noted by Janvier and Martin (1979), with the implication that the path taken by the mandibular canal may vary. Thus, in *D. ? heilegenstockiensis* and *D.* sp., the line of pores rises posteriorly, suggesting that the mandibular canal entered the angular shortly behind the articular facet. In *D. kayseri* and *N. schmidti* the line of pores has been restored as running at the ventral margin. Janvier and Martin (1979: 500) suggest that the ventral path exhibited by the two last-mentioned species corresponds to the subopercular branch of *Latimeria* (p. 99) and that this represents the original path of the main canal. The more dorsally inclined path taken by the canal in the first two species mentioned would be a derived condition and comparable to the path of the main canal in *Latimeria*, which must therefore be considered as showing a modified path of the main canal.

An opposite view is adopted here, whereby it is considered that the subopercular branch is a derived condition having arisen in comparatively cladistically derived coelacanths. Our present knowledge suggests that *Diplocercides* (including *Nesides*) has only one row of pores, which must represent the main canal. Thus, there is no evidence of a separate subopercular branch in *N. schmidti* and *D. kayseri*. The posterior end of the angular of the only specimen of *N. schmidti* is not known, while in the relevant specimens of *D. kayseri*, the path of the mandibular canal of the right angular is indicated by Stensiö (1937: pl. 1) as running posteriorly to leave the angular immediately behind the articular facet. The true ventral edge of the angular is obscured by the overlying gular plate. Thus, the path of the mandibular canal may be similar to that in

D. ? heilegenstockiensis and *D.* sp. and not as indicated by Stensiö (1937: fig. 7A) in his restoration.

The ornament upon the angular, dentary and splenial consists of elongate ridges and matches that upon the gular plate.

In summary, the lower jaw is plesiomorphic in having a long pit line reaching on to the dentary in at least one species, no dentary pore, fused dentary teeth, simple dentition upon the fourth coronoid (Fig. 5.2, Co.4), fused articular–retroarticular and ridged ornament.

Allenypterus

The lower jaw has been described by Lund and Lund (1985) and only a few additional comments are needed here. My observations are based are based on three specimens (FMNH PF10939, PF10940, PF10942), one of which is illustrated in Fig. 5.3(B) and forms the basis for the reconstruction as shown in Fig. 4.6.

The lower jaw is unusual in the shape of the dentary. Overall the jaw is relatively short and deep, and the jaw articulation is located beneath the hind margin of the orbit. Lund and Lund (1985: fig. 62) record that the articular and retroarticular fuse in large individuals. Compared with many other coelacanths, this articular–retroarticular is a very large ossification. A mentomeckelian has not yet been recorded.

As usual the angular is the largest dermal bone of the lower jaw, although it is unlikely to have reached forward to the symphysis as shown by Lund and Lund (1985: fig. 57). The coronoid expansion is low and ill defined. Most of the lateral surface is ornamented with prominent ridges which tend to run longitudinally. The oral margin immediately behind the dentary is devoid of ornamentation and this, together with the smooth dentary, represents the area of insertion of the lip fold. The splenial is short and deep and bears ornament.

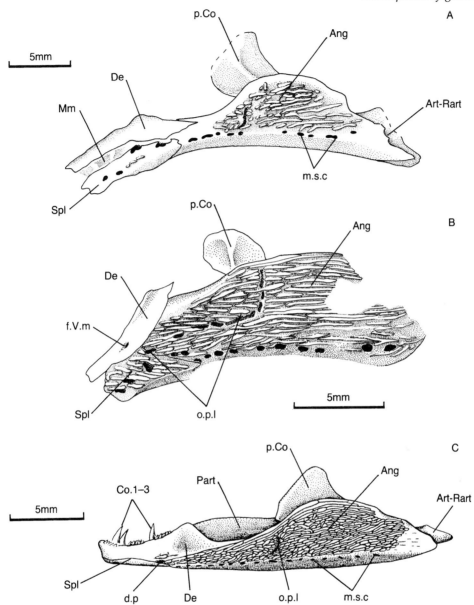

Fig. 5.3 Lower jaws of plesiomorphic coelacanths. (A) *Caridosuctor populosum* Lund and Lund, left lower jaw in lateral view. Camera lucida drawing of latex cast, FMNH PF31505a. (B) *Allenypterus montanus* Melton, left lower jaw in lateral view. Camera lucida drawing of latex cast, FMNH PF10939. (C) *Rhabdoderma elegans* (Newberry), restoration of left lower jaw in lateral view, based on several specimens in BMNH collection.

The dentary is much longer but narrower than the splenial. It is concave dorsally and is set at an angle to both the splenial and the angular. It is not hook-shaped, although the posterior aspect may be roughened for the insertion of the lip fold. No dentary teeth have been found.

The anterior coronoids are very poorly known. FM PF10942 shows faint impressions of a coronoid with large teeth lying mesial to the posterior end of the dentary but this observation needs confirmation from better specimens. Lund and Lund (1985: 44) refer to a 'thin precoronoid continuous between the posterior end of the dentary and the prominent coronoid [principal coronoid in this work]'. This implies that there is no modified member of the anterior coronoid series. In the data matrix this character is scored as unclear. The principal coronoid is relatively small. My observation (Fig. 5.3) suggests that the dorsal margin of this element is rounded (cf. Lund and Lund 1985: fig. 62). The prearticular is shown by Lund and Lund (1985: fig. 62) to be covered with rows of tiny teeth.

The mandibular sensory canal opens within the angular and the splenial through a series of large pores. There is no dentary sensory pore. The oral pit line is long and broken into short sections. It passes forward and just runs onto the splenial (Fig. 5.3).

In summary: the lower jaw is plesio-morphic in showing a single articular–retro-articular ossification, long oral pit line, no hook-shaped dentary, no subopercular branch of the mandibular sensory canal, no dentary pore and ridged ornament upon the splenial and angular. It is derived with respect to the shape and orientation of the dentary, which also lacks ornament.

Hadronector

The lower jaw has been described by Lund and Lund (1985) but remains very poorly known. Salient features include the follow-

ing. The angular is relatively shallow with only a low coronoid expansion. The short oral pit line marks the bone at its greatest depth. The dentary and splenial are of equal size and the dentary is simple without a hook-shaped process. All three outer dermal bones are heavily ornamented. The mandibular sensory canal opens through a series of many small pores. There is no specialized dentary pore. One further fact mentioned by Lund and Lund is that the articular and retroarticular are fused together but I was unable to confirm this.

Caridosuctor

The lower jaw has been described by Lund and Lund (1985). Three specimens were examined here (FM PF12920, FM PF13505a,b and a specimen in a private collection of Bill Hawes, number 82-80302B) and a camera lucida drawing of one of these is shown in Fig. 5.3(A).

The jaw is relatively long and shallow and the jaw articulation lies beneath the level of the posterior margin of the skull. The angular shows only a narrow overlap for the gular plate and there is a low, rounded coronoid expansion. The oral pit line is short and vertical and located anterior to the deepest level of the angular. The mentomeckelian is poorly ossified, probably as a perichondral shell only. The posterior end of Meckel's cartilage is ossified as an articular–retro-articular.

Both dentary and splenial are shallow bones: the dentary does not have a hook-shaped profile but the posterior end is slightly expanded. The dentary tooth plates are probably separate. Lund and Lund (1985: 23) suggest that the dentary bears two or three large teeth near its posterior end. These almost certainly belong to the coronoid series. The coronoid series consists of three or, more probably, four anterior coronoids, the posterior member bears the large teeth. The principal coronoid has a curved anterior margin but straight dorsal and posterior margins. It is inclined anterodorsally.

Ornament upon the lower jaw is confined to a few ridges and tubercles on the central area of the angular and to a few large tubercles on the splenial. The mandibular sensory canal enters the jaw immediately behind the level of the quadrate–articular jaw joint and runs the length of the angular and splenial opening through a series of small pores. The dentary pore is either located wholly within the splenial (Fig. 5.3(A)) or developed equally between the dentary and the splenial (Lund and Lund, 1985: fig. 18).

In all, the lower jaw of *Caridosuctor* is relatively primitive: features derived relative to the presumed primitive coelacanth conditions include the presence of a modified coronoid opposite the posterior end of the dentary, separate dentary teeth, ornament absent from the dentary, short oral pit line and a dentary sensory pore.

Polyosteorhynchus

The lower jaw was described by Lund and Lund (1985: figs 46, 54, 56) and in many features it is similar to *Lochmocercus*. The jaw is shallow and the dentary and splenial are about the same length and equal to approximately one-third of the total jaw length. The dentary is particularly shallow and there is no hook-shaped process. Lund and Lund record that in small specimens, the articular and retroarticular are separate, but that in large individuals a single articulo-retroarticular is present. The dentition consists of tiny villiform teeth, although it is not known if the dentary teeth are fused or separate from the supporting bone. The mandibular canal opens through a series of tiny pores, and Lund and Lund (1985: fig. 46) show a dentary pore in their restoration of the whole fish. The oral pit line is short and located beneath the level of the quadrangular principal coronoid.

Lochmocercus

The lower jaw was described by Lund and Lund (1985: figs 69, 70). The jaw is shallow throughout most of its length but there is a well-developed coronoid expansion located relatively far posteriorly. The dentary and splenial are of similar length and occupy a little less than one-third the total length of the jaw. The dentary is shallow with no hook-shaped process but it does bear large, pointed teeth which appear to be fused to the bone. A dentary sensory pore is not described or figured by Lund and Lund. The mandibular sensory canal opens by many small pores posteriorly but these pores become large towards the front of the angular and within the splenial. Little is known of the inner jaw bones.

Rhabdoderma

Within the genus *Rhabdoderma* the anatomy of the lower jaw is known in detail only in *R. elegans* (Fig. 5.3(C)) and *R. tingleyense*, and it appears to be similar in these two species. The jaw morphology of other species, such as that of *R. huxleyi*, *R. (?) alderingi* and *R. ardrossense*, is known to differ only in the ornament pattern upon the angular and these details are mentioned in the diagnoses (p. 333). The lower jaw of *R. elegans* has been described by Moy-Thomas (1937) and Forey (1981).

The mentomeckelian is small and there is a single articular–retroarticular forming both the quadrate/articular joint and the articulation with the symplectic. The angular has a gently rounded coronoid expansion but is otherwise shallow. The oral pit line is well developed and found anterior to the maximum depth of the angular as a short, posteriorly convex groove. The angular is heavily ornamented.

The dentary is longer than the splenial as in *Diplocercides* and *Caridosuctor*. It is posteriorly expanded but lacks the distinctive hook-shaped process seen in many coelacanths. A few small dentary tooth plates lie separate from the dentary. A few tubercles of ornament are present on the dentary. The splenial is very narrow and unornamented.

The mandibular sensory canal opens through a series of 16–20 small and regularly spaced pores running along the ventral edge of the angular and through the centre of the splenial. A single dentary pore opens at the junction of the dentary, angular and splenial. There is no evidence of a subopercular branch of the mandibular canal.

The prearticular is shallow and covered with tiny villiform teeth. The anterior coronoids bear fang-like teeth (Fig. 5.3(C)), substantially longer than those in most other coelacanths, but there is no evidence of these being folded as are the fangs of porolepiforms and some 'osteolepiforms'. The principal coronoid is triangular and the dentition on the medial surface is divisible to small isolated teeth anteriorly and crenulate ridges posteriorly.

In the shapes and relative proportions of the outer dermal bones, the lower jaw of *Rhabdoderma* is similar to that of *Caridosuctor* (e.g. Lund and Lund, 1985: fig. 29). The dentition upon the principal coronoid is also similar and the species of both genera have large fang-like teeth upon the anterior coronoids.

Coelacanthus granulatus

The lower jaw has been described by Moy-Thomas and Westoll (1935: 450) and, more completely, by Schaumberg (1978: 182). I have seen only one additional specimen (SM D.435) and can confirm most of the points contained in previous descriptions. Therefore only a few comparative remarks are necessary.

The jaw is long and relatively shallow. The angular is produced as a low coronoid expansion such that the proportions are simi-

lar to those of *Spermatodus* and *Sassenia*. Both the dentary and splenial are particularly shallow, the former being almost rod-like (Fig. 5.4, De). The dentary bears no posterior excavation or hook-shaped dorsal process and is therefore primitive in these respects. The posterior end of Meckel's cartilage is ossified as a single articulo-retroarticular and there is a small mentomeckelian (Schaumberg, 1978: fig. 12) wedged between the dentary and the splenial.

The principal coronoid is of similar shape to that in *Spermatodus*: it is quadrangular and inclined anterodorsally. The report that the external surface is ornamented (Moy-Thomas and Westoll, 1935: 450) is incorrect: it is smooth, marked only by the usual 'T'-shaped thickening (Fig. 5.4, p.Co). The dentition upon the principal coronoid consists of fine concentric rows of teeth forming tooth ridges. The prearticular dentition remains largely unknown, but the extreme anterior and posterior ends can be seen in SM D.435, where the teeth are tiny, irregular in size and shape and do not appear to be arranged into rows. The anterior dentition consists of four coronoids: the anterior three (Dentalplatte of Schaumberg, 1978) are rounded tooth plates, each carrying 40–50 acutely conical teeth. Most of the teeth are small, but those towards the lateral margin tend to be slightly enlarged. The fourth coronoid (CO_2 of Schaumberg, 1978) is considerably larger, stouter and bears an outer row of five or six prominent teeth (BMNH P.3339 and HM 926.52); and in terms of the characterization of this coronoid amongst coelacanths, it is recorded as 'derived' (Table 9.1).

The mandibular sensory canal can be traced as series of medium-sized pores. There are approximately 15 within the angular and about six or seven within the splenial, all regularly spaced. The observations of Schaumberg (1978: figs 6, 11, 12) suggest that the mandibular canal ran posteriorly to the level of the jaw articulation, turned abruptly

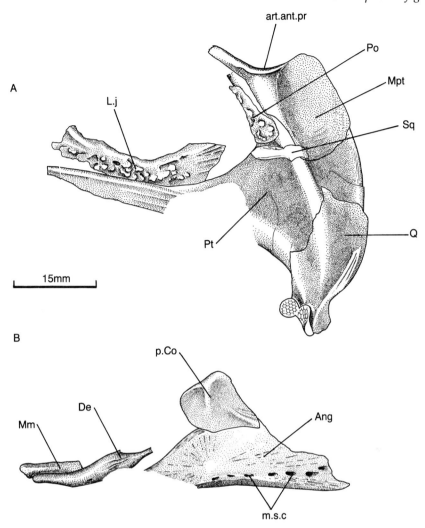

Fig. 5.4 *Coelacanthus granulatus* Agassiz. Camera lucida drawings of SM D.435. (A) Palate and cheek bones of left side in lateral view. (B) Lower jaw in left lateral view. The principal coronoid appears to have been displaced during preservation to lie outside of the angular.

dorsally to join the preopercular canal and sent off a subopercular branch which left the angular posteroventrally. Moy-Thomas and Westoll (1935: fig. 3) suggest that the main canal turned gently posterodorsally and ran near to the level of the jaw articulation without giving off a subopercular branch. This observation was based on a single specimen (HM 926.52) but can now be confirmed in SM D.435 (Fig. 5.4(B)). I therefore favour the view that the mandibular canal described a very oblique path across the posterior half of the angular, much like that in *Spermatodus*, and that there was no subopercular branch. Ornament is totally absent from the lower jaw and there is no pit line marking the angular (if present in life it must have lain superficially).

Laugia

The lower jaw (Figs 4.10, 5.5, 5.6) was briefly mentioned by Stensiö (1932) in his original description of the genus, but no further notes or comments have been published. This description is based on many specimens.

The meckelian cartilage is usually ossified in two portions; anteriorly there is a small, laterally flattened mentomeckelian while posteriorly there is a large and robust articular–retroarticular. In one small specimen (MGUH no. 82) of unknown angular length, the articular and retroarticular are separate bones (Fig. 5.5(B)) and this presumably represents an early ontogenetic stage. Comparison between this specimen and others, where there is a fused articular–retroarticular, demonstrates that most of the articular fossa lies within the articular ossification. The retroarticular is very similar to that of *Whiteia* in that there is a prominent process upon the lateral surface, separating the point of entry of the mandibular sensory canal above from the point of entry of the external ramus of mandibular VII below. There is no clearly defined facet for articulation with the symplectic, but articulated specimens (e.g. MGUH VP2311) show the usual coelacanth tandem joint.

The angular is shallow throughout much of its length but it deepens considerably immediately anterior to the articular facet where the dorsal margin is very rounded. The angular is the only bone within the lower jaw to show ornament. Even here it is absent from most specimens; but occasionally it is represented as a few rounded tubercles located within the area of greatest depth. The oral pit line is 'L'-shaped (Fig. 5.5(A)); it lies at the centre of growth which is anterior to the greatest depth of the bone. Ventrally the overlap area for the gular (Fig. 5.5(A), o.Gu) is rather narrow.

The dentary and splenial are each rather small bones. The dentary bears a pronounced embayed area which is marked with promi-nent ridges (Fig. 5.5(A)) for the insertion of the lip ligament. It is, however, a relatively shallow bone compared with the dentary of more derived coelacanths. Instead, the proportions of the dentary are more like those of *Rhabdoderma*.

The coronoid series is, as usual, divided into an anterior series of four bones and a single large principal coronoid. The anterior series consists of three small tooth plates followed by a long shallow tooth plate which is rolled over the dorsal surface of the posterior end of the dentary (Fig. 5.5(C), Co.4). The dentition upon the anterior coronoids consists of a shagreen of small villiform teeth, there are no enlarged teeth. The principal coronoid is large rectangular, with the longer axis running anterodorsally. The outer surface bears the usual 'T'-shaped strengthening ridge while the inner surface is covered with minute teeth running in concentric rows, parallel to the margin.

The prearticular is a long, shallow element covered sparsely with tiny teeth (Fig. 5.5(D)). These teeth are arranged mostly in lines, forming tooth ridges which tend to lie parallel to the bone margin (MGUH VP2316a). The posterior limit of the prearticular remains unknown but in the restoration of the inner face of the jaw it is assumed that it resembled that of other coelacanths.

The mandibular sensory canal entered the angular immediately behind the level of the articular fossa. From here the canal followed its usual course obliquely across the posterior part of the angular and then followed the ventral margin to penetrate the splenial. There is no indication that a separate subopercular branch was developed. The canal is relatively large, as indicated by Stensiö (1932: fig. 19(B)). The openings of the mandibular canal show some interesting variations (Fig. 5.6). In small specimens the angular is pierced by about 10 large pores. In larger specimens the pores are considerably smaller and far more numerous, there being about 40 in the largest specimens. In some specimens, such as MGUH VP2308B

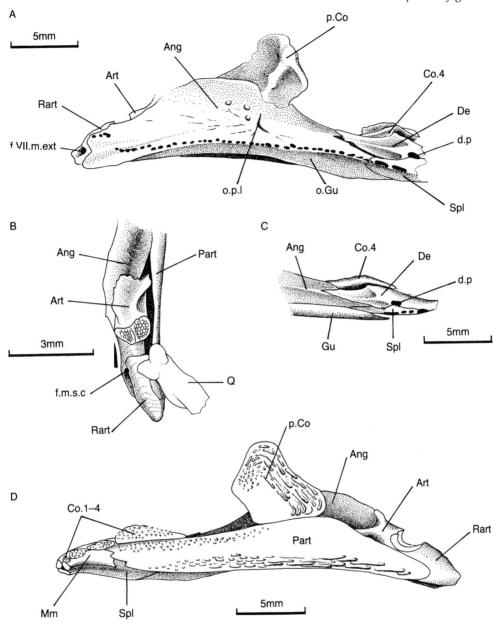

Fig. 5.5 *Laugia groenlandica* Stensiö, lower jaw. (A) Lateral view, camera lucida drawing of MGUH VP2308a. (B) Dorsal view of jaw joint, camera lucida drawing of MGUH VP12. (C) Anterior end of right jaw, camera lucida drawing of MGUH VP2311. (D) Reconstruction of right jaw in medial view, based on MGUH VP2308a, VP2316, VP2312, VP2317.

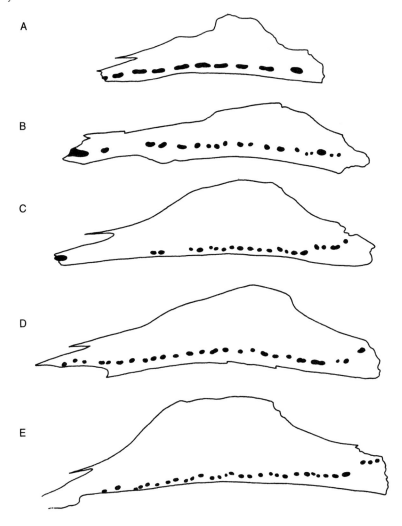

Fig. 5.6 *Laugia groenlandica* Stensiö. Camera lucida drawings of angulars of several specimens drawn to same scale and showing ontogenetic increase in the number of mandibular sensory pores associated with decrease in the relative size of pores. Some drawings have been reversed for easier comparison. (A) MGUH VP2309 (13.5 mm); (B) MGUH VP2312A (16.6 mm); (C) MGUH VP2311 (18.1 mm); (D) MGUH VP2316B (19.8 mm); (E) MGUH VP2308B (22.2 mm).

(Fig. 5.6(E)), groups of small pores can be recognized, implying that the increase in pore number has come about by the subdivision of the larger pores of younger growth stages (see also Stensiö, 1932: fig. 19). A series of five different-sized angulars is illustrated in Fig. 5.6 to support this idea.

It is further noticeable that the pores towards the anterior end of the angular, and the four or five pores on the splenial are substantially larger than those posteriorly (Fig. 4.10). This pattern matches a similar pattern affecting the supraorbital and infra-orbital canals over the snout (p. 62). In one specimen (MGUH VP2317), a medium-sized individual with an angular length of 19 mm, there are three groups of double pores within the splenial, suggesting that subdivision has occurred exceptionally here. The dentary pore is large and lies along the dentary/splenial suture.

Coccoderma

The anatomy of the lower jaw of *Coccoderma* (Figs 4.11, 5.7) is imperfectly known. Only the outer aspect of the jaw is known and even this imperfectly so. Two other species (*C. bavaricum* Reis and *C. gigas* Reis), previously referred to the genus, are no longer considered to belong here, and because part of the reason for this evaluation concerns lower jaw morphology, some comparative notes are given below.

The angular is produced as a prominent dorsal expansion marked by a shallow notch along the posterodorsal margin opposite the preoperculum. None of the available specimens show the anterior suture with the splenial, but it is assumed to lie well anteriorly as in other coelacanths.

The dentary is relatively shallow and is without the hook-shaped process seen, for instance, in *Latimeria* or *Macropoma*. It does, however, bear a few ridges for the insertion of the lower lip fold. The anterior coronoid series consists of three small equidimensional plates, each carrying about 20 bluntly conical teeth. This is followed by an elongate coronoid which is rolled over the posterodorsal aspect of the dentary. The teeth within the outer row upon this coronoid are slightly larger and more acutely pointed than neighbouring teeth but they cannot be described as fangs. Coronoid 4 is considered to be of the derived type (Table 9.1). The principal coronoid is distinctively shaped: it is rectangular and considerably deeper than wide and is inclined anterodorsally. In shape and position it is most closely similar to the principal coronoid of *Laugia*.

The mandibular canal appears to open to the surface within the angular through slit-like pores (Fig. 5.7) which are notched in such a way to suggest that they are the result of ontogenetic subdivision (cf. *Laugia*, p. 140). There is no evidence of a subopercular branch of the preopercular sensory canal. The anterior pores within the splenial are considerably larger than those posteriorly. There is a single large dentary pore. The chief difference between the lower jaw of *C. suevicum* and the formerly recognized *C. nudum* is claimed to be the absence of ornament from the latter. However, the ornament upon the angular of *C. suevicum* is very sparse (Fig. 5.7) and at least one specimen of *C. nudum* (JME 1956.2) shows a few characteristic spicules of ornament. In sum, there is nothing in the lower jaw to refute the possibility that *C. nudum* is synonymous with *C. suevicum*.

Observations on the lower jaw of two further nominal species of *Coccoderma* are pertinent here. *Coccoderma gigas* is based on two syntype specimens showing the remains of large jaws (Reis 1888: pl. 3, figs 17–19). The differences between these jaws and those of *C. suevicum* and *C. nudum* are quite profound. In *C. gigas* the angular is long and low and bears granular ornament (BSM 1870-XIV-25). The principal coronoid is also long and low with the posterodorsal angle produced as a blunt process. The dentary (BSM 1870-XIV-26) is hook-shaped, unlike that of *C. suevicum*. BSM 1870-XIV-25 also shows an elongate dermal bone lying near the jaws. This element is poorly preserved, but the curved groove crossing the expanded end, which may be interpreted as a sensory canal, suggests that Reis (1888) was correct in identifying this as a preoperculum and it invites comparison with the elongate preoperculum of *Undina* species.

Reis (1888: 60) erected another poorly defined species, *C. bavaricum*, based on a single specimen (BSM 1870-XIV-24 a,b) which, amongst other remains, shows part of the lower jaw (Reis, 1888: pl. 5, fig. 2). The angu-

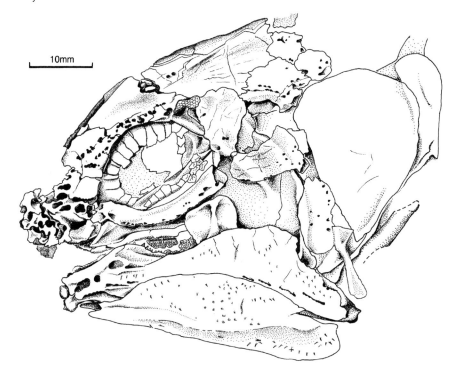

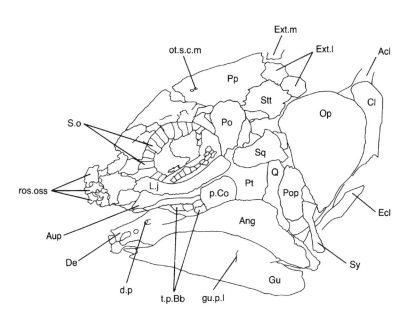

Fig. 5.7 *Coccoderma suevicum* Quenstedt. Camera lucida drawing of the head and shoulder girdle as preserved in BMNH P.8356.

lar is long and shallow, without the prominent dorsal expansion seen in the type species. The principal coronoid is also long and low and may be produced posterodorsally as a rounded process. The sensory canal opens by a regular series of very small openings. All of these features are unlike the corresponding areas in *C. suevicum* and *C. nudum* and show certain similarity with those in *Undina*.

Sassenia

The lower jaw of *Sassenia* (Figs 4.13, 5.8(B)) is very imperfectly known from the type species (*S. tuberculata* Stensiö, 1921). That of *S. groenlandica* is much better known and is described here on the basis of three specimens (MGUH VP2326, VP3258, VP2327a). In many respects the lower jaw of *Sassenia* is similar to the lower jaw of *Spermatodus*.

The jaw is long with an elongate but shallow coronoid expansion developed on the angular. The overlap surface for the lateral gular (o.Gu) is very narrow and, unlike *Spermatodus*, there is no pronounced groove beneath the sensory canal. The angular also appears to be of more constant thickness (cf. *Spermatodus*, below). The ornament on the angular is developed as a uniform covering of closely packed hemispherical tubercles which cover most of the angular above the line of the sensory canal. Ornament is absent from the dentary and splenial. There is no trace of an oral pit line in any of the specimens examined here.

The dentary and the splenial are both shallow. The dentary does not bear a hook-shaped process and the dentary sensory pore (MGUH VP2326) is rather small. The retroarticular is large and robust but the articular cannot be seen in available material and it is therefore impossible to say whether it is fused or separate.

The prearticular remains unknown in *S. groenlandica* but a very imperfect prearticular is known in *S. tuberculata* (Stensiö 1921: pl.

10, fig. 4 – intercoronoideo-prearticular). In the type species and in *S. groenlandica*, there is a modified fourth coronoid lying opposite the dentary. The principal coronoid is subrectangular and tilted anteriorly. The dentition is very poorly known, but where evident, it is composed of a covering of small villiform teeth.

The mandibular sensory canal opens through a series of many tiny pores in the angular and slightly larger pores in the splenial. At the posterior end of the angular, the path of the mandibular sensory canal is marked by two clusters of three pores, showing that the canal crossed the angular obliquely without sending off a subopercular branch.

Particular points of similarity with the lower jaw of *Spermatodus* include the overall length of the jaw compared with the length of the head, the similar shapes of the angular and the principal coronoid, the extreme shallowness of the jaw anterior to the coronoid expansion of the angular, and the pattern of tiny openings along the mandibular sensory canal.

Spermatodus

The lower jaw is very incompletely known (Fig. 5.8(A)). Previously, it has been briefly described by Westoll (1939) and the lack of suitable material means that only a little more information can be added. In several respects (see below) the lower jaw of *Spermatodus* is very similar to that of *Sassenia*.

The lower jaw is long, compared with that of most coelacanths, in keeping with the relatively posterior location of the jaw articulation (Westoll, 1939: fig. 2a). The angular (angulo-surangular of Westoll) is distinctively shallow except for a pronounced and well-rounded coronoid expansion which occurs midway along its length. The angular also varies considerably in thickness: the portion immediately surrounding the sensory canal is very thick, but the coronoid expansion and

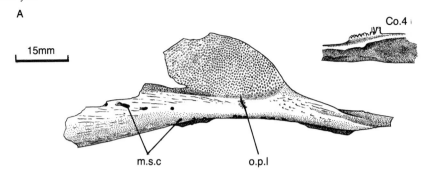

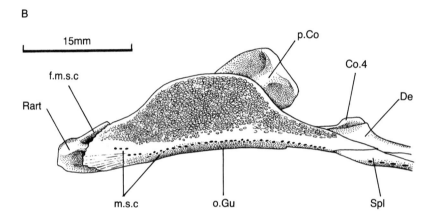

Fig. 5.8 (A) Angular and coronoid 4 (inset) of *Spermatodus pustulosus* Cope, lateral views of right elements, UM 16165. Camera lucida drawing. (B) *Sassenia groenlandica* sp. nov. Right lower jaw in lateral view, MGUH VP3258. Camera lucida drawing.

the ventromesial edge forming the overlap surface for the gular plate are both very thin. On the lateral face, there is a very deep groove immediately below the sensory canal.

The ornament upon the angular consists of many closely packed hemispherical tubercles as on other parts of the head. It is similar to that in *Sassenia* except that in *Spermatodus* the ornament ends abruptly at the ventral end of the coronoid expansion, and gives way to rugose bone (Fig. 5.8(A)). In *Sassenia* the ornament continues over most of the angular (Fig. 5.8(B)).

The posterior end of Meckel's cartilage is ossified as a very stout articulo-retroarticular (UM 10308) which is tightly sutured with the angular laterally and the prearticular medially. The dentary and splenial and the mentomeckelian remain unknown and the prearticular is known only from fragments (AMNH 4612, UM 1038, 11660). These fragments suggest that the prearticular must have been very shallow and elongate, and the dentition consists of a shagreen of irregularly sized teeth, each of which is marked by striations radiating from the crown. A similar type of tooth surface is seen in *Mawsonia* and *Axelro-*

dichthys. Towards the ventral margin and the posterior end of the prearticular, the dentition becomes still more irregular and several teeth appear to have grown together forming short tooth ridges.

The coronoid series is very robust in *Spermatodus*. The three anteriormost coronoids are represented as rectangular tooth plates each bearing 30–40 rounded teeth, each of which is marked with fine striations. The fourth coronoid (Westoll, 1939: fig. 1a Vo?) is elongate and modified (Fig. 5.8(A)). It is deeply grooved on its lateral face and the dentition is confined to a small raised area along the oral margin, where there is a gradation of large teeth situated anteriorly to small teeth posteriorly. The principal coronoid is very large, deeper than long and rectangular in outline. The articulated specimen examined here (AMNH 4612) suggests that, in life, this coronoid was inclined forward.

The path of the mandibular sensory canal is traced posteriorly as a line of four or five well-spaced pores traversing the angular from a level opposite the jaw articulation anteroventrally to the ventral edge of the angular beneath the coronoid expansion. Anterior to this level the mandibular canal opens by a series of tiny, closely spaced pores. There is no subopercular branch of the preopercular canal. The oral pit-line is well marked as a near-vertical groove lying just anterior to the greatest depth of the bone.

The lower jaw of *Spermatodus* is similar to that of *Sassenia*; the particular points of similarity are listed on p. 337.

Wimania

The lower jaw is very poorly known. A few aspects have been described and figured by Stensiö (1921) for the type species (*W. sinuosa*). The angular (supraangulo-angular) is long and low and the coronoid expansion is very weakly developed; the centre of ossi-

fication lies at the centre of the bone. The dentary is shallow with only a slight expansion posteriorly and there does not appear to be a hook-shaped process. The principal coronoid is triangular and large. The coronoid series is also well developed. Of the two elements labelled as precoronoids by Stensiö (1921: pl. 6, fig. 1), the posterior is probably coronoid 4 and this is low with a few slightly enlarged teeth; it may be coded as a modified coronoid 4. The anterior of the coronoid labelled by Stensiö is probably coronoid 3 and this bears a few large fangs also figured for *W. multistriata*.

Axelia

Stensiö (1921) provides the only description. The lower jaw is known only in *A. robusta* where the angular (supraangulo-angular) is particularly deep posteriorly and the centre of ossification lies anterior to this level (Stensiö, 1921: pl. 11, pl. 15, fig. 2). The dentary and splenial appear to be shallow but they are very incompletely known. The prearticular (intercoronoideo-prearticular) is of an unusual shape. It is deep at the level of the principal coronoid and tapers rapidly posteriorly.

The dentition on the prearticular and principal coronoid is formed by many rounded flat teeth and these bear fine striations radiating from the crown.

The mandibular sensory canal opens within the angular through five large pores (Stensiö, 1921: pl. 15) and at least one pore within the splenial. There may also be a dentary sensory pore present.

It is worth remarking here that the angulars of both *Mylacanthus lobatus* and *Scleracanthus asper*, which are two species very poorly distinguished from *Axelia robusta*, are similar in shape to that of *Axelia robusta* and that the mandibular sensory canal opens through five large pores. The dentition is also described as being similar in all three species.

Chinlea

The lower jaw is poorly known from the outer aspect. Several observations have been made by Schaeffer and Gregory (1961), Schaeffer (1967) and Elliott (1987). The angular is very similar in shape to that of *Whiteia*; that is, relatively shallow with a weakly developed coronoid expansion midway along the bone. The dentary bears a marked hook-shaped process and the narrow splenial is strongly inturned at the symphysis (Elliott, 1987). The outer dermal bones are marked by rugose ornament, similar to that in *Mawsonia* and *Axelrodichthys*, and there is no record of a pit line. Similarly, the condition of the sensory canal is unknown and this may mean that the openings are very small and obscured by the rugose ornament. There is a modified coronoid 4 opposite the dentary and this bears three or four prominent tusks.

Whiteia

The lower jaw of *Whiteia* (Figs 4.15, 5.9) was described by Moy-Thomas (1935b: figs 1, 3) and again by Lehman (1952: figs 4, 5, 10c, 13; pls 1, 2). Both descriptions contain errors and omissions and therefore a redescription is given below including information from a third species, *W. nielseni*, recognized in this work (p. 344). In all aspects of jaw anatomy the three species are very similar to one another.

The meckelian cartilage is ossified as three portions. Anteriorly the mentomeckelian is very small (Fig. 5.9(C,E)). Posteriorly there are separate articular and retroarticular bones which together make up the articular fossa. The retroarticular (Fig. 5.9(B)) bears a prominent lateral process which separates two foramina entering the overlying angular. Above there is the foramen for the entry of the mandibular sensory canal (f.m.s.c) and this means that the canal entered the jaw behind the articular facet and not in front as

shown by Lehman (1952: fig. 5). The more ventral foramen (f.VII.m.ext) allowed the external ramus of the mandibular VII to enter the jaw. In some specimens, such as MGUH VP2335a (*W. nielseni*), there is a small area posterodorsally which is devoid of perichondral bone. This represents the point of articulation with the symplectic (Fig. 4.15). Some specimens (e.g. BMNH P.17205 – *W. woodwardi*) also show that the ventral edge of the retroarticular is unfinished and presumably passed into cartilage as in embryo *Latimeria*.

The angular is a relatively shallow bone with no obvious coronoid expansion and bears only a narrow overlap for the gular plate. As in most coelacanths, the anterior end is pointed where it forms a complicated overlapping suture with both the dentary and the splenial. The external surface is marked by a vertical oral pit line at the centre of growth and lies immediately below the greatest depth of the angular.

The dentary and splenial are both well developed in *Whiteia*. The dentary, in particular, shows a large posterior embayed area with a rugose surface and this marks the position of insertion of the lip ligament. Thus, the dorsal surface of the dentary may be described as hook-shaped. The posterior end of the dentary reaches back considerably further in *W. woodwardi* than in *W. groenlandica* and probably *W. tuberculata*. The dentary is perforated by an oblique canal which marks the path of a nerve supplying the lower lip (Fig. 5.9(C), f.V.m). It is not definitely known if the dentary bore teeth: certainly they were not fused to the bone. Some specimens (e.g. BMNH P.16636) show very small tooth plates lying near the dentary and these may represent dentary tooth plates. The splenial is strongly recurved both medially and ventrally.

The prearticular is covered with a uniform shagreen of tiny villiform teeth. There are four anterior coronoids (Fig. 5.9(E)). The three lying in front of the prearticular are approximately equidimensional. Coronoid 4

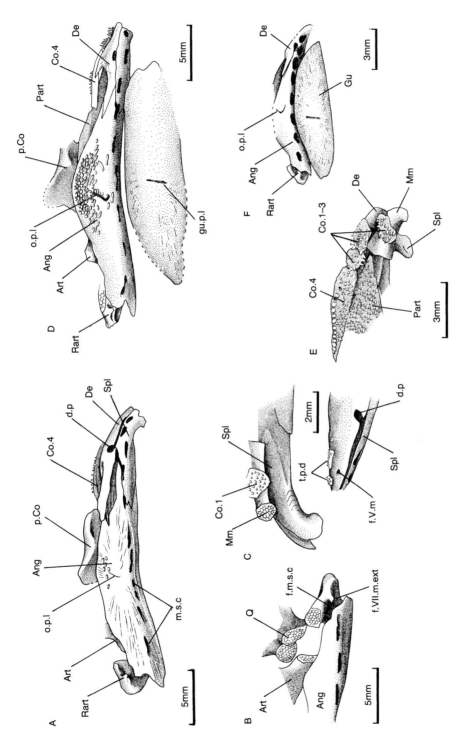

Fig. 5.9 *Whiteia* spp. lower jaw. (A–C) *Whiteia woodwardi* Moy-Thomas: (A) right lateral view of BMNH P.17169; (B) right lateral view of jaw joint of BMNH P.19501; (C) symphysial regions of left and right jaws of BMNH P.16636. (D and E) *Whiteia nielseni* sp. nov.: (D) right lower jaw and gular plate as preserved in MGUH VP2335a; (E) medial view of symphysis of MGUH VP2335a. (F) *Whiteia tuberculata* Moy-Thomas, right lower jaw and gular plate in lateral view as preserved in BMNH P.17214. All diagrams are camera lucida drawings.

within the anterior series is very elongate and shallow. Like the other coronoids it bears a uniform covering of small villiform teeth, but there is also a marginal series of slightly enlarged teeth (Fig. 5.9(E), Co.4). The principal coronoid is large, longer than deep and is saddle-shaped with a prominent dorsal thickening.

The mandibular sensory canal runs its usual course along the ventral edge of the angular and splenial. Within the angular it opens to the surface through six pores; occasionally one or two of those pores may be subdivided (P.17169). In *W. woodwardi* and *W. groenlandica* the pores are elongated. In *W. tuberculata* they are relatively larger and more rounded (cf. Fig. 5.9(A,D) with (F)). The splenial usually contains four pores. In two of the specimens of *W. woodwardi* examined here (P.17169 and P.17167) it appears as though the anterior two pores are out of line (Fig. 5.9(A)). This feature was also figured by Lehman (1952: fig. 10) but its significance is unclear. This feature does not appear to be developed within either *W. groenlandica* or *W. tuberculata*. The dentary sensory pore, omitted in earlier restorations, is well developed in *Whiteia*. It is unlikely that a subopercular sensory canal was present in *Whiteia*.

The three species considered here differ slightly in aspects of ornamentation of the lower jaw. *W. tuberculata* appears to be without any ornament, while ornament in the other species is confined to the angular. *W. groenlandica* shows a dense aggregation of tubercles located in the centre of the angular while the ornament of *W. woodwardi* is confined to widely spaced tubercles scattered around the oral pit line.

Diplurus

The lower jaw (Fig. 4.16(A)) is highly apomorphic in its shape. It has been described by Schaeffer (1952a) but there are a few points of emendation to be made.

The lower jaw is relatively short and shal-low with the quadrate/articular joint lying beneath the centre of the orbit. No mento-meckelian has been found and the posterior ossifications of Meckel's cartilage were described by Schaeffer (1952a: 44) as a single ossification (called the articular) fused with the angular and the prearticular. However, several specimens of *D. newarki* (PUGM 14943a, 14944; BMNH P.62104) show the articular/retroarticular clearly separate from the angular, although it is by no means clear if there are separate articular and retro-articular. A specimen of *D. longicaudatus* (PUGM 14955) shows separate articular and retroarticular and shows clearly that these lie free from both the angular and the prearti-cular. Because the observations are more reli-able for *D. longicaudatus* I have used this to score the condition of the articular and retro-articular for the genus (Table 9.1). The retro-articular is particularly short and there is no evidence of a symplectic articulation. In fact, a symplectic has not been described for *Diplurus*. However, considering the conserva-tism of coelacanths, it is unlikely to have been absent.

The angular (Fig. 4.16, Ang) is shallow and parallel-sided – an unusual shape for a coela-canth angular. It is smooth, without orna-ment or a pit line. The lower half of the bone is much inflated to contain the large mandi-bular sensory canal. This inflation is particu-larly marked on the inside of the dentary and exactly matches the area designated by Schaeffer (1952a: fig. 7B) as the prearticular. Anterior to the angular lie the splenial and dentary, arranged in typical coelacanth fash-ion. These have been well described by Schaeffer but it is worth noting that the dentary is narrow throughout, without a hook-shaped process, and markedly angled downwards at the level of the dentary sensory pore. The dentary is also strongly incurved at the point so that the two dentar-ies form a scoop-shaped profile to the lower jaw. At least one specimen (PUGM 14944) shows a group of fine needle-like teeth which lie medial to the dentary; these are tentatively identified as anterior coronoid teeth. None of the outer dermal bones of the jaw bears ornament. The principal coronoid may be seen in BMNH P.41632 as a very elongate and shallow bone barely protruding above the dorsal margin of the angular (Fig. 4.16). The prearticular is a separate element developed as in *Latimeria*. It bears tiny villiform teeth, developed within the dorsal half of the prearticular.

The mandibular sensory canal opens through four posteroventrally directed pores within the angular and through two pores within the splenial, in addition to opening through the offset dentary pore. There is no evidence of a separate subopercular branch of the preopercular sensory canal.

Ticinepomis

The lower jaw (Fig. 4.16(B)) was described by Rieppel (1980). As noted by that author, it is rather unusual in shape and shares similarities with the lower jaw of *Diplurus*.

The angular is long and low. The dentary is splint-like and angled at its mid length; a hook-shaped process is absent. The shape suggests that the symphysis was strongly inturned as in *Diplurus*. The splenial is much deeper than the dentary. A foramen between the dentary and the angular is probably the dentary sensory pore. Reippel (1980: fig. 3) describes and figures two anterior coronoid tooth plates and these bear stout teeth (not seen in *Diplurus*). The principal coronoid is small and triangular.

Mawsonia and *Axelrodichthys*

The lower jaw is so similar in species of these two genera that they can be described together. The complete jaw morphology is best known for *Axelrodichthys araripensis* (Figs 4.17, 5.10).

The meckelian cartilage is ossified as a small mentomeckelian (unknown for species of *Mawsonia*) and separate articular and retroarticular ossifications which share in the formation of a particularly deep glenoid articulation with the quadrate (Fig. 5.10(B)).

The angular is differently shaped in *Axelrodichthys* and *Mawsonia* and there are claimed to be differences in the shape of the angular between species of *Mawsonia* (but see below). In *Axelrodichthys* the angular reaches its deepest point near to the anterior margin of the bone and the suture with the principal coronoid lies well behind this level. In *Mawsonia*, the deepest point of the angular lies approximately midway along the length and at this level there is a marked anterodorsally directed process which sutures with the principal coronoid. In *Mawsonia* there is a long overlap surface with the dentary; in *Axelrodichthys* this is much shorter. In both genera the centre of ossification lies well behind the level of the greatest depth (Figs 5.10, 5.11). The angular is much inflated where the mandibular sensory canal passes through the bone. Subtle differences in the shapes of the angulars of different species of *Mawsonia* were noted (Tabaste, 1963) but, as Wenz (1975) argues, these differences blend into one another when larger samples are taken. For instance, the dorsal process of *Mawsonia tegamensis* was originally described from a small specimen as being ill defined (Wenz, 1975: fig. 6) but a larger specimen described later (Wenz, 1981: pl. 2C) shows a prominent process. Angulars of four species of *Mawsonia* are illustrated in Fig. 5.11. The inner surface of the angular shows prominent ridges which delimit the adductor fossa and the path taken by the external mandibular VII (Fig. 5.10). These ridges are developed exactly the same as in *Macropoma* and probably reflect general conditions in coelacanths.

The dentary in *Mawsonia* and *Axelrodichthys* is characteristic. It is markedly incurved anteriorly, it bears a well-developed, hook-shaped process, the depression

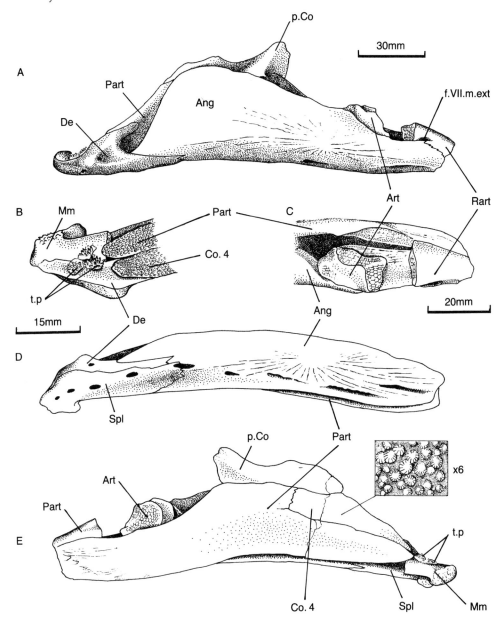

Fig. 5.10 *Axelrodichthys araripensis* Maisey. Lower jaw; camera lucida drawings of FMNH PF10726, a left lower jaw. (A) Lateral view. (B) Dorsal view of extreme anterior end showing the lateral swelling on the dentary. (C) Dorsal view of articular region. (D) Ventral view. (E) Medial view with inset showing form of teeth covering coronoid 4. Note the overlap of coronoid 4 with the principal coronoid, which is a derived feature.

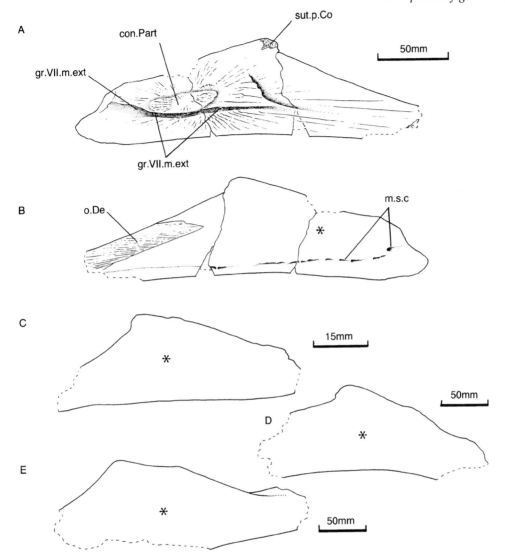

Fig. 5.11 *Mawsonia* angulars to show subtle differences in shapes between species. Asterisk shows centre of ossification. (A) *Mawsonia gigas* Woodward, left angular medial view, BMNH P.10360. (B) *Mawsonia gigas* Woodward, left angular lateral view, BMNH P.10360. (C) *Mawsonia tegamensis* Wenz, medial view outline, from Wenz (1981). (D) *Mawsonia lavocati* Tabaste, lateral view outline, from Tabaste (1963). (E) *Mawsonia libyca* Weiler, lateral view outline, from Weiler (1935).

for the pseudomaxillary fold is deep, and immediately in front of this level the dentary bears a lateral swelling (Fig. 5.10) which is not found in other coelacanths except *Luala-baea*. The posterior extension of the dentary

of *Mawsonia* is relatively longer (BMNH P.10363) than that of *Axelrodichthys*.

The splenial of *Axelrodichthys* is only visible in a ventral view of the jaw (Fig. 5.10(D)) and is shallow. The splenial of *Mawsonia* has not certainly been identified but a small isolated fragment (BMNH P.7120) may be a splenial of *M. gigas* (it is associated with clearly identifiable remains of this species). This specimen is very similar to the splenial of *Axelrodichthys*.

The prearticular reaches nearly the entire length of the jaw and is covered with tiny hemispherical teeth, each marked with fine radiating striations. The coronoid series is unusual in at least *Axelrodichthys* (it remains very poorly known in *Mawsonia*). A large coronoid (Fig. 5.10, Co.4) which, in position, corresponds to coronoid 4 of other coelacanths is sutured with the prearticular. It reaches from the anterior end of the prearticular posteriorly to the midpoint of the principal coronoid. In most other coelacanths there is a distinct gap between coronoid 4 and the principal coronoid. One other exception may be found in *Macropoma*, where coronoid 4 appears to be fused with the prearticular, in which case the true limits may not be seen. The dentition upon this coronoid lies in continuity with the prearticular dentition, there are no enlarged teeth. Anterior to the prearticular and coronoid 4 there are three tiny tooth plates which carry small needle-like teeth. These needle-like teeth may also be seen associated with the isolated splenial of *Mawsonia* mentioned above. These small tooth plates may be the anterior three coronoids and they are considered as such here, but it remains possible that they may be dentary tooth plates.

The principal coronoid of both *Mawsonia* and *Axelrodichthys* is long, low and saddle-shaped (Woodward, 1907). The posterior end rises higher than the anterior end. There is a clear overlap surface along the ventral edge receiving the prearticular and coronoid 4. The dorsal edge of the lateral surface is rolled over. In *Mawsonia* this has a small sutural contact surface with the dorsal process of the angular. In *Axelrodichthys* the principal coronoid is produced as a stout process which sutures with the angular (Fig. 4.17). The sutural contact between the principal coronoid and the angular is a synapomorphy of these two genera.

Ornament is absent from the dermal bones but the angular is marked by rugosities. In some species of *Mawsonia*, especially *M. ubangiana* and *M. lavocati* this is particularly coarse and has been used as a point of specific distinction (Tabaste, 1963).

The mandibular sensory canal enters the jaw immediately behind the quadrate/articular joint and from a relatively small opening it expands rapidly to a large lumen. The main canal of *Axelrodichthys* opens ventrally to the surface through five large openings (Fig. 5.10(D)). The posterior three or four are long and slit-like and may be multiple openings. The anterior openings, as well as the four or five on the splenial, are simple pores. The openings are very similar in *M. lavocati* (Tabaste 1963: pl. 13, fig. 5). However, in *M. tegamensis* and *M. gigas* there appear to be many small openings (Wenz, 1975: fig. 6).

Lualabaea

The lower jaw of *Lualabaea lerichei* is strikingly like that of *Mawsonia* and *Axelrodichthys* and has been illustrated by Saint-Seine (1955: pls 1 and 2). Particular points of similarity are the shape of the angular (spléniosurangular), which has a small anterodorsally directed process approximately midway along its length as in *Mawsonia*. The dentary bears a prominent hook-shaped process and extends posteriorly as a splint-like process developed to about the same degree as in *M. gigas*, and there is a prominent dentary swelling. As in both *Mawsonia* and *Axelrodichthys*, the principal coronoid is saddle-shaped although it is unclear if it is sutured with the angular. As in some species of *Mawsonia* and in *Axelrodichthys*, the mandibular sensory canal opens within the angular through approximately five slit-like pores. The angular is marked by coarse rugosities which radiate from a point posterior to the deepest part of the bone.

Indocoelacanthus

Jain (1974) has described the lower jaw. Features of relevance to the characters recognized here are the relatively long angular with a weakly developed coronoid expansion; the short splenial and dentary which are of approximately equal length; a hook-shaped process on the dentary; ornamentation present on the splenial and the angular but absent from the dentary. The principal coronoid is described by Jain (1974) as being small but it is clearly incomplete in the only specimen (the holotype) showing the jaw.

Libys

The lower jaws of the two species of *Libys* are identical, as far as known. Some aspects of the jaw have been described by Reis (1888); much remains unknown.

The angular and the principal coronoid are each distinctively shaped in *Libys*. The angular rises to a rounded coronoid expansion and behind this level the dorsal margin is slightly concave and matches exactly the ventral margin of the preoperculum. This is similar to conditions in *Macropoma*, *Holophagus* and *Undina*. The angular remains relatively deep at the level of the jaw joint. In keeping with the rest of the sensory canals, the mandibular canal opens through several very large pores (Reis, 1888: pl. 2, fig. 2) and there is a subopercular branch developed.

The principal coronoid has a broad base, a steeply inclined anterior margin and a shallow excavated posterior margin. It is produced posterodorsally as a finger-like process and the overall shape is similar to the principal coronoid of *Macropoma*. The dentary and splenial remain incompletely known but one specimen of *L. superbus* (Reis, 1888: pl. 2, fig. 2) shows that the dentary bears a hook-shaped process. The anterior

coronoid series consists of three small quadrangular tooth plates followed by an elongated fourth coronoid sutured firmly with the prearticular. The teeth upon the coronoids are small and rounded. Those upon the anterior end of the prearticular are sightly elongated anteroposteriorly. The dentition within the rear half of the jaw remains unknown and details of ossification within Meckel's cartilage are unknown.

Undina and *Holophagus*

The lower jaws belonging to species of these two genera are so similar that a combined description is appropriate. The lower jaw of *Undina penicillata* has been described by Reis (1888) and Aldinger (1930), that of *Undina cirinensis* by Saint-Seine (1949), while those of *Holophagus gulo*, *U.* (?) *barroviensis* and *U. purbeckensis* remain to be described. The following description is based primarily upon *Holophagus gulo* (Fig. 5.12(A,B)) and *Undina penicillata* (Fig. 5.12(C)).

The jaw is long and in the reconstructed head would reach back to the level of the occiput. It is shallow throughout. The mentomeckelian is short (known only in *U. penicillata* – BMNH P.10884) and there are separate articular and retroarticular elements. The angular is particularly shallow, with an ill-defined coronoid expansion, and the greatest depth occurs immediately in front of the articular, giving the angular a distinctively elongate shape among coelacanths. The ventral margin of the angular in *H. gulo* is deeply emarginated and roughened for the insertion of fibres of the posterior intermandibularis muscles (cf. *Latimeria*). This area of the angular of *U. penicillata* is poorly known. An oral pit line marks the angular at the centre of ossification and this lies well posteriorly, particularly so in *U. penicillata*.

The dentary is narrow and relatively long, equal to approximately 40% of the total jaw length. There is a prominent hook-shaped process upon the dentary. The splenial is also

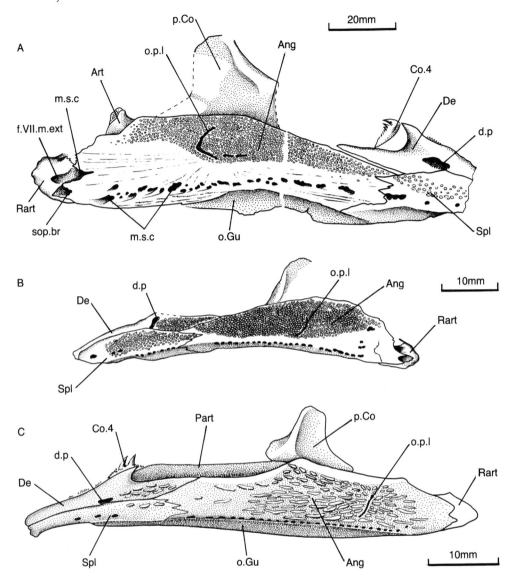

Fig. 5.12 *Holophagus* and *Undina*, lower jaws. A. *Holophagus gulo* Egerton, left lateral view as preserved in BMNH P.7795, camera lucida drawing. (B) *Holophagus gulo* Egerton, right lateral view as preserved in BMNH P.2022. (C) *Undina penicillata* Münster, reconstruction in left lateral view, based on BMNH 37032, P.7182, P.10884, Munich 1870.XIV.516, 1870.XIV.517.

narrow in *U. penicillata* but in *H. gulo* it is substantially deeper than the dentary (Fig. 5.12(B)). This appears to be the case also in *U. cirinensis* (Saint-Seine, 1949: fig. 34).

The coronoid series is very poorly known. Both *H. gulo* and *U. penicillata* show a modified fourth coronoid, bearing one or two enlarged teeth, lying opposite the hook process upon the dentary. Details of the more anterior coronoids are very unclear although several specimens of *U. penicillata* and at least one specimen of *H. gulo* (BMNH P.7795) show patches of tiny villiform teeth lying medial to the dentary. The principal coronoid is best seen in *U. penicillata* (Reis, 1888: pl. 1, fig. 21). It is broad ventrally and developed as a narrow dorsal process. The overall shape is similar to that in *Libys* and *Macropoma*. The prearticular is known only in *U. penicillata* and is, as usual, long and shallow. The prearticular dentition is weakly developed as tiny villiform teeth near the dorsal margin and with teeth arranged in rows towards the ventral margin.

The path of the mandibular sensory canal can be traced as a series of tiny pores which are arranged regularly along the ventral edge of the angular. The pores continue into the splenial where, in *H. gulo*, the anterior pores are larger. The anterior end of the canal remains unknown in *U. penicillata*. Posteriorly, the canal turns dorsally to leave the angular opposite the glenoid articulation. In *H. gulo* there is a large pore (Fig. 5.12(A), sop.br) lying immediately beneath the foramen for the external ramus of mandibular VII. This pore provides evidence of a subopercular branch of the preopercular sensory canal. This has not been seen in *U. penicillata* but this area of all specimens examined here is very badly preserved. A large dentary pore is present in both *U. penicillata* and *H. gulo*.

Ornament is present on the angular, dentary and splenial although it is developed differently in the two species. In *H. gulo* it consists of many tiny rounded tubercles and the ornament cover extends up to the dorsal margin of the angular. In *U. penicillata* the ornament is of sparse, elongate tubercles. There is a smooth, unornamented area along the dorsal edge of the angular and this is the overlap area for the elongate preoperculum. This overlap area appears to be absent from *H. gulo* even though the preoperculum is of similar shape.

A comment about *Coccoderma bavaricum* Reis is appropriate here because the shape of the angular, the principal coronoid and the pattern and distribution of the mandibular sensory pores are all very similar to that seen in *U. penicillata* (see also p. 347).

Macropoma

The lower jaw of *M. lewesiensis* and *M. precursor* has been described previously by Huxley (1866), Woodward (1909) and Watson (1921). That of *M. speciosum* has been briefly mentioned by Tima (1986). These authors agree on the morphology of the jaw although bone terminology varies. The acid-prepared material used here allows a far more detailed description to be made and allows clarification of earlier descriptions.

Restorations of the lower jaw and constituent bones of *M. lewesiensis* are shown in Figs 4.19 and 5.13 and based on several specimens (BMNH 4237, 28388, 39070, 49094, 49833, 49836, 43851, 49887, P.742, P.9114, P.12886, P.33373, P.33540). Several specimens of *M. precursor* (BMNH 35700, P.3353, P.6453, P.10916, P.10917, P.11002) show that, except for details of ornamentation, the structure of the lower jaw is very similar to that of the type species.

Meckel's cartilage is ossified as three separate bones. Anteriorly there is a very small mentomeckelian. It is laterally compressed, curved markedly inwards anteriorly and becomes expanded, forming the mandibular symphysis. Posteriorly there are separate articular and retroarticular ossifications. The articular is small and rests upon a swelling on the inside of the angular. Both Huxley (1866: 38) and Woodward (1909: 176) considered that the angular and articular were fused. This is not the case, although it is true that the suture between the bones can be traced for only a short distance (e.g. BMNH P.33540). The articulatory facet is, as usual, a double concave facet. Both parts of the facet are inclined anteromedially. The retro-

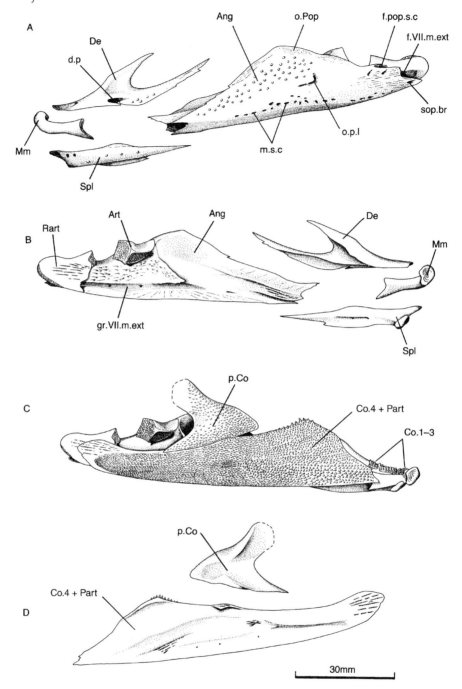

Fig. 5.13 *Macropoma lewesiensis* (Mantell). Restoration of the lower jaw with bones pulled slightly apart to show nature of sutural overlap. Based mainly on BMNH P.33540. (A) Lateral view, showing endochondral bones plus outer dermal bones. (B) Medial view of same. (C) Medial view of inner dermal bones. (D) Mesial surface of inner dermal bones. For a restoration of the lateral surface of the jaw see Fig. 4.19.

articular is narrow and anteriorly, ventrally and posteriorly it lacks perichondral bone and must therefore have been capped with cartilage as in *Latimeria*. The anterior face could only have contributed little to the articular facet. The dorsal edge of the retroarticular bears a small facet for articulation with the symplectic and this too must have been cartilage capped. Longitudinal ridges and grooves mark the mesial face of the retroarticular (Fig. 5.13(B), Rart) and this roughened area is the overlap surface of the prearticular. In *Latimeria* the internal ramus of mandibular VII penetrates the mandible between the retroarticular and the prearticular, and there is a distinct foramen between these bones. I have been unable to find a comparable foramen in *Macropoma*. However, there is a distinct groove on the inside of the prearticular (Fig. 5.13(D)) which marks the path of this nerve within the jaw.

The dentary and splenial are both slender. The dentary is deeply embayed posteriorly and bears a hook-shaped process which received the lip fold. There is a long overlapping suture with the angular. The dentary sensory pore is large and lies along the suture with the splenial. Jarvik (1947: fig. 5(B)) interpreted this foramen to carry nerves and vessels but it was almost certainly for the sensory canal as it is connected directly with the main mandibular canal.

The angular (supra-angulo-angular of Stensiö 1932) is strongly curved ventrally, with a large and well marked overlap area for the gular plate. The oral pit line is 'L'-shaped and lies slightly behind the deepest point of the angular. Along the dorsal edge of the angular, there are three areas devoid of ornament. The anterior probably received the labial fold of skin while the middle and posterior area was overlapped by the preoperculum.

The points of entry of the mandibular sensory canal and the external ramus of mandibular VII are seen as two notches, corresponding with grooves upon the outer face of the retroarticular. As usual, the mandibular sensory canal entered the angular immediately behind the level of the articulatory facet. At this point the surface of the angular is marked by two or three distinct pits directed anteroventrally. They may have contained minor bands of connective tissue. They are absent from *Latimeria*.

The notch for the external ramus of mandibular VII is well marked and the margin of the angular at this point is roughened where it must have received part of the retroarticular–hyomandibular ligament. The internal face of the angular is considerably thickened at the level of the articular and there is a prominent medial process for contact with the prearticular. Anteriorly, there is another contact area for the prearticular marked by ridges and grooves.

The prearticular, as usual, covers most of the inside of the jaw. It is deepest within the anterior third where it rises to a crest opposite the embayment of the dentary. In *Latimeria* (Fig. 5.1(B)) and most other coelacanths this crest is represented as a separate coronoid (the fourth in the anterior series). It is possible that the conditions in *Macropoma* represent an adult fused condition and that younger individuals would show the usual separate coronoid 4. Anterior to the prearticular there are three anterior coronoids developed as small tooth plates. Dentary tooth plates have not been found. The principal coronoid is large and distinctively shaped. It has a broad elongate base and a relatively narrow, posterodorsally directed process which ends in a rounded tip. The dentition on the coronoids and the prearticular consists of tiny villiform teeth arranged in a regular shagreen. The teeth are of roughly equal size and some are marked with fine radiating ridges.

The mandibular sensory canal is a wide canal opening to the surface through many small openings in both angular and splenial. The anterior two pores within the splenial are enlarged. As in *Latimeria*, there is a subopercular branch of the preopercular canal.

This branch leaves the angular immediately beneath the opening for the external ramus of mandibular V.

In *M. lewesiensis* and *M. speciosum* the ornamentation of the jaw is present as rounded tubercles extending over the central area of the angular, the splenial and a narrow ventral area of the dentary. There is no ornament upon the lower jaw of *M. precursor*.

5.3 DISCUSSION OF LOWER JAW

Within coelacanths the jaw anatomy varies considerably and some of this variation has been used by authors in constructing classifications (Forey, 1981, 1991; Lund and Lund, 1985; Cloutier, 1991a,b). Cloutier (1991b) has used two characters considered here, numbered 57 and 67 below. We disagree on some of the codings for character 57 for some taxa and this is pointed out when discussing that character. Cloutier (1991a,b) also used two further characters which are not considered here. One of these is the shape of the principal coronoid which he considered to be either 'sub-triangular' or 'sub-quadrilateral'. In some cases (e.g. *Rhabdoderma*, Fig. 5.3) it is clearly triangular whilst in others it is clearly quadrilateral (e.g. *Laugia*, Fig. 5.5). But in many coelacanths it is difficult to assign the principal coronoid to one or other of the specified shapes (e.g. *Axelrodichthys*, Fig. 5.12(C), *Undina*, Fig. 5.10C or *Macropoma*, Fig. 5.13). Indeed Cloutier (1991b, appendix 1) found this difficult by coding the condition in *Axelrodichthys* as 'unclear'. However, it may be significant that Cloutier was dealing with Palaeozoic coelacanths in which the shapes of the principal coronoids more clearly fall into one description or another. The other character used by Cloutier (1991a,b) and Lund and Lund (1985) concerns the orientation of the dentary relative to the occlusal margin of the jaw. In some coelacanths the oral margin of the dentary lies in perfect continuity with the angular behind (e.g. *Lochmocercus*). In others

the dentary is orientated diagonally with respect to the angular and "results in a concave lateral profile of the lower jaw" (Cloutier, 1991a: 407). I agree that the offset dentary is a derived feature but, in several cases, have found difficulty in describing particular species as having one or other condition, and in some cases disagree with the specific codings that might be ascribed to certain taxa. For example Cloutier (1991a) describes the dentary of *Allenypterus* as being 'straight' while that of *Diplocercides kayseri* is described as 'diagonal'. Yet, the orientation of these two is rather similar and, indeed, it may be said that the dentary of *Allenypterus* is even more angled. And those coelacanths with hook-shaped dentary must also show a concave dorsal margin (but see Cloutier, 1991b, codings for *Chinlea sorenseni*). In view of these difficulties I have not used this character.

Lower jaw characters

The jaw characters used in this classification of coelacanths are listed below and are coded for taxa in the combined data matrix (Table 9.1): in most instances the derived condition is coded as 1, 2 etc. and given in parentheses.

53. Retroarticular and articular co-ossified (0), separate (1). The posterior end of Meckel's cartilage primitively ossifies from two centres in bony fishes, articular and retroarticular (Nelson, 1973). In *Latimeria* and many fossil coelacanths, these ossifications remain separate, while in others they fuse to form a compound articular–retroarticular. It is assumed here that a single ossification is the plesiomorphic adult condition, even though it may develop from two centres and be represented as two bones in young individuals (e.g. *Laugia*). Persistence of two centres into the adult is regarded as derived. It is recognized that

this can provide some difficulty in coding for certain taxa, particularly those known only from few specimens, because there is no guarantee that they are adults.

54. Dentary teeth fused to dentary (0), separate from dentary (1). Teeth firmly attached to the supporting dentary are here regarded as a plesiomorphic feature of adult bony fishes because this is the common condition in outgroups. An edentulous dentary or a dentary bearing separate tooth plates as the adult condition is regarded as derived. In most fossil coelacanths it is impossible to distinguish between an edentulous dentary and one having separate tooth plates because the tooth plates drift away during fossilization. There may have been edentulous coelacanths (e.g. *Diplurus*).

55. Number of coronoids, coded as integers. The coronoid series of coelacanths is divisible into an anterior series of three or four elements lying medial to the dentary and separated by a diastema from the highly distinctive principal coronoid which is located immediately anterior to the jaw articulation. There may be either three or four coronoids in the anterior series. Which of these numbers is regarded as the plesiomorphic condition is uncertain. Actinopterygians have a variable number, but primitive members of this group (*Mimia* and *Moythomasia*) show four while osteolepiforms, porolepiforms and tetrapods consistently show three plus an adsymphysial plate which is regarded here as a modified coronoid. Here the number of anterior coronoids is recorded.

56. Coronoid opposite posterior end of dentary not modified (0), modified (1). Within the anterior series some coelacanths show an enlarged coronoid, usually the fourth coronoid closely associated with the dentary, bearing enlarged teeth and often with its lateral edge rolled over (e.g. *Spermatodus*). This is here regarded as a derived condition and, in all probability, is associated with the development of enlarged teeth on the opposing dermopalatine and ectopterygoid.

57. Dentary simple (0), dentary hook-shaped (1). The dentary of all coelacanths is a relatively small element, reaching only a small distance along the jaw margin. As in other bony fishes, it is regarded as being primitively shallow. The hook-shaped posterior end of the dentary, as seen in many coelacanths, is regarded as a derived condition (1) and is no doubt associated with the elaboration of the muscular lip fold as seen in *Latimeria* and which inserts at this point.

It was pointed out above that Cloutier (1991b) has also used this character and that our codings differ slightly. The particular points are that Cloutier claims that the dentary in *Diplurus* and *Ticinepomis* is hook-shaped whereas the dentary of *Whiteia* is not. My codings are the reverse and I have justified these in the relevant descriptions and illustrations.

58. Oral pit line confined to angular (0), oral pit line reaching forward to the dentary and/or the splenial (1). The plesiomorphic condition appears to be the presence of a short pit line on the angular with no extension on to the dentary or splenial. And this appears to be the usual condition in coelacanths. Exceptionally, in the primitive actinopterygian *Mimia* (Gardiner, 1984a: fig. 90) there may be a separate dentary pit-line. But a long pit line extending across both bones is here regarded as derived.

59. Oral pit line located at centre of ossification of angular (0), removed from centre of ossification (1). The oral pit line is usually located at the centre of ossifica-

tion of the angular and, in turn, this is usually the deepest point of the angular. Thus, the location of the pit line away from the deepest point is considered derived (1).

60. Subopercular branch of the mandibular sensory canal absent (0), present (1). This feature is developed in some coelacanths where the branch leaves the mandibular canal within the angular and sweeps back over the subopercular flap of skin (see *Latimeria*, Fig. 4.3). Its presence is regarded as derived.

61. Dentary sensory pore absent (0), present (1). This pore, which is regarded as a derived feature (1), is located near the dentary/splenial suture. It is located above the trajectory of the mandibular sensory canal but is connected with the main canal and usually is substantially larger than the pores leading from the main line of the mandibular sensory canal. Some porolepiforms, *Youngolepis*, *Powichthys* and the osteolepiform *Gogonasus* (Fox *et al.*, 1995) have a series of three large infradentary foramina within the lower jaw and these are also located above the trajectory of the mandibular sensory canal. Because coelacanths have only the anterior two infradentaries, it remains possible that the single pore in these fishes is homologous with the most anterior of the infradentary foramina in porolepiforms etc. The presumed primitive member of the porolepiforms (e.g. *Porolepis*) does not show these enlarged pores. Ahlberg (1991) concluded that the presence of infradentary foramina is a synapomorphy of *Youngolepis*, *Powichthys* and porolepiforms and that the absence of such foramina from *Porolepis* is secondary. The presence of a single foramen in some coelacanths might reinforce that view. But it should be pointed out that the single pore in coelacanths occurs between the splenial and the dentary and not between infradentary

bones. Also, the infradentary foramina in porolepiforms, *Youngolepis* and *Powichthys* do not connect with the mandibular sensory canal (Ahlberg, pers. comm.). For these reasons I consider the presence of a dentary pore to be derived within coelacanths.

62. Ridged (0) or granular ornament (1). The form of the ornament varies considerably within coelacanths, although there do appear to be two basic types. In some the ornament is arranged in ridges of enamel which usually anastomose. In others the ornament is made exclusively of enamel-capped tubercles (1). Within the lower jaw it is of one type or another. It is difficult to be certain of the polarity of this character but I will accept the ridged ornament as plesiomorphic based on the presence of this kind of ornament in primitive actinopterygians (Gardiner, 1984a). Primitive lungfishes, 'osteolepiforms' and porolepiforms show a continuous covering of cosmine and thus are uninformative about the polarity of this character. In coelacanths which show no ornament (the bone is either smooth or rugose, e.g. *Latimeria*), this character is scored as 'not applicable'.

63. Dentary with ornament (0), without ornament (1). It is assumed here that a complete ornament cover on the outer dermal jaw bones is primitive. This is justified by the observation that these bones are completely covered with cosmine in primitive lungfishes, 'osteolepiforms' and porolepiforms. Absence of ornament from the dentary is derived and may mean that the lip fold was more extensive in these coelacanths.

64. Splenial with ornament (0), without ornament (1). See arguments used for the previous character. Absence of ornament is regarded as derived.

65. Dentary without prominent lateral swelling (0), with swelling (1).

66. Principal coronoid lying free (0), sutured to angular (1). In most coelacanths the principal coronoid lies free from the angular although there is, in *Latimeria*, a strong ligament which runs from the dorsal tip of the principal coronoid to the dorsal edge of the angular (Millot and Anthony, 1958: fig. 19, *lig. cor. a.*) and presumably was present in many fossil coelacanths. An elaboration of this support is provided by sutural connection between these two bones (1).

67. Coronoid fangs absent (0), present (1). In most coelacanths the coronoid series bears a shagreen of many tiny villiform teeth and this is regarded as the plesiomorphic condition, taken from the condition in primitive actinopterygians. The presence of enlarged teeth lying alongside a replacement socket upon at least some of the anterior coronoids is accepted as the derived condition.

68. Prearticular and/or coronoid teeth pointed and smooth (0), rounded and marked with fine striations radiating from the crown (1). This character implies a crushing dentition and is considered derived relative to the grasping and holding type of dentition signified by small villiform smooth teeth.

6

NEUROCRANIUM, PARASPHENOID AND VOMER

6.1 INTRODUCTION

In this chapter, the endochondral ossifications of the neurocranium are described and discussed together with the dermal parasphenoid and vomers, which are associated with the ethmosphenoid portion.

The coelacanth neurocranium is completely divided into ethmosphenoid and otico-occipital portions which articulate one with the other through the intracranial joint. The joint is much better developed in coelacanths than in other sarcopterygian fishes reflecting great mobility at this point of the skull (p. 202). The ethmosphenoid encloses the nasal sacs, the median rostral organ and the eyes. The otico-occipital portion encloses the brain and the labyrinth. In *Latimeria* the brain is confined to the otico-occipital region and thus lies wholly behind the joint. The extent of the brain in the extinct taxa may have been different (p. 171). As in other sarcopterygian fishes and presumably primitive tetrapods, the notochord extended through the otico-occipital region unconstricted, to end at the level of the ventral fissure, where it rested against the posterior surface of the basisphenoid within a shallow notochordal pit. As it passes through the otic region, the notochord is surrounded dorsally by one anazygal, which articulates with the sphenoid condyles on the basisphenoid, and ventrally by two catazygals which lie free from each other and from other neurocranial bones and occupy the basicranial fenestra.

6.2 DESCRIPTIONS OF GENERA

Latimeria

The neurocranium with the parasphenoid and the paths of cranial nerves of *Latimeria* has been described and illustrated on several occasions (Smith, 1939c, Millot and Anthony, 1958, 1965; Bjerring, 1967, 1971, 1972, 1973; Bemis and Northcutt, 1991; Northcutt and Bemis, 1993). With respect to the understanding of fossil neurocrania, these published descriptions give sufficient information and will not be duplicated here. However, I have provided illustrations (Fig. 6.1), modified after Millot and Anthony (1958) to exemplify landmark features and give a basis for comparison between the extinct and modern species. Only a few features need to be emphasized.

Much of the neurocranium remains cartilaginous, more so than in most fossil coelacanths. In the ethmosphenoid there are paired lateral ethmoids (ectethmoids) and a large median basisphenoid. Each lateral ethmoid (L.e) forms the floor and part of the lateral wall of the nasal capsule. It passes into cartilage both dorsally and posteriorly. The lateral wall is marked by a triangular depression – the ventrolateral fossa (v.l.fo) – which receives the autopalatine. Anterior to this there is a swelling which articulates with the ventral process of the lateral rostral. The anterior and posterior nostrils mark the lateral ethmoid as notches along the dorsal edge of the lateral wall, and between these two notches there is a shallow excavation for the buccal foramen. All three foramina are closed dorsally by cartilage. The endoskeletal openings leading into and out of the rostral organ lie entirely within cartilage.

The basisphenoid is relatively small, compared with that in many other coelacanths. Prominent antotic processes (ant.pr) and the paired processus connectens (pr.con), which are capped by cartilage, are developed to articulate with the metapterygoid and the otic shelf of the prootic respectively. Paired sphenoid condyles, which articulate with the anazygal, are found medial to the posterior ends of the processus connectens and dorsal to the notochordal pit. Immediately in front of the antotic process on the lateral face of the basisphenoid there is a deep suprapterygoid fossa (spt.fos) which is continued onto the ethmosphenoid cartilage. This fossa is the site of origin for the adductor palatini muscle. The lateral face of the basisphenoid is also marked by a deep groove above the processus connectens and this carries the jugular vein. In front of this there is a short, poorly developed ridge, marking the site of origin for the superior and the external rectus eye muscles. The inferior and internal rectus muscles originate on the ethmosphenoid cartilage immediately anterior to the foramen for the internal carotid artery. The superior and inferior oblique muscles originate on the ethmosphenoid cartilage between the orbitonasal canal and the anterior opening for the profundus (Fig. 6.1(A), V1) at the point where that nerve pierces the postnasal wall. In *Latimeria* the basisphenoid is pierced by the profundus and oculomotor nerves, while the superficial ophthalmic and trochlear nerves loop over the antotic process and pierce the descending process of the parietal before piercing the ethmosphenoid cartilage.

The parasphenoid (Par) underlies the basisphenoid, ethmosphenoid cartilage and the posterior part of the lateral ethmoids. The area covered by teeth is restricted to the anterior third of the parasphenoid and this reflects the forward insertion of the basicranial muscle (Millot and Anthony, 1958; Bjerring, 1967). The tendons from the basicranial muscle reach forward to insert on the lateral face of the parasphenoid, ending at a level where the toothed area begins. Anteriorly there are small paired vomers which lie beneath the lateral ethmoids and just meet in the midline. The vomers are rarely preserved in fossil coelacanths, having presumably a loose connection with the lateral ethmoids.

The otico-occipital portion of the neurocranium is mostly cartilaginous. The paired prootic is the largest ossification. It forms the lateral commissure, which carries a large cartilage-capped double hyomandibular facet (f.Hy) which is orientated vertically; a prominent otic shelf (ot.sh) which reaches forward to form the groove component of the track-and-groove intracranial joint, the track being the processus connectens of the basisphenoid; the anterior wall of the otic capsule; and is developed posteriorly and ventrally as a wing which forms part of the lateral wall of the notochordal canal as well as suturing with the basioccipital.

The basioccipital (Boc) is largely embedded in cartilage and consists of a narrow 'U'-shaped ossification underlying the notochord. Above the notochord there are small, paired opisthotics (exoccipitals), each of which is pierced anteriorly by the glossopharyngeal foramen (IX). The supraoccipital is a median ossification embedded within cartilage above the foramen magnum.

In *Latimeria* the anterior end of the prootic is capped with cartilage. The neurocranium is securely attached to the roofing bones through the descending processes of the postparietal and the supratemporal. There is, in addition, a strong ligament stretching between the descending process of the postparietal to insert to a small prefacial eminence (pr.lig) lying immediately ante-

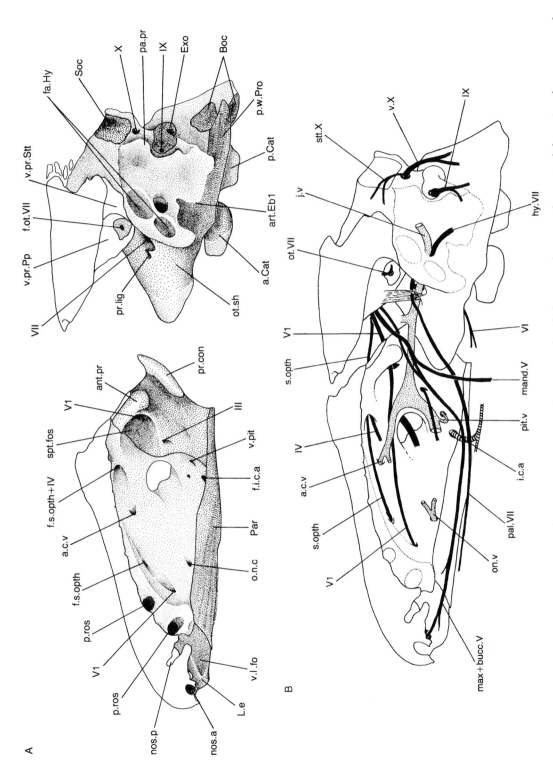

Fig. 6.1 Neurocranium and parasphenoid of *Latimeria*. In the top figure (A) the skeleton is shown and principal foramina for nerves and blood vessels labelled. Bones indicated in darker stipple. In the lower figure (B) the paths taken by the principal cranial nerves and blood vessels are shown. This figure is a compilation of several in Millot and Anthony (1958, 1965) with information from Northcutt and Bemis (1993).

riorly to the facial foramen (VII). The zygal plates (one anazygal and two catazygals – a.Cat, p.Cat) are formed mostly of cartilage (Millot and Anthony, 1958; Bjerring, 1973). The shape of the anazygal suggests that it grows as a paired structure.

The paths taken by the nerves through the braincase have been described by Millot *et al.* (1978) and especially by Bemis and Northcutt (1991) and Northcutt and Bemis (1993). The paths taken by some of the major nerves are shown in Fig. 6.1(B). Because of the previous extensive descriptions, only a few comments are necessary. It has been pointed out that *Latimeria* and other coelacanths are unusual amongst sarcopterygians in that the intracranial joint marks the point of exit of the main trigeminal branches (Jarvik, 1942) – the profundus (V1), maxillary (max+bucc.V), mandibular (mand.V) and the anterodorsal lateral line nerve (s.opth). Several authors have pointed out that this arrangement is different from that in most other sarcopterygians (except *Powichthys* and *Youngolepis*) in which the trigeminal opens entirely within the otico-occipital and the profundus opens through the intracranial joint (Jarvik, 1942; Bjerring, 1973; Ahlberg, 1991; Cloutier and Ahlberg, 1996). This led Bjerring (1973) to propose that the intracranial joint of coelacanths was non-homologous with that of other sarcopterygians.

The abducens (VI) takes a rather unusual path through the anterior part of the otic capsule before leaving the neurocranium through the ventral surface of the prootic to innervate the basicranial muscle and the lateral rectus eye muscle (Bemis and Northcutt, 1991). Because the innervation of the basicranial muscle is via the abducens, there is no need to postulate a special branch of the vagus (the nervus rarus as proposed by Bjerring, 1978: see Bemis and Northcutt, 1991, for discussion).

In most other respects the cranial nerve distribution and blood supply (Millot *et al.*, 1978; Northcutt and Bemis, 1993) follows a standard pattern for sarcopterygian fishes.

Diplocercides

The neurocranium of *Diplocercides* (usually referred to as *Nesides*) has been described and figured on numerous occasions and it is unnecessary to give a lengthy description here. Apart from the first rather sketchy description and illustration (Stensiö, 1922a,b), the many subsequent descriptions, based on a wax model made from a grinding series of the holotype of *Nesides schmidti* (Jarvik, 1980: fig. 206B), agree in most essential details (Holmgren and Stensiö, 1936; Stensiö, 1937; Jarvik, 1954, 1980; Bjerring, 1967, 1972, 1973, 1977). The snout has been particularly thoroughly described by Jarvik (1942). An illustration based on that given by Jarvik (1980) is included here (Fig. 6.2).

The ethmosphenoid and otico-occipital portions of the neurocranium are each ossified as single units. There are no traces of separate prootic, opisthotic, basisphenoid ossification etc. although the topographic areas can easily be compared with those in *Latimeria* (cf. Fig. 6.1 with 6.2), emphasizing the extreme conservatism of coelacanth braincase morphology. Endochondral and perichondral bone is present throughout the otico-occipital and in the basisphenoid region of the ethmosphenoid. But anteriorly, in the interorbital and ethmoid regions there is only thin perichondral bone (Jarvik, 1942). This differential degree of ossification is very similar to that seen in primitive actinopterygians (Gardiner, 1984a) and lungfishes (Miles 1977). The anterior tip and the roof of the nasal capsules, as well as the extreme posterior margin of the occiput, may have been formed by cartilage only (they are shown fully restored in some restorations: Bjerring, 1972, 1973, 1977).

In the ethmoid region the lateral, ventral and posterior walls of the nasal capsule are ossified. The position of the endoskeletal narial openings is not known for certain but

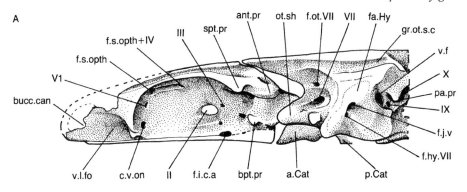

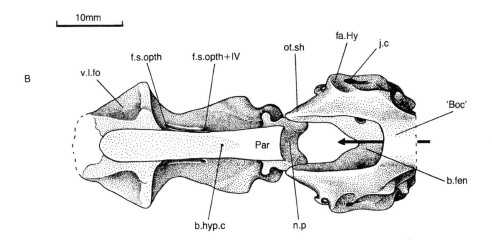

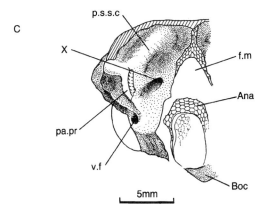

Fig. 6.2 Braincase of *Diplocercides kayseri* (von Koenen) = *Nesides*. (A) Restoration in lateral view. (B) Restoration in ventral view (catazygals omitted). Simplified after Jarvik (1980). (C) Posterior view of left side of occipital region as preserved in specimen 'd' of Stensiö (1937). Arrow indicates notochordal canal passing dorsal to basicranial fenestra.

there was presumably a single fenestra endonarina communis. The lateral wall of the nasal capsule is notched by the buccal foramen (bucc.can). There is a very deep ventrolateral fossa which received the autopalatine. This is said to be absent from *Latimeria* but this is only a matter of degree. There is a depression in *Latimeria* formed by the lateral ethmoid and the anterior ascending wing of the parasphenoid. The depression is admittedly much shallower than in *Diplocercides* or, especially, *Euporosteus*, but it serves the same function: that is, to receive the autopalatine. The wall lining the ventrolateral fossa of *Diplocercides* shows a prominent ridge which may have received ligaments attached to the autopalatine.

The dorsolateral wall of the nasal capsule was initially restored as being entire (e.g. Stensiö, 1932) but after the discovery of *Latimeria* the presence of the two posterior openings from the rostral organ is usually included (Bjerring 1972, 1973, 1977). The restoration given by Jarvik (1980) is probably the most faithful to real conditions, where there is a notch in the ossificiation of the nasal capsule which marks the ventral edge of the inferior posterior opening of the rostral organ. This notch lies directly beneath the anterior opening in the overlying preorbital (Jarvik 1942: fig. 77a).

The orbital face of the postnasal wall is shown to be pierced by three foramina (Jarvik, 1942: fig. 75B) and all three have homologues in *Latimeria*. The most dorsal marks the passage of the superficial ophthalmic branch of the anterodorsal lateral line nerve (f.s.opth), the middle gave passage to the profundus and the ventral contained the orbitonasal artery and vein (c.v.on). Jarvik (1942) also suggested that the ventral opening was also an anterior ventral myodome containing the inferior oblique eye muscle. Comparison with *Latimeria* suggests that the inferior oblique originates above the orbitonasal canal (Millot and Anthony, 1965).

The basisphenoid region is somewhat different from that in *Latimeria* because in *Diplocercides*, there are suprapterygoid and basipterygoid processes in addition to the antotic process. The suprapterygoid process (spt.pr) lies immediately anterior to the antotic process but is separated from it by a distinct notch. The suprapterygoid process is considered to be the primitive dorsal articulation between the braincase and the metapterygoid (Stensiö, 1932; Jarvik, 1954). Thus the antotic process is a coelacanth synapomorphy. The suprapterygoid process is absent in *Latimeria* and most other coelacanths (except *Euporosteus* and *Sassenia*) in which this region of the braincase is ossified, but it should be noted that most coelacanth braincases are cartilaginous at this point and are not preserved. The basipterygoid process (bpt.pr) is well developed and lies beneath the antotic process: it is clearly separated from the processus connectens. The distribution of the foramina within the ethmosphenoid agrees well with that in *Latimeria* and allows clear homology statements to be made (cf. Fig. 6.1(A) with 6.2).

The ventral surface of the ethmosphenoid is covered by an elongate parasphenoid which reaches well forward beneath the nasal capsules. The toothed area extends nearly the entire length of the parasphenoid (Jarvik, 1954: fig. 4c) and this implies that the basicranial muscle inserted well posteriorly on the ethmosphenoid portion of the braincase (Bjerring, 1967: figs 4B, 5D). There is a narrow ridge stretching between the basipterygoid process and the anteroventral end of the processus connectens which may have received the basicranial muscle. Jarvik (1954: fig. 4A) identifies an ascending process on the parasphenoid but this is hardly recognizable: it does not show a spiracular groove and is in no way comparable with the ascending process as identified in actinopterygians (Gardiner, 1984a), *Eusthenopteron* or porolepiforms (Jarvik, 1954). An ascending process is here regarded as being absent from *Diplocercides* and other coelacanths.

The otico-occipital portion of the neurocranium is very similar in shape to that of *Latimeria* but there are some interesting differences. The otic shelf which reaches forward to form a groove in the 'track-and-groove' endoskeletal intracranial joint, is relatively shorter in *Diplocercides*.

The lateral wall of the prootic region is developed entirely as bone and this extends up to the undersurface of the skull roof meaning that the temporal excavation is lined entirely with bone. There is also a well-marked groove for the palatine ramus of VII, leading from the facial foramen (VII) and running across the otic shelf. Such a groove is absent in *Latimeria* and most other coelacanths. There is a small prefacial eminence implying that, as in *Latimeria* a prefacial ligament is present. There is also a small afacial eminence which probably supported a small spiracular tooth plate.

The hyomandibular facet is not completely known, probably because much of it was capped by cartilage. That which remains is very similar to that in *Latimeria* except that the entire facet appears to be orientated more horizontally. As usual in sarcopterygians the large jugular canal pierces the lateral commissure to run horizontally and emerge (f.j.v) at the mid level of the hyomandibular facet. Posterior to the hyomandibular facet, the lateral wall of the saccular chamber, which is relatively large in *Diplocercides*, is pierced by a large fenestra (also seen in *Laugia* and *Sassenia*). This fenestra lies separate from but adjacent to the glossopharyngeal foramen (cf. *Laugia*). It is interpreted as the vestibular fontanelle (v.f) which is perichondrally lined.

Immediately above the glossopharyngeal foramen, there is a well-developed parampullary process to which the first suprapharyngobranchial articulates. Ventral to the parampullary process there is a strong vertically orientated ridge, also found in *Latimeria* and which receives epaxial trunk musculature. The vagus foramen (X) is found high up on the posterior face of the neurocranium.

A portion of an ethmosphenoid with attached parietals has been described from the Fammenian of Morocco (Lelièvre and Janvier, 1988, see p. 363). The specimen shows the postnasal wall, most of the interorbital septum and the upper portion of the basisphenoid region and is roofed by the parietal. The neurocranium is completely ossified and this is why it is discussed here with *Diplocercides*. The broken lateral edges are contoured in a way that suggests that both an antotic and suprapterygoid process were developed. The specimen is interesting for a number of reasons. The interobital septum is very wide as in *Euporosteus* and houses an anterior extension of the cranial cavity, the roof of which contains a deep pineal recess (Lelièvre and Janvier, 1988: fig. 1A) which lies well forward in a position comparable to that in *Eusthenopteron*. This may mean that the brain continued well forward in front of the intracranial joint. There is no external opening of the pineal. The position of the foramina in the rear wall of the orbit make interpretation difficult. A small foramen pierces the cranial cavity to pass anterolaterally into the orbit and this corresponds to the oculomotor foramen. There is, however, no obvious foramen for the profundus, which usually exits the cranial cavity at the same level and lies close to the oculomotor foramen. Instead, a large foramen occupying the usual profundus position leads into a canal which passes posteriorly through the antotic process and opens on the lateral edge of the posterior face (Lelièvre and Janvier, 1988: fig. 1C). Any nerve which penetrated this canal must have left the cranial cavity behind the level of the ethmosphenoid, either within or behind the intracranial joint. In *Latimeria* the superficial ophthalmic branch of the anterodorsal lateral line nerve would follow this course. Lelièvre and Janvier (1988) suggest that the profundus must also have accompanied this nerve (they usually lie quite separate), implying that the profundus left the cranial cavity within the intracranial joint, as in most other sarcopterygians, but unlike other coelacanths.

The postnasal wall is pierced by an anteriorly directed foramen and canal, which presumably carried the profundus. Immediately behind this level, there is a large posterolaterally directed foramen which leaves the anterior limit of the cranial cavity. Lelièvre and Janvier (1988) suggest that this is a foramen for the anterior cerebral vein, which certainly has this position and course in other sarcopterygians. It is, however, very much larger than would be expected. It is also curious that there is no obvious foramen for the optic tract. Because so little is known it would be unwise to draw too many conclusions from this one specimen but it may show that there may have been considerable species variation between the braincases of Devonian coelacanths.

Euporosteus

This taxon is represented by a single specimen of an ethmosphenoid portion of a braincase (Fig. 6.3) and has been described by Jaekel (1927), Stensiö (1937) and Jarvik (1942). Jarvik's description of the snout is comprehensive but the remainder of the ethmosphenoid remains poorly known. The ethmosphenoid is completely ossified, at least as perichondral bone. As Jarvik remarks (1942: 556), the nasal capsules form nearly half the length of the ethmosphenoid. The postnasal wall is particularly thick and ventrally there is a deep ventrolateral fossa which received the autopalatine. Jarvik (1942: fig. 79) has restored a median rostral cavity and in this he is probably correct because there are faint indications of posterior openings (p.ros) from the rostral organ (Jarvik, 1942: pl. 14, fig. 2, *f*). There is a single large fenestra endonarina communis (fen) which may also have contained a buccal foramen because there is an anterodorsally directed

groove which notches the edge of the fenestra as in *Latimeria*. Immediately in front of the fenestra endonarina communis, there is a groove within the perichondral bone marking the path of the ethmoid commissure (gr.eth.-com). The ventral surface of the nasal capsule shows paired scars showing where the vomers attached (the vomers are unknown). They must have been very large, unlike those of most other coelacanths, and probably separated from each other in the midline.

The postnasal wall is pierced by at least three foramina (Jarvik, 1942: fig. 75A; pl. 16, fig. 1). Ventrally there is a large orbitonasal canal which Jarvik suggests also was an anterior myodome (see also *Diplocercides* and comments). There is some variation from one side of the specimen to the other. On the left there are two foramina very close together. Presumably the more dorsal is for the superficial ophthalmic branch of the anterodorsal lateral line nerve while the more ventral is for the profundus. On the right side these two foramina lie at some distance from each other (Jarvik, 1942: pl. 14, fig. 2) and are accompanied by a small third foramen which may have carried the anterior cerebral vein.

The interorbital septum is very broad dorsally as it flares out on either side to underlie the skull roof and it is very likely that the anterior part of the brain was located here as in *Diplocercides*. This idea is further supported by the fact that there is a very weakly developed dorsum sellae and the opening of the cranial cavity through the basisphenoid is very large as in *Diplocercides*. In the dorsal half of the interorbital septum there is a pronounced longitudinal groove with a foramen at either end. The posterior foramen marks the point at which the superficial ophthalmic left the cranial cavity and the anterior foramen where it re-entered. The posterior foramen of the right side is considerably larger than that of the left, where there is a small associated foramen. It is likely that the trochlear (IV) left the cranial cavity with the superficial ophthalmic on the

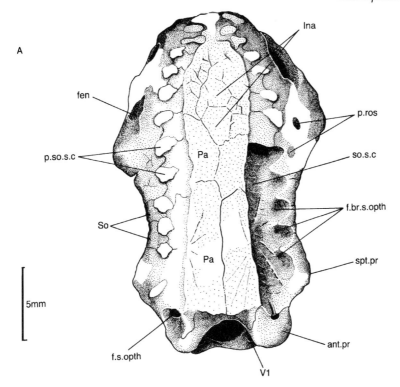

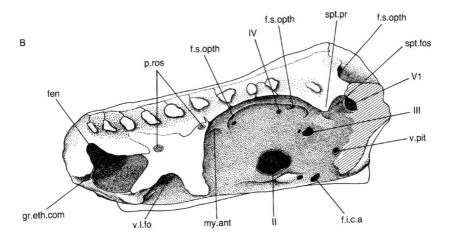

Fig. 6.3 *Euporosteus eifeliensis* Jaekel. Ethmosphenoid portion of neurocranium, UB, camera lucida drawings. (A) Dorsal view. The roof of the supraorbital sensory canal has broken away on the right side to show paths of the nerves in the floor of the canal. (B) Left lateral view. Note the relatively long nasal region which is nearly equal to half the total length. Hatching indicates a broken surface.

right side and exited separately on the left. The centre of the interorbital septum is perforated by the large optic foramen (II) and immediately beneath this there is a small but pronounced swelling developed as a longitudinal ridge. This ridge is not seen in other coelacanths. In position it corresponds to a small isolated section of an anteriorly elongated basipterygoid process as seen in *Glyptolepis* (Jarvik 1972: fig. 21A). And it has the same topographic relationship to a small foramen for the ophthalmic artery which lies immediately below the ridge. However, in comparsion with the baspterygoid process of *Diplocercides*, it lies well anteriorly to the expected position. Unfortunately the palate remains unknown and therefore it is impossible to confirm if there was a basal articulation as in *Diplocercides*. An alternative explanation is that this process marks the site of origin of the inferior rectus muscle. Dorsally the basipterygoid region shows well-developed suprapterygoid and antotic processes which lie in tandem, separated one from another by the insertion point for the adductor arcus palatini. A large profundus foramen lies anteroventrally to the antotic process and just in front of this level there is the foramen for the oculomotor (III) which may (left side) be accompanied by a separate foramen for the ciliary artery. The processus connectens is quite long and reaches the level of the parasphenoid. Stensiö (1937: pl. 11, fig. 4; pl. 12, fig. 2) labels a basipterygoid process, but this is not obviously separate from the processus connectens as in *Diplocercides* and the presence of a basipterygoid process must remain in doubt. In posterior view (Stensiö 1937: pl. 11, fig. 3) it is worth noting that the sphenoid condyles are very poorly developed and the dorsum sellae is very small. This, together with the broad nature of the dorsal part of the interorbital septum implies that the brain extended into the ethmosphenoid and may, in turn, mean that movement of the intracranial joint was very limited.

The parasphenoid is unusually broad throughout its length. The buccohypophysial canal pierces the parasphenoid close to the centre of ossification. The dentition is formed by tiny villiform teeth which are arranged in whorls around the opening of the buccohypophysial canal.

Rhabdoderma

The neurocranium has been described by Moy-Thomas (1937) and Forey (1981). Moy-Thomas' description was strongly influenced by earlier restorations of the neurocrania of *Macropoma* (Watson, 1921), *Wimania* (Stensiö, 1921) and *Undina* (Aldinger, 1931). All of these coelacanth templates show fragmented otico-occipital moieties with small ossifications separated in life by large areas of cartilage. Moy-Thomas' restoration of *Rhabdoderma* is similarly restored (Moy-Thomas, 1937: fig. 3). Moy-Thomas also accepted the braincase described by Aldinger (1931) as *Coelacanthus* sp. as belonging to *Rhabdoderma* and this is the source of the opinion that there is a basipterygoid process present. Aldinger's *Coelacanthus* sp. should probably be referred to *Diplocercides* (Forey, 1981: 219).

Forey (1981) redescribed the neurocranium of *Rhabdoderma elegans* and provided a restoration, reproduced here as Fig. 6.5. This description was based on evidence from three specimens (Fig. 6.4), one used by Moy-Thomas (BMNH P.7912), SM E.169 and BMNH P.10473, all referable to *Rhabdoderma elegans*. This remains the only species of the genus in which articulated neurocrania are known, although isolated basisphenoids and parasphenoids of other species are frequently found.

The ethmosphenoid is ossified as paired lateral ethmoids which are little more than ossifications in the floor of the nasal capsule. The basisphenoid shows pronounced anterior laminae forming a partly ossified interorbital septum and is pierced by a large optic fora-

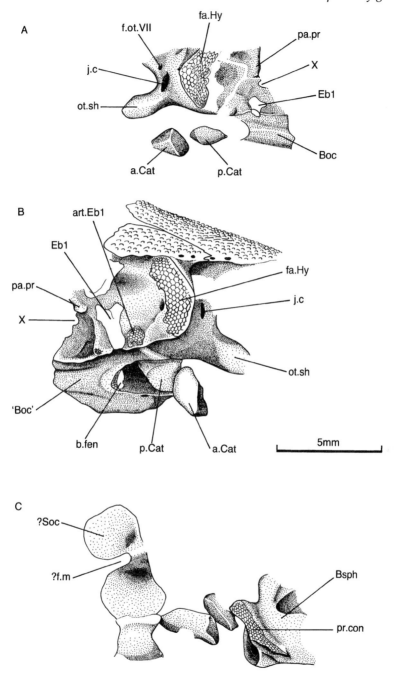

Fig. 6.4 *Rhabdoderma elegans* (Newberry). Camera lucida drawings of casts of braincase specimens. (A) Left lateral view of braincase as preserved in SMC E.169, a small individual. (B) BMNH P.10473 right lateral view. (C) BMNH P.7912, right lateral view.

men, behind which lies a small oculomotor foramen positioned at the lower rim of the suprapterygoid fossa. Ventral to the optic foramen, there is a small foramen for the pituitary vein located at the contact between the basisphenoid and the parasphenoid. The antotic processes are narrow and only partly overlapped by the descending processes of the parietal. There is no obvious foramen for the passage of the superficial ophthalmic and this must have passed from the cranial cavity to the orbit lateral to the parietal descending process (cf. *Latimeria*). The sphenoid condyles are poorly developed as small, cartilage-capped protuberances. The processus connectens reaches far ventrally to end at the level of the parasphenoid (Fig. 6.5). Moy-Thomas restores it as turning horizontally where it meets the parasphenoid and he interprets this horizontal portion as the basipterygoid process. I cannot confirm this observation on the specimen used by Moy-Thomas, nor on any other specimen examined here. In *Diplocercides* and Aldinger's

Coelacanthus sp., the basipterygoid process is quite separate.

The parasphenoid is closely applied but never fused (cf. Moy-Thomas, 1937) to the basisphenoid. Anteriorly it expands but remains a flat bone without the dorsal processes seen in *Latimeria* and more derived coelacanths. Teeth are borne over most of the oral surface, extending posteriorly to nearly reach the ventral fissure.

The otico-occipital portion of the neurocranium is ossified as a large single unit corresponding to combined prootic, opisthotic and basioccipital areas (cf. Moy-Thomas, 1937) in addition to the two catazygals and a possibly separate 'supraoccipital' (see below). Of course it is very likely that the otico-occipital ossified from several centres because it is difficult to imagine how else it may have grown but no evidence of separate ossification centres remains. In the small individual (SMC E.169, Fig. 6.4(A)) examined here, ossification of the otico-occipital is incomplete immediately behind the level of the hyoman-

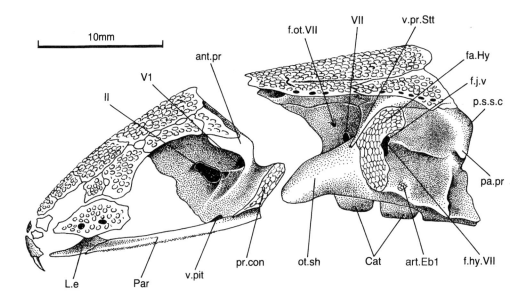

Fig. 6.5 *Rhabdoderma elegans* (Newberry). Restoration of the braincase in left lateral view based mainly on specimens illustrated in Fig. 6.4. From Forey (1981).

dibular facet. In this respect this corresponds to the conditions seen in large specimens of *Laugia*. The prootic region of the braincase is well developed as a prominent otic shelf which articulates with the processus connectens. Above this level the prootic reaches far dorsally to contact the undersurface of the postparietal (there is no descending process of the postparietal). This extensive ossification of the prootic means that the temporal excavation is completely lined with bone as in *Diiplocercides*. The anterior opening of the jugular canal and exit for the palatine nerve is large, and the small, dorsally directed foramen for the exit of the otic lateral line nerve (f.ot.VII) lies immediately above.

The hyomandibular facet must have been completely lined with cartilage and there is only a slight constriction at the level where the jugular canal opens posteriorly. As in *Laugia*, the posterior opening of the jugular canal is bilobed, reflecting the passage of both the jugular vein and the hyomandibular ramus VII. The site of articulation of epibranchial 1 (art.Eb.1) can be recognized as an area of exposed endochondral bone posterior to the lower end of the hyomandibular facet. Posterior to the dorsal half of the hyomandibular facet the opisthotic region is excavated and provided the site of origin of the adductor hyomandibulae. This excavation is bordered posteriorly by a swelling for the posterior semicircular canal (p.s.s.c) which is continuous ventrally as a small parampullary process. The vagus foramen is located immediately below the process. Ventral to the parampullary process, there is a pronounced vertical ridge which ends in a poster-oventrally directed process, which received epibranchial 2. This vertical ridge is very similar to that in *Latimeria* and corresponds to the rather oblique ridge seen in *Laugia*.

The basioccipital region forms the posterior limit of the basicranial fenestra. The fenestra is filled with two catazygals which are of approximately equal size. An anazygal has not been identified, but was presumably present because sphenoid condyles are present on the basisphenoid.

None of the specimens shows the position of the supratemporal branches of either IX, or X but the material is rather poorly preserved. There is no evidence of the vestibular fontanelle opening into the otic capsule.

The posterior face of the neurocranium is largely unknown but in BMNH P.7912, the specimen used by Moy-Thomas, there is a large butterfly-shaped median bone (Fig. 6.4(C), ?Soc) identified as a supraoccipital (Moy-Thomas, 1937: fig. 3). Accepting this interpretation, the deep median notch would be the foramen magnum (?f.m). It is very much larger than the bone of the same name in *Holophagus*, *Macropoma* and *Latimeria*, and the life position of this element is uncertain. No other Palaeozoic coelacanth shows such a bone and that of more derived coelacanths is considerably smaller. I remain uncertain that this bone is a supraoccipital, more material may clarify this uncertainty.

Laugia

The neurocranium of *Laugia* has been briefly described by Stensiö (1932). The description given here is based on acid-prepared specimens, in particular MGUH VP2308a, VP2312a, VP2315a/b, VP2316a/b, VP2317 and VPspec 82.

The ethmosphenoid and otico-occipital are approximately of the same length and both are more extensively ossified than the neurocranium of most other Mesozoic coelacanths. The ethmosphenoid is ossified as paired lateral ethmoids and a basisphenoid, which is co-ossified with an orbitosphenoid. In the otico-occipital region there are paired prootics and opisthotic/basioccipital ossifications as well as the zygal plates.

In the ethmosphenoid the lateral ethmoids are small, forming the floor of the nasal capsules (Fig. 6.7, L.e). The posterior edge of the lateral ethmoid passed into cartilage. Small, perichondrally lined tubes are found within the nasal capsule of MGUH VP2312a and these probably describe the paths taken by minor branches of the profundus on their way to innervate the snout. These would correspond to branch 'p.10' of Northcutt and Bemis (1993).

The interorbital ossification is a very thin double sheet of bone variously developed in different specimens. The most completely developed is seen in MGUH VP2317 where it lies in continuity with the basisphenoid posteriorly. It appears to be unfenestrated. The basisphenoid (Fig. 6.6(B)) is, as usual, a robust element. The anterior limits of the basisphenoid ossification vary from specimen to specimen. The antotic process is very slender and the anterior edge passes into cartilage which formed the roof of the orbit. Immediately anterior to the antotic process, the basisphenoid shows a deep suprapterygoid fossa at the base of which lies the large foramen for profundus V. This suprapterygoid fossa is the point of origin for the adductor palatini muscle. Immediately beneath the antotic process the basisphenoid is pierced by the oculomotor foramen, and ventral to this level is the foramen for the pituitary vein.

The processus connectens is developed as an oblique cartilage-capped ridge, in front of which the surface of the basisphenoid is roughened for the attachment of the external and superior rectus eye muscles. On the posterior surface of the basisphenoid, the sphenoid condyles are well separated from one another.

The parasphenoid is shallow over its entire length. It expands anteriorly but there are no dorsal wings (cf. *Macropoma*, *Latimeria*). The buccohypophysial canal pierces the parasphenoid and small, villiform teeth are present over the anterior 80% of the palatal surface.

The otico-occipital is composed of two large paired ossifications (Figs 6.6(A), 6.7): anteriorly there is a large prootic which contains the jugular canal, hyomandibular

facet and the anterior half of the saccular chamber; posteriorly a large ossification houses the posterior half of the auditory capsule and the occipital region. With the exception of the overlapping suture between the prootic and opisthotic/basioccipital ossification, the posterior margin of the prootic and the anterior margin of the opisthotic ossifications end in cartilage surfaces and no doubt were joined by cartilage in life. The prootic is very large and reaches far dorsally to suture with the undersurface of the postparietal along a broad area. This dorsal extension is excavated deeply immediately in front of the descending process of the supratemporal, and this temporal excavation is the site of origin for the palatal levator muscle. This dorsal extension of the prootic is pierced by a small, upwardly directed foramen (f.ot.VII) which carried the otic lateral line nerve dorsally to innervate the anterior part of the otic sensory canal through three small foramina which pierce the undersurface of the postparietal (Fig. 3.9(B)). The dorsal extension also bears a deep notch which is open anteriorly into the intracranial joint. In *Diplocercides*, *Sassenia* and *Latimeria*, the prefacial commissure is ossified and therefore VII leaves the cranial cavity through a foramen rather than a notch as here. The lateral aspect of the prootic is swollen as the otic shelf, which reaches far forwards to lie laterally to the processus connectens of the basisphenoid. The anterior extension of the prootic is much longer in *Laugia* than in *Diplocercides* (cf. Fig. 6.7(A) with 6.2) and is comparable with that in *Latimeria* (Fig. 6.1(A)). The otic shelf is developed as a prominent ridge which broadens posteriorly to form the anterior limit of the hyomandibular facet. There is no affacial process or prefacial eminence upon the otic shelf. Nor is there any groove marking the path of the palatine ramus of VII. Presumably, this nerve coursed along the dorsal edge of the otic shelf. The hyomandibular facet is very large and orientated vertically. It is distinctly bilo-

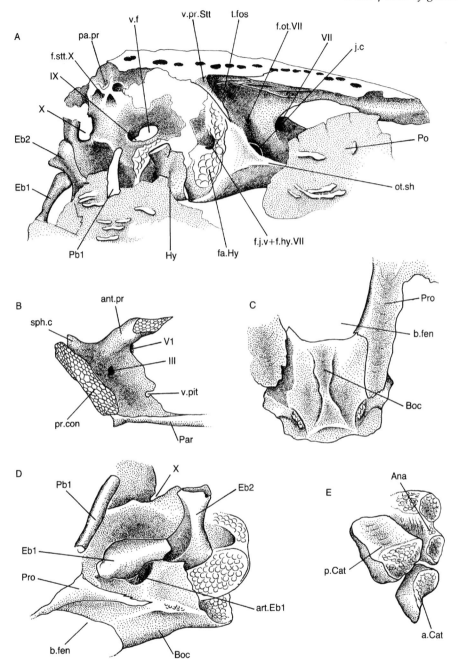

Fig. 6.6 Camera lucida drawings of parts of the braincase of *Laugia groenlandica* Stensiö, forming the basis on which the restoration in Fig. 6.7 is made. (A) Right lateral view of braincase as preserved in MGUH VP2316a. (B) Basisphenoid in right lateral view as preserved in MGUH VP2315a. (C) Basioccipital region as preserved in MGUH VP.spec 82 (anterior towards top). (D) Ventrolateral view of left side of basioccipital region of MGUH VP.spec82. (E) Zygal plates seen in anterolateral view as preserved in MGUH VP2315a.

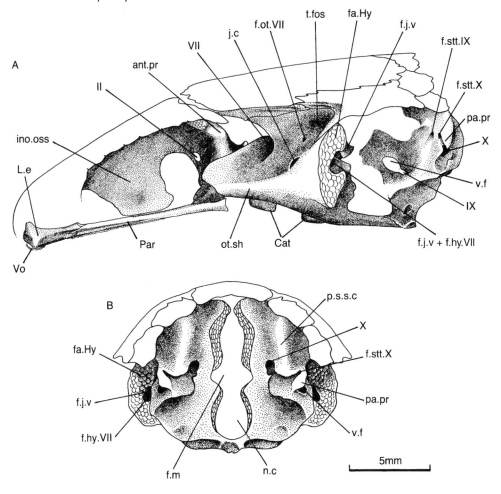

Fig. 6.7 Restoration of the braincase of *Laugia groenlandica* Stensiö. (A) in left lateral view and (B) in posterior view. Based mainly on specimens illustrated in previous figure.

bed with the central constriction lying at the level of the jugular canal. However, it must have been completely covered by cartilage because the facet shows no perichondral bone surface (MGUH VP2316a, VP2317). The facial foramen lies above the otic shelf and leads into the jugular canal which, in turn, opens behind the hyomandibular facet. The posterior opening is distinctly double (Fig. 6.6, f.j.v+f.hy.VII), with a smaller canal above for the passage of the jugular vein and a posteroventrally orientated canal below marking the exit for the hyomandibular

trunk of the facial. The ventral edge of the prootic continues posteriorly as a wing to overlap the basioccipital region. The ventral surface of the prootic is concave and is the site of origin of the basicranial muscle.

The occipital ossification is large and covers the area which in later coelacanths ossifies as the opisthotic, supraoccipital and basioccipital. This large ossification is clearly paired except basally, where the ossification is continuous beneath the notochordal canal (Fig. 6.7(B)). The anterior edge is irregular where it must have passed into cartilage.

In lateral aspect, the contour of the dorsal half of the ossification (the area occupied by the opisthotic) is deeply excavated, at the base of which is a large vestibular fontanelle. The depression provided an insertion area for the adductor hyomandibulae muscle, and below this there is a very pronounced ridge (Fig. 6.7(A)) which curves posterodorsally to end in a well-developed parampullary process. The parampullary process receives suprapharyngobranchial 1 (Fig. 6.6(D), Pb1) and the surface of the bone is roughened just above this level for the insertion of the branched levator muscles 1–4. The vestibular fontanelle is perichondrally lined and is also present in *Diplocercides*, but absent in *Latimeria*, in which the wall is entirely cartilaginous at this point. In one specimen (MGUH VP2316a, Fig. 6.6(A)), the fenestra is notched along the posterior border by the glossopharyngeal foramen, and from this a shallow groove leads posterodorsally and is directed towards a foramen (Fig. 6.7(A), f.stt.IX) at the anterior edge of the parampullary process. The groove probably contained the middle lateral line nerve which, in *Latimeria*, exits with the glossopharyngeal, passes dorsally, and pierces the parampullary process to emerge on the upper surface of that process before running with the posterior lateral line nerve to innervate the neuromasts and the middle pit line.

Immediately behind the parampullary process, there is a large vagus foramen which is best seen in the posterior view of the skull (Fig. 6.7(B)). The foramen leads to a prominent groove which runs posteroventrally carrying the branches of the vagus. There is a small dorsolaterally directed groove from the foramen which leads to a foramen (Fig. 6.7(B), f.stt.X) high up on the parampullary process. This groove contained the posterior lateral line nerve and it probably linked up with the middle lateral line nerve within the parampullary process to emerge dorsally and innervate the posterior pit line and the neuromasts of the supra-

temporal commissure. The path taken by the posterior semicircular canal can be seen as a pronounced ridge (Fig. 6.7(B), p.s.s.c) which curves dorsally above the parampullary process.

The posterior face of the neurocranium shows that the margins of the foramen magnum and the notochordal canal are cartilage capped. That is, the posterior face of the neurocranium is unfinished. This may imply that the apparently unfinished posterior end of the neurocranium of *Diplocercides* (often shown in restorations) is real and not an artefact of the specimen. Certainly, there is no trace of any occipital nerve foramina in *Laugia* or indeed other coelacanths, such as are present in actinopterygians (Gardiner 1984a) or *Eusthenopteron* (Jarvik, 1980). Also, there is no trace of an occipital fissure although the ventral end persists as the vestibular fontanelle. The cartilage-capped surface surrounding the foramen magnum (Fig. 6.7(B)) appears to lie in continuity with the anterior surface of the first neural arch, which is also cartilage capped. Between the foramen magnum and the notochordal canal, there is another laterally expanded foramen corresponding to the restored separate foramen in *Diplocercides* (Bjerring, 1973) and the ventral expansion of the foramen magnum of *Latimeria*. This may be evidence of the endolymphatic occipital commissure.

The basioccipital region of the occipital ossification shows a median ventral ridge flanked by a roughened area (Fig. 6.6(C)) immediately below the ridge running from the parampullary process. This area is the site of articulation with an epibranchial element (Fig. 6.6(D), Eb1).

The three zygal plates are seen in MGUH VP2315a (Fig. 6.6(E)). The two catazygals are very unequal in size. The anterior is small and saddle-shaped while the posterior is more elongate with a rounded posterior margin which reflects the concave anterior margin of the basioccipital (Fig. 6.6(C)). The anazygal (Fig. 6.6(E)) bears two deep concavities along the anterior edge to match the sphenoid condyles with which it articulates.

Sassenia

There are several specimens of *Sassenia groen-landica* which show the neurocranium: MGUH VP2327a (holotype), MGUH VP2336, VP2337 and VP2326, the last mentioned being the best preserved, most complete and illustrated here (Fig. 6.8). The braincase of *Sassenia* is extensively ossified and very much like that of *Diplocercides*.

The ethmosphenoid is nearly completely ossified, with just the tip of the snout and the anteroventral corner of the interorbital septum remaining as presumed cartilage areas. It is possible that the lateral ethmoids are separate ossifications, because on the right side of MGUH VP2326 there is a clear line of separation between the ossifications covering the anterior limit of the interorbital septum and the posterior edge of the nasal capsule. The lateral ethmoid is extensive and forms part of the lateral wall and the floor of the nasal capsule. It shows a prominent dorsal process forming the posterior border

to the common anterior and posterior endos-keletal openings into the nasal capsule. The posterior endoskeletal opening from the rostral organ could not be seen in this speci-men, but the position of the overlying preor-bital, which is pierced by relevant openings (Fig. 4.13) suggests that the endoskeletal openings would have lain more dorsally than in *Latimeria*. The lateral ethmoid is consider-ably thickened along the intersection of the lateral wall and the floor, and this thickening provides a shelf on which the autopalatine sits, delimiting a rather weakly developed ventrolateral fossa.

There is an extensively ossified interorbital septum which flares anteriorly to form a partially ossified postnasal wall. Posteriorly the interorbital septum is continuous with the basisphenoid region. Within the roof of the interorbital septum there are two oppo-site-facing foramina linked by a deep groove, and similar to conditions in *Diplocercides* but located more centrally within the orbit. In this latter respect, *Sassenia* is more like *Lati-meria* where, however, the septum is cartila-ginous. These foramina and groove mark the course of the superficial ophthalmic branch

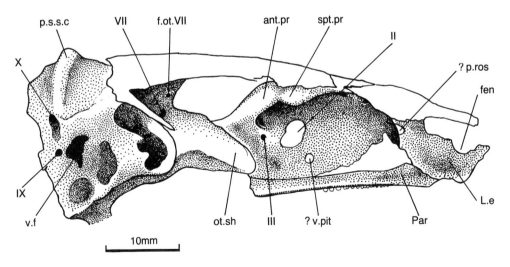

Fig. 6.8 *Sassenia groenlandica* sp. nov. Drawing of braincase in right lateral view as preserved in MGUH VP2326.

of the anterodorsal lateral line nerve as it left the cranial cavity posteriorly, to run beneath the roof of the orbit and penetrate the rostral organ and nasal capsule. The posteriorly placed exit foramen is slightly the larger and it may also have allowed the trochlear nerve to pass into the orbit as suggested for *Diplocercides* by Jarvik (1954). In *Latimeria* there are separate but closely spaced foramina for the lateral line nerve and the trochlear. Anteroventrally to the anterior foramen there is a larger foramen marking the entry for the profundus nerve into the nasal capsule. This foramen is located as in *Diplocercides* but more dorsally than in *Latimeria*.

The interorbital septum is pierced by two very obvious foramina. The larger one lies immediately in front of the basisphenoid region and is for the passage of the optic tract. The smaller lies anteroventral to the optic foramen and may correspond to that for the internal carotid artery, or the pituitary vein (Fig. 6.8, ?v.pit), or the ophthalmica magna artery of *Latimeria*.

The main body of the basisphenoid is narrow across the level of the antotic processes, and the suprapterygoid fossa is very deep with well-marked rims. An anteroventrally directed foramen lies along the lower rim and is interpreted here as the exit for the profundus. A small oculomotor foramen lies below the profundus foramen. The processus connectens reaches far ventrally to end near the contact with the parasphenoid. Immediately in front of the antotic process, there is a well-developed suprapterygoid process as in *Diplocercides* and *Euporosteus*.

The otico-occipital is completely ossified except for the zygal plates. The prootic region is particularly thick, with the otic shelf well demarcated from the hyomandibular facet. Beneath both the shelf and the hyomandibular facet the ventral wall of the prootic region is deeply grooved as the site of insertion of the basicranial muscle. Above the otic shelf, the temporal excavation is completely lined with bone and is partly overlapped by the well-developed descending process of the supratemporal.

The facial foramen is large and gave exit to both the hyomandibular and palatine trunks of the facial nerve, the former passing back directly into the jugular canal and the latter running along the top of the otic shelf to the orbit. MGUH VP2336 shows that the dorsal rim of the facial foramen is contoured into a dorsally directed groove and this may indicate the path of the middle lateral line nerve in this individual in which there is no separate foramen for the middle lateral line nerve. The jugular canal runs through the lateral commissure and emerges posteriorly at mid-level along the vertically inclined facet. The posterior exit is distinctly bilobed for the jugular vein above and the hyomandibular branch of VII below. Ventral to the hyomandibular facet, there is a foramen in the roof of the groove for the basicranial muscle. It is obscured in lateral view by the ventral expansion of the hyomandibular facet. The foramen is directed dorsally and enters the labyrinth cavity. This is presumably a foramen for the abducens (VI), which in *Latimeria* shows a circuitous route through the otic capsule before penetrating the prootic to innervate both the basicranial muscle and the external rectus eye muscle (Bemis and Northcutt, 1991).

Posterior to the hyomandibular facet the side wall of the otic capsule shows a deep excavation which runs dorsally and then posteriorly where it becomes much shallower. The deepest part of the excavation is the site of origin of the adductor hyomandibulae while the posterodorsal part contained the adductor operculi. The posterior rim of the deep excavation is formed by a very well-developed vertical crest. This crest is presumably comparable to the ridge seen in other coelacanths where it runs vertically or obliquely to a level beneath the parampullary process. Because the dorsal end of the crest is broken in MGUH VP2326 its true extent is unknown for *Sassenia*. In *Latimeria* the posterior surface of the ridge receives epaxial musculature. Further comparisons with other coelacanths would suggest that the second epibranchial articulated with the ventral end of the crest. Posterior to the crest, there is a small but deep excavation in the side wall of the basioccipital region. I do not know the function of this excavation. In *Latimeria*, epaxial muscles insert to this area of the neurocranium but it is not so markedly depressed.

In MGUH VP2326 the side wall of the otic capsule posterior to the dorsal half of the hyomandibular facet is badly broken as a large broken-edged fenestra. However, the anterior rim of this fenestra tips into the cavity as smooth perichondral bone and it is therefore very likely that there was a vestibular fontanelle opening in the wall of the labyrinth cavity (cf. *Diplocercides* and *Laugia*). The glossopharyngeal foramen lies immediately behind the vestibular fontanelle. The swelling for the posterior semicircular canal is obvious in the specimen. The vagus foramen, which is strongly directed ventrally, lies posterior to the ventral end of the swelling but there is no clearly defined parampullary process. A supraoccipital appears to be absent and the posterior face of the occiput resembles that of *Laugia*.

Wimania

The neurocranium has been described by Stensiö (1921), who used this taxon as a reference to make general comments about the coelacanth neurocranium. Considering subsequent discoveries of the braincases of other coelacanths, that of *Wimania* is relatively poorly known. Much of the neurocranium must have been cartilaginous: the ethmosphenoid is represented by lateral ethmoids (pre-ethmoids of Stensiö) and a basisphenoid; the only known otico-occipital ossification is the prootic (prootico-opisthotic of Stensiö). The prootic is similar to that of most other Mesozoic coelacanths in that it

forms the anterior wall of the saccular chamber and bears a posterior wing (process 'c' of Stensiö) which extends posteriorly and presumably sutured with a basioccipital, although the latter bone remains unknown. Above the otic shelf, the prootic extends dorsally as a broad flange (process f_1 of Stensiö) to meet the undersurface of the postparietal. Thus the inner wall of the temporal excavation is lined with bone. The parasphenoid of *Wimania* expands anteriorly from about one-third of the distance from the posterior end (Stensiö, 1921: fig. 23) and bears small teeth throughout the anterior two-thirds.

Whiteia

The neurocranium is partly known in *Whiteia woodwardi* and *W. nielseni* and the basisphenoid has been described for *W.* sp. from British Columbia (Schaeffer and Mangus, 1976). Beltan (1968) gave a description of the neurocranium of *W. woodwardi* but, unfortunately, the material examined was seriously misinterpreted: the otico-occipital portion was interpreted upside down and the elements described as belonging to the nasal capsule region represent the basibranchial and overlying tooth plates (Fig. 7.6(D)).

The ethmosphenoid is ossified as lateral ethmoids and the basisphenoid with the underlying parasphenoid and vomers. The lateral ethmoids are particularly small and, as in *Laugia*, they form only an ossified floor to the nasal capsule which must therefore have been formed mostly by cartilage. The basisphenoid is also small and very short, with strongly divergent antotic processes. Each antotic process is perfectly smooth and is not perforated by the superficial ophthalmic nerve. Beltan (1968) records anteroventral processes in *Whiteia*, and she cites the presence of these as differences from *Latimeria*. However, these processes are dorsal and they are the antotic processes. The processus connectens is short and fails to reach the parasphenoid, and the sphenoid condyles are placed close together near the midline. The parasphenoid bears teeth over most of the ventral surface. The teeth towards the edges are slightly larger than those centrally placed. The buccohypophysial canal fails to penetrate the parasphenoid. The paired vomers are very small and are separate from each other in the midline.

The otico-occipital portion of the braincase is ossified as the prootics, the basioccipital and the zygal plates. The prootic is developed as an ossification which reaches to the underside of the postparietal but fails to form a completely ossified lining to the temporal excavation (Bang, 1976: pl. 1A). Otherwise it is very similar to that in *Latimeria*, in that the ossified support for the hyomandibular articulation is poorly defined and it must have passed into cartilage to form the sharp contours of the facet. Posterior to the ventral tip of the facet, there is a prominent swelling for articulation with infrabranchial 1. The prootic reaches posteriorly and ventrally beneath the saccular chamber to show a complex overlap surface with the basioccipital. There is some individual variation in the positions of the foramina that pierce the prootic. Most specimens show the usual large facial foramen which lies adjacent to the jugular canal. The facial foramen conveyed both the hyomandibular trunk and the palatine nerve: the latter passed anteroventrally over the surface of the otic shelf. However, in at least one specimen of *W. nielseni* (MGUH VP2323A) there is a separate foramen for the palatine nerve which lies anteroventral to the jugular canal and faces anteroventrally. In this respect this specimen is similar to conditions in *Macropoma*.

The basioccipital is represented as a 'U'-shaped ossification with a roughened inner surface where it lay against the notochord. Beltan recorded a supraoccipital (1968: 114) but this is the basioccipital. The anazygal is very short and broad and dumb-bell-shaped in dorsal view. There are two large catazy-gals which occupy the basicranial fenestra: the rectangular anterior catazygal is considerably larger than the semicircular posterior catazygal. There are some specimens (BMNH P.17205, MNHN IP 230 *W. wood-wardi*, MGUH VP96) in which there is evidence of further ossification in the exoccipital region, but none of the specimens shows any detail.

The saccular chamber is often wholly occupied by the very large saccular otolith, which is approximately disc-shaped and very similar in size and shape to that in *Latimeria*. The otolith is externally slightly concave and medially more strongly convex. The outer surface is smooth while the inner surface is smooth except for a rim of roughened calcification which is separated from the central area by a narrow but well-marked groove. The medial protuberance, present in the otolith of *Latimeria* (Nolf, 1985: fig. 80), is absent in *Whiteia*. The otolith was incorrectly identified as the opisthotic by Beltan (1968: pl. 47).

Undina

The neurocranium has been described by Reis (1888) and Aldinger (1930). It is very similar to that of *Macropoma*. Ossifications within the ethmosphenoid include the lateral ethmoids and basisphenoid, and in the otico-occipital region there are prootic, basioccipital and supraoccipital ossifications. The lateral ethmoid forms the anterior half of a well-marked ventrolateral fossa and anteriorly the bone is pierced by a foramen for the ventral branch of the naso-basal canal. The prootic shows a well-developed posterior wing which sutures with the basioccipital (Reis, 1888: pl. 1, fig. 22). Dorsally the prootic is extended as a process to brace against the descending process of the postparietal. A small supraoccipital has been described by Aldinger (1930). The parasphenoid is narrow posteriorly but broadens considerably in the anterior half, where there are prominent ascending wings forming part of the ventrolateral fossa. Teeth are confined to the anterior half of the ventral surface.

Holophagus

Only parts of the otico-occipital are known (Fig. 6.9) and in all details it is similar to *Macropoma*. The prootic shows a prominent posterior wing which is sutured with a large basioccipital. There is a small supraoccipital which, except for its small size, is similar to that of *Latimeria*.

Macropoma

The neurocranium of *Macropoma* has been described and commented on by Huxley (1866), Woodward (1909), Watson (1921) and Stensiö (1932). It is known by many specimens of both *M. lewesiensis* and *M. precursor* in which details are very similar. As in most Mesozoic coelacanths, the neurocranium is developed as a few well-separated ossifications; a composite restoration is shown in Fig. 6.10.

The ethmosphenoid is ossified as paired lateral ethmoids (pre-ethmoid of Stensiö,

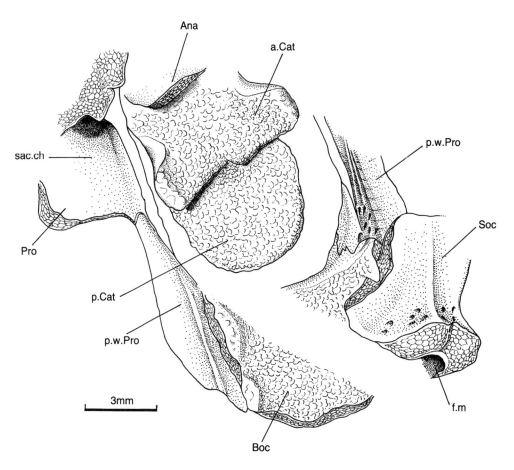

Fig. 6.9 *Holophagus gulo* Egerton. Camera lucida drawing of the posterior part of the braincase as preserved in BMNH P.875 showing a well-ossified supraoccipital and the complex suture between the prootic and basioccipital.

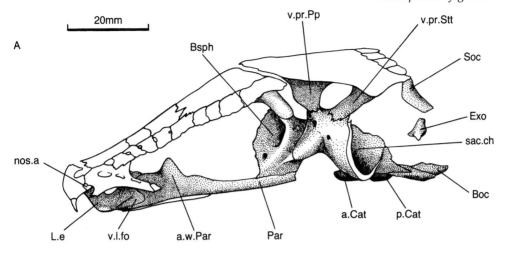

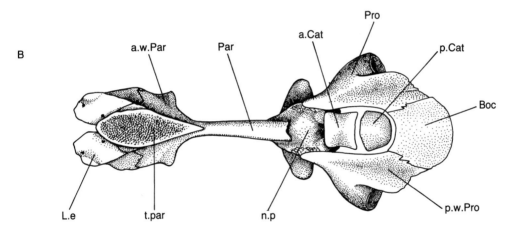

Fig. 6.10 *Macropoma.* Restoration of the braincase based on composite of *M. lewesiensis* (Mantell) and *M. precursor* Woodward. (A) Left lateral view; (B) ventral view. From specimens in BMNH collection.

1921; part of prevomer of Watson, 1921) and the basisphenoid with the dermal para-sphenoid and vomers. The lateral ethmoid forms the floor and part of the lateral wall of the nasal capsule. The medial and posterior margins passed into cartilage but the dorsal margin of the lateral wall is perfectly finished in perichondral bone (Fig. 6.12(C)). The lateral face is marked anteriorly with a roughened area which receives the ventral process of the lateral rostral (Fig. 6.11, v.pr.L.r). Anterior to this roughened area, there is a notch marking the position of the anterior nostril (Figs 6.10, 6.11, nos.a) and behind the roughened area there is a larger notch for the buccal canal. Posteriorly the lateral wall of the lateral ethmoid rises stee-ply and is notched for the posterior nostril.

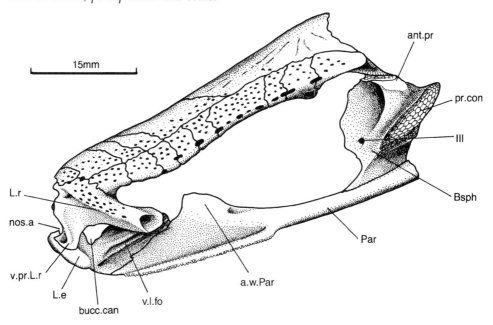

15mm

Fig. 6.11 Restoration of the ethmosphenoid portion of braincase of *Macropoma precursor* Woodward in left lateral view, to show especially the relationship between the lateral rostral, lateral ethmoid and parasphenoid.

The ventrolateral fossa (Fig. 6.11, v.l.fo) is shallow but it is clearly demarcated dorsally by an oblique ridge which matches the dorsal contour of the autopalatine in articulated specimens. The ventral margin of the ventrolateral fossa is pitted and rugose, suggesting the site of insertion of ligaments which bound the palate to the braincase. Several foramina pierce the floor of the lateral ethmoid. A large anteroventrally directed foramen pierces the anterior tip and corresponds to a similarly orientated foramen in *Eusthenopteron* (Jarvik, 1972: Fig. 56) for the ventral branch of the naso-basal canal, allowing anterior branches of the profundus to innervate the snout. Other foramina are not consistent from specimen to specimen and must have conveyed minor branches of the buccal nerve.

The basisphenoid (Fig. 6.12(A,B,D)) is very similar to that of *Latimeria*. Both the antotic processes and the sphenoid condyles are well developed. The entire anterior face passed into cartilage. The processus connectens ends anteroventrally well short of the parasphenoid. The foramen for the profundus lies in the usual position, deep within the suprapterygoid fossa, and the oculomotor foramen lies anteroventrally. There is no separate foramen for the pituitary vein and this probably entered the pituitary fossa through cartilage. The anterior face of the antotic process (Fig. 6.12(A), ant.pr) bears strong ridges and these may have provided anchorage for the adductor palatini muscle. The antotic process is not perforated and the superficial ophthalmic and trochlear nerves almost certainly passed above, within a prominent lateral groove in the parietal descending process.

The parasphenoid is very narrow and tubular beneath the orbit (Fig. 6.12(C,D)): it is open posteriorly, at the level of the ventral fissure. The anterior end of the parasphenoid

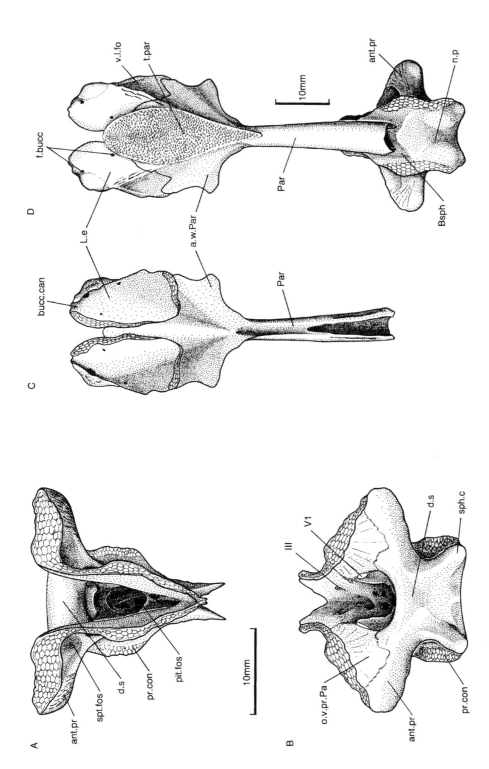

Fig. 6.12 *Macropoma lewesiensis* (Mantell). Individual braincase elements: (A) anterior and (B) dorsal views of basisphenoid of BMNH P.33540; (C) dorsal and (D) ventral views of parasphenoid, lateral ethmoids and basisphenoid. Based on BMNH 4237, P742, P.33540.

expands rapidly and is upturned as prominent lateral wings (a.w.par) which partly embrace the posterior ends of the lateral ethmoids. Small villiform teeth cover the anterior half of the parasphenoid. Immediately anterior to the parasphenoid, there are small paired vomers which fail to meet one another in the midline (Watson, 1921, has described the vomers – his prevomers – but he failed to distinguish the dermal vomers from the overlying lateral ethmoids).

The otico-occipital portion of the neurocranium shows prootics, opisthotics and median supraoccipital, basioccipital, and zygal plates. The prootic (Fig. 6.13) has a short, steeply inclined otic shelf (ot.sh), the inner surface of which is roughened to form the groove receiving the processus connectens (Fig. 6.13(B)). The dorsal edge of the prootic passed into cartilage as the inner wall of the jugular canal and, as in *Latimeria*, it fails to meet the skull roof. There are two roughened areas on the lateral wall of the jugular canal. The anterior area, which is developed as the prefacial eminence, receives the postparietal descending process (Fig. 6.13(A), o.v.pr.Pp) and the posterior area is overlapped by the descending process of the supratemporal (o.v.pr.Stt). The inner wall of the jugular canal is pierced by a ventrolaterally directed facial foramen, which allowed the hyomandibular trunk and the palatine rami to pass from the cranial cavity. In *Macropoma* the palatine nerve pierced the floor of the jugular canal (f.pal.VII) to emerge on the lateral face of the otic shelf ventral to the prefacial eminence. There is an additional foramen beneath the overlap with the supratemporal, which gave passage to the orbital artery (Fig. 6.13(A), f.o.a) into the base of the jugular canal. Neither of these two foramina is present in *Latimeria*. The supratemporal excavation was largely cartilage-lined and, as in *Latimeria*, the middle lateral line nerve must have passed out through cartilage. One specimen of a prootic (Fig. 6.13, f.ot.VII) shows a small foramen posterodorsal to the

facial foramen and this may represent a bone-enclosed middle lateral line nerve. This appears to be an exceptional condition for *Macropoma*.

Posterior to the otic shelf, the prootic is inflated to form the medial, lateral and ventral walls of the saccular chamber. The lateral wall passed into cartilage which formed the hyomandibular facet, the medial wall is more complete and separated the saccular chamber from the notochordal canal. The ventral floor passes posteriorly as a process which shows a complex interdigitating suture with the basioccipital.

The semicircular basioccipital (Fig. 6.10, Boc) is open ended posteriorly and laterally, suggesting that it passed to cartilage. It is perichondrally lined anteriorly as it formed the posterior boundary of the basicranial fenestra. The dorsal surface is roughened where it lay against the notochord.

The supraoccipital is very poorly known (Watson, 1921: 324) and may have been ossified in large specimens only. BMNH P.63999 shows this bone to be trapezoidal with the wider ventral edge notched for the foramen magnum. Most of the bone is very thin and all edges of the bone passed into cartilage. Only the rim of the foramen magnum is thickened to match the enlarged first neural arch. Behind this level, the first three neural arches are enlarged and closely articulated to one another through cartilaginous interfaces. There is no neural spine on the first neural arch: a clearly developed spine is only apparent on the third and succeeding neural arches.

Watson (1921) described an opisthotic (exoccipital) for *Macropoma lewesiensis* but I have been unable to confirm his observations. However, one is developed in *M. precursor*, where it is rather differently shaped from the square bone described by Watson. In BMNH P.11002 it is a bone of complex shape, more easily illustrated than described. Although the bone appears to have shifted from its original position in this

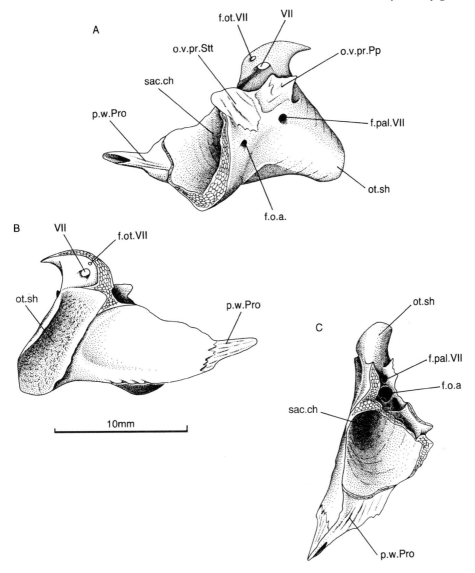

Fig. 6.13 *Macropoma precursor* Woodward. Restoration of isolated prootic based on BMNH P.10916, P.11002, P.12887. (A) right lateral view. (B) Medial view. Note prominent groove on inside of the otic shelf which receives the processus connectens of the basisphenoid. (C) Dorsal view.

specimen, an approximate orientation remains and it has been restored in Fig. 6.10. It has a lateral and a posterior face, both of which pass into cartilage except ventrally, where the ventral margin is perichondrally lined and strongly excavated immediately below a prominent parampullary process. In this position, the excavation corresponds to the dorsal half of the glossopharyngeal foramen, but it appears much larger than would be expected for such a nerve foramen. It is therefore possible that an open vestibular fontanelle was present, including passage for the glossopharyngeal. The parampullary process bears complex roughened areas: an anterior area marks the site of insertion of the first suprapharyngobranchial and behind this there are roughened areas for the origination of the branchial elevator muscles 2, 3 and 4.

The zygal plates are developed as in other coelacanths. The anazygal is short and broad, with closely spaced anterior facets articulating with the sphenoid condyles. The anterior catazygal is long and saddle-shaped, in contrast to the very small hemispherical posterior catazygal.

6.3 CHARACTER DESCRIPTIONS

The following neurocranial characters potentially useful for establishing relationships between coelacanths have been recognized here. The codings for each taxon are given in Table 9.1.

69. Orbitosphenoid and basisphenoid regions co-ossified (0), separate (1).
70. Basisphenoid extending forward to enclose the optic foramen (0), optic foramen lying within separate interorbital ossification or cartilage (1).
71. Processus connectens meeting parasphenoid (0), failing to meet parasphenoid (1).
72. Basipterygoid process absent (0), present (1).
73. Antotic process not covered by parietal descending process (0), covered (1).
74. Temporal excavation lined with bone (1), not lined (0).
75. Otico-occipital solid (0), separated to prootic/opisthotic (1).
76. Supraoccipital absent (0), present (1).
77. Vestibular fontanelle absent (0), present (1).
78. Buccohypophysial canal opening through parasphenoid (1), closed (0).
79. Parasphenoid without ascending laminae anteriorly (0), with ascending laminae (1).
80. Suprapterygoid process absent (0), present (1).
81. Vomers not meeting in the midline (0), meeting medially (1).
82. Prootic without complex suture with the basioccipital (0), with complex suture (1).
83. Superficial ophthalmic branch of anterodorsal lateral line nerve not piercing antotic process (0), piercing antotic process (1).
84. Process on braincase for articulation of infrabranchial 1 absent (0), present (1).
85. Separate lateral ethmoids absent (0), present (1).
86. Separate basioccipital absent (0), present (1).
87. Dorsum sellae small (0), large and constricting entrance to cranial cavity anterior to the intracranial joint (1).

PALATE, HYOID AND GILL ARCHES

7.1 PALATE

The palate, hyoid arch and the gill arches are all very distinctive in coelacanths, although usually only the palate is well preserved in fossil coelacanths, together with the ceratohyal, ceratobranchials and the urohyal, which are discussed below.

The palate (Fig. 7.1) is characteristic in shape, usually being described as triangular. The endoskeletal part of the palate consists of an autopalatine (Aup) anteriorly, a quadrate (Q) posteroventrally and metapterygoid (Mpt) dorsally. The autopalatine is partly ossified but in *Latimeria* continues posteriorly as a narrow tongue of cartilage (Fig. 7.1, Aup). This is probably also the case for fossil coelacanths because there is usually a small triangular ossification anteriorly which is open ended along the posterior margin (Fig. 7.2). The metapterygoid and quadrate are linked by cartilage in *Latimeria*, and in some coelacanths the ventral end of the metapterygoid and the dorsal end of the quadrate are similarly open ended (Fig. 7.2), suggesting a similar cartilage infill. However, in *Latimeria* the anterior autopalatine cartilage and the posterior cartilage do not join each other, and therefore the original palatoquadrate cartilage must form from two centres, anterior and posterior.

The autopalatine articulates against the ethmoid region along its dorsal edge, which sometimes bears a small process. However, the articulation is always a simple contact, unlike the complex head seen in osteolepiforms and onychodonts, and unlike the condition in porolepiforms in which the autopalatine is received within a deep fossa autopalatine (Jarvik, 1972). In primitive coelacanths such as *Diplocercides*, and especially *Euporosteus*, the autopalatine fits into a deep ventrolateral fossa on the floor of the nasal capsule. In more derived coelacanths, the lateral wall of the lateral ethmoid is marked by a shallow trough rather than a deep fossa. In *Latimeria*, even though there is no deep ventrolateral fossa, the contact between the autopalatine and lateral ethmoid is very tight, with a strong ligament connecting the two, and there is very limited movement at this point.

The metapterygoid has a complex articulatory dorsal surface which is usually saddle-shaped. In *Latimeria* this articulates (Fig. 7.1, art.ant.pr) against the ventral surface of the antotic process and there is some lateral movement at this point. This appears to be the condition in most fossil coelacanths. In some primitive coelacanths (*Diplocercides*, *Euporosteus* and *Sassenia*), there is an additional suprapterygoid process developed on the neurocranium which is the original metapterygoid–braincase contact seen in other primitive sarcopterygians. In those coelacanths that have both processes, the suprapterygoid process lies anteriorly to the antotic process and the dorsal tip of the metapterygoid. Presumably the metapterygoid was connected to the suprapterygoid process by ligaments. In *Diplocercides* the metapterygoid is very broad antero–posteriorly (Fig. 7.1, Mpt), and towards the anterior margin there is a well-developed pit

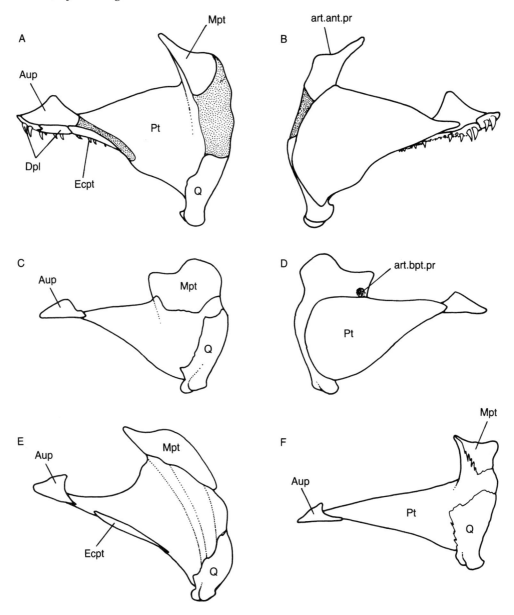

Fig. 7.1 Outline drawings of left palates of four coelacanth species to show variation in shapes and proportions. (A and B) *Latimeria chalumnae* Smith; (A) lateral view, cartilage stippled; (B) medial view, cartilage stippled; both based on Millot and Anthony (1958). (C and D) *Diplocercides kayseri* (von Koenen): (C) lateral view; (D) medial view; both based on Jarvik (1980). (E) *Rhabdoderma elegans* (Newberry), lateral view. (F) *Axelrodichthys araripensis* Maisey, lateral view.

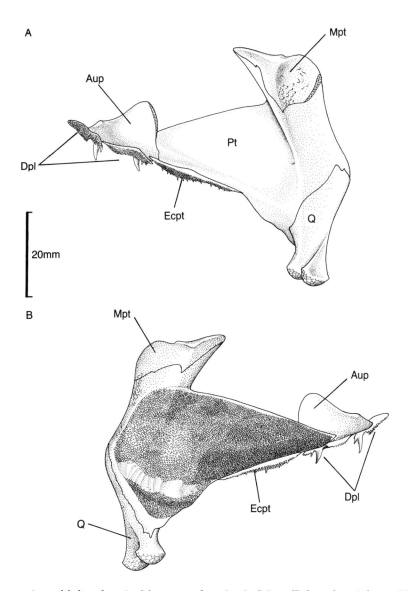

Fig. 7.2 Restoration of left palate in *Macropoma lewesiensis* (Mantell) based mainly on BMNH P.33540: (A) lateral view; (B) medial view. Note the swollen contour of the ventral pterygoid margin seen also in *Latimeria*. The band devoid of teeth on the pterygoid is probably restricted to this specimen.

(art.bpt.pr) for the articulation with the basipterygoid process. *Diplocercides* is the only coelacanth known to have this basal articulation, but it is suspected that other less well-known coelacanths (e.g. *Euporosteus*) had such an articulation. An articulation between the metapterygoid and the otic shelf

– the paratemporal articulation of *Eusthenopteron* (Jarvik, 1954, 1980) – is absent in coelacanths. This absence of a paratemporal articulation might be expected in fishes showing a mobile intracranial joint.

The quadrate (Q) is always a robust bone orientated vertically and ending in a large double condyle for articulation with the lower jaw. The two halves of the condyle are offset such that the outer condyle lies posterior to the inner condyle. This asymmetry matches the glenoid on the lower jaw.

The dermal palate is formed by the pterygoid (Pt), dermopalatines (Dpl) and an ectopterygoid (Ecpt). The pterygoid (homologous with the posterior ectopterygoid of actinopterygians; Gardiner, 1984a) is the most distinctively shaped element in the coelacanth palate. It is usually described as being triangular (Cloutier, 1991a,b) and it is certainly of a very different shape from that seen in actinopterygians and other sarcopterygians. But within coelacanths the shape varies considerably (Figs 7.1, 5.4). In *Latimeria*, *Macropoma* (Fig. 7.2), *Holophagus* and *Undina* it is almost as deep as long, whereas in coelacanths belonging to the *Mawsonia* clade (p. 235) it is very shallow and elongate (Fig. 7.1(F)), reflecting differences in the proportions of the head. Further, in *Latimeria* and *Macropoma* (Fig. 7.2) the ventral margin of the pterygoid is swollen immediately in front of the quadrate. The significance of this is uncertain. In others (e.g. *Rhabdoderma*, Fig. 7.1(E)) it is more cleaver-shaped. To some extent the shape of the palate can be used to recognize coelacanth genera, but it is difficult to quantify the subtle differences.

Anteriorly and lying along the lateral edge of the autopalatine, there are two dermopalatines (prevomer of Smith, 1939c) (Fig. 7.1, Dpl) set in tandem. Each of these bones in *Latimeria* carries a covering of small teeth as well as large fangs together with replacement sockets (Millot and Anthony, 1958: pl. 38). Dermopalatines are present in many fossil coelacanths (Fig. 7.2), and are probably universally present, but they are only loosely associated with the autopalatine and are often missing in specimens. This contrasts with the palate of osteolepiforms and porolepiforms in which the dermopalatines are tightly sutured to the pterygoid. In *Macropoma* there are also fangs present (Fig. 7.2).

Posterior to the second dermopalatine, there is an elongate ectopterygoid which in *Latimeria* and *Macropoma* is covered by a shagreen of small teeth. Schaeffer (1941) suggested that some Palaeozoic coelacanths showed fusion between the pterygoid and the ectopterygoid, but this appears not to be the case. In some early literature (Huxley, 1866; Nielsen, 1936) the dermopalatines were identified as the maxilla.

7.2 HYOID ARCH

The hyoid arch and associated musculature of *Latimeria* has been described by Millot and Anthony (1958) and Lauder (1980). The arch consists of hyomandibular, interhyal, ceratohyal, hypohyal and a symplectic. Only the ceratohyal and symplectic show any ossification, and in fossil coelacanths these are usually the only elements preserved. Therefore we can say very little about variation in the hyoid arch.

The hyomandibular lies free from the palate and this is a derived condition amongst osteichthyans. It is also very short compared with that in most sarcopterygian fishes. In *Latimeria* the hyomandibular is quadrangular (Fig. 7.3, Hy; Millot and Anthony, 1958: pl. 43) and articulates along its anterodorsal edge with the hyomandibular facet. The posterodorsal edge is produced to a small and ill-defined opercular process. An opercular process is usually absent from the sarcopterygian hyomandibular but is well developed in *Polypterus* and actinopterans except some palaeoniscids (Gardiner, 1984a). Such a distribution would suggest that it has arisen on more than one occasion. The medial surface is pierced by the hyomandibular canal, which carries the hyomandibular trunk of the facial nerve to emerge onto the outer face. All branches that make up the hyomandibular trunk pass through this single canal, but as soon as the nerve exits on the lateral face it divides to the internal and external mandibular rami as well as the hyoidean

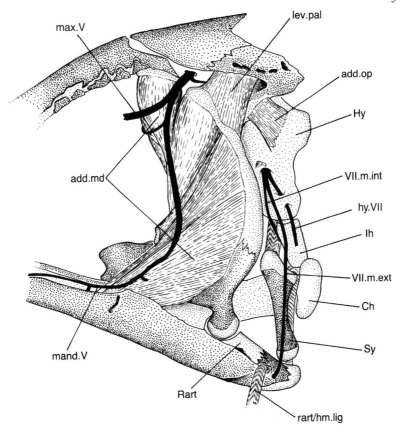

Fig. 7.3 *Latimeria chalumnae* Smith. Drawing of dissected BMNH 1976.7.1.16 in lateral view to show relationship between hyoid arch, palate and lower jaw and associated muscles, ligaments and principal nerve rami. Cartilage shown as light stipple, bones as dark stipple. Note the separate divisions of the adductor mandibulae which are seen in lateral but not medial view (Fig. 7.4). For scale see Fig. 2.2

ramus (Fig. 7.3). The last mentioned then passes through a narrow bridge of cartilage.

The ventral end of the hyomandibular articulates with a small interhyal cartilage (Fig. 7.3, Ih) which is sometimes called the stylohyal. This, in turn, articulates with the ceratohyal (Ch), which is very long and curves forward to the level of the anterior end of the basibranchial cartilage, with which it articulates via a small cartilaginous hypohyal. Ossification within the ceratohyal is confined to the central part of the ceratohyal and there are large areas of cartilage

at either end. The posterior cartilaginous tip is particularly large and swollen and Millot and Anthony (1958) distinguish this as a stylohyal. It is, however, part of the ceratohyal and serves as the site of origin of part of the geniohyoideus muscle. The ventral margin of the ceratohyal is swollen at about two-thirds distance from the anterior end but this provides no special site of muscle insertion. However, the swelling gives a very characteristic shape to the ceratohyal, allowing fossils to be instantly recognized as belonging to coelacanths. The dorsal surface of the ceratohyal is covered by skin in which are embedded tiny tooth plates.

The symplectic (Fig. 7.3, Sy) is developed as a perichondral ossification around a rod-shaped cartilage which is orientated vertically from the retroarticular to articulate with both the ceratohyal and the interhyal. The dorsal and ventral tips are capped with cartilage, and because the dorsal end is expanded to form a double articulation, the ossified portion as preserved in fossils has a characteristic shape. The symplectic is a synapomorphy of coelacanths even though similarly named bones occur in neopterygians and Recent chondrosteans. Patterson (1982) has shown that the bone named symplectic has different topological relationships in members of each of these groups and, furthermore, the symplectic of chondrosteans may be a modified interhyal (Patterson, 1982) whereas the neopterygian and coelacanth symplectics are independently derived neomorphs.

As mentioned above, the ceratohyal is usually preserved in fossil coelacanths as a long curved bone with a prominent ventral expansion (Fig. 7.7, Ch). The anterior and posterior ends were capped in cartilage as was the ventral edge of the expansion. An ossified hypohyal has never been found in coelacanths. The symplectic is usually preserved in fossils. Although the shape is constant, the length varies considerably from species to species. However, it is difficult to know what this might mean in functional terms, because the ends were usually cartilaginous (an exception might be *Laugia*, Fig. 4.10) and the more dorsal ends of the hyoid arch were not ossified. In *Diplocercides*, the hyomandibular and the interhyal (stylohyal of Jarvik, 1980) are each partly ossified and, in all important respects, are similar to those of *Latimeria*. *Laugia* also shows a partly ossified hyomandibular. Here, at least the dorsal half of the hyomandibular is perichondrally ossified and this passes into a cartilage-capped articulatory head (Fig. 6.6, Hy). In

Laugia there also appears to be a pronounced opercular process (Fig. 6.6(A)).

The topological relationship between the hyoid arch and the palatoquadrate and the lower jaw is shown in Fig. 7.3 together with some of the muscle and ligaments associated with jaw opening and closing. This figure shows the tandem jaw articulation which is a synapomorphy of coelacanths (p. 299).

The adductor mandibulae muscles are the most prominent and in lateral view (Fig. 7.3, add.md) they are arranged in three units which have been called antero-superior, middle vertical, and superficial postero-inferior. The fibres within each of these divisions do show slightly different orientations, but the distinction between these muscle bundles is not so evident in medial view (Fig. 7.4). The antero-superior division takes origin from the cartilage of the antotic process and the dorsal edge of the metapterygoid. The fibres pass vertically down and there is some fusion with fibres of the middle vertical division. In medial view the fibres of the antero-superior division converge to a tendon (Fig. 7.4) which inserts on to the medial edge of Meckel's cartilage just behind the principal coronoid.

The fibres of the middle division originate high up on the posterior edge of the cartilage, which flanks the metapterygoid, and from here they run anteroventrally to join distally with the fibres of the postero-inferior division. The postero-inferior division originates along a broad posterior edge of the palatoquadrate cartilage and it can be divided to three groups of fibres. Dorsally there are a few fibres which run to insert into the maxillary fold (these fibres are shown cut in Fig. 7.3). A middle group of fibres converge to a thin strand of muscle which enters the meckelian fossa. Ventrally there is a broad sheet of fibres which insert to the meckelian fossa at an oblique angle.

The palatal levator muscle (Fig. 7.3, lev.pal) takes origin within a deep temporal

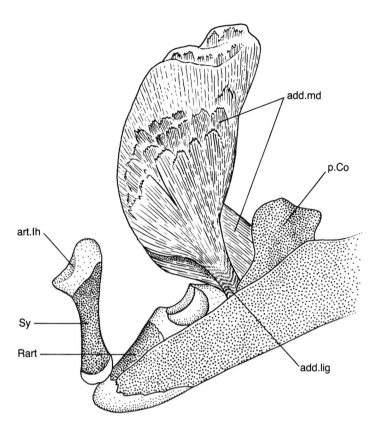

Fig. 7.4 *Latimeria chalumnae* Smith. Drawing of dissected BMNH 1976.7.1.16 in medial view to show relationship between adductor mandibulae and lower jaw. Cartilage shown as light stipple, bones as dark stipple. For scale see Fig. 2.2.

excavation and passes anteroventrally to insert along the posterodorsal edge of the metapterygoid. The paired basicranial muscle originates on the undersurface of the basioccipital and prootic and passes forward to insert at the base of the ascending wing of the parasphenoid.

In addition there are some muscles associated with the lower jaw and ceratohyal important for the jaw functioning. In particular, the sternohyoideus extends from the ceratohyal to the pectoral girdle, and the coracomandibularis (geniohyoideus of Thomson, 1973) reaches from the symphysis of the mandible to the pectoral girdle. There are also powerful ligaments which link together the various parts of the hyoid arch and

symplectic with the palate and the lower jaw. Chief among these are the retroarticular/hyomandibular ligament (anterior mandibulohyoid ligament of Lauder, 1980) (Fig. 7.3, rart/hm.lig), which stretches between the lateral face of the hyomandibular and the lateral face of the retroarticular; the posterior mandibulohyoid ligament (inferior quadrate–hyoidean ligament of Millot and Anthony, 1958), which runs on the inner side of the symplectic between the interhyal and the inner surface of the retroarticular; and the inferior quadratohyoid ligament, which attaches the posterior end of the ceratohyal to the inner face of the quadrate. These ligaments are shown as hatched lines in Fig. 7.5(A).

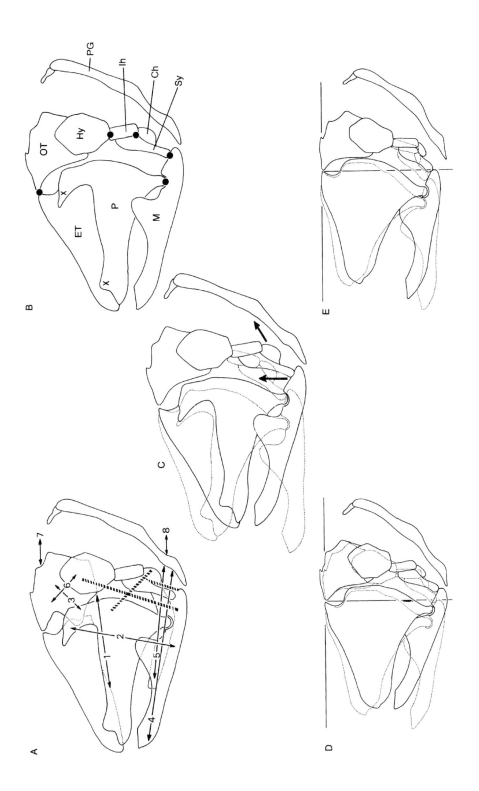

Fig. 7.5 Diagrams to illustrate model of jaw mechanics (after Lauder, 1980).

(A) Diagrammatic head and shoulder girdle showing lines of action of principal muscles involved in opening and closing the jaw and intracranial joint, together with ligaments that bind together the mandibular and hyoid arches. The hypaxial muscles have been categorized as abductors but their role is probably to stabilize actions of both abductors and adductors. Key to muscles: 1, basicranial muscle; 2, adductor mandibulae; 3, levator arcus palatini; 4, coracomandibularis; 5, geniohyoideus; 6, levator hyomandibularis; 7, epaxial muscles; 8, hypaxial muscles.

(B) Diagrammatic head showing skeletal elements and rotational joints (solid circles). The palate is securely tied to the ethmosphenoid through the autopalatine and the metapterygoid (X). These positions allow lateral movement only. Abbreviations: Ch, ceratohyal; ET, ethmosphenoid; Hy, hyomandibular; Ih, interhyal; M, mandible; OT, otico-occipital; P, palate; PG, pectoral girdle; Sy, symplectic.

(C) The skull in resting position (solid black line) and with joint and jaw abducted (dotted line). During mouth opening, the intracranial joint moves through a maximum of 10°, the lower jaw through 30°. The diagonal arrow shows the direction of movement of the posterior end of the ceratohyal following contraction of the sternohyoideus. This movement is probably also accompanied by a lateral force directed at the symplectic/interhyal joint. The posterodorsal movement of the ceratohyal is translated to the symplectic/retroarticular joint where the reroarticular is pulled dorsally, causing the mandible to be depressed. The most important vector of this movement is that shown by the vertical arrow.

(D and E) Two stages in the model of jaw opening sequence. (D) From the resting stage (solid lines), contraction of the epaxial muscles (7) lifts the entire head and moves the lower end of the palate forward. At the same time, contraction of the sternohyoideus (5) causes the vertical vector ((C), shown in above) to lower the jaw (dotted line). (E) From a partially open jaw (solid line corresponding to the dotted line in (D)), the symplectic/retroarticular joint moves just anterior to the vertical level of the intracranial joint and the vertical vector now pushes the palate dorsally and opens the intracranial joint. Throughout these movements the hypaxial muscles probably stabilize the shoulder girdle. Closure of the intracranial joint and jaws is effected by contraction of the adductor mandibulae (2), levator arcus palatini (3) and basicranial muscles (1).

7.3 JAW MECHANISM

The relationships of skeletal elements, the direction of contraction of the muscles and the path of the ligaments are shown in schematic form in Fig. 7.5(A), together with moveable articulations (Fig. 7.5(B)). The palate is secured to the ethmosphenoid portion of the braincase through an ethmoid articulation anteriorly and a metapterygoid articulation posterodorsally (Fig. 7.5(B), X). The palate may move laterally relative to the ethmosphenoid, but independent movement is impossible in the sagittal plane. Therefore the palate and the ethmosphenoid move up and down together. The only relative movement between these two parts is lateral movement of the palate upon the ethmosphenoid.

The lower jaw articulates with both the palate and the hyoid arch in a distinctive tandem jaw articulation which is one of the synapomorphies of coelacanths. The usual sarcopterygian articulation is via a biconcave ball-and-socket joint between the quadrate on the palate and the articular and retroarticular in the lower jaw. The symplectic joint occurs more posteriorly between the tip of the retroarticular and the ventral end of the symplectic (Fig. 7.5(B), Sy). This joint is very simple, with two juxtaposed convex cartilaginous surfaces coming together and bound by connective tissue. This joint appears to be able to resist only tensional forces. The dorsal end of the symplectic articulates with the interhyal (Fig. 7.5(B), Ih) and with the posterior end of the ceratohyal. The plane of articulation with the interhyal (Fig. 7.4, art.Ih) is oblique and seems to allow the symplectic to flex laterally relative to the interhyal. Further dorsally there is a diarthrosis between the interhyal and the hyomandibular which allows limited antero–posterior rotational movement of the interhyal on the hyomandibular. In turn, the hyomandibular articulates with the braincase via a double articulation. The principal movement of the hyomandibular on the braincase is lateral, but because the plane of this articulation is very oblique, this has the effect of moving the distal end of the hyomandibular forward.

The intracranial joint between the ethmosphenoid and the otico-occipital is also part of this jaw mechanism. It consists of a dorsal part between the dermal bones of the skull roof, as well as an endochondral part between the basisphenoid and the prootic portion and anazygal within the otico-occipital portion of the neurocranium. Movement about the intracranial joint is kept strictly in the vertical plane by the presence of the track and groove between the prootic and the basisphenoid.

Several authors have examined cine film (Thomson, 1973) or manipulated freshly dead or thawed specimens (Thomson, 1966, 1973; Alexander, 1973; Robineau and Anthony, 1973). All have noted that depression of the lower jaw is followed by a raising of the intracranial joint. Thomson (1970) noted that the first few degrees of jaw abduction occur without any obvious movement of the intracranial joint. Thereafter, for every 3° that the symphysis of the lower jaw is depressed, then the snout is elevated through 1°. This will happen until the intracranial joint moves through about 10° and the lower jaw through 30°. It was also noted by Thomson (1966) that the tip of the lower jaw moves forward considerably during abduction. There appears to be no detectable cheek movement during mouth opening.

Several attempts have been made to explain the cycle of mouth opening and closing. As yet there have been no electromyographic studies such as have been performed on actinopterygian fishes (Lauder, 1980) and therefore all proposed mechanisms are based on indirect evidence.

Adduction of the jaw and lowering of the snout (adduction of the intracranial joint) can be accounted for by the contraction of the adductor mandibulae (Fig. 7.5, 2), levator arcus palatini (Fig. 7.5, 3) and basicranial muscles (Fig. 7.5, 1). But the opening of the lower jaw and raising of the snout (abduction of the intracranial joint) has been difficult to explain because there are no muscles obviously placed to perform this function.

Early hypotheses (Millot and Anthony, 1958; Thomson, 1967, 1970; Cracraft, 1968; Robineau and Anthony, 1973; Adamicka and Ahnelt, 1976) suggested that the contraction of the coracomandibularis muscles (geniohyoideus of Thomson, 1970) depressed the lower jaw and resulted in a posterodorsal force behind the quadrate/articular joint. This force was assumed to be translated through the symplectic and the hyoid arch into an anterior movement of both the lower jaw and the palate, thereby raising (abducting) the intracranial joint. The hypothesized details of how this force was translated vary from author to author (review: Lauder, 1980), but the entire system has been interpreted as a four-bar linkage system.

As Lauder (1980) has pointed out there are four major criticisms of these suggested mechanisms. First, measurements taken during manipulated mouth opening imply that the coracomandibularis (Fig. 7.5(A), 4), which runs from the shoulder girdle to the mandibular symphysis, must increase in length while it is contracting. (Thomson explained this anomaly by suggesting that the shoulder girdle moved forward to compensate.) Second, the line of action of the coracomandibularis appears to pass above the level of the quadrate/articular joint (Fig. 7.5(A)) and therefore contraction of this muscle should act to close the jaw. Third, the proposed mechanisms imply that the symplectic and hyoid arch form a rigid stucture, able to withstand compression. In fact this is not the case: there are several joints and points of contact, at any one of which there may be movement. And in particular, the symplectic/retroarticular joint does not appear well enough defined to be able to transmit a compressional force. Fourth, some of the proposed mechanisms (e.g. that of

Robineau and Anthony, 1973) depend on ligaments acting as compression members. All of these criticisms suggest that jaw opening does not involve the coracomandibularis muscle.

Lauder (1980) has outlined a different sequence of events, based on the assumption that jaw opening is initiated in fundamentally the same way as in primitive actinopteryians. This model, outlined in Fig. 7.5, proposes that the initial phase of jaw opening is brought about by the contraction of epaxial muscles (Fig. 7.5(A), 7), which raise the entire head and thrust both the palate and jaws forward. It is possible that contraction of the hypaxial muscles (Fig. 7.5(A), 8) stabilizes movement of the shoulder girdle. At the same time, contraction of the sternohyoideus (Fig. 7.5(A), 5) pulls the ceratohyal posterodorsally and probably laterally. The posterior end of the ceratohyal is attached to both the symplectic and the interhyal, and this force is translated to a tensional force, which acts through the symplectic and the posterior mandibulohyoid ligament to lift the retroarticular process and depress the mandible. At this stage, the forward thrust of the lower jaw brings the symplectic/retroarticular joint anterior to the level of the intracranial joint. The dorsal vector of the force applied by the posterodorsal movement of the ceratohyal now acts to raise the palate and ethmosphenoid and to abduct the intracranial joint. Lauder (1980) has calculated that the length of the moment arm around the jaw joint means that the symphysis will move three times as far as the retroarticular/symplectic joint. There is therefore the potential for very rapid jaw opening.

The hyomandibular of *Latimeria* is assumed to be relatively immobile, unlike that of many fishes. Instead, the key joint lies between the distal end of the hyomandibular and the interhyal (Alexander, 1973). But it is probable that the hyomandibular is capable of some movement because there is a sliver of levator hyomandibularis muscle (Fig.

7.5(A), 6). Contraction of this muscle would cause the distal end of the hyomandibular to move medially. This would probably act to reinforce the action of the sternohyoideus.

7.4 BRANCHIAL ARCHES AND UROHYAL

The branchial arches of *Latimeria* have been described by Millot and Anthony (1958) and Nelson (1969), and commented on by Rosen *et al.* (1981). Little remains of the gill arches in fossil coelacanths. In some respects the gill arches of *Latimeria*, and presumably other coelacanths, are unusual amongst osteichthyans.

In *Latimeria* five arches are present (Fig. 7.6(A)). Each arch is represented as a ceratobranchial, the first four arches show epibranchials, and pharyngobranchials are associated with the first two arches. There are no separate hypobranchials and in this respect *Latimeria* is similar to lungfishes (Rosen *et al.*, 1981, suggested that the shapes of the ventral ends of the third and fourth ceratobranchials, as well as a small independent cartilage on one side of one specimen of *Neoceratodus*, imply that hypobranchials may be present in at least some of the arches).

The basibranchial is large and diamond-shaped and formed mostly of cartilage with a single embedded central ossification (Fig. 7.6(B)). The cartilage is single but in AMNH 32949SW the dorsal surface appears to show a partial division (Fig. 7.6(A)) into a large anterior portion articulating with the first arch and a smaller posterior portion receiving the ventral ends of arches 2–4. In some other sarcopterygian fishes (the lungfishes *Griphognathus*, *Chirodipterus* – Miles, 1977, the osteolepiforms *Medoevia* – Lebedev, 1995 and *Mandageria* – Johansen and Ahlberg, in press) there appears to be a large anterior basibranchial articulating with arches 1 and 2, followed by a much smaller basibranchial supporting the third arch. This is a pattern seen also in *Eusthenopteron*, in which the basibranchial series (incompletely known) is more elongated than in most other sarcopterygians. In primitive actinopterygians (*Mimia* and *Moythomasia*) there is a single basibranchial supporting the first four arches. However, Gardiner (1984a) maintains that this single basibranchial is the result of ossification from three separate centres. In elasmobranchs there are primitively two basibranchials. Thus, a reduction to a single basibranchial in coelacanths may be a derived feature. In fossil coelacanths only, the central ossification is seen as a rounded element and this shows a dorsal and ventral perichondral surface surrounded by a perimeter of cartilage surfaces (Figs 7.6(B), 7.7, Bb). The dorsal surface is flat or gently rounded. The ventral surface is marked by pits and by an articulation surface (Fig. 7.6(B), art.Uhy) for the urohyal (Uhy). The endochondral urohyal is a characteristic bone in coelacanths. It is dorsoventrally flattened (the urohyal of other sarcopterygians is compressed from side to side), narrow anteriorly and expanded posteriorly where it ends in a bifid tip. It is very constant in shape in coelacanths.

The tooth plates associated with the basibranchial in *Latimeria* (Fig. 7.6(A), t.p.Bb, Fig. 7.6(I)) are represented as two paired plates with a small diamond-shaped plate wedged in the midline. Surrounding these large tooth plates there are many small and irregularly sized tooth plates which lie free in the skin (Nelson, 1969: pl. 81, fig. 1). The tooth plates bear tiny villiform teeth. Nelson (1969) demonstrated that the primitive osteichthyan basibranchial dentition consisted of many small paired plates and that there was a trend towards consolidation into larger tooth plates and, in some, a fusion in the midline to form median plates. Basibranchial tooth plates are often preserved in fossil coelacanths, but it is rare to find them in place and a clear pattern cannot easily be recognized. However, the basibranchial dentition of some taxa is shown in Fig. 7.6(C–I). These variations in basibranchial dentitions, when placed against a coelacanth phylogeny (Chapter 9), justify Nelson's claim. In *Laugia* and *Whiteia*, there are three pairs of large paired plates bordered by symmetrically arranged small paired plates. In *Diplurus* (three pairs) and *Axelrodichthys* (two pairs), only the large paired plates remain, while in *Undina* and *Latimeria* there is a small median tooth plate in addition to the consolidated paired plates. *Macropoma* appears to be unusual in that there is a single large median basibranchial tooth plate (Figs 7.6(H), 7.7).

The ceratobranchials are large, curved elements and the first four each bear a deep groove on the ventral surface to receive the afferent branchial arteries. In *Latimeria* the first four arches articulate with the single basibranchial while the fifth ceratobranchial is much smaller, lacks a groove and articulates with the base of the fourth ceratobranchial (Fig. 7.6(A), Cb). Rosen *et al.* (1981) suggested that a pattern where the posterior arches articulated with the bases of those more anteriorly was a character of coelacanths, porolepiforms, lungfishes and tetrapods (hynobid salamanders). It now appears that this is a more universal character which may be found in all sarcopterygians (the branchial arches of *Eusthenopteron*, which had formerly been used as the sarcopterygian 'model', are poorly known). The chief difference between *Latimeria* on the one hand and fossil lungfishes, *Medoevia*, *Glyptolepis* and *Mandageria* on the other is that in these latter fishes, the fourth arch is supported by the fourth and, if a fifth were present, that must be supported by the fourth arch or lie free. The distal tips of the ceratobranchials are capped by a large cartilage head. On arches 1–4 these cartilages have two prominent swellings (Fig. 7.6(A)); one of these swellings articulates with the adjacent epibranchial, the other receives the branchial elevator muscle related to that arch. The first four branchial elevators converge anterodorsally to the posterior face of the neurocranium immediately behind the parampullary process. There

is in addition a large muscle attached to the dorsal end of the fourth ceratobranchial, which passes posterodorsally to insert onto the anocleithrum. This may be a modified fifth branchial elevator which has switched insertion to the preceding arch. The cartilage head of the fifth ceratobranchial is much simpler. It does not articulate with an epibranchial but there is a small sliver of muscle connecting the head with the fourth ceratobranchial. The dentition on the ceratobranchials is confined to three rows of small tooth plates, each bearing tiny villiform teeth (Fig. 7.6, t.p).

In fossil coelacanths the relationships between the ceratobranchials cannot be seen but it is assumed that they were like those of *Latimeria*. Small tooth plates cover the ceratobranchials and in some (e.g. *Rhabdoderma*, Forey, 1981), these plates can be seen to be arranged in three rows. Rarely, there may be modifications of this primitive dentition. In *Undina penicillata* (Reis, 1888: pl. 1, fig. 21) and in *Undina* sp. from the Upper Jurassic of Turkey (Forey *et al.*, 1985), one row of tooth plates carries large teeth. In *Axelrodichthys*, there is an enlarged tooth plate lying along the base of the first two ceratobranchials.

The dorsal parts of the gill arches of *Latimeria* are reduced in comparison with the assumed primitive osteichthyan condition. There are epibranchials (Fig. 7.6(A), Eb) associated with the first four arches and there is a gradation in size from a small rod-like element on the first arch to a large and swollen element on the fourth arch. This is the reverse size-gradation to those of *Acanthodes*, actinopterygians, *Griphognathus*, *Neoceratodus* and *Eusthenopteron* (as restored by Jarvik, 1954), all of which are assumed to show the primitive osteichthyan condition. The fourth epibranchial is pierced by a foramen transmitting the efferent branchial artery. In osteichthyans the pharyngobranchials are developed as infrapharyngobranchials and suprapharyngobranchials, which each articulate with prongs on the distal ends of the

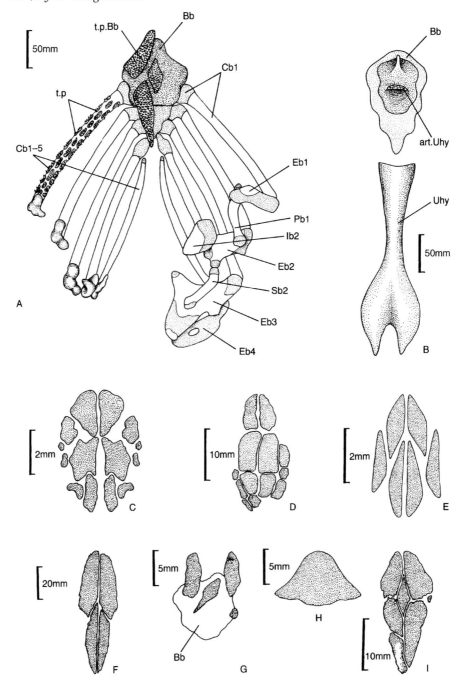

Fig. 7.6 Gill arches and basibranchial tooth plates. (A) *Latimeria chalumnae* Smith. Gill arches in dorsal view. From Millot and Anthony (1958) and Nelson (1969) and AMNH 32949SW. The dorsal arch elements are indicated on the right side, the basibranchial tooth plates and the tooth plates associated with first ceratobranchial on the left side. Note the complex cartilaginous caps at the posterior ends of the

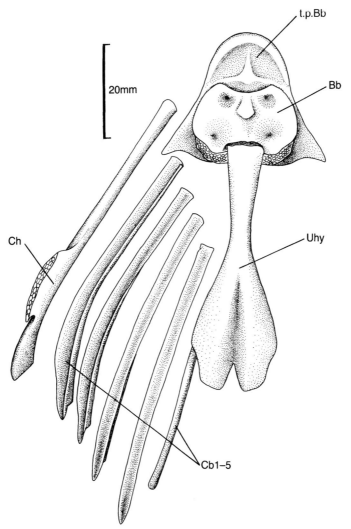

Fig. 7.7 *Macropoma lewesiensis* (Mantell). Restoration of the ceratohyal, gill arches and urohyal in ventral view. Based on several specimens in the BMNH collection.

ceratobranchials. These serve to articulate with the epibranchials as well as to provide insertion for the branchial elevator muscles. (B) *Latimeria chalumnae* Smith. Ventral aspect of the basibranchial and urohyal. The elements have been pulled apart. The basibranchial is represented as a central ossification (dark stipple) surrounded by cartilage. (C–I) Patterns of basibranchial tooth plates: (C) *Laugia groenlandica* Stensiö, MGUH VP.2316b; (D) *Whiteia woodwardi* Moy-Thomas, MNHN IP.230a,b; (E) *Diplurus newarki* (Bryant) after Schaeffer (1952a); (F) *Axelrodichthys araripensis* Maisey, FMNH PF11856; (G) *Undina penicillata* Münster, BSM 1870.xiv.517; (H) *Macropoma lewesiensis* (Mantell), BMNH P.64001; (I) *Latimeria chalumnae* Smith, AMNH 32949SW.

epibranchial. In *Latimeria* the first epibranchial carries a single rod-like pharyngobranchial element which, because it is orientated posterodorsally, is comparable to a suprapharyngobranchial. At the dorsal end, this element articulates with the parampullary process on the neurocranium. This interpretation implies that an infrapharyngobranchial is missing from the first arch. However, it is possible that this is a fused infra- and suprapharyngobranchial (Nelson, 1969), as has been restored in *Eusthenopteron* (Jarvik, 1954). In Fig. 7.6(A) it has simply been termed a pharyngobranchial. The second arch is more normal in the sense that there are both an infrapharyngobranchial (Fig. 7.6(A), Ib2) and suprapharyngobranchial (Sb2). The infrapharyngobranchial is swollen while the suprapharyngobranchial is rod-like. Pharyngobranchials are not clearly recognizable on the third and fourth arches, but may be represented as small cartilages attached to the tips of the epibranchials.

In fossil coelacanths virtually nothing is known of the dorsal gill arch elements. In *Laugia* (Fig. 6.6) there is evidence of the first pharyngobranchial (Fig. 6.6, Pb1) and epibranchials 1 and 2. Furthermore, in *Rhabdoderma* (Fig. 6.5, art.Ib1) there is, on the side wall of the neurocranium, evidence of an articulation which may have been for infrapharyngobranchial 1 (by comparison with conditions in primitive actinopterygians). It is possible, therefore, that the single pharyngobranchial on the first arch of *Latimeria* is a derived condition within coelacanths.

With respect to the structure of the palate, hyoid arch and gill arches, so few coelacanths are known that it is difficult to quantify any significant changes that might be useful in a phylogenetic analysis of coelacanths. Consolidation of the basibranchial dentition is seen as a general trend but the taxon sample is so limited that an unequivocal statement is not possible.

8

POSTCRANIAL SKELETON AND SCALES

8.1 INTRODUCTION

The postcranial skeleton of coelacanths is both conservative in structure throughout the group and highly apomorphic amongst osteichthyans. Both of these circumstances helped Smith (1939a,b) to recognize *Latimeria* as a Recent member of the group. The skeleton of *Latimeria* is shown in Fig. 8.1. The body is usually plump, with a deep caudal peduncle which passes into a broad trilobed diphycercal tail consisting of dorsal and ventral lobes and a horizontal extension of the body axis forming a small supplementary lobe carrying a fin tuft (Millot and Anthony, 1958, 1965). The notochord and the lateral line extends right to the tip of the supplementary lobe (Millot and Anthony, 1958, 1965). There are two dorsal fins, a condition which is plesiomorphic for sarcopterygians. However, unlike most other fin-bearing sarcopterygians, the anterior dorsal fin (D_1) is located well anteriorly, usually within the anterior half of the body. The second dorsal fin (D_2) is also located relatively more anteriorly than in most other sarcopterygians and this fin lies opposite, or nearly opposite, the anal fin. The position of the paired fins is broadly similar to that in other sarcopterygians. The pectoral fin is held vertically on the flank (pressed against the body); the level of insertion varies between 50% and 70% of the depth of the body, taken from the dorsal midline (although it needs to be stressed that the flattened preservation of such plump-bodied fishes makes it difficult to be precise about such statements in fossils). This relatively high insertion is more like that of lungfishes and porolepiforms than it is that of 'osteolepiforms'. The pelvic fins are held horizontally and are inserted close to the ventral midline at a vertical level close to or shortly behind the insertion of the first dorsal, although there are some exceptional taxa (e.g. *Laugia*, *Coccoderma*).

This description of the general coelacanth body shape needs to be modified by adding that exact proportions do vary among the different taxa. A range of body shapes is shown in Fig. 8.2, where it can be seen that some taxa such as *Laugia* (Figs 8.2(J), 11.10) are decidedly more elongate than others such as *Holophagus* (Figs 8.2(M), 11.8), *Macropoma* (Fig. 11.11) and *Latimeria*. There are some very differently shaped coelacanths such as *Miguashaia* (Fig. 11.13), which is distinctly more like other plesiomorphic sarcopterygians than other coelacanths, and *Allenypterus* (Figs 8.2(H), 11.2), which has a very unusual body shape in which the ventral profile is straight and the dorsal profile is markedly arched, passing into a long, tapering tail. In most coelacanths the dorsal and ventral lobes of the caudal fin are approximately of equal size, but in many there is a slight asymmetry in the numbers of caudal fin rays and internal radial supports. In these cases the dorsal lobe is longer than the ventral lobe (cf. plesiomorphic sarcopterygian conditions also seen in *Miguashaia*) and

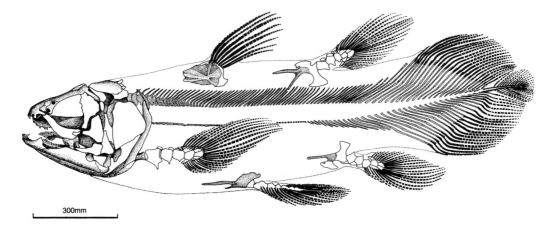

300mm

Fig. 8.1 *Latimeria chalumnae* Smith. Skeleton of entire fish, based on Millot and Anthony (1958: frontispiece). Note the single basal plate supporting the first dorsal fin; the endoskeleton of the second dorsal and the anal fins are mirror images of each other and the pattern of mesomeres resembles that in the paired fins. The notochord tapers rapidly at the junction between abdominal and caudal divisions.

begins in advance of the ventral lobe (this asymmetry is seen exaggerated in *Allenypterus*, Fig. 11.2). The supplementary lobe which is developed as a symmetrical arrangement of fin rays around the terminal end of the notochord is, with the exception of *Miguashaia*, always separate; there is a clear gap between the dermal fin rays of the principal caudal lobes and the fin rays within the supplementary lobe.

8.2 VERTEBRAL COLUMN

The axial skeleton (vertebral column and median fins) is very conservative. Although it is usually seen in coelacanth fossils, it is rarely well enough preserved for detailed description. This is partly because there are no ossified centra, the neural and haemal spines are usually only perichondrally ossified and there must have been a considerable amount of cartilage present. The neural spines are median throughout the column and this is presumably a derived condition amongst sarcopterygians. The neural arches immediately behind the head are broader

antero–posteriorly than those further back down the column, and in some taxa the first few neural arches abut one another leaving nerve foramina between them. Lund and Lund (1985: 14) refer to these expanded neural arches as constituting a 'functional cervical region' but this is not elaborated and here they are simply referred to as expanded neural arches. No doubt they would have served to strengthen this part of the column and, at the same time, decrease flexibility. The anterior neural spines are short and they gradually increase in length further down the column. The neural spines within the anterior part of the caudal region are the longest. All except the first few neural spines are capped with cartilage (in fossils they remain open ended).

On the ventral side of the notochord, there are small parapophyses throughout the abdominal region. Many coelacanths do not show such structures, which may have been entirely cartilaginous in these taxa. In the caudal region, co-ossified haemal arches and median haemal spines are developed; again these are longest in the anterior part of the

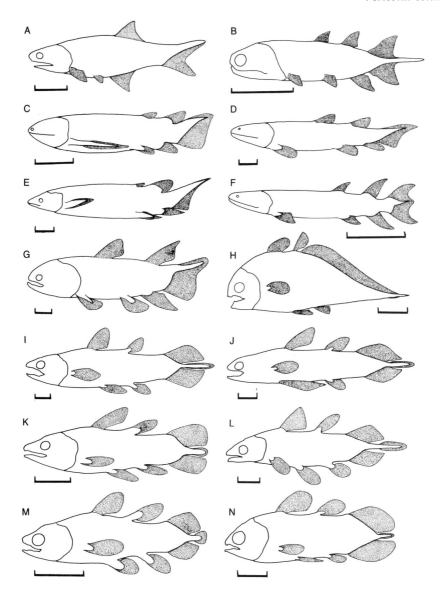

Fig. 8.2 Outline drawings of a variety of osteichthyan fishes to show variation in the body form and fin positions. (G–N) are coelacanths. (A) *Mimia toombsi* Gardiner and Bartram, a primitive actinopterygian (from Gardiner, 1984a). (B) *Strunius walteri* Jessen, an onychodont (from Jessen, 1966). (C) *Glyptolepis paucidens* (Agassiz), a porolepiform (from Ahlberg, 1989a). (D) *Osteolepis macrolepidotus* Agassiz, an 'osteolepiform' (from Jarvik, 1948). (E) *Dipterus valenciences* Sedgwick and Murchison, a primitive lungfish (from Ahlberg and Trewin, 1994). (F) *Eusthenopteron foordi* Whiteaves, an 'osteolepiform' (from Jarvik, 1980). (G) *Miguashaia bureaui* Schultze (from Cloutier, 1996). (H) *Allenypterus montanus* Melton. (I) *Rhabdoderma elegans* (Newberry). (J) *Laugia groenlandica* Stensiö. (K) *Whiteia woodwardi* Moy-Thomas. (L) *Diplurus newarki* (Bryant) (from Schaeffer, 1952a). (M) *Holophagus gulo* Egerton. (N) *Nacropomoides orientalis* Woodward. Scale bar for most fishes equals 20 mm, except (C), (F) and (M), where scale equals 200 mm.

caudal region. Both neural and haemal spines in the caudal region lie in series with dorsal and ventral radials, which are also formed only in perichondral bone and are cartilage capped. A few coelacanths show ossified ribs throughout the abdominal region, but the majority show ribs confined to the posterior third of the abdominal segments.

There is considerable variation in the number of neural arches/spines and haemal arches/spines, as well as the spacing throughout the abdominal and caudal regions. Such variations may well have significance for locomotory adaptations (Forey, 1991, and see p. 239).

Because there is so little variation in the structure of the vertebral column of different taxa, and because many are known in only superficial detail, separate descriptions for most coelacanths are unnecessary. *Latimeria* is, of course, the best known and some descriptive comments are necessary before comparative comments about different fossil coelacanths are given.

Latimeria

The vertebral column (Fig. 8.1) has been described by Millot and Anthony (1956, 1958) and, especially, Andrews (1977). The notochord extends from the tip of the supplementary lobe of the tail to reach well forward, through the otic region of the neurocranium where it may be seen ventrally through the basicranial fenestra to abut the posterior face of the basisphenoid within the notochordal pit. As it passes through the otic region it tapers only slightly and is surrounded by one anazygal dorsally and two catazygals (Fig. 6.1(A)) ventrally, the latter being set in tandem. The zygal plates are median but the anazygal is clearly bilobed and probably arises as a paired structure. The anazygal is therefore perfectly comparable with the zygal plates in primitive actinopterygians (Gardiner, 1984a). Actinopterygians lack catazygals because the

basioccipital region is completely ossified in the ventral midline to occlude the basicranial fenestra.

The notochord is very broad. It is, however, not as broad relative to body depth as in some other fishes such as sturgeon (*Acipenser*) or the Australian lungfish (*Neoceratodus*). In two respects *Latimeria* and many other coelacanths are unusual among 'notochordal fishes'. First, the notochord maintains its diameter as far as the posterior end of the abdominal region and thereafter tapers very rapidly into the caudal region. Second the abdominal segments are very short: i.e. there are many segments over a short body length (Forey, 1991). The detailed structure of the notochord, including the contained viscous jelly-like fluid, has been described by Millot and Anthony (1958), Andrews (1977), Griffith (1980) and Locket (1980).

There are about 115 neural arches throughout the column, and all except the first two are co-ossified with median neural spines (which are absent from the first two segments). The neural arches are deep anteriorly and decrease in height to about the 20th vertebra, thereafter remaining approximately the same height. Between successive anterior neural arches there are notches allowing the dorsal and ventral roots to pass through. These notches disappear by about the 20th vertebra, when the neural arches are much more slender and lie at some distance from one another. Throughout most of the column, from about the level of D_1, there are small cartilages wedged between the bases of several neural arches. These cartilages have been identified as pleurocentra (Andrews 1977). They are never seen in fossil coelacanths. The cartilaginous haemal arch bases are also swollen and thought to represent intercentra. The pattern of vertebral components of *Latimeria* is very similar to the caudal vertebra of *Neoceratodus*, with the notable absence of supraneurals from *Latimeria* and other coelacanths.

Within the supplementary lobe the regular

segmental architecture breaks down and there is a series of cartilaginous neural and haemal arches and, occasionally, ossified neural and haemal spines. Within this region there can be a variable number of neural and haemal arches per muscle segment (Millot and Anthony, 1958; Andrews, 1977). In fossil coelacanths no endoskeletal ossifications are seen within the supplementary lobe.

As mentioned above, supraneurals are absent from coelacanths. However, in the caudal region there is a single series of dorsal radials (epineural spines) which lie in series with the neural arches and support the dorsal caudal fin rays. They are perichondral ossifications around cartilages distinct from, but in series with, the neural spines. Ventrally the haemal spines also articulate with a single series of ventral radials which support the ventral caudal fin rays. In *Latimeria* and most fossil coelacanths, there are one or two neural and haemal spines lying immediately anterior to the caudal fin which support radials but which do not carry fin rays. The most anterior of these vertebrae is taken as the beginning of the caudal region in vertebral counts (Chapter 11, diagnoses).

Ventral radials are common in both actinopterygians and sarcopterygians, where there is usually more than one series (i.e. proximal and distal radials). Therefore coelacanths may be uniquely derived in having only one series of ventral radials.

Dorsal radials appear to be absent from other sarcopterygians and, if an epichordal lobe is developed, the fin rays articulate directly with the neural arches (e.g. *Eusthenopteron*, Andrews and Westoll, 1970a). It needs to be noted here that the endoskeleton of the caudal region of most sarcopterygians remains unknown.

Comparative comments on the vertebral column of other coelacanths

The remarks given for the vertebral column of *Latimeria* apply also to the vast majority of coelacanths. In *Miguashaia* and *Diplocercides* the neural arches and spines and the haemal arches and spines in successive segments of the caudal region are very broad and may abut one another distally (Fig. 11.13). Outgroup comparison with porolepiforms and 'osteolepiforms' (but not actinopterygians) suggests this condition to be plesiomorphic.

In some coelacanths (e.g. *Diplurus*, *Axelrodichthys*, *Chinlea* and *Changxingia*) long, ossified ribs are present throughout the abdominal region and encircle the abdominal cavity. In many coelacanths there are short ribs developed towards the posterior end of the abdominal region but they must have had only limited supporting function. In the data matrix (Table 9.1), only those taxa that have long ribs are scored.

With regard to the number of neural arches and relative spacing, there is considerable variation amongst coelacanths. With some exceptions there are more neural arches in more derived coelacanths (this ranges from about 45 in the relatively plesiomorphic *Rhabdoderma* to about 110 in *Latimeria*) and furthermore, the neural arches are more closely spaced in the abdominal region than those in the caudal region. Also, there are relatively more abdominal vertebrae in more derived coelacanths. Such variation in the numbers and proportions of abdominal/caudal neural arches may well have implications for functional evolution in coelacanths (Forey, 1991 and see p. 239).

8.3 THE MEDIAN FINS

The anterior dorsal fin is usually composed of a few unbranched rays (8–15) which are stouter than those of all other fins except the anterior rays of the caudal fin. The exceptional taxon is *Miguashaia*, in which there are more than 20 rays (exact number unknown), some of the more posterior of which are branched. The basal (Fig. 8.5, $D_1r.b$), which supports the fin rays, is usually a plate-like bone strengthened by thickened ridges. In most coelacanths the basal plate lies completely above the level of the neural arches and shows an entire ventral margin to an essentially triangular plate (Schaeffer, 1941). Some coelacanths (e.g. *Rhabdoderma*, Fig. 11.14) show a rather differently shaped basal in which the plate is kidney-shaped and the ventral margin is emarginated to receive the tips of adjacent neural spines.

The second dorsal (D_2) and the anal fins of *Latimeria* and other coelacanths are uniquely constructed amongst Recent fishes. In most sarcopterygians there is a basal plate (sometimes called a proximal radial or simply fin support) supporting two or more radials which, in turn, support the fin rays. The endoskeletons of the dorsal and anal fins are closely comparable with each other but completely different from the single-axis paired fins. In *Latimeria*, D_2 (Fig. 8.3(A)) and the anal (Fig. 8.3(B)) are each supported by a single basal plate, more complicated in shape than the simple plate of other sarcopterygians. More distally the fin rays are supported by a single axis of at least four mesomeres (*ax.mes.*) which are almost identical to the axis of the paired fins. The proximal three mesomeres have processes (Fig. 8.3(A,B)) comparable with the dorsal and ventral processes in the paired fin mesomeres (Fig. 8.4). The D_2 and anal fin axes are practically identical to each other and are held in mirror image position (cf. Fig. 8.3(A) with

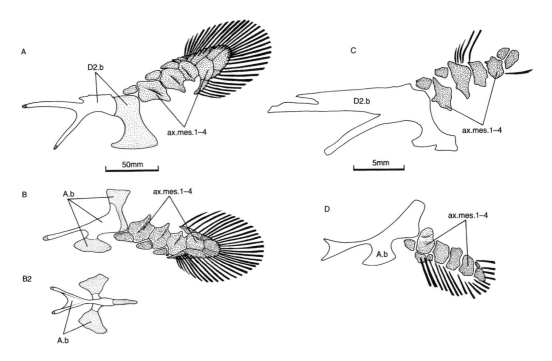

Fig. 8.3 Median fin skeletons. (A) *Latimeria chalumnae* Smith, second dorsal fin in left lateral view based on Millot and Anthony (1958), stipple indicates cartilage. (B) *Latimeria chalumnae* Smith, anal fin skeleton in left lateral view: B2, basal of the anal fin in dorsal view, stipple indicates cartilage. (C) *Laugia groenlandica* Stensiö, second dorsal fin basal and mesomeres in left lateral view, from MGUH VP3174. (D) *Laugia groenlandica* Stensiö, anal fin basal and mesomeres in left lateral view, from MGUH VP3174. The bases of the fin rays are shown in black. Note the marked asymmetry of insertion in *Laugia* as compared with *Latimeria*.

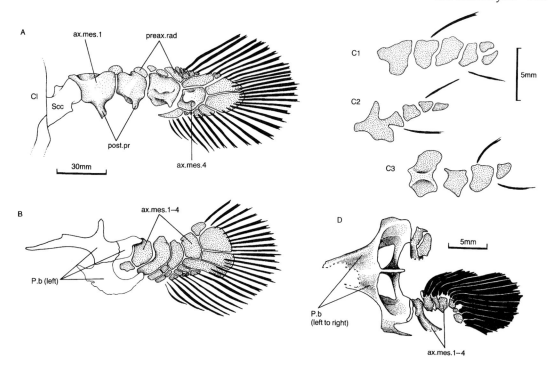

Fig. 8.4 Paired fin skeletons. (A) *Latimeria chalumnae* Smith, left pectoral fin as preserved in AMNH 32949SW. (B) *Latimeria chalumnae* Smith, left pelvic girdle and fin skeleton in dorsal view as preserved in AMNH 32949SW. (C) *Laugia groenlandica* Stensiö, pectoral fin mesomeres as preserved in C1 – MGUH VP3170a, C2 – MGUH VP3175b, C3 – MGUH VPspec4. The curved black lines indicate the level of insertion of the leading fin rays. (D) *Laugia groenlandica* Stensiö, pelvic girdle and fin skeleton as preserved in MGUH VP1017, seen in ventral view. Note that the position of the fin rays, shown in black, suggests that the fin has shifted through 180° as a preservational artefact.

(B)). The mesomeres are surrounded by smaller radials which, as in the paired fins, are arranged asymmetrically with those radials corresponding to the preaxial radials extending further down the fin. The exact arrangement of the radials is more like that of the pelvic fins than that of the pectorals, in that the radials are arranged in one-to-one correspondence right down to the proximal mesomere (compare Fig. 8.3(A,B) with Fig. 8.4(A,B)). Towards the distal end of the axis the exact correspondence between radial and mesomere breaks down. The similarity between the paired fin skeletons and D_2 and anal extends to the musculature and the innervation (Millot and Anthony, 1958). This is undoubtedly of some functional significance because the paired fins, especially the pectorals, are capable of considerable rotational movements. A high degree of rotation is also seen in the median fins (p. 23). Ahlberg (1992) also noted the great similarity between the median fins and, especially, the pelvic fin, and noted that this similarity extended to the construction of the pelvic girdle and the basal plates of D_2 and the anal. He noted that if we imagine a 90° rotation of the pelvic bone we see the shapes of the basal plates of the D_2 and anal fins (cf. Fig. 8.4(B) with Fig. 8.3(A,B)). Ahlberg (1992) suggests that the D_2 and anal fins of *Latimeria* are new developments which have replaced the usual sarcopterygian D_2 and anal by a once-off developmental genetic change causing paired fin morphology to be expressed in median fin positions. This is something that, hopefully, in the future might be tested.

The skeleton of the D_2 and anal fins of fossil coelacanths is very poorly known. Rarely is the endoskeletal axis preserved, although the basal plates are often seen. The basal plates of these fin skeletons in *Latimeria* end posteriorly in cartilage and this was probably so in most fossil coelacanths. In *Latimeria*, there is a prominent posteroventral process upon the basal plate of D_2 which is formed entirely of cartilage (Fig. 8.3(A)). This is rarely seen in fossil coelacanths and the recording of this process as a potential character (Cloutier, 1991a) leads to difficulty. *Laugia* is one rare fossil coelacanth showing a highly ossified fin endoskeleton, and in this taxon there is considerable individual variation between those that show a process and those that do not. There is a difference between those taxa that show a deeply bifurcated D_2 basal plate and those few taxa (*Allenypterus*, *Lochmocercus* and *Polyosteorhynchus*) that have a simple plate.

The axis skeleton of the D_2 and anal fins of fossil coelacanths is very rarely seen and presumably was represented as cartilage only. Exceptions are *Laugia* where the entire skeleton is seen (Fig. 8.3(C,D)). Further, the internal supports for D_2 have been figured for *Hainbergia granulatus* (Schweizer, 1966: fig. 5d) and *Coccoderma suevicum* (Schweizer, 1966: fig. 5e). In both cases they are very similar to both *Latimeria* and *Laugia* in showing an axis of four elements with small radial-like elements along the leading edge in a one-to-one correspondence.

Comparison of the fins of *Latimeria* and *Laugia* shows some striking similarities but also some interesting differences, the most notable of which is the relative position of the insertion of the fin rays. In *Latimeria* the fin rays of both D_2 and the anal fin are inserted in almost symmetrical fashion around the tip of the fin axis (Fig. 8.3(A,B)) whereas in *Laugia* there is a marked asymmetry, with the leading-edge ray being inserted well down the axis and associated with the second axial mesomere in D_2 or the first in the anal fin (Fig. 8.3(C,D)). The relationship between the fin ray insertion and the endoskeleton is not known for other coelacanths but there is evidence that most of the plesiomorphic coelacanths were like *Laugia* in this respect, because the preservation of the fin rays shows markedly asymmetrical arrangements. A range of fin ray distributions was recorded by Forey (1991). *Caridosuctor* and *Rhabdoderma* are taxa showing marked asymmetry, more extreme than that of *Laugia*, while taxa such as *Macropomoides* (Fig. 11.21), *Undina* and *Holophagus* (Fig. 11.11) show a nearly symmetrical arrangement like that of *Latimeria*. All coelacanths with the possible exceptions of *Miguashaia* and *Allenypterus*, show a lobed D_2 fin even though the degree of lobation may vary. And the fact that all coelacanths known in this respect show the unusually shaped basal plates for the D_2 and the anal fins suggests that the basic structure of the coelacanth D_2 and anal has been similar since the origin, or near to the origin of the group. As yet the endoskeleton of the fins in *Miguashaia* is unknown.

The caudal fin of all coelacanths except that of *Miguashaia* is also highly apomorphic amongst sarcopterygians. As mentioned above, the fin is usually developed as nearly equal dorsal and ventral lobes, which are structural mirror images of each other, and a separate supplementary lobe. The shape of the tail varies from being relatively elongate in forms such as *Coelacanthus* and *Laugia* to short and square-cut as in *Macropomoides* and *Holophagus*. Most coelacanths have between 16 and 24 rays in the dorsal lobe, usually one or two fewer in the ventral lobe. Asymmetry is extreme in the tail of *Allenypterus* (Fig.

11.2). The rays are segmented but unbran-ched (except in *Miguashaia*) and usually show a one-to-one relationship with the supporting radials. Some more plesiomorphic taxa show a greater than one-to-one ratio between the radials and fin rays, and this is presumed to be the plesiomorphic condition. Only *Migua-shaia* among coelacanths shows branched caudal fin rays (Schultze, 1973; Cloutier, 1996). The supplementary fin varies in length both between adults of different species and in different growth stages where it is often longer in the young (*Rhabdoderma exiguum* Eastman). It is rarely preserved intact and it is difficult to quantify relative lengths of the supplementary lobe. The function of this lobe, which is so characteristic of coelacanths, is unknown. Fricke's underwater observa-tions of *Latimeria* show that the supplemen-tary lobe is flexed from side to side, particularly when the animal performs its head-standing manoeuvres, but the signifi-cance of this is not known.

8.4 SHOULDER GIRDLE AND FIN

The shoulder girdle of coelacanths is remark-ably conservative. With the exception of that of *Miguashaia*, the variations seen concern only minor differences in proportions of the cleithrum and clavicle and in the shape of the anocleithrum. The shoulder girdle (Fig. 2.5) is usually narrow throughout its entire length and consists of an unornamented anocleithrum (Acl), a cleithrum (Cl), clavicle (Cla) and, probably primitively, an inter-clavicle (see below). There is also an extra-cleithrum (Ecl) which is a coelacanth synapomorphy. This narrow bone lies ventral to the level of insertion of the pectoral fin and is sutured to the posterior edges of both the clavicle and the cleithrum. Sometimes it is more closely associated with the cleithrum (e.g. *Macropoma*, Fig. 4.19; *Laugia*, Fig. 4.10), sometimes with the clavicle (e.g. *Whiteia*, Fig. 4.15) or equally spaced between the two (*Latimeria*, Fig. 2.5). Subtle differences in

these mutual relationships, which depend on the relative sizes of cleithrum and clavicle, are difficult to code as discrete character states and consequently are not used in the cladistic analysis.

The positional relationships of the different bones within the girdle are constant. The clavicle overlaps the ventral part of the clei-thrum as is usual in sarcopterygians (except rhizodonts, which have a characteristic reverse overlap – Andrews and Westoll, 1970b). In many cases the clavicle wraps completely around the leading edge of the cleithrum, and in members of the clade lead-ing to *Axelrodichthys* and *Mawsonia*, both the clavicle and the cleithrum are broadly expan-ded medially into a broad wing (Fig. 4.17). The anocleithrum is always without orna-ment and this means that, as in *Latimeria* and lungfishes, it is completely subdermal. In *Latimeria* the anocleithrum receives a large muscle attached to the dorsal end of the fourth ceratobranchial. This muscle presum-ably helps to elevate the posterior branchial arches and may be a coelacanth synapomor-phy. The shape of the anocleithrum varies from those that are blade-like and forked anteriorly (Figs 4.10, 5.7, 11.11) to a simple sigmoid element (Figs 8.5, 11.8).

The shoulder girdle lies free from the skull and there is no post-temporal or supraclei-thrum. In this coelacanths are similar to Recent lungfishes (and tetrapods which have also lost the anocleithrum). However, because primitive lungfishes such as *Chir-odipterus* have these bones, this loss of dorsal bones in the shoulder girdle must be homo-plasious.

Coelacanths are generally assumed to lack an interclavicle (Andrews, 1973) and indeed such an element is missing in *Latimeria*. In most sarcopterygians and primitive acti-nopterygians (Gardiner, 1984a), the inter-clavicle is a median dermal bone, usually ornamented, which lies in the ventral midline and is partly overlapped by the clavicles of either side. At least two fossil coelacanths

(*Whiteia* and *Laugia*) show the presence of an interclavicle (Fig. 4.15, Icla), which is a small subdermal bone capped with cartilage and therefore presumably now of endochondral origin.

The scapulocoracoid (Fig. 8.4, Scc) is a single element as in all sarcopterygian fishes and it articulates with the cleithrum and clavicle over a broad area. It is mostly cartilaginous in *Latimeria* with only the articulatory tip being ossified, and this was probably the case in most fossil coelacanths because the scapulocoracoid is usually a small ossification lying free from the dermal shoulder girdle (Fig. 4.17). It is not perforated by any nerve or vascular foramina and in this feature it resembles those of porolepiforms and modern lungfishes (some fossil lungfishes show a perforated scapulocoracoid – Rosen *et al.*, 1981: fig. 40). The imperforate girdle is probably a derived character within sarcopterygians. The glenoid surface is convex, as in lungfishes and porolepiforms (although porolepiforms show an unusual strap-shaped glenoid – Ahlberg, 1989a,b).

As in all sarcopterygians, the paired fins are single-axis fins, the axis representing the original metapterygial axis (Rosen *et al.*, 1981). The pectoral fin endoskeleton (Fig. 8.4(A)) is slightly longer than the pelvic fin, although in both there are four large axial mesomeres (ax.mes.), each bearing ridges for the attachment of segmental muscles as described by Millot and Anthony (1958). In the pectoral fin these mesomeres are equidimensional or slightly longer than wide. The basal mesomere (humerus) articulates with the scapulocoracoid through a concave articulatory head. At least the first three mesomeres have both dorsal and ventral processes: the ventral processes (corresponding to the postaxial processes (post.pr) of most sarcopterygians) are much better developed than the dorsal (preaxial), and there is a series of preaxial radials (preax.-rad), dorsal in position, associated with each

mesomere. However, unlike the condition in porolepiforms and lungfishes, there is no clear one-to-one correspondence between mesomere and radial throughout the entire length of the fin: such correspondence breaks down at the third mesomere. Furthermore, the radials are not obviously segmented. A ventral (postaxial) radial is associated with the fourth mesomere but these are lacking on the first three mesomeres. In *Latimeria* the fin rays are arranged symmetrically around the tip of the fin, with both the leading preaxial and postaxial rays being associated with the radials attached to the fourth mesomere.

The pectoral fin endoskeleton is rarely seen in fossil coelacanths. *Laugia* (Fig. 8.4(C)) is one of the few exceptions (traces have been seen in *Coccoderma*). Here there are four axial mesomeres with, occasionally, one or two distal radials also preserved, and we must assume that the remainder of the skeleton was cartilaginous. There are two differences between the fin endoskeleton of *Laugia* and *Latimeria*. First, in *Laugia* the basal axial mesomere is substantially larger than those succeeding and both the dorsal and ventral processes are very well developed (Fig. 8.4(C2)). Second, in *Laugia* the insertion of the fin rays is very asymmetrical, with the dorsal (preaxial) rays being inserted much further forward than the ventral (postaxial) rays. The anteriormost preaxial rays are associated with the second or third mesomere, the anteriormost postaxial rays are associated with the fourth (as in *Latimeria*). This might imply that the coelacanth pectoral fin ray insertion was primitively asymmetrical, as we would expect through outgroup comparison, and that symmetry is derived within the group and comparable with that seen in the D_1 and the anal fins. Many other coelacanths also appear to show an asymmetrical arrangement, but because the relationship between fin rays and endoskeleton is unknown, this inference must be regarded with caution.

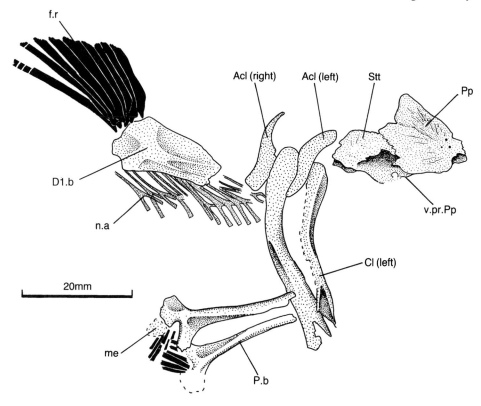

Fig. 8.5 *Undina penicillata* Münster. Camera lucida drawing of the posterior part of the skull roof, shoulder girdle, anterior dorsal fin and pelvic girdle as preserved in BSM 1899.I.40.

8.5 PELVIC GIRDLE AND FIN

The pelvic girdle of *Latimeria* (Figs 8.1, 8.4(B)) appears to conform with that expected in a primitive sarcopterygian and lies in an abdominal position (Fig. 8.1). There are some coelacanths in which the pelvic girdle is located immediately below and behind the pectoral, and in at least two of these (*Laugia* and *Coccoderma*) the girdle is more solidly constructed than in most coelacanths (see below). The pelvic girdle includes posteriorly diverging pelvic bones. The posterior end of each pelvic bone/cartilage (Fig. 8.4(B), P.b) is expanded as the acetabulum, alongside which there is a lateral expansion or process and, usually, a medial

expansion or process which is attached to the pelvic bone of the opposite side through connective tissue. The lateral expansion (corresponding to the ilium of a tetrapod) is large and cartilaginous. The medial process (corresponding to the ischium of a tetrapod) is narrow, while the acetabular surface is also cartilaginous with a convex articulating surface. The pelvic bones of fossil coelacanths (Fig. 8.5) are often seen, and when they are preserved they are fundamentally like those of *Latimeria*. Variations have been noted by Schaeffer (1941), but it needs to be noted that posterior surfaces are often open ended, suggesting that they were cartilage-capped and that the true shape may not always be seen. The shapes of the ossified portions do vary between species. Many of the more plesiomorphic coelacanths show an anterior process (pubis of tetrapod terminology) which is broad and divided to two or three processes, and posteriorly the medial process often shows an interdigitating connection with the pelvic bone of the opposite side. In more derived coelacanths the pelvic bone is simpler, with just a single anterior process. *Laugia* (Fig. 8.4(D)) and *Coccoderma* are derived among coelacanths in that the pelvic girdles of either side are joined in the ventral midline throughout most of their length.

The pelvic fin skeleton of *Latimeria* (Fig. 8.4(B)) is constructed as a structural reverse of the pectoral with the preaxial edge as the leading edge of the fin. The four axial mesomeres (ax.mes.) are short, resulting in a slightly shorter fin, and the preaxial radials are more obviously segmented and more clearly associated with the mesomeres. There are one or two postaxial radials. The fin rays are distributed in a more asymmetric fashion around the fin axis than in the pectoral skeleton.

As with the pectoral fin, the pelvic fin endoskeleton is known in any detail only in *Laugia*. Here the mesomeres are very compressed antero–posteriorly and only one or two radials are seen at the tip of the axis. The fin rays are arranged in very asymmetric fashion (in Fig. 8.4(D) the fin has rotated as a preservational artefact because other specimens show a 'reversed' asymmetry matching that in *Latimeria*).

It therefore appears that the pelvic fin shows more plesiomorphic sarcopterygian conditions (high degree of asymmetry, preaxial side forming the fin leading edge, short mesomeres).

The paired fin rays are usually slender, more so than those of D_1 and the caudal fin. However, some taxa show markedly expanded pelvic fin rays (occasionally, as in *Libys*, all fin rays are expanded). A survey of specimens of *Laugia* suggests that not all individuals have expanded pelvic rays and it is therefore possible that this is a feature of maturity or a difference between the sexes.

8.6 SCALES

The scales of coelacanths are circular or subcircular and deeply embedded such that the exposed portion, which is always ornamented (reports of the absence of ornament in *Chagrinia* probably reflect preservational artefact), is less than one-third of the total surface area. No comparative histological work has been done on coelacanth scales, but there is little to suggest that they vary in any significant way from the structure seen in *Latimeria*. *Latimeria* scales have been described by J.L.B. Smith (1939c), Roux (1942), Millot and Anthony (1958), M.M. Smith *et al.* (1972), Millot *et al.* (1978) and M.M. Smith (1979). The scales consist of a basal layer of cellular isopedine which is laid down as a series of layers with successive layers enclosing collagen fibres in different orientations. The surface layer consists of cellular bone, dentine and enamel or enameloid (the distinction between these two tissues which depends on developmental criteria – Reif, 1982 – cannot be seen here). There is no evidence of a pore canal system and therefore these scales cannot be said to be cosmoid. The ornament pattern does vary considerably within coelacanths and, with minor individual variation and variation over the body, can be useful for identification. Pertinent notes and some illustrations are given in Chapter 11. The more plesiomorphic coelacanths tend to show many crowded tubercles or ridges of enamel, whereas more derived taxa show a sparse covering of larger denticles as in *Latimeria*. In a few coelacanths, the scale ornament on any one scale is clearly differentiated into large central denticles and much smaller marginal denticles (Figs 11.7, 11.12). In most coelacanths the scales in front of the level of the pelvic fin usually show much denser ornament than those more posteriorly, but in species that have scales with clearly differentiated ornament (e.g. *Macropoma*), these scales predominate on the posterior part of the body. The lateral line may open through the scales via a single pore (Fig. 11.19(B)) or the canal may ramify through tubes within the lateral line scales and open through several small pores (Fig. 11.12(B)).

8.7 CHARACTER DESCRIPTIONS

The following characters relating to the post-cranial skeleton and scales are used in Table 9.1.

88. Extracleithrum absent (0), present (1).
89. Anocleithrum simple (0), forked (1).
90. Posterior neural and haemal spines abutting one another (0), not abutting (1).
91. Occipital neural arches not expanded (0), expanded (1).
92. Ossified ribs absent (0), present (1).
93. Diphycercal tail absent (0), present (1).
94. Fin rays more numerous than radials (0), equal in number (1).
95. Fin rays branched (0), unbranched (1).
96. Fin rays in D_1 > 10 = (0), 8–9 = (1), < 8 = (2).
97. Caudal lobes symmetrical (0), asymmetrical (1)
98. D_1 without denticles (0), with denticles (1).
99. Paired fin rays not expanded (0), expanded (1).
100. Pelvics abdominal (0), thoracic (1).
101. Basal plate of D_1 with smooth ventral margin (0), emarginate and accommodating the tips of adjacent neural spines (1).
102. D_2 basal support simple (0), forked anteriorly (1).
103. Median fin rays not expanded (0), expanded (1).
104. Scale ornament not differentiated (0), differentiated (1).
105. Lateral line openings in scales single (0), multiple (1).
106. Scale ornament of ridges or tubercles (0), rugose (1).
107. Swimbladder not ossified (0), swimbladder ossified (1).
108. Pelvic bones of each side remain separate (0), pelvic bones of either side fused in midline (1).

INTERRELATIONSHIPS OF COELACANTHS AND EVOLUTIONARY TRENDS

9.1 INTRODUCTION

Ideas about the pattern of coelacanth evolution have been offered on several occasions (Schaeffer, 1952a,b; Forey, 1984, 1988, 1991; Cloutier 1991a,b). Coelacanths have been used as an example of a group of organisms that show rapid morphological evolution in the early history of the group followed by near stasis (Schaeffer, 1952a,b; Cloutier, 1991b) and in this respect are often thought to be similar to lungfishes (Westoll, 1949). Both might therefore be considered to concur with the punctuationists' view of evolutionary change. Any estimation of such changes in rate must be based on a given phylogeny plotted against time. In this chapter, a phylogeny is proposed from which deductions about rates of taxonomic and morphological change can be made as well as deductions about evolutionary trends.

9.2 PREVIOUS IDEAS ON COELACANTH PHYLOGENY

Coelacanth phylogeny has either been expressed overtly in diagrams (Schaeffer, 1941; Forey, 1981, 1984, 1988, 1991; Cloutier, 1991a,b; Maisey, 1991) or in Linnaean rank classifications (Berg, 1955; Lehman, 1996; Vorobjeva and Obruchev, 1967; Moy-Thomas and Miles, 1971; Andrews, 1973; Lund and Lund, 1985; Schultze, 1993). The latter classi-

fications have implied phylogenies that are based on grades of organization with one or two monophyletic groups distinguished on clear apomorphies. Although these Linnaean classifications vary in detail, particularly with respect to included genera, they are all based on the idea that the primitive evolutionary 'grade' consists of Devonian and Carboniferous coelacanths, which show a fully ossified braincase with a basipterygoid process and a skull roof consisting of many bones. In the classifications of Berg (1955), Vorobjeva and Obruchev (1967), Andrews (1973) and Lund and Lund (1985), shown in Fig. 9.1, fishes assigned to the families Diplocercididae, Hadronectoridae and Rhabdodermatidae belong to this primitive grade. Schultze (1993) added the monospecific family Miguashaiidae (Fig. 9.1). A more advanced grade, usually categorized as the family Coelacanthidae or suborder Coelacanthoidei, shows coelacanths in which the braincase is divided to separate ossifications, there is no basipterygoid process, and there is a preorbital pierced by the posterior openings of the rostral organ. This family includes most of the Mesozoic coelacanths. The Recent *Latimeria* is usually separated as a monospecific family, representing the final, most advanced grade, on the basis that the premaxillae are represented as a series of small tooth-bearing splints together with the absence of a preorbital. The family Laugiidae, originally erected

Berg 1955
Superorder Coelacanthi Berg
Order Coelacanthiformes Berg
Suborder Diplocercidoidei Berg
Suborder Coelacanthoidei Berg
Suborder Laugioidei Berg

Vorobjeva & Obruchev 1967
Order Coelacanthida (Actinistia Cope) Vorobjeva & Obruchev
Suborder Diplocercidoidei Berg
Family Diplocercidae Stensiö
Family Rhabdodermatidae Berg
Suborder Coelacanthoidei Berg
Family Coelacanthidae Agassiz
Family Latimeriidae Berg
Suborder Laugioidei Berg
Family Laugiidae Berg

Andrews 1973
Order Coelacanthida Vorobjeva & Obruchev
Family Diplocercidae Stensiö
Family Coelacanthidae Agassiz
Family Laugiidae Berg

Lund & Lund 1985
Order Coelacanthiformes Berg
Suborder Hadronectoroidei Lund & Lund
Family Hadronectoridae Lund & Lund
Suborder Coelacanthoidei Berg
Family Diplocercidae Stensiö
Family Rhabdodermatidae Berg
Family Coelacanthidae Agassiz
Family Laugiidae Berg

Schultze 1993
Infraclass Actinistia Cope
Suborder Diplocercidoidei Berg
Family Miguashaiidae Schultze
Family Diplocercidae Stensiö
Suborder Hadronectoroidei Lund & Lund
Family Hadronectoridae Lund & Lund
Family Rhabdodermatidae Berg
Suborder Coelacanthoidei Berg
Family Laugiidae Berg
Family Whiteiidae Schultze
Family Coelacanthidae Agassiz
Suborder Latimeroidei Schultze
Family Mawsoniidae Schultze
Family Latimeriidae Berg

Fig. 9.1 Linnaean classifications of coelacanths. These classifications reflect the concept of grades of organization. *Latimeria* is usually placed in its own family to represent a distinct grade or simply because it is the only living representative. *Laugia* is usually treated as a separate monophyletic family because of the anterior position of the pelvic fins.

to contain only the genus *Laugia*, was separated on the basis that this coelacanth showed the anterior position of the pelvic fins. The classification proposed by Schultze (1993) is slightly different from most in that, although the basic grade arrangement is maintained, a few genera are placed in separate Mesozoic families, thought to be monophyletic (families Whiteiidae, Mawsoniidae). The family Laugiidae was expanded to include *Synaptotylus* and *Coccoderma* as well as *Laugia*.

These Linnaean classifications are difficult to compare in detail because the generic assignments are not always equivalent and, because they are acknowledged to contain paraphyletic groups (Schultze 1993), they are of limited value in using them for ideas of evolutionary trends and rates of evolution.

The phylogenies proposed as diagrams are shown in Fig. 9.2. These are much more explicit statements. The earliest of these is from Schaeffer (1941) and to some extent reflects the concepts of grades of organization inherent in the Linnaean classifications discussed above. *Diplocercides* and *Euporosteus* were grouped on the basis of primitive characters (node 1, Fig, 9.2(A)) such as the possession of a solidly ossified braincase and a basipterygoid process. All other coelacanths were regarded as derived, in having a fragmented braincase and having lost or reduced the basipterygoid process (at the time Schaeffer wrote *Rhabdoderma* was considered to have reduced the size of the basipterygoid process and to have a fragmented braincase), and this characterized node 2 in Fig. 9.2(A). All other coelacanths (node 3) were assumed to be most closely related to each other because they had lost both a basipterygoid process and the suprapterygoid process. At this level this division of coelacanths follows the ideas of grade evolution outlined above. Schaeffer went further than this by suggesting that there were two monophyletic groups of Mesozoic coelacanths which could be recognized. The first (*Wimania* to *Scler-*

acanthus) grouped together at node 4 (Fig. 9.2(A)) the forms that Stensiö (1921) had described from the Lower Triassic of Spitzbergen with *Whiteia*. The chief evidence for this was the shape of the pterygoid, described as long and low in these coelacanths. The second monophyletic group (*Macropoma* to *Coccoderma*), grouped at node 5, was recognized on the basis that the metapterygoid had partially or completely fused with the pterygoid and that there was an independent dermopalatine. Both of these notions are now known to be incorrect. There is no fusion, and an independent dermopalatine is a universal character in coelacanths. The relationships of these two proposed monophyletic groups to each other and to five other genera (*Coelacanthus*, *Diplurus*, *Laugia*, *Spermatodus* and. *Latimeria*) could not be specified more precisely. Despite the shortcomings of this classification, which was due mainly to incomplete information, it provided the basis on which subsequent classifications were expanded.

Later attempts to classify coelacanths were based on cladistic methodology, which attempts to recognize monophyletic groups based on the parsimonious distributions of characters. These hypotheses are shown here (Figs 9.2(B), 9.3, 9.4). Those of Forey (1984, 1988) and Maisey (1991) are very limited in the numbers of taxa chosen and the numbers of characters used, and contain much the same information as those more extensive analyses of Cloutier (1991a,b) and Forey (1991). There is a considerable degree of similarity between these hypotheses. Most importantly, those taxa that had been included in the primitive grade of coelacanths (usually as the suborder/family Diplocercoidei and the family Hadronectoridae) in previous classifications are now resolved as a series of successively more derived coelacanths from *Miguashaia* to *Lochmocercus* or *Allenypterus*. Within the more derived coelacanths, then all these classifications recognize a monophyletic group which includes genera

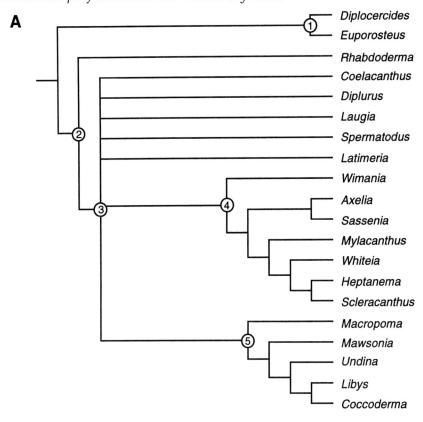

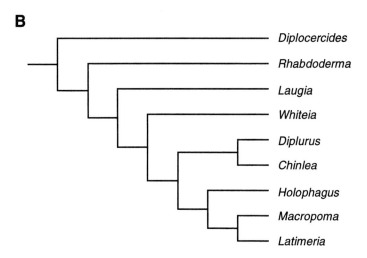

Fig. 9.2 Classifications of coelacanths proposed by (A) Schaeffer (1941) and (B) Forey (1981). See text for discussion.

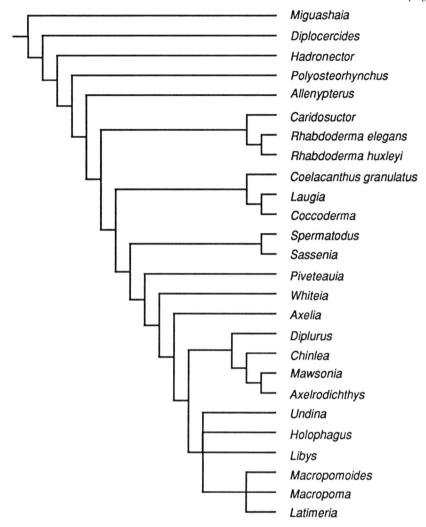

Fig. 9.3 Classification of coelacanth genera as proposed by Forey (1991). This cladogram shows areas of ambiguity among the cladistically most derived coelacanths.

such as *Diplurus, Chinlea, Mawsonia* and *Axelrodichthys*, as well as a variable number of less well-known genera and another monophyletic group including *Undina, Holophagus, Macropoma* and *Latimeria*. Other genera are scattered between these groups and the most primitive coelacanths, and there is some disagreement in the placing of certain taxa such as *Whiteia* and *Coelacanthus*. These differences are expanded upon below, after

the findings of the analysis resulting from the anatomical descriptions given in previous chapters are presented.

9.3 RELATIONSHIPS OF COELACANTH GENERA

In Chapters 3 to 8, all of the better-known coelacanths have been described and the potentially phylogenetically useful characters

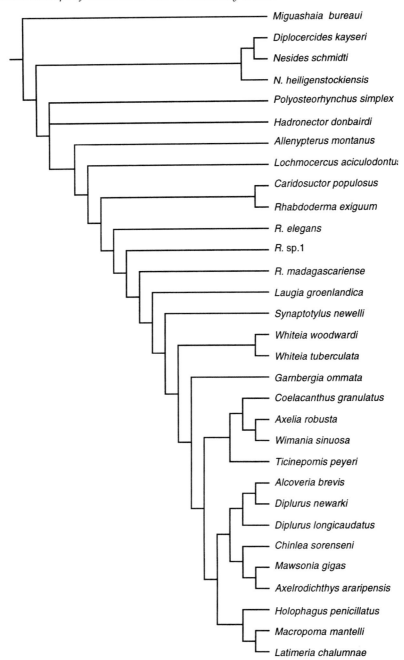

Fig. 9.4 Classification of coelacanth genera as proposed by Cloutier (1991b). See text for discussion.

Table 9.1 Data matrixx for 30 coelacanth taxa plus porolepiforms and actinopterygians. The 108 characters used here are described and numbered in Chapters 3–8. The 'N' coding is used for illogical scores: that is, character state cannot be coded because the structure does not exist in that particular taxon. For Computational purposes, 'N' and '?' are treated equally

```
                                    111111111122222222223333333333444444444455555555556666666666777777777788888888889999999999000000000
TAXA                      123456789012345678901234567890123456789012345678901234567890123456789012345678901234567890123456789012345678
Diplocercides             001??2210000000000100010100000000000101010000001101101?010101010000?111001101?01010000??2000??
Rhabdoderma               10000020011001001010000010000001010000000110010001011101?00000101011100011110000110000011000010
Caridosuctor              100002101?0?20011000010000010?2000000001010001011000141000011010111010111011101101000110010010
Hadronector               00100021?1?0?20000100001010010?2000000007?1?010100?200100100014100001010001110?0111000001100?2010
Polyosteorhynchus         00?00021?2??100?20101?2?00010010?200001010?2000000?00010?200100?????????????1011000010??7010
Allenypterus              0??????2001?0??0000100000000?201000000001001?2101010101?20100?????????????7?1101011?70001000??000
Lochmocercus              ?????????????????????0?010000?2000010101?00??200010?200?7?0001?20100?????????711010100001?00??2020
Coelacanthus              00?11?2100100100011?1110?20?010100?20000101?2100141000?0110014?2001??00210?????20110?00107?010
Spermatodus               1000002011001101011001000101000001100000011?011141100?00110110001110101101011?12010001001000
Whiteia                   11?0?2?011001011100001000001100000001010110000100010114101?1110000011120100001001?01?1110010071000
Laugia                    10??????201100100?001?020100001110000001114001?700?10011?201?11001017?111?100011010011011010100?011
Sassenia                  ?????7?201101?0211?10027?200??????????????????????????????????????????????????????????????????
Axelia                    ?????1??7?1001?????????1007?20??????????????????????????????????????????????????????????????????
Wimania                   ?0?012000102?0011?110??2?201100??????????????????????????????????????????????????????????11210
Chinlea                   00012110101012111100071100?1100010000000?7110101?7000011?20??7111011111110100010120?0
Holophagus                102012110101117111120701001020100101101??00??2111?10110?10?????????????????????????????????????
Indocoelacanthus          0??????????????????????????????71100??0?00000??7?10?20010?7?????7?7?711?7110?11010110?10?010
Undina                    0020021101011102117270110?1007?000000011410001110?011?7111017?0111017?111101100011001010
Coccoderma                1001210010010110027010?0?01110100014?0001?110?7?????????????????????????????????????????????10
Libys                     1????12107?10011?7?11117?2027120?1210010?2?00010017?7?11101?0100?????7?11101110011101010
Mawsonia                  1?20?21001001117112027121001?2?00017?01?11101110?01?1?101117?1007??1111011701000101?120
Macropoma                 01207211010110211111207100107?200100100101101114?100?71011?0111111011112100010011010
Latimeria                 001111211010111070111100102110101001000111111011N11001010010100110111011?120100010010
Miguashaia                002000170201??00000000020200000210?2000000?71072020007?102000070?2000007?0000?2000007?
Axelrodichthys            1000112100101001111?10207112010?000101100100120101140?70111210011?207110711111010010
Ticinepomis               0?07??????207?72120717?2?7120?272110?27?000?2120?27120071111007?2011101270071020711102
Lualabea                  07?7?7?7207?7?0???7??7??1?7?77177001100?2?712072??2?????????????????????????????????????
Garnbergia                ?????????21?07?7??7???177??0???7?17?2007??1702?7?17?0?010?2?0?77????????????????????????
Euporosteus               0?1????207102?7?7772?7207?207?200114?07?111010?20101N00?21000N0001007?070000N000000
Porolepiforms             00100N1??000000000000210?0001N111NN0000000000011NN0?00NNNNN?NN01N000?1N004000N0000000
Actinopterygians          N0000N1N?001000N0000021100000011N00011N?00NNNNN?NNN01oN10N100?10001000010001N00000000N0000
```

to be used in the current analysis have been listed. Genera have been used here as the terminal taxa. Ideally, in any cladistic analysis terminal taxa would be species. However, for coelacanths it is all too common that either the genera are monospecific or only one species is adequately known (exceptions are *Whiteia*, *Macropoma*) for sufficient characters to be coded (Chapter 11). Difficulties in substituting genera for species would only present a problem if the codings for the characters were polymorphic (i.e. if one species had a particular state of a character different from the state(s) in other species). Another potential problem may arise when genera cannot be shown to be monophyletic: that is, when there is the possibility that some species of the genus are more closely related to species of other genera. For instance, Cloutier (1991a,b) considers that the genera *Nesides* and *Rhabdoderma* are paraphyletic (Fig. 9.4).

The data matrix as shown in Table 9.1 contains 32 taxa and 108 characters. Of these, 30 are coelacanth genera, which form the ingroup, and there are two taxa coded as outgroups – porolepiforms and actinopterygians (*Mimia* has been used to code for the character states). In those cases where there are states of structures that are not represented in actinopterygians (e.g. conditions of the processus connectens which is found only in coelacanths and porolepiforms in this analysis), then an 'N' has been scored in the matrix. However, computer parsimony programs treat these as question marks and the same as genuinely missing data which are always encountered in fossils. There is a considerable literature on the problems associated with coding in this way (Pimentel and Riggins, 1987; Maddison, 1994). As far as this analysis is concerned, the fact that question marks (which stand for 'not applicable') are located in the outgroup should not make much difference, because it has been noted by Maddison (1994) that question marks in taxa placed at the base of the tree (in the

outgroup position) have little influence on the internal relationships of ingroup taxa.

The size of the matrix allowed a heuristic search only and this was carried out using both the PAUP and Hennig86 parsimony programs, which gave similar results. PAUP results are discussed here. The result of running all taxa, using actinopterygians and porolepiforms as combined outgroup, gave 486 equally parsimonious trees (length = 251, $CI = 0.450$, $RI = 0.706$). Using either actinopterygians or porolepiforms as a single outgroup taxon resulted in no change in the topological relationships of ingroup taxa and therefore for all subsequent analyses only porolepiforms were used. This is justified on the grounds that actinopterygians could not be scored for several of the characters that are restricted to sarcopterygians (e.g. states of the intracranial joint, some neurocranial characters).

Despite this plethora of trees, the retention index (RI) suggested that there is considerable internal structure that is being masked by some taxa with high percentages of missing values due to incompleteness. Missing values are known to generate spurious trees in both PAUP and Hennig86 parsimony programs (Platnick *et al.*, 1991a). A series of 'experiments' was tried by deleting some of the taxa with many question marks (*Euporosteus*, 84%; *Wimania*, 81%; *Axelia*, 78%; *Lualabaea*, 94%; *Indocoelacanthus*, 81%; and *Ticinepomis*, 65%). This resulted in relatively few parsimonious trees for the remaining taxa.

Deletion of these six coelacanth taxa originally coded for, but with many question marks, resulted in nine equally parsimonious solutions, the strict consensus of which is shown in Fig. 9.5. However, it should be noted that there is no direct relationship between the amount of missing data and any disruptive influence a particular taxon may have. For instance the inclusion of *Euporosteus* resulted in 18 parsimonious trees. However, the inclusion of *Ticinepomis*,

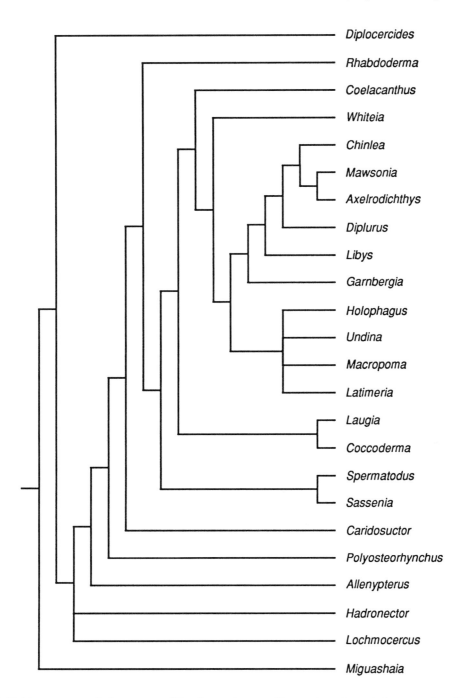

Fig. 9.5 Strict consensus cladogram resulting from nine equally parsimonious cladograms (length = 236, consistency index = 0.465, retention index = 0.692) using data from Table 9.1 and PAUP 3.1.1. (Swofford, 1993) parsimony program. In this analysis, certain taxa were omitted because of the disruptive influence caused by many question marks. All characters were run unordered.

a taxon with considerably more real data, resulted in 171 equally parsimonious trees and complete instability of relationships between taxa which are cladistically more derived than *Laugia + Coccoderma* in Fig. 9.5. It is probable, therefore, that the disruptive influence brought about by the inclusion of *Ticinepomis* is caused not only by the question marks but also by the alternative resolutions caused by the real data of *Ticinepomis*. We may here be running into a problem recognized by Nixon and Wheeler (1992), that the disruptive influence of a taxon may be caused, not by the number of missing entries, but rather by the distribution of the few real values possessed by that taxon. In relative terms, the inclusion/exclusion of *Euporosteus* has little impact on the overall relationships of coelacanth taxa. *Ticinepomis* is highly influential and therefore this might be an argument for inclusion, because any analysis cannot be considered comprehensive if such an influential taxon is excluded. Conversely, we must acknowledge that were more real data known for *Ticinepomis*, they might place this taxon unequivocally. In terms of overall phylogenetic hypothesis of coelacanths, it is advantageous to omit this taxon at this stage. Like other omitted taxa, it may be inserted to the tree(s) between nodes based on positive statements that can be made about the morphology (see below).

The consensus cladogram of nine equally parsimonious solutions (Fig. 9.5) shows two points of ambiguity. There is a polytomy among four of the more derived taxa (*Holophagus*, *Undina*, *Macropoma* and *Latimeria*) and another involving cladistically more plesiomorphic taxa (*Lochmocercus*, *Hadronector* and *Allenypterus*).

For the more derived taxa, there are three alternative solutions (Fig. 9.6(A–C)) among the nine trees. Of these, only two can be justified by synapomorphies (Fig. 9.6(A,B)) and the difference between these reflects alternative optimizations of zero-length branches. The third alternative (Fig. 9.6(C)) can only be supported by homoplasy.

For the trichotomy between *Lochmocercus*, *Hadronector* and *Allenypterus* + cladistically more derived taxa, then this arises because of question marks about the dentition in *Hadronector* (character 54) and the condition of the anterior pit-line in *Lochmocercus* (character 24). A stratigraphic arbiter is of no use in this case because all taxa are of the same age.

A further approach to selecting one tree in order to discuss patterns of evolution is to apply an a posteriori weighting procedure ('reweight characters' option in PAUP, successive approximation in Hennig86). This technique reweights the characters according to the contribution that each character makes to group taxa in the initial analysis. The technique has been applied very effectively by Platnick *et al.* (1991b) in work on spider interrelationships. A posteriori weighting is considered justified here because the overall retention index was quite high (0.692) compared with the overall consistency index (0.465). The 'reweight character' option was used, setting the maximum to 100 – meaning that any synapomorphy identified in the initial run was assigned a weight of 100 and other characters were weighted less according to their individual rescaled consistency index ($RI \times CI$).

The result of a single reweighted analysis gave a single tree and this is shown in Fig. 9.7. This is nearly identical to one of the nine trees of the initial unweighted analysis. It resolves the ambiguous relationships among the cladistically more derived coelacanths in favour of Fig. 9.6(A). However, for the ambiguity amongst more plesiomorphic taxa it suggests that *Hadronector* is cladistically more derived than either *Allenypterus* or *Lochmocercus*. This topology was not among the alternatives suggested by the unweighted analysis. However, considering the numbers of question marks against each of these taxa, we must admit this to be an equally likely solution.

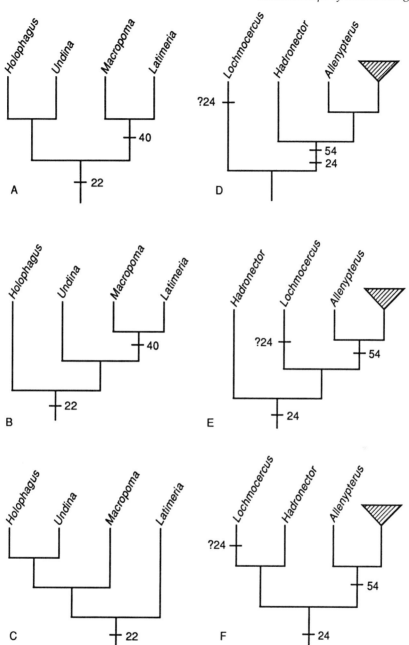

Fig. 9.6 The two areas of ambiguity resulting in the nine equally parsimonious cladograms concern, on the one hand, the taxa *Holophagus*, *Undina*, *Macropoma* and *Latimeria*, and on the other the taxa *Lochmocercus*, *Allenypterus*, *Hadronector* and cladistically more derived coelacanths (hatched triangle). The three alternative resolutions are shown for each of these areas of ambiguity (A–C and D–F). Numbers indicate characters judged as synapomorphies in alternative resolutions. In both instances, question marks in taxa cause the alternative solutions. See text for discussion.

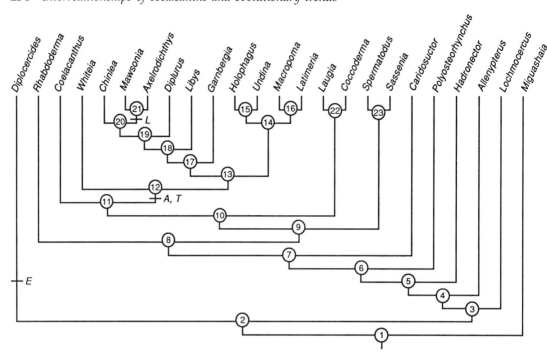

Fig. 9.7 Result of re-analysis of the nine equally parsimonious solutions arrived at in Fig. 9.5. Here characters have been reweighted (see text) and a PAUP 3.1.1. analysis carried out. All multistate characters were run unordered under the heuristic search option (TBR). A single cladogram was found and this is accepted as a phylogenetic tree (length = 8171, consistency index = 0.708, retention index = 0.871). Several taxa with high percentages of question marks and left out of the analysis have been added to this tree and placed at modes or along branches commensurate with their real data: E, *Euporosteus*; A, *Axelia*; T, *Ticinepomis*; L, *Lualabaea*. Characters were optimized to the tree using the ACCTRAN option, and those relating to the numbered nodes are as follows (those starred are resolved as synapomorphies):

Node 2. $7(1\rightarrow2)$, $12(1\rightarrow0)^*$, $10(0\rightarrow1)$, $19(0\rightarrow1)^*$, $25(0\rightarrow1)$, $58(0\rightarrow1)$, $71(0\rightarrow1)$, $74(0\rightarrow1)$, $81(0\rightarrow1)^*$, $93(0\rightarrow1)^*$, $95(0\rightarrow1)^*$.

Node 3. $11(0\rightarrow1)^*$, $14(0\rightarrow1)$, $23(2\rightarrow0)$, $24(1\rightarrow0)^*$, $34(1\rightarrow0)$, $45(0\rightarrow1)$, $55(3\rightarrow4)^*$, $63(0\rightarrow1)$, $69(0\rightarrow1)$, $72(1\rightarrow0)^*$, $73(0\rightarrow1)^*$, $80(1\rightarrow0)$, $85(0\rightarrow1)^*$, $87(0\rightarrow1)^*$, $90(0\rightarrow1)^*$.

Node 4. $54(0\rightarrow1)^*$, $97(0\rightarrow1)$.

Node 5. $23(0\rightarrow1)$, $58(1\rightarrow0)$, $94(0\rightarrow1)^*$, $102(0\rightarrow1)$, $107(0\rightarrow1)$.

Node 6. $3(1\rightarrow0)$, $17(0\rightarrow1)$, $61(0\rightarrow1)^*$, $101(0\rightarrow1)$.

Node 7. $1(0\rightarrow1)$, $20(0\rightarrow1)^*$, $23(1\rightarrow0)$, $56(0\rightarrow1)$, $59(1\rightarrow0)$, $67(0\rightarrow1)$.

Node 8. $64(0\rightarrow1)$, $97(1\rightarrow0)$.

Node 9. $26(0\rightarrow1)$, $29(0\rightarrow1)$, $62(0\rightarrow1)^*$, $101(1\rightarrow0)$, $105(0\rightarrow1)^*$.

Node 10. $5(0\rightarrow1)$, $6(0\rightarrow1)$, $30(1\rightarrow0)$, $44(0\rightarrow1)$, $47(0\rightarrow1)$, $48(0\rightarrow1)$, $67(1\rightarrow0)$, $75(0\rightarrow1)^*$.

Node 11. $1(1\rightarrow0)$, $10(1\rightarrow0)$, $18(0\rightarrow1)$, $21(0\rightarrow1)$, $59(0\rightarrow1)$, $70(1\rightarrow0)$, $71(1\rightarrow0)$, $76(0\rightarrow1)^*$, $77(1\rightarrow0)$, $78(1\rightarrow0)$, $82(0\rightarrow1)^*$, $86(0\rightarrow1)^*$.

Node 12. $15(0\rightarrow1)$, $53(0\rightarrow1)^*$, $57(0\rightarrow1)$, $60(0\rightarrow1)$, $96(0\rightarrow1)$, $98(0\rightarrow1)$.

Node 13. $9(0\rightarrow1)$, $13(0\rightarrow1)^*$, $17(1\rightarrow2)$, $35(0\rightarrow1)^*$, $51(0\rightarrow1)$, $52(1\rightarrow0)$, $91(0\rightarrow1)$.

Node 14. $3(0\rightarrow1)$, $22(0\rightarrow1)^*$, $23(0\rightarrow2)$, $67(0\rightarrow1)$, $74(1\rightarrow0)$, $79(0\rightarrow1)^*$.

Node 15. $64(1\rightarrow0)$.

Node 16. $39(0\rightarrow1)^*$, $40(0\rightarrow1)^*$, $96(1\rightarrow2)$.

Node 17. $27(0\rightarrow1)$, $31(0\rightarrow1)$, $32(1\rightarrow0)$, $50(0\rightarrow1)$, $59(1\rightarrow0)$, $68(0\rightarrow1)$.

Node 18. $49(1\rightarrow0)$.

This reweighted solution is chosen as the framework to discuss the evolution of coelacanths. The behaviour of all characters on this tree (optimized under the ACCTRAN option) is given in the legend to Fig. 9.7.

There are strong points of similarity between this tree and those of Forey (1991) and Cloutier (1991a,b). In particular all these hypotheses recognize an essentially pectinated tree throughout Palaeozoic coelacanths with two major clades of Mesozoic coelacanths. One of these Mesozoic clades includes such taxa as *Mawsonia*, *Axelrodichthys*, *Diplurus* and *Chinlea*. The other includes *Holophagus*, *Undina*, with *Latimeria* as the sister taxon to *Macropoma*. The mutual relationships of *Whiteia* and *Laugia* remain the same and, with minor differences so do the relationships of *Rhabdoderma/Allenypterus/Hadronector* and *Diplocercides*. Beyond this, comparisons are difficult to make with Cloutier (1991a,b) because the taxon sampling is different.

One of the major differences between the trees produced by Cloutier (1991a,b; Fig. 9.4) and Forey (1991; Fig. 9.3) is the phylogenetic position of *Coelacanthus*: this taxon is relatively derived in Cloutier's trees and relatively plesiomorphic in Forey's tree. This new analysis favours the derived position in placing *Coelacanthus* as the sister group to the two dominant Mesozoic clades. The analysis of Forey (1991) also suggested that *Coelacanthus* is the sister group to *Laugia* and *Coccoderma*, but this node in that earlier analysis was supported entirely by homoplasy.

It can be seen from the character distribution across this tree that a number of the nodes (numbered 8, 17, 18, 23 in Fig. 9.7) are supported entirely by homoplasy and must therefore be regarded as tentative hypotheses. Node 23 groups *Spermatodus* and *Sassenia* and this is probably realistic. These two taxa are very similar to each other in the shapes of the cheek and opercular bones, the proportions of the various components of the skull roof and the lower jaw, and the juxtaposition of the posterior openings from the rostral organ.

The node suggesting that *Rhabdoderma* is more derived than *Caridosuctor* is supported by two characters (loss of splenial ornament and asymmetrical caudal lobes in *Rhabdoderma*), both of which show reversals in other parts of the tree (interestingly, the only character suggesting that *Holophagus* and *Undina* are sister taxa is the 'reacquisition' of splenial ornament). This hypothesis must therefore be open to question. Cloutier (1991b) reached the same systematic conclusion based on relative size of the spiracular bone. This character was not used here because of the difficulty of precisely coding this attribute.

The two successive nodes (17, 18) suggesting that *Garnbergia* and *Libys* are related to the *Diplurus–Mawsonia* clade are particularly problematical. All changes associated with these nodes are resolved as parallelisms. However, in nearly all these characters *Garnbergia* is unknown. *Libys* is placed with these coelacanths in part on the reduction of ornament and the absence of a subopercum; both of these characters show a very homoplasious distribution among coelacanths.

Some of the taxa originally coded for but excluded from the analysis can be inserted to the tree at the most inclusive hierarchical level justified by their real data. As far as known character states are concerned then

---

Node 19. 14(1→0), 30(0→1), 56(1→0), 60(1→0), 92(0→1)*, 104(0→1).
Node 20. 9(1→0), 15(1→0), 17(2→1), 23(0→2), 27(1→2)*, 43(0→1)*, 45(1→0), 47(1→0), 49(0→2)*, 50(1→0), 106(0→1).
Node 21. 1(0→1), 16(0→1)*, 30(1→0), 38(0→1), 41(0→1)*, 65(0→1)*, 66(0→1)*.
Node 22. 32(1→0), 45(1→0), 91(0→1), 99(0→1), 100(0→1)*, 108(0→1)*.
Node 23. 23(0→2), 31(0→1), 34(0→1), 68(0→1), 80(0→1), 84(0→1).

Euporosteus is interchangeable with *Diplocercides*. *Axelia* (and probably *Wimania*) and *Ticinepomis* can be placed at node 12. *Lualabaea* is very poorly known but it does show a dentary swelling (character 65) which places it with *Mawsonia* and *Axelrodichthys* (node 21) and no other known features would deny this position.

A Linnaean classification that attempts to incorporate the phylogenetic ideas advocated here is given in Chapter 11.

9.4 EVOLUTIONARY TRENDS

Several trends in morphological evolution may be deduced from the phylogenetic tree although it is to be admitted that many of these show secondary reversal. The behaviour of individual characters is given in the legend to Fig. 9.7. Here, an attempt is made to summarize some of the more obvious general patterns of change and any correlated changes.

The evolution of the skull of coelacanths shows some marked trends in numbers of bones and in proportions. In contrast to most other sarcopterygian lineages, the skull roof is always divided by the dermal intracranial joint, reflecting the universal presence and presumably the functional importance of the endoskeletal joint. The postparietal shield is flat in plesiomorphic coelacanths but becomes more convex in many Mesozoic taxa (node 12), although it is difficult to suggest a functional explanation for this phenomenon. The pattern of the major bones of the postparietal shield became fixed very early in coelacanth history (node 2). All coelacanths with the exception of *Miguashaia* show the loss/fusion of the intertemporal. This bone persists in a primitive coelacanth and in primitive porolepiforms such as *Powichthys* and this suggests that the similarity between porolepiforms and coelacanths (Andrews, 1973) is convergent. There was also a relatively early shift in the position of the ossification centre of the postparietal from a posterior to an anterior position (node 2) although *Allenypterus* also shows a primitive position. This shift in ossification centre is evidenced by the position of the middle and posterior pit lines.

The extrascapular series shows marked evolutionary trends, with successive increases in the numbers of bones (nodes 6 and 13) and in most Mesozoic coelacanths and *Latimeria* they are contained within a deep embayment of the postparietal shield, which gives a rather characteristic shape to the posterior margin of the skull roof. As part of this process, the lateralmost extrascapular becomes 'fused' with the supratemporal (node 11). As a specialization amongst a small group of derived coelacanths (*Mawsonia* lineage), the extrascapulars become part of the skull roof and come to overlie the neurocranium.

The history of the anterior part of the skull roof is more complicated. The median internasals (postrostrals) were lost in most coelacanths relatively early (node 6) to reappear much later. Otherwise there appears to have been a history of fragmentation. Thus the number of tectals/supraorbitals increases in most coelacanths above node 13. The premaxilla generally becomes smaller and fails to enclose the anterior opening to the rostral organ (node 10) except in *Whiteia*. Furthermore, in two taxa (*Coelacanthus* and *Latimeria*) the premaxillae are fragmented to several small tooth-bearing plates (although Schultze, 1991, records that *Latimeria* may be polymorphic for this character).

The evolution of the skull roof is also marked by the development of ventral processes which grow down to strengthen the contact with the underlying neurocranium. At first these processes appeared on the parietal and the supratemporal (node 3) and then, much later, upon the postparietal (node 13). It is possible that the development of the postparietal descending process is associated with a reduction in the ossification of the neurocranium.

The general trend of neurocranial evolution is towards the 'fragmentation' of a primitively solidly ossified braincase into discrete ossifications separated by cartilage. This history matches that seen in actinopterygians (Patterson, 1975; Gardiner, 1984a) but it occurred independently. Most of the separate ossifications remaining in derived coelacanths occur at obvious points of mechanical stress (e.g. articulation points with the palate and the hyomandibular) and these might be expected to ossify first in the ontogeny of the primitive coelacanth braincase. Evolution may then be marked by delay in ossification of these separate centres in successive ontogenies. More needs to be known about the development of *Latimeria*'s braincase before such an idea may be carried further. A similar process has been suggested for the evolution of the actinopterygian braincase (Gardiner, 1984a).

In primitive coelacanths, with the possible exception of *Miguashaia* in which the neurocranium may have been cartilaginous, both ethmosphenoid and otico-occipital parts of the neurocranium are each solidly ossified. The sequence of 'fragmentation' appears to have been (1) the separation of lateral ethmoids and separation of basisphenoid and interorbital ossifications (somewhere between nodes 3 and 8, the neurocraniam of intervening taxa are unknown), (2) separation of prootic and combined opisthotic/basioccipital ossifications, (3) separation of the basioccipital ossification which develops a complex suture with the prootic, together with the appearance of a median supraoccipital ossification (node 11).

Within the ethmosphenoid the principal changes affect the basisphenoid, where there is the early loss of the basipterygoid process (node 3) and the development of a distinct dorsum sellae. The latter structure restricts the size of the cavity leading forward into the anterior part of the cranial cavity in most coelacanths. This may signify that the brain was restricted to the otico-occipital portion of the neurocranium as in adult *Latimeria*. In *Diplocercides* and *Euporosteus* there is some evidence that the brain was larger and occupied part of the interorbital region (p. 172).

Trends in the evolution of the parasphenoid concern the reduction in the area of the covering dentition as well as the anterior expansion to form ascending wings (node 14). The reduction of the toothed area appears to have been a gradual process and because this signifies a forward migration of the insertion of the basicranial muscle, this may have functional significance. This change of muscle insertion may, in turn, be correlated with the change in the relative sizes of the two halves of the braincase as measured by the lengths of the parietal and postparietal shields. As Forey (1991) and Cloutier (1991b) pointed out, there is a general trend across the phylogeny towards a relative decrease in the length of the postparietal shield. In more plesiomorphic coelacanths such as *Rhabdoderma* and *Sassenia*, the postparietal shield equals about 75% of the length of the parietonasal shield (exceptionally this figure is 50% in *Allenypterus* and 65% in *Diplocercides*). In taxa above node 12, the postparietal shield is usually less than half the length of the parietal shield. The inferred position of the anterior insertion of the basicranial muscle also changes (Forey, 1991: fig. 8) such that more derived coelacanths probably had a longer basicranial muscle.

These relative changes in the proportions of the two halves of the braincase may have functional implications. A relatively longer anterior moiety may potentially increase the gape of the mouth. It may also be significant that the relative length of the processus connectens and the otic shelf increase (although accurate measurements are very difficult to make). The probable increase in the length of the basicranial muscle presumably matched the mechanical advantage gained by greater length of the ethmosphenoid.

Within the cheek bones the dominant evolutionary trend has been a reduction of the total cheek cover. This has taken place by reduction in the sizes of individual bones (usually the squamosal and preoperculum are the bones affected) or by loss (suboperculum, preoperculum and spiracular). However, this loss and reduction has taken place in homoplasious fashion across the phylogeny above node 9. One of the more significant changes took place at node 13 where the lachrymojugal expanded anteriorly and became grooved to receive the soft-tissue tubes leading from the posterior openings from the rostral organ. This expansion followed earlier loss of the preorbital through which the rostral organ opened in more plesiomorphic coelacanths (node 11 – reversal in *Whiteia*).

The lower jaw is highly distinctive in coelacanths, even in the most plesiomorphic taxa (p. 132). Subsequent evolution chiefly concerned (1) separation of dentary teeth from the supporting bone (node 4), (2) the development of the hook-shaped process on the dentary (node 12) which provided a pocket receiving the pseudomaxillary fold, (3) 'fragmentation' of the articular/retroarticular (node 12).

The gill arches are virtually unknown in fossil coelacanths and few evolutionary trends can be recognized, with the exception of a possible consolidation of basibranchial dentition. In some more plesiomorphic taxa, this dentition consists of many small tooth plates which in more derived taxa become consolidated to larger plates and in some there is the appearance of median plates (p. 204, Fig. 7.6).

The evolution of the laterosensory system consisting of pit lines and sensory canals shows some obvious trends. The pit lines show reduction both in their extent as well as in the degree to which they mark the bones. Only *Diplocercides* may show an anterior pit line, which is a plesiomorphic attribute. The oral pit line is long in some

plesiomorphic coelacanths and reaches well forward to mark the dentary. At node 5 this line becomes much reduced, and in most coelacanths is restricted to a small 'comma-shaped' line lying at or near the centre of ossification of the angular. In plesiomorphic coelacanths, the middle and posterior pit lines on the postparietal and the jugal pit line which marks the squamosal are usually conspicuous. In most Mesozoic coelacanths (except *Whiteia* and *Coccoderma*), these pit lines fail to mark these bones (node 10). However, the facts that middle and posterior pit lines are present in *Latimeria* and that the cheek line lies superficially within the skin suggest that pit lines are universally present.

In the cephalic lateral line system, there is an elaboration of the otic, mandibular and preopercular canals and the supratemporal commissure. Relatively early in coelacanth history, there is a development of a short dorsal branch of the mandibular sensory canal which opens through a dentary pore (node 6) and the appearance of a median branch of the otic sensory canal at the level of the middle pit line (node 7). In some coelacanths such as *Whiteia* and *Mawsonia*, the ipsilateral medial branches join in the midline and, as further elaboration, the otic branch may be subdivided (*Latimeria*). In most Mesozoic taxa the preopercular canal sends out a subopercular branch (node 12), which in *Latimeria* is elaborately branched within the soft-tissue subopercular flap (Fig. 2.2). And as a final elaboration there is, within the *Holophagus–Latimeria* clade, the development of anterior branches of the supratemporal commissure. Elaboration of the sensory system is also reflected in the body lateral line openings where, in each scale, there are multiple openings in more derived taxa (node 9).

In the postcranial skeleton there was an early change from a heterocercal tail to a diphycercal tail (node 2). This arose by the development and elongation of epaxial radials. The caudal skeleton also shows the

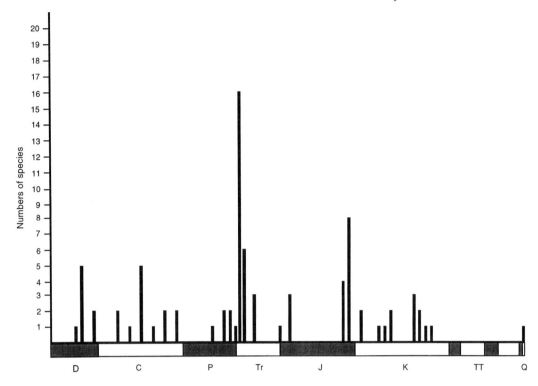

Fig. 9.8 Plot of numbers of coelacanth species through time showing a maximum diversity in the Lower Triassic (Tr) and another lesser peak in the Upper Jurassic. The species used to calculate this plot are only those which can be recognized as distinct species even though not all are able to be placed unequivocally within a phylogenetic tree.

reduction in the number of fin rays so that there is a one-to-one ratio between the radials and the fin rays, a feature that is inherited by most coelacanths.

Within the vertebral column there is a general trend towards an increase in the numbers of vertebrae in more derived coelacanths, although there is no absolute correspondence with the phylogeny. However, there is a more obvious trend in the increase in number and relatively closer spacing of the abdominal vertebrae (Forey, 1991: fig. 6). This may be explained as a trend towards less flexibility of the abdominal region. There is also an increase in the lobation of the second dorsal and the anal fins. This increase in lobation is accompanied by a tendency for

the fin rays to become symmetrically arranged around the tips of the second dorsal and anal lobes (Forey 1991: fig. 7). It is possible that this 'stiffening' of the abdominal region and increased lobation of the second dorsal and anal fins is associated with an emphasis on these fins to produce the locomotory thrust during slow locomotion (p. 22).

9.5 PATTERNS OF COELACANTH EVOLUTION

Several aspects of coelacanth evolution are relevant to general questions of vertebrate evolution and to the particular patterns within the coelacanth lineage itself. General questions concern the relationship between morphological and species diversity, the

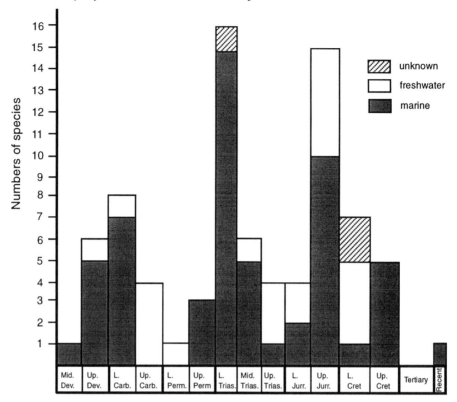

Fig. 9.9 Plot of coelacanth species through time differentiating environments in which they have been found. As can be seen, most coelacanths were marine. In this diagram, only species considered valid are included: those records simply noted as coelacanth indet. or sp. are omitted. M, marine record; FW, freshwater; ?, unknown environment or at present debated. The records are as follows. Middle Devonian: *Euporosteus eifeliensis* [M]. Upper Devonian: *Chagrinia enodis* [M], *Coelacanthus welleri* (M), *Diplocercides kayseri* [M], *Diplocercides jaekeli* [M], *Diplocercides heiligenstockiensis* [M], *Miguashaia bureaui* [FW]*. Lower Carboniferous: *Allenypterus montanus* [M], *Diplocercides davisi* [M], *Hadronector donbairdi* [M], *Lochmocercus aciculodontus* [M], *Polyosteorynchus simplex* [M], *Rhabdoderma ardrossense* [M], *Rhabdoderma* (?) *alderingi* [FW], *Rhabdoderma huxleyi* [M]. Upper Carboniferous: *Rhabdoderma elegans* [FW], *Rhabdoderma tinglyense* [FW], *Rhabdoderma exiguum* [FW], *Synaptotylus newelli* [FW]. Lower Permian: *Spermatodus pustulosus* [FW]. Upper Permian: *Coelacanthus granulatus* [M], *Changxingia aspratilis* [M], *Youngichthys xinhuanisis* [M]. Lower Triassic: *Axelia robusta* [M], *Axelia elegans* [M], *Laugia groenlandica* [M], *Mylacanthus spinosus* [M], *Mylacanthus lobatus* [M], *Piveteauia madagascariensis* [M], *Whiteia* sp. of Gardiner (1996) [M], *Sassenia tuberculata* [M], *Sassenia groenlandica* [M], *Scleracanthus asper* [M], *Sinocoelacanthus fengshanensis* [?], *Wimania sinuosa* [M], *Whiteia nielseni* [M], *Whiteia tuberculata* [M], *Whiteia woodwardi* [M], *Wimania* (?) *multistriata* [M]. Middle Triassic: *Alcoveria brevis* [M], *Garnbergia ommata* [M], *Hainbergia granulata* [M], *Heptanema paradoxum* [M], *Moenkopia wellesi* [FW], *Ticinepomis peyeri* [M]. Upper Triassic: *Chinlea sorenseni* [FW], *Coelacanthus lunzensis* [FW], *Diplurus newarki* [FW], *Graphiurichthys callopterus* [M]. Lower Jurassic: *Diplurus longicaudatus* [FW], *Holophagus gulo* [M], *Indocoelacanthus robustus* [FW], *Undina* (?) *barroviensis* [M]. Upper Jurassic: *Coccoderma bavaricum* [M], *Coccoderma gigas* [M], *Coccoderma seuvicum* [M], *Coccoderma* (?) *substriolatum* [M], *Libys polypterus* [M], *Libys superbus* [M], *Lualabaea henryi* [FW], *Lualabaea lerichei* [FW], *Macropoma willemoesi* [M], *Undina cirinensis* [M], *Undina purbeckensis* [M], *Undina penicillata* [M]. Lower Cretaceous: *Axelrodichthys araripensis* [?], *Macropoma lewesiensis* [M], *Mawsonia gigas* [?], *Mawsonia lavocati* [FW], *Mawsonia libyca* [FW], *Mawsonia tegamensis* [FW], *Mawsonia ubangiensis* [FW]. Upper

pattern of species diversity through time, and the concordance or otherwise of phylogenetic hypotheses with the stratigraphic record, all of which may be summed up by Simpson's book title *Tempo and Mode in Evolution* (1944).

Coelacanths are usually described as forming a slowly evolving lineage (Simpson, 1944) similar to lungfishes, and as following a pattern of evolution that shows rapid change in the early history of the group followed by stasis (Schaeffer, 1952a), a phenomenon which has also been claimed for lungfishes (Westoll, 1949). Such a pattern may favour a punctuationist view of evolutionary history. Also, coelacanths have been used as one classic case of bradytely. These ideas were formulated before a phylogeny of coelacanths was proposed and therefore it is pertinent to ask if such a reputation can be maintained. In many ways coelacanths are not particularly good to address questions of morphological and species evolution, because the taxon sampling through time is very poor and the records are usually only 'spot records', being known from only one horizon and locality. However, the common citations to coelacanth evolution in textbooks (Simpson, 1944, 1953) demand some comments here.

Diversity through time

A simple plot of genera or species through time has been attempted on several occasions (Schaeffer, 1952b; Forey, 1984, 1988, 1991; Cloutier and Forey, 1991; Cloutier, 1991a) and another is given here (Fig. 9.8). The precise way in which these plots have been made is different and slight variations occur

depending on the current assessment of species status. Here a total of 83 species are plotted and these correspond to those included in categories A and B of Cloutier and Forey (1991): that is, those species which can be placed unequivocally within a phylogenetic tree (46 species) and those which are good species but the relationships of which cannot be accurately determined. Despite the differences in how these plots have been made, the pattern is very similar. The peak of diversity occurs in the Lower Triassic (16 species) with minor peaks in the Upper Devonian (5 species in the Frasnian), Lower Carboniferous (5 species in the Namurian) and the Upper Jurassic (8 species in the Kimmeridgian/Tithonian). The distribution of coelacanths through time against environments in which they are found is shown in Fig. 9.9. Most are found in marine deposits. The 'freshwater peaks' in Upper Carboniferous – Lower Permian and in Lower Cretaceous may do no more than reflect the general pattern of fish-bearing deposits. Consideration of those coelacanths that cannot be regarded as valid species (either because of poor material or owing to incomplete study) would not significantly alter this overall pattern. There are about 10 species names for Triassic taxa that need validation. The only possible change may influence the number of Upper Carboniferous species.

There are two major gaps shown in the species plot (Fig. 9.8): Santonian → Recent (83 million years) and throughout much of the Jurassic (Pleinsbachian → Oxfordian, 40 million years). The Jurassic 'gap' is false in the sense that indeterminate coelacanth remains are known from the intervening

---

Cretaceous: *Macropoma lewesiensis* [M], *Macropoma precursor* [M], *Macropoma speciosum* [M], *Macropomoides orientalis* [M], *Megalocoelacanthus dobei* [M]. Recent: *Latimeria chalumnae* [M].

*Subsequent to the generation of this plot, earlier suspicions (Schultze, 1972) that the Escuminac Formation of Quebec, containing *Miguashaia*, may represent a marine environment of deposition have been altered (Schultze and Cloutier, 1996). This would mean that all Devonian coelacanths are marine.

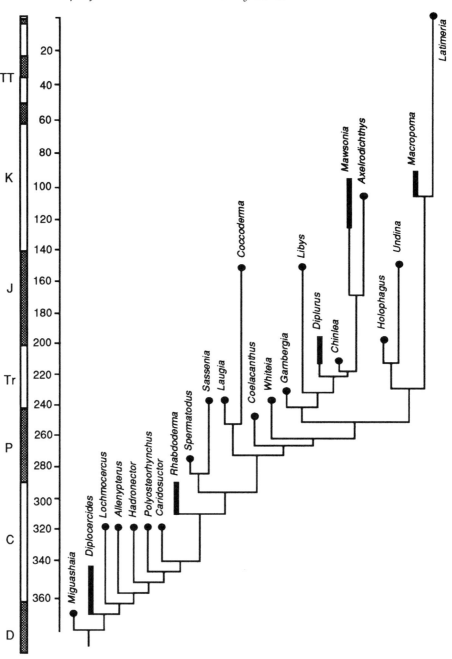

Fig. 9.10 Plot of the phylogenetic tree against stratigraphy to show overall good correlation between stratigraphy and phylogeny. The black circles denote a single time at which a genus is known; thick vertical bars represent time ranges for the genera. It can be seen that most genera are restricted to one time level. The major stratigraphic range extension required throughout the Tertiary remains unexplained. Absolute time (million years) shown on left-hand side.

Bajocian and Oxfordian (Wenz 1979) but these remains cannot be shown to be distinct species or referable to named species. The Santonian → Recent gap is more difficult to explain. Some coelacanth remains from the Maastrichtian of New Jersey have doubtfully been referred to *Megalocoelacanthus* (Schwimmer *et al.*, 1994). Although these remains cannot be positively identified as a particular species they are coelacanth remains and show that coelacanths can be traced in the fossil record until 75 million years ago. The only younger record of coelacanths in the fossil record is that of a piece of bone bearing tubercles from the Palaeocene of Sweden (Ørvig, 1986) but this is not determinable as a coelacanth fish. Thus, to date, there are no records of coelacanths for the past 75 million years.

Stratigraphic fit

The concordance between the phylogeny given here and the stratigraphic occurrence of taxa is very good (Fig. 9.10). At all nodes except two, either the sister taxa are of the same age or the more derived taxon is younger. One exception is *Libys*, which is Tithonian, whereas the more derived sister group is at least Scythian. Thus a range extension for *Libys* of at least 85 million years must be assumed. It is possible that some of the omitted taxa may fit along this branch but it must also be admitted that the phylogenetic position is also problematical (see above). The other incongruence between the phylogeny and stratigraphy concerns *Coelacanthus* (Upper Permian–Guadaloupian), which occurs approximately some 10 million years earlier than its plesiomorphic sister group, *Laugia + Coccoderma* (Lower Triassic–Scythian). Given the overall quality of the fossil record of coelacanths, this mismatch is here regarded as insignificant. Other hypotheses of range extensions or ghost lineages necessary are those leading to *Coccoderma* (90 million years), *Rhabdoderma* and cladistically

more derived coelacanths (50 million years), *Chinlea* and *Mawsonia + Axelrodichthys* and, of course, that between *Macropoma* and *Latimeria*. Some of the taxa excluded from this analysis may occupy these time intervals if their relationships could be more accurately specified. For instance, *Lualabaea* (Upper Jurassic) may well fit along the internode leading to *Mawsonia + Axelrodichthys* because it has a very similar lower jaw structure. The fact that there are many excluded taxa with appropriate stratigraphic positions may well mean that some of the 'gaps' are exaggerated by our inability to accurately assess the relationships of all coelacanth taxa.

Morphological evolution

Accepting this as a representative sample of coelacanth diversity through time, and further accepting that the species considered in the construction of the phylogenetic tree (Fig. 9.7) are a representative sample of the morphological pattern of diversity, there are a number of deductions to be made about the mode and tempo of coelacanth evolution.

Morphological disparity can be measured by the numbers of cladogenetic events at selected intervals as well as the number of character changes (steps on the phylogenetic tree) occurring between taxa at specified time intervals (Cloutier, 1991b). Both methods have been outlined by Smith (1994). Five intervals are selected here: Frasnian (two taxa), Namurian (six taxa), Scythian (four taxa), Kimmeridgian/Tithonian (three taxa) and Aptian–Cenomanian (three taxa). The mean number of nodes separating contemporaneous taxa increases to the Kimmeridgian/Tithonian followed by a slight decline (Fig. 9.11(B)) into the Cretaceous. Similarly the mean number of character changes separating contemporaneous taxa increases to maxima in the Scythian and Kimmeridgian/Tithonian followed by a small decline (Fig. 9.11(A)). Therefore the time of maximum cladogenetic disparity (number of nodes)

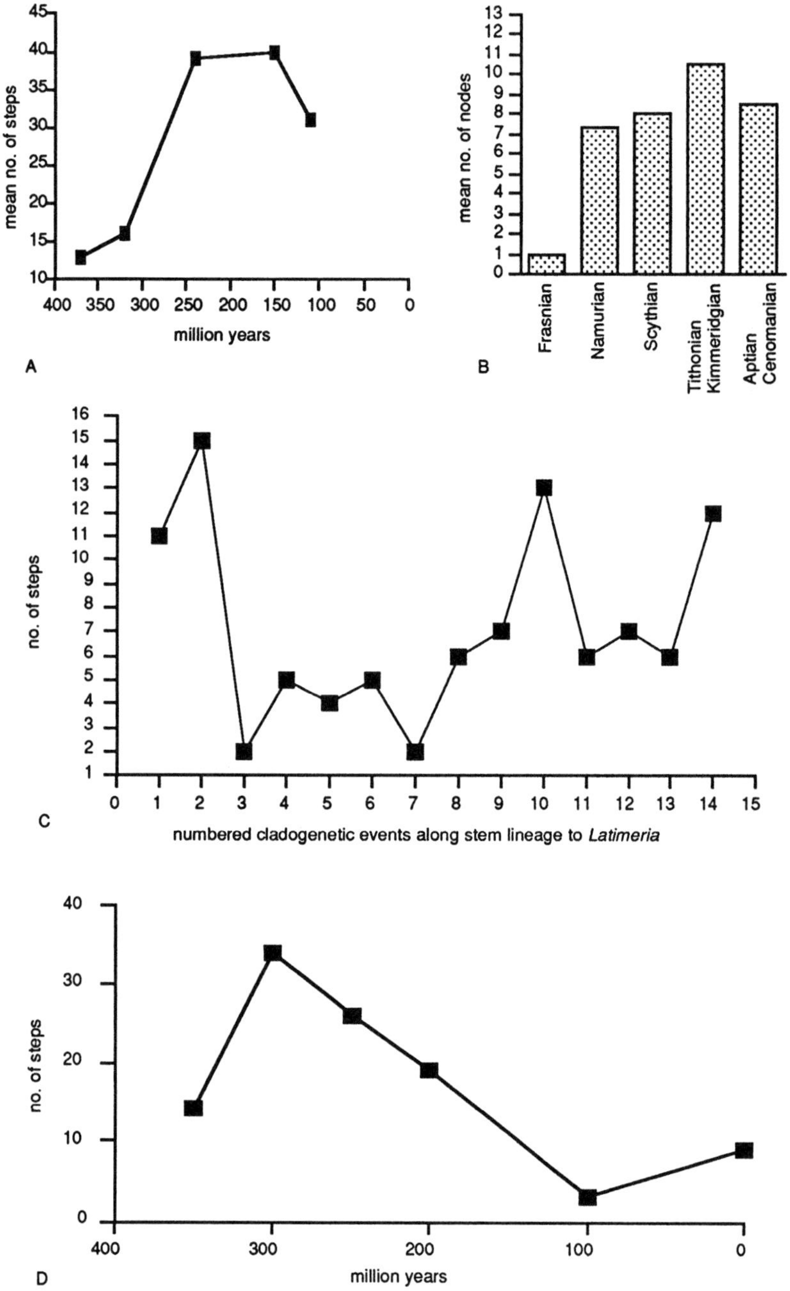

Fig. 9.11 Measures of morphological evolution through time. (A) Plot of mean number of steps between any two contemporaneous taxa at given times (Frasnian, Namurian, Scythian, Tithonian/Kimmeridgian, Aptian/Cenomanian). (B) Mean number of nodes separating taxa at given times. (C) Number of character changes occurring between the numbered nodes (horizontal axis) on the phylogenetic tree shown in Fig. 9.7. This pathway is the direct phylogenetic path leading to *Latimeria*. (D) Plot of the number of character changes on the tree shown in Fig. 9.7 occurring between arbitrarily selected geological times.

postdates the time of maximum species diversity. However the rate of increase in morphological diversity is coincident with maximum species diversity but, using this measure, the rate is maintained well into the Upper Jurassic. This kind of measure compares most closely to the measure of morphological evolution made by Forey (1988), who concluded that the maximum rate of morphological evolution was coincident with maximum species diversity. It does not support a punctuationist view of coelacanth evolution, but rather suggests a gradual increase in cladogenetic and morphological diversity followed by near extinction in the late Cretaceous.

However, the measure of morphological evolution used by Schaeffer (1952b) suggested rapid initial evolution followed by near stasis from the Permian onwards. Schaeffer's approach, like that of Westoll (1949), who investigated the rate of morphological change in lungfishes, established a series of evolutionary steps for certain characters. Each step was given a score and the scores were summed by contemporaneous taxa for various time intervals. No phylogenetic tree was used. Cloutier (1991a) used a phylogenetic tree which was broadly similar to that used here (see above) and agreed with Schaeffer's conclusion. However, Cloutier (1991a) was concerned only with changes that took place along the stem lineage leading to *Latimeria*. In the context of the current tree, those changes are those occurring at nodes 1–14, 16, and along the terminal branch leading to *Latimeria*. Measured in this way, and corresponding to Cloutier's rate 'f' then there is indeed an initial rise in the rate

of character acquisition per unit time followed by a steady decrease (Fig. 9.11(D)). The difference between Cloutier's plot and the one given here concerns the rate at which character changes decrease. Because these studies are based on different character/taxa data sets, the differences cannot be compared easily and may not be significant. The rate of morphological evolution (character changes) between successive nodes along the stem lineage leading to *Latimeria* (Cloutier's (1991b) rate 'c') shows exactly the same pattern (Fig. 9.11(C)) in the initial phases of diversification. There is an initial rise, with many changes taking place between the plesiomorphic *Miguashaia* and Namurian taxa, followed by a low but steady rate of change per cladogenetic event. But this is succeeded by a subsequent rise to a second peak during the evolution of *Coelacanthus granulatus* and most Mesozoic taxa + *Latimeria*.

In sum, the rate of morphological evolution along the stem lineage leading to *Latimeria* measured against absolute time increased rapidly and thereafter decreased gradually, not rapidly as has been previously assumed. At no time can there be said to have been a stabilized and low rate of change, although it has fluctuated with time. The rate of change between cladogenetic events shows two peaks: once in the early evolution of the group and another broadly coincident with the species 'flowering' in the Triassic. Finally, morphological evolution over the entire coelacanth tree shows a gradual increase to a maximum which lasted from the Lower Triassic to the Upper Jurassic.

RELATIONSHIPS OF COELACANTHS

10.1 INTRODUCTION

The relationships of coelacanths to other fishes and to tetrapods are clearly important in our assessment of *Latimeria* as a surviving 'missing link' between fishes and tetrapods and whether it is entitled to be called 'Old Fourlegs' (Smith, 1956). In a wider context, understanding the relationships of coelacanths is part of theories concerned with the origin of tetrapods, which involve modern actinopterygians and lungfishes as well as a suite of fossil forms. Interrelationships amongst these animals has been one of the most frequently debated theories in lower vertebrate systematics. Early ideas on the relationships of coelacanths centred on the notion that coelacanths were modified rhipidistians (osteolepiforms and porolepiforms). These extinct taxa, in turn, were thought to be ancestral to tetrapods and therefore *Latimeria* was regarded as the nearest living relative of modern tetrapods. As has been pointed out (Patterson, 1980; Rosen *et al.*, 1981, Forey, 1984, 1987), the idea that rhipidistians were ancestral to tetrapods was flawed by a false expectation that ancestors could be recognized. Instead, for the past 15 years or so the problem of the relationships of coelacanths in particular, and the origin of tetrapods in general, has been approached through cladistic systematic methods. That is: an attempt has been made to recognize shared derived characters which may be useful in recognizing sister groups. If *Latimeria* is to uphold its reputation as a surviving missing link, then it must be shown that the most likely theory of relationship amongst living taxa places *Latimeria* and tetrapods as sister groups. Any complete assessment about the phylogenetic position of coelacanths must take into account the relationships of the many kinds of sarcopterygian fishes, both Recent and fossil. However, the kinds of data that have been used to assess the relationships of coelacanths have been rather different.

10.2 THEORIES OF RELATIONSHIPS AMONG RECENT FISHES AND TETRAPODS

Evidence for the one or more theories shown in Fig. 10.1 has come from comparative anatomy, physiology and molecular studies. The last category of evidence is dealt with separately here (p. 250).

One of the first cladistic studies was that by Løvtrup (1977), who suggested that *Latimeria* and Chondrichthyes were sister groups (Fig. 10.1(A)). Løvtrup listed eight similarites between *Latimeria* and cartilaginous fishes: (a) fatty liver, (b) structure of the eye, (c) structure of the thymus, (d) structure of the thyroid, (e) shape of the grey matter (flattened in cross section within the spinal cord), (f) absence of Mauthner's fibres, (g) first pair of nerves in telencephalic vesicle, and (h) large eggs. Lagios (1979) suggested the following *Latimeria* + chondrichthyan characters: (i) osmoregulation by urea retention, (j) rectal gland, (k) structure of the pancreas and (l) the structure of the pituitary. To this list we may add some similarities between these two groups noted by Lemire and Lagios (1979) but not stated as synapomorphies: (m) presence of cerebellar auricles, (n) absence of a valvula, (o) large renal glomeruli, (p) ovoviviparity. This is an impressive list but they have all been discussed by Forey (1980). Many of them can be judged as primitive vertebrate characters (c, d and see discussion on osmoregulation, p. 27, relating to i, j, p), some are known in other fishes (e, n, o) and one at least is not universally present in chondrichthyans (f). However, the structure of the pituitary is very similar between the two groups (p. 19) and remains a putative synapomorphy (Lagios, 1982).

Other theories (Fig. 10.1(B–F)) accept *Latimeria* as an osteichthyan and there are many characters that can be cited in evidence (Forey, 1980; Maisey, 1986a), among which are the presence of large dermal bones that can be referred to a common pattern investing the skull, lower jaw and covering the gills and the lateral face of the scapulocoracoid; marginal replacement teeth associated with premaxilla and dentary; swim bladder (or lung), endochondral bone, pharyngobranchials which are directed anteriorly, an interhyal in the hyoid arch, separate branchial levator muscles, consolidated tooth plates upon the copula and the palate, lepidotrichia supporting the fin rays.

The theory shown in Fig. 10.1(B) was suggested by von Wahlert (1968) and is based on the idea that there was a sequential acquisition of the upper jaw and palatal bones. Modern lungfishes and *Latimeria* lack the maxilla while actinopterygians and tetrapods have both. *Latimeria* is claimed to be more derived than lungfishes because, like actinopterygians and tetrapods, it shows dermopalatines and an ectopterygoid. This theory is based on an assumed polarity of character change (simple to

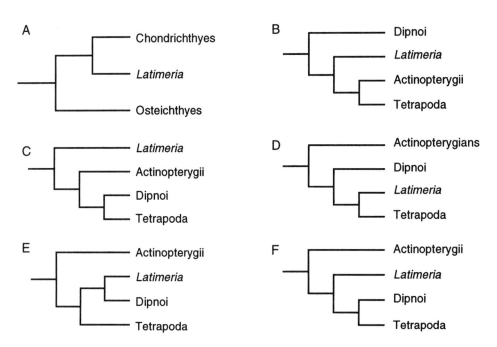

Fig. 10.1 Theories of the relationship of *Latimeria* to other modern groups which have been proposed using information from recent comparative anatomical and physiological studies. See text for discussion.

complex) and must be judged against other character distributions (Miles, 1977: 315) as well as acknowledging that fossil lungfishes may have extra palatal bones (see below). In fact, there is little other morphological evidence to support this theory but some molecular studies have concurred with this theory.

Wiley (1979) suggested that *Latimeria* is the most primitive osteichthyan (Fig. 10.1(C)). This theory is based on the anatomy of the ventral gill arch muscles which are relatively simple and undifferentiated in *Latimeria* in comparison with other bony fishes and tetrapods.

The idea that *Latimeria* is the sister group of tetrapods (Fig. 10.1(D)) has been based on almost purely palaeontological arguments (p. 1). Such an idea has been difficult to substantiate using evidence exclusively from the anatomy/physiology of the modern animals. Northcutt (1987) notes that the spinal cord of *Latimeria* shows enlargements at the levels of the pectoral and pelvic fins. This may have some implication for the tetrapodous coordination of fin movement (p. 23). However, this feature needs to be checked in lungfishes, which also show the same kind of fin coordination. The structure of the inner ear is similar in *Latimeria* and tetrapods in showing a perilymphatic connection between left and right ears and in showing a basilar papilla (p. 20).

A theory suggesting that lungfishes and coelacanths are sister groups (Fig. 10.1(E)) has been advocated by Northcutt (1987) based on similarities in brain anatomy (p. 18) and the suggestion that the rostral organ of *Latimeria* is homologous with the the labial cavity of lungfishes (p. 20).

One final theory: a modern version of 19th-century work which reported the great similarity between lungfishes and tetrapods (review: Patterson, 1980) suggests that lungfishes and tetrapods are sister groups with the coelacanth more distantly related (Fig. 9.1(F)). Amongst the modern fauna this theory has been supported by a large number of characters shared by lungfishes and tetrapods. Some of these emphasize the similar way in which the palate is associated with the braincase and a decreasing importance of the gills for gaseous exchange: (a) palate fused with the braincase, (b) abbreviated palatine region involving loss of the autopalatine, (c) fusion or suturing of right and left pterygoids in midline, (d) choana or internal nostril present, (e) hyomandibular decoupled from ceratohyal, (f) loss of interhyal, (g) levator hyoideus muscle inserts directly to ceratohyal (usually in fishes it attaches to the hyomandibular), (h) loss of pharyngobranchials, (i) efferent branchial arteries join one another before joining to form dorsal aorta, (j) depressor mandibulae muscle present. In the postcranial skeleton: (k) pectoral and pelvic fins of equal size, (l) second axial segment of the limb formed by two subequal bones (e.g. radius and ulna) with the postaxial element articulating against the postaxial side of the proximal segment (humerus), (m) fusion of right and left pelvic bones to form ischial and pubic processes, (n) rib gradient, where the largest ribs are found immediately behind the head, thereafter decreasing in size. Some similarities are associated with the lungs and air breathing: (o) similar internal structure of the lung, (p) pulmonary circulation with both a pulmonary artery and a pulmonary vein, (q) divided auricle, (r) valves in the truncus arteriosus providing a separation of oxygenated and deoxygenated blood, (s) glottis and epiglottis present providing a guarded passage to the lungs. Other shared features are: (t) non-compartmentalized adenohypophysis, (u) mesotocin present, (v) lens proteins D_1, D_2 and D_5. Most of the features listed above have been discussed at greater length by Rosen *et al.* (1981).

Of these different theories the two that have received most evidence are Figs 10.1(E) and 10.1(F), and the theory suggesting that lungfishes are the Recent sister group of tetrapods is by far the best supported.

10.3 MOLECULAR SEQUENCING AND RELATIONSHIPS OF COELACANTHS

The analysis of molecular data for phylogenetic reconstruction is viewed by many workers to have distinct advantages over the use of traditional morphological data (Hillis, 1987). First, there is a potentially very rich source of characters available in molecular sequences when those sequences are analysed with parsimony techniques, in which each base position is regarded as a separate and independent character. In the genome of some eukaryote cells there may be as many as 4×10^{11} base pairs (Hillis, 1987); although so far only a very small sample of this genome has been sampled. Second, DNA and RNA is universally present in living organisms and therefore there is little difficulty in being able to compare molecular sequences over a broad taxonomic range. In this sense problems of homology are less than those with morphological structures where some initial estimate of similarity is necessary before a theory of homology may be suggested. Third, all nuclear genomic sequence information is inherited and is less prone to environmental modification than morphological structure (Hillis, 1987), which is often described as ecophenotypic variation and can lead to polymorphic codings for characters. Fourth, molecular data are much closer to the genetic code and therefore might be expected to reveal more faithful phylogenetic relationships. The physicochemical interactions between genes and gene products are far better understood than between the genetic code and morphological structures (although the increasingly intensive studies of Hox genes are making significant contributions to this understanding). This means that we are better able to set constraints to any molecular analysis with a high degree of empirical justification.

The analysis of molecular data is, however, limited by its own set of problems. First, sampling taxa, both in number of species and number of individuals, is rather limited and this can cause problems of long branch attraction in phylogenetic analysis leading to false theories of relationship. With increasingly sophisticated techniques this sampling problem will rapidly recede. Second, there will remain many palaeontological species for which there are morphological data but for which molecular sequences will be unavailable, notwithstanding some conspicuous claims where sequences have been obtained (Golenberg *et al.*, 1990). Third, molecular sequence data deal with variation in only four character states, or 20 amino acids, and this leads to problems in trying to align sequences such that we are not always sure that we are comparing like positions within the genome between different taxa. Alignment is probably the single most important problem currently facing molecular systematists. Fourth, molecular sequences do not have ontogenies which, for morphological characters allow us to polarize characters (Nelson, 1978) and to root trees. Rooting molecular trees is known to be problematical with the choice of the root significantly affecting the resulting topology (Smith, 1989; Stock *et al.*, 1991a; Stock and Whitt, 1992). Lastly, there are problems associated with recognizing homologous gene sequences (Patterson, 1987).

The analysis of molecular data has been used on several occasions to attempt to discover the relationships of coelacanths, lungfishes and tetrapods. The results of these analyses give a far from clear phylogenetic signal. This may be a reflection of the methods of analysis, the unsuitability of the genes/molecules sequenced so far, inadequate or inappropriate selection of taxa, or a combination of these factors. Some of these problem areas are discussed below with respect to individual studies. The rapid development of techniques designed to obtain sequence data (PCR, automatic sequencing, production of primers) means that molecular sequences will become available that are more complete and known for a greater variety of taxa than is currently available. We are far from reaching the limits of molecular analyses for this systematic problem and any conclusions reached here will surely be tested with more data in the near future.

Two kinds of sequence data have so far been gathered bearing on the problem of coelacanth relationships. First, nucleotide sequences from a variety of nuclear and mitochondrial genes have been obtained. Second, amino acid sequences for several proteins have been analysed. Sometimes these sequences have been determined directly (Jauregui-Adell and Pechere, 1978; Pechere *et al.*, 1978; Maeda *et al.*, 1984). More usually the amino acid sequence is inferred from the DNA sequence within the coding gene (Gorr *et al.*, 1991; Normark *et al.*, 1991; Yokobori *et al.*, 1994) according to the parameters of the genetic code (Osawa *et al.*, 1992).

DNA sequences and nuclear genes

Nuclear genes are usually considered more useful for relationships between lineages known to be of great antiquity (>100 million years). The most frequently sequenced genes are those coding for large (28S) and small subunit (18S) ribosomal RNA. These genes generally have large sections of nucleotide sequences which are closely similar between distantly related organisms and this makes alignment of one sequence with another a relatively easy operation. Between these highly conserved regions there are also very variable regions, called divergent domains (Hillis and Dixon, 1989). Although they may be difficult to align, they are potentially useful for detecting phylogenetic relationships between closely related species. Thus, the nuclear genes are often thought to be

very useful for a variety of systematic problems.

Nucleotide sequences from both large and small subunit ribosomal genes have been obtained and used in analyses designed to discover the relationships of the coelacanth. Unfortunately, most of the studies include only the coelacanth or lungfishes. Joss *et al.* (1991) analysed a sequence of about 500 nucleotides in their study which included only the Australian lungfish *Neoceratodus*. The result was inconclusive and they were unable to suggest whether lungfishes or actinopterygians were closer to tetrapods. During the course of their study, a partial sequence of 18S rRNA became available for *Latimeria* (Hedges *et al.*, 1990) which they incorporated but with no further resolution. One unsuspected grouping of *Calamoichthys* (an actinopterygian) with *Hydrolagus* (a chimaeroid) appeared consistently but, as the authors acknowledged, the *Calamoichthys* sequence was incomplete. Also this study used the very distantly related *Branchiostoma* and *Styela* to root the tree.

In a study including both a lungfish (*Neoceratodus*) and the coelacanth, Stock *et al.* (1991b) were able to provide a much larger sequence of 18S rRNA for a variety of craniates. They were able to align 1668 base positions. The outcome of parsimony analysis, using a lamprey as outgroup, is shown in Fig. 10.2(A) and can be seen to give very unexpected results (e.g. the grouping of *Rhinobatus* with *Homo* as well as the cladistically derived position of the actinopterygian *Atractosteus*). They tried a number of analytical 'experiments' (deleting taxa with long branch lengths, adding taxa to attempt to break up long braches, weighting transversions, comparing separate analyses of paired stem regions of the gene with unpaired loop regions, removing sites from analysis which were variable amongst acknowledged monophyletic taxa). None of these 'experiments' resulted in any significant improvement in arriving at a solution which came anywhere

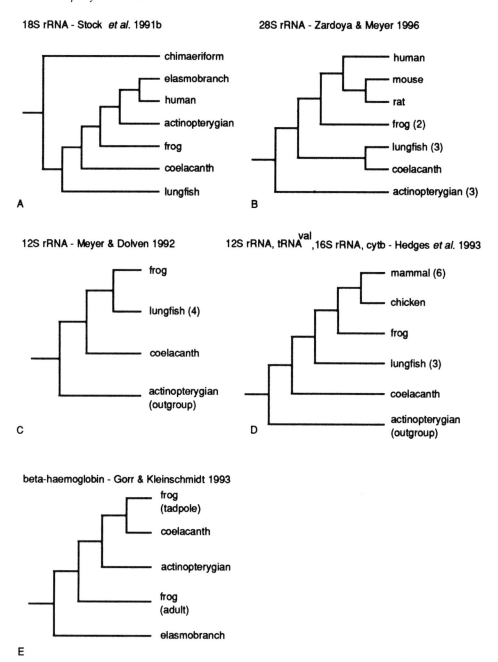

Fig. 10.2 In recent years a number of studies using molecular sequences have resulted in many different hypotheses of relationships between the coelacanth and other vertebrates. A sample is shown here with the genes analysed indicated in the heading to each tree. Apart from the different phylogenetic conclusions, the studies differ considerably in the molecules used, the number of characters and the methods of analysis. See text for discussion.

near matching solutions arrived at by analyses using morphological data. They concluded that the cladogenetic events leading to the subdivision of osteichthyan groups may have happened in a relatively short time interval, such that very few informative substitution events took place, and that these were probably swamped out by parallel substitutions in other lineages (see also Stock *et al.* 1991a). This must be considered a real problem with slowly evolving genes.

In another study of 18S rRNA which was designed to address the classification of agnathans (Stock and Whitt, 1992), the coelacanth grouped with tetrapod (*Xenopus*) rather than the chondrichthyan (*Rhinobatos*). Because neither a lungfish nor an actinopterygian was included, this study has limited significance to this systematic problem.

Analyses of 28S rRNA relevant to the relationships of lungfishes, coelacanths and tetrapods have been carried out by Le *et al.* (1993), Hillis and Dixon (1989) and Hillis *et al.* (1991). The studies of Le *et al.* included a lungfish (*Protopterus*) but not the coelacanth, and were designed to discover relationships between a dense sampling of actinopterygian fishes. Sequences of about 500 base pairs were used. The lungfish grouped with anurans rather than with chondrichthyans. However, the two species of urodeles were resolved as sister taxa to the five species of chondrichthyans, implying that lissamphibia were non-monophyletic.

The studies of Hillis and Dixon (1989) and Hillis *et al.* (1991) included the coelacanth but no lungfishes amongst seven gnathostome taxa; 1989 base positions within the 28S rRNA genetic code could be aligned and the resulting most parsimonious tree placed the coelacanth as the sister group to tetrapods, cladistically more derived than actinopterygians.

The most complete sequence of 28S rRNA has been analysed by Zardoya and Meyer (1996). These authors sequenced the entire 28S gene and, after deleting regions which

could not be easily aligned, they were able to analyse 4786 base pairs (bp) for 12 vertebrate species. In all, 807 sites were phylogenetically informative and after using a transition/transversion weighting ratio of 2:1 they obtained a single tree, a simplified version of which is shown in Fig. 10.2(B), showing coelacanths to be most closely related to lungfishes. This tree resulted with three different kinds of tree-building procedure.

To summarize, the evidence from the nuclear genes analyses so far is inconclusive. In those studies which considered both lungfishes and coelacanths (Joss *et al.*, 1991; Stock *et al.*, 1991a,b; Zardoya and Meyer, 1996), there is either no clear signal as to which of the two fish taxa is nearer to tetrapods, or lungfishes and coelacanths are resolved as sister groups. It should be emphasized that taxon sampling for each gene is very different between the studies, a deficiency which can surely be corrected within the next few years.

Mitochondrial genes

A number of mitochondrial genes have been examined in separate studies. Meyer and Wilson (1990) sequenced 664 bp of the 12S rRNA and cytochrome b genes for *Latimeria*, an actinopterygian and the South American lungfish. Meyer and Dolven (1992) added 12S sequences from the African lungfishes (*Protopterus annectens* and *P. aethiopicus*). Both of these studies used the actinopterygian as the outgroup and published sequences for the tetrapod (*Xenopus*) and resulted in a sister-group pairing between lungfishes and tetrapods (Fig. 10.2(C)). Bootstrap values for the nodes in the tree were very low, except for that linking the two African lungfish species. Normark *et al.* (1991) sequenced about 300 bp of the cytochrome b (cyt b) gene which they translated to amino acids before analysis. They used about 25 taxa – a much denser taxon sampling than that used by Meyer and Dolven (1992) – and suggested that lungfishes and tetrapods were sister groups. However, *Latimeria* was resolved as the sister group of sturgeons and paddle-fishes within the actinopterygians. Once again the support for the node linking lung-fishes and tetrapods was low.

One of the more extensive studies of the mitochondrial genome is that of Hedges *et al.* (1993), who were able to sequence nearly 3000 bp from several genes (12S, 16S, cyto-chrome b, tRNA^{val}). Their analysis, using an actinopterygian outgroup, agreed with the studies mentioned above: lungfishes and tetrapods were resolved as sister groups. And further study (Yokobori *et al.*, 1994) of sequences from the cytochrome oxidase subunit I gene (CoI), added to the 12S and cyt b data already available gave mixed results. Analysis of the CoI amino acid sequence by parsimony, maximum likelihood or neighbour joining methods of analysis suggest that lungfish and coelacanths are sister groups. Combining several genes toge-ther gave a data set that could not distin-guish between the hypotheses expressed in Fig. 10.1(E) and (F).

Thus, the majority of analyses of mito-chondrial genes suggest that lungfishes and tetrapods are sister groups. The nearest rival theory suggests that lungfishes and *Latimeria* are sister groups.

Amino acid data

Analysis of amino acid data (either by direct sequencing or translating DNA sequences) have also given inconclusive results. An early analysis of α and β parvalbumins (calcium-binding proteins common in white muscle) suggest that teleosts are more closely related to tetrapods than either are to coelacanths (Maeda *et al.*, 1984). Lungfishes were not included in this study.

In an analyisis of α and β haemoglobin amino acids, Gorr *et al.* (1991) and Gorr and Kleinschmidt (1993) conclude that *Latimeria* and tetrapods are sister groups. Their study

and the critical reviews which it received (Meyer and Wilson, 1991; Sharpe *et al.*, 1991; Stock and Swofford, 1991; Meyer, 1993; Mili-nkovitch, 1993) encapsulate many of the problems faced by molecular systematists. Gorr *et al.* (1991) started from the evolu-tionary scenario that both α and β haemoglo-bin had changed by gene duplication to adult and larval haemoglobin which are functional at different stages of the life history. Tracing this history is not always easy because it involves inferring history from phylogenetic trees – which is what we are trying to discover – and inferences about homology between different haemoglobins. In other words, gene trees and species trees may or may not be the same.

Gorr *et al.* (1991) and Gorr and Kleinsch-midt (1993) suggest that the fish haemoglo-bin (only one of each of α and β is present) is homologous with the larval tetrapod haemo-globin and therefore these are the sequences that should be compared. This deduction was made using overall similarity measures (the numbers of amino acids in common). Further, the analysis used to build the phylo-genetic tree was a distance method rather than a parsimony method and subsequent workers who have re-analysed the data using parsimony arrive at different phylogenetic conclusions which were inconclusive (Meyer and Wilson, 1991; Stock and Swofford, 1991; Milinkovitch, 1993). Gorr *et al.* (1991) and Gorr and Kleinschmidt (1993) justified the use of a distance method (UPGMA) because the method provides an automatic root for the tree. This was thought desirable because it is known that the choice of different outgroups, as in strict parsimony analyses, can substantially alter the topological rela-tionships of the ingroup (Stock and Whitt, 1992). However, most distance methods root the tree between the most divergent taxa and assume a constant rate of nucleotide substi-tution within all lineages. This has been shown to be an unwarranted assumption (Sharpe *et al.*, 1991; Stock *et al.*, 1991). Within this one study, the problems of paralogy vs homology, the use of distance vs parsimony methods of analysis and the acceptance or rejection of a molecular clock were all exposed as potential influences in inferring phylogeny from molecules.

Summary

This survey of molecular sequence data relevant to the problems of relationships between lungfishes, coelacanths and tetrapods demonstrates the lack of current consensus. Ideally we would like to be able to combine all data together into one analysis. Unfortunately it is not yet possible because different studies have used different taxa and the use of 'hybrid' taxa brings with it another class of problems. Furthermore, we need to recognize that the divergence between lungfishes, coelacanths and tetrapods must have occurred at least 390 million years ago. Such antiquity of lineages poses problems for the analysis of rapidly evolving molecules, or parts of molecules, unequal rates of molecular evolution between different lineages and possible instances of gene duplication. All of these facts will confuse any phylogenetic signal by potentially mistaking homoplasy for homology (long branch attraction). One common solution to circumvent this problem of long branch attraction is to add more taxa of the particular lineages. For coelacanths and lungfishes this is impossible because most are fossil taxa. Molecular data may therefore be inappropriate for this systematic problem. There is, however, much to be done before we reluctantly accept this conclusion. Nearly all studies done so far have used an actinopterygian as an outgroup, and usually the particular species is quite derived within this lineage (e.g. a carp). A wider survey of taxa, including several chondrichthyans (some morphological studies have suggested that coelacanths or lungfishes may be most closely related to cartilaginous fishes), as well as combining information for many genes of the same species, may well reveal stronger phylogenetic signals. At present we can say that parsimony analysis for most of the molecular data at hand suggests that two theories demand more attention: either that lungfishes and tetrapods are sister groups with *Latimeria* more distantly related or that lungfishes and *Latimeria* are sister groups with tetrapods more distantly related.

10.4 STUDIES INCLUDING FOSSIL TAXA

Most investigations of the origin of tetrapods include fossil taxa in the analyses or deal exclusively with fossil taxa. There have been many attempts but it is only within the last 15 years that these have been conducted using cladistic methodology. But even with a common methodology there has been a variety of hypotheses proposed and it cannot be said that an overall consensus has emerged. Figure 10.3 illustrates most of the topological relationships which have been proposed so far. They are not easy to compare one with another, although Schultze (1994) has made some attempt.

The analyses differ in many ways from one another. Some include more taxa than others. For instance, some (Fig. 10.3(D,F–H,M)) include onychodonts, others panderichthyids (Fig. 10.3(F–H,K,L,M)), *Youngolepis* and/or *Powichthys* (Fig. 10.3(C,E,F–H,I,M)). In some cases the taxa used to code for a group were not the same (e.g. Rosen *et al.*, 1981, relied to some extent on Recent amphibians to code for tetrapods, while many of these studies concentrate entirely on fossil lungfishes and tetrapods rather than acknowledging the Recent genera). Many of these analyses relied on accepting a presumed primitive member as representative of the group. Very few of the trees are computer generated (Fig. 10.3(F,K,L,M)) and they are based on very different numbers of characters. Thus, Schultze (1994) considered 216 characters, Cloutier and Ahlberg (1995, 1996) 140 characters. Some classifications are limited in

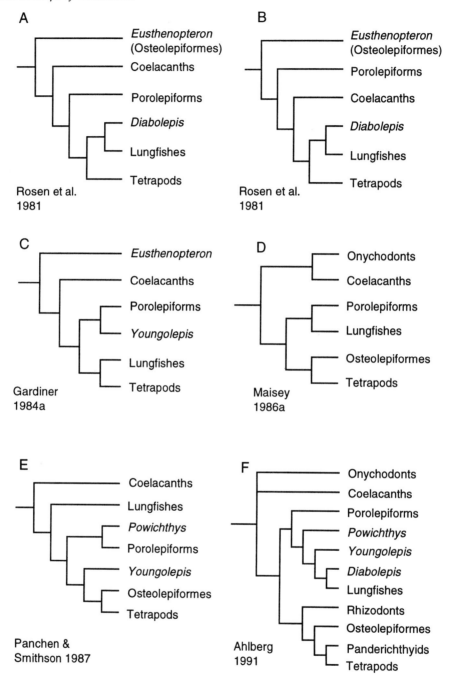

Fig. 10.3 During the last 15 years there have been many attempts to discover the relationships of sarcopterygians using fossils as the principal source of data. This figure illustrates the individual results of those attempts. (A) to (F) conclude that lungfishes are the living sister group of tetrapods, (G) and (H) resolve coelacanths as the living sister group to tetrapods, (I) to (L) suggest that coelacanths are the

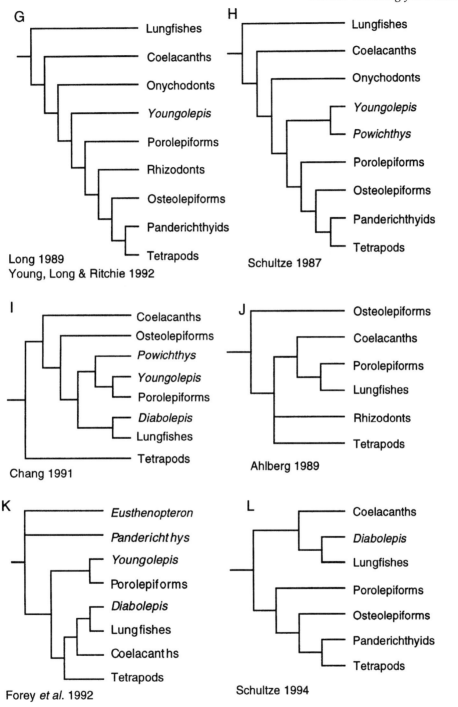

G
Lungfishes
Coelacanths
Onychodonts
Youngolepis
Porolepiforms
Rhizodonts
Osteolepiforms
Panderichthyids
Tetrapods
Long 1989
Young, Long & Ritchie 1992

H
Lungfishes
Coelacanths
Onychodonts
Youngolepis
Powichthys
Porolepiforms
Osteolepiforms
Panderichthyids
Tetrapods
Schultze 1987

I
Coelacanths
Osteolepiforms
Powichthys
Youngolepis
Porolepiforms
Diabolepis
Lungfishes
Tetrapods
Chang 1991

J
Osteolepiforms
Coelacanths
Porolepiforms
Lungfishes
Rhizodonts
Tetrapods
Ahlberg 1989

K
Eusthenopteron
Panderichthys
Youngolepis
Porolepiforms
Diabolepis
Lungfishes
Coelacanths
Tetrapods
Forey *et al.* 1992

L
Coelacanths
Diabolepis
Lungfishes
Porolepiforms
Osteolepiforms
Panderichthyids
Tetrapods
Schultze 1994

sistergroup of lungfishes with or without intervening fossil groups. (I) is interesting to the extent that it is the only one to suggest that there is a monophyletic group which includes sarcopterygian fishes. (M) is the most comprehensive analysis undertaken in terms of numbers of characters and numbers of taxa.

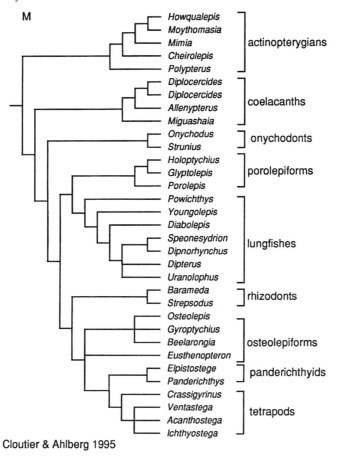

Cloutier & Ahlberg 1995

Figure 10.3 continued

either the numbers of characters (e.g. Chang, 1991a–17 characters) or the scope of the anatomy considered (Ahlberg, 1989a,b). More significantly there have been considerable differences in coding characters for the same organisms. In particular, certain structures in lungfishes such as premaxillae, maxillae, dermal bones of the lower jaw and skull roof, and the choana were regarded as homologous throughout Osteichthyes by some authors (e.g. Rosen *et al.*, 1981) and hence any characters relating to these structures were coded for all taxa. However, some analyses (e.g. Schultze, 1994) denied the initial theory of homology between anatomi-

cal structures and therefore lungfishes were left 'uncoded'. Classifications that accept these assumptions result in lungfishes as the most plesiomorphic of sarcopterygians (Fig. 10.3(G,H)).

Nearly all of these attempts have considered only osteological characters in order to incorporate as many fossil taxa as possible. The most comprehensive analysis, in terms of density of taxon sampling, is that by Cloutier and Ahlberg (1995, 1996).

Throughout the many analyses there is some consensus that a few recognizable monophyletic sarcopterygian groups exist which may be used as terminal taxa.

Onychodontiformes (onychodonts)

This is a small group of sarcopterygians which are very poorly known. *Strunius* and *Onychodus* are the best-known genera but many details of their anatomy remain unpublished. Of all sarcopterygians they are, superficially, most like actinopterygians with large eyes and fusiform body. The most distinctive feature is the development of a large paired symphysial whorl of teeth in the lower jaw which is accommodated in a large cavity behind the premaxilla of the upper jaw. Unfortunately, so little is known of these fishes that these are omitted from this analysis.

Rhizodontiformes (rhizodonts)

This is a group of Devonian and Carboniferous, often large sarcopterygians including such 'better-known' genera as *Barameda, Rhizodus, Screbinodus, Strepsodus, Notorhizodon*; although it is doubtful if any of these can be described as well known. These fishes are usually cosmine covered. They have an unusual posterior skull roof pattern (Andrews, 1973) in which the ossification centres are placed well forward and distant from the pit lines. The dermal shoulder girdle is unique in showing a reverse overlap relationship between the cleithrum and clavicle (Andrews and Westoll, 1970b). The paired fins show long unjointed lepidotrichia which are mostly covered by scales and this gives the overall aspect of the fin a distinctive appearance of a narrow fringe of fin rays. The group has most recently been discussed by Young *et al.* (1992).

Porolepiformes

These are exclusively Devonian sarcopterygians of which the best-known genera are *Porolepis, Glyptolepis, Holoptychius* and *Laccognathus*. These are plump-bodied fishes with small eyes and a large cheek containing more

bones than all other sarcopterygians (except some early lungfishes). They have folded teeth with a unique pattern or infolding (dendrodont) and long biserial fins somewhat similar to those of lungfishes. The group has recently been revised by Ahlberg (1989a,b, 1991).

Dipnoi (lungfishes)

This very large group of sarcopterygians is represented today by six species of exclusively freshwater fishes belonging to the genera *Neoceratodus* (Australian lungfish), *Protopterus* (African lungfish – four species) and *Lepidosiren* (South American lungfish) which have large durophagus tooth plates. Lungfishes are represented in the Devonian by a diverse assemblage of species, some of which show dentitions very different from the modern species.

The Devonian representatives (most completely known are *Griphognathus, Chirodipterus, Scaumenacia* and *Uranolophus*) show a much greater degree of ossification than Carboniferous or later species. Lungfishes are recognized by a characteristic pattern of skull roof bones, a palate which is fused with the braincase, reduced hyoid and gill arches, specialized extensions of the inner ear to form large cavities in the skull above the brain, a highly characteristic lower jaw which has large labial pits near the symphysis and a snout which is penetrated by many small tubules (rostral tubuli) associated with the laterosensory system (some porolepiforms may also have had such a system).

Panderichthyidae

This is a group of distinctive Upper Devonian fishes sometimes known as Elpistostegidae. The included genera are *Panderichthys* and *Elpistostege*, the former is much better known than the latter. *Panderichthys* is unusual amongst bony fishes in that there are no dorsal or anal fins. Both have a characteristic head shape which is decidedly tetrapod-like. There are also low brow-crests above the eyes. The latest summary of the structure of panderichthyids is that of Vorobjeva and Schultze (1991).

'Osteolepiformes'

This is the one of largest and the most problematic of all the sarcopterygian fish 'groups'. *Eusthenopteron* is the best-known member and from a skeletal point of view is probably better-known than most Recent fishes (Jarvik, 1980). Alongside *Eusthenopteron* there are many other genera ranging in time from the Devonian to Lower Permian. They are very diverse and it is unlikely that they form a monophyletic group. However, some authors (Long, 1985) have suggested that they all share a characteristic cheek bone pattern as well as a large basal scute at the base of the paired fins. Unfortunately both of these features are known to occur outside the group and not all osteolepiforms show basal scutes. Many forms such as *Osteolepis* and *Thursius* are relatively poorly known. However, during the last few years there have been some spectacular finds of 'osteolepiform' fishes such as *Gogonasus* (Long, 1985) and *Medoevia* (Lebedev, 1995) and it is expected that we will learn much more about the diversity of these fishes. In most analyses mentioned above, *Eusthenopteron* is synonymous with 'osteolepiforms' but it is generally recognized that this form is assumed to be specialized in having 'lost' the cosmine cover seen in most members. For any systematic analysis it is advisable to use several different genera (Cloutier and Ahlberg, 1995) as representative of 'osteolepiforms'.

Actinistia (coelacanths)

See p. 299 for diagnosis of this monophyletic group.

Other sarcopterygians

There are, in addition, several genera of sarcopterygians which have been discovered within the last 20 years and which show some but not all of the characters of the above groups. In many ways these taxa are interesting as potentially providing an insight into the early evolution of sarcopterygian groups but, at the same time, they are rather incompletely known and difficult to place easily. Genera such as *Powichthys*, *Youngolepis* and *Diabolepis* are thought to be important because of the mixture of plesiomorphic and apomorphic characters. This is particularly true of *Diabolepis*, which shows many lungfish characters in the skull roof and lower jaw, yet suggests that primitive lungfishes had a palate free from the braincase.

10.5 DISCUSSION OF MORPHOLOGICAL CHARACTERS

In the following section an attempt is made to construct a data matrix from knowledge of skeletal anatomy of actinoperygians and sarcopterygians.

Skull roof and extrascapular series

Skull roof

The general shape of most fish skulls shows a smoothly convex transverse profile in which the eyes are placed laterally on the head. This contrasts with the primitive tetrapod head which is usually flattened, particularly above the brain, and in which the eyes are positioned close together and face upwards rather than laterally. In *Panderichthys* and *Elpistostege* the eyes are set in tetrapod-like fashion and there is even the development of 'eyebrow' ridges (Vorobjeva and Schultze, 1991). The snout of tetrapods and some sarcopterygian fishes such as *Panderichthys* and *Elpistostege* is acute rather than rounded.

Patterns of bones within the skull roof of sarcopterygian fishes have been used on many occasions to suggest relationships between one or other group and tetrapods. Discussions have been lengthy, detailed and confused. A summary of the arguments is provided by Rosen *et al.* (1981). The discussions concern the homologies between the bones in the tetrapod skull roof, which tend to show a relatively stable pattern of few bones, with those in sarcopterygians which display a variety in both the number and positional relationships to underlying cranial structures. A particular point of contention is the homology of the median series of large paired bones in tetrapods, the frontals, parietals and nasals, and the consequent names which we apply to these bones in fishes. These names are applied to tetrapod bones with little difficulty, all workers agreeing on their identity between different tetrapods. But which of these bones is represented in fishes is contentious. Schultze (1994) suggests that only tetrapods have nasals and only panderichthyid fishes and tetrapods have frontals (Schultze and Arsenault, 1987). This stands in contrast to Jarvik (1972) and most other fish workers who accept that fishes have nasals but there is dispute over whether they have frontals. For instance, Cloutier and Ahlberg (1996) accept, like Schultze, that only panderichthyids have frontals, but they also recognize frontals in the actinopterygian *Polypterus*.

The pattern of these median bones in most but not all sarcopterygian fishes (e.g. lungfishes) includes two large pairs of bones covering the posterior part of the orbito-temporal region and the otic region of the underlying neurocranium. There is usually no difficulty in recognizing the positional homologies of these two pairs and in most the anterior pair carries the anterior pit line and is often associated with the pineal opening, if present, while the posterior pair carries the middle and posterior pit lines. The dermal part of the intracranial joint, where present, separates the anterior from the posterior pair. One style of skull roof nomenclature holds that the anterior pair are the frontals and posterior pair are parietals (Watson and Day, 1916) and this is the usual terminology applied to the skull roofing bones of actinopterygians (Gardiner, 1984a). The other, adopted here, calls the anterior pair the parietals (the parietals of tetrapods surround the pineal eye) and the posterior pair the postparietals (Westoll, 1938). The arguments that have been used for and against one nomenclature or the other depend on different observations and different assumptions about which parts of the skull are stable and which are labile. These arguments have been adequately reviewed by Rosen *et al.* (1981), Borgen (1983), Panchen and Smithson (1987) and Ahlberg (1991).

The parietonasal series is the dominant series in the skull roof above the eye and nasal capsules of osteichthyans. In forming a continuous series there is no clear distinction between parietals and nasals (see also Westoll, 1936: 164), but parietals are accepted here as those bones positioned above the eye and nasals above the nasal capsules. This is in agreement with many workers except Westoll (1938, 1943), who would refer to the bones above the eye as the parietals; the frontals would therefore lie above the nasal capsules. In the majority of coelacanths there are two pairs of bones (parietals) above the eye and three or four pairs of bones (nasals) above the nasal capsules. As a deduction from the coelacanth cladogram (Fig. 9.7) this would appear to be the primitive condition for the group. Thus *Laugia*, *Wimania* and *Axelia* are derived within coelacanths in showing only a single pair of frontals above the eye. It may be possible to meaningfully distinguish between parietals and nasals in coelacanths because in a few species the suture between the posterior two pairs of bones (parietals) is far more complicated, often irregular and shows a greater degree of overlap than those separating the more anterior pairs. Moy-Thomas (1935b: 217) noted this in *Whiteia* and it is also well developed in *Latimeria*, *Spermatodus* and *Macropoma*.

This means that the skull roof of a primitive coelacanth contains at least three pairs of large bones: two pairs lying above the eye (parietals) and one pair above the otic region (postparietals). The separation between the parietals and postparietals is here taken as the position of the dermal intracranial joint in coelacanths, most 'osteolepiforms' and porolepiforms or a position above the exit of the trigeminal nerve. This pattern of three large paired bones is similar to the pattern seen in panderichthyids and primitive tetrapods. The pattern in porolepiforms, *Powichthys*, *Youngolepis* and lungfishes is difficult to compare with that in other bony fishes. In these there are at least one pair of enlarged bones (C bones in lungfishes) above the eye but anteriorly the roofing bones generally pass into a mosaic where any attempt at one-to-one matching becomes very difficult. In *Eusthenopteron*, *Eusthenodon* and primitive actinopterygians there is only one pair of large bones above the eye which, in area, cover the two pairs of panderichthyids, coelacanths and tetrapods.

From this discussion I suggest that three patterns might be distinguished: the first involving two large paired bones in the skull roof – one above the eye and one above the otic region (exemplified by *Eusthenopteron*); the second consisting of three large paired bones (coelacanths, panderichthyids and tetrapods); the third exemplified by porolepiforms and lungfishes in which there is at least one pair of large bones above the eye and an anterior mosaic. The dermal bone pattern of the otic region of porolepiforms and lungfishes differs from one another. In lungfishes the pattern of bones covering the head has always been difficult to reconcile with the pattern in other fishes and tetrapods, and the bones are usually given a number notation (Forster-Cooper, 1937; Westoll, 1949; White, 1965). This numbering

system is adequate for comparing patterns within lungfishes but fails to accommodate the question of relationships of lungfishes to other fish groups. One apomorphic and constant feature of lungfishes is the presence of a median 'B' bone above the occiput which, as Rosen *et al.* (1981) also point out, occurs, exceptionally, in *Ichthyostega* among tetrapods. Anterior to this 'B' bone there are, in lungfishes, large paired bones (C bones) while lateral to the 'B' bone there is a series of paired bones (Y_1, Y_2, X) carrying the otic sensory canal and which might be compared with supratemporal and intertemporal.

The nasal capsule is covered by at least three nasals in coelacanths. This is a feature also seen in osteolepiforms (including, as far as we know, those with cosmine-covered snouts), panderichthyids, porolepiforms, primitive lungfishes with *Polypterus* among actinopterygians. In most actinopterygians there is a single nasal generally taken to correspond to the three nasals in *Polypterus* (Jollie 1985). Because of its limited distribution, the condition of a single nasal is probably a derived feature of most actinopterygians (Gardiner, 1984a: 270; Jollie, 1985: 368) and, by inference, independently derived in tetrapods.

The position of the nasal bone and the contained supraorbital sensory canal varies with respect to the nostrils amongst bony fishes and is of potential use for character analysis. In more primitive actinopterygians the supraorbital sensory canal, enclosed by the nasal(s), passes between the incurrent and excurrent nostrils (for *Mimia* and *Moythomasia* see Gardiner, 1984a; for *Acipenser* see Pehrson, 1944; for *Polypterus* and *Lepisosteus* see Jollie, 1984c; for *Acentrophorus*, *Lepidotes*, *Dapedium* see Patterson, 1975). In many it is common for the nostrils to notch the nasal, or even pierce the bone (e.g. *Pholidophorus macrocephalus*, Patterson, 1975: Fig. 145). In *Amia* the supraorbital sensory canal ends in a fork above the incurrent (anterior) nostril (Allis, 1889) and cannot therefore be said to pass between the nostrils. The branch of the sensory canal that clearly passes between the nostrils of *Amia* is developed as an outgrowth of the infraorbital canal as judged by innervation (Allis, 1889). In none of these instances is there a connection between the supraorbital and infraorbital canal (or between the supraorbital canal and the ethmoid commissure), although in *Polypterus* the anterior angle formed by the highly recurved canal lies very close to the ethmoid commissure (Allis 1900). In Recent teleosts the nasal, with enclosed sensory canal runs anterior to (above) both nostrils but the supraorbital canal remains separate from the ethmoid commissure, which is usually only represented as a pit line. Even in those teleosts with a well-developed and bone-enclosed ethmoid commissure ('pholidophorids', 'leptolepids', Elopiformes and Albuliformes), there is no anterior connection with the supraorbital canal.

It therefore appears that the primitive condition of the nostrils and nasals in actinopterygians involves a nasal bone that separates the two nostrils and carries the supraorbital sensory canal. The canal terminates within the nasal and fails to connect with either the infraorbital canal or the ethmoid commissure. Any connection between the supraorbital and infraorbital canals in the snout region of actinopterygians is therefore considered a derived condition within this group. Such connections do exist in *Amia*, where the antorbital branch of the infraorbital canal is hypertrophied to join the supraorbital canal (Allis, 1889), and in *Lepisosteus*, where the supraorbital and infraorbital canals join between the nostrils and where there is an additional connection between these canals behind both nostrils (Hammerberg, 1937). Because of the difference in details of these connections they are here regarded as independently derived in *Amia* and *Lepisosteus*.

In sarcopterygians most of the ethmosphenoid is roofed by two longitudinally paired series, the parietonasal series medially and the tectal–supraorbital series laterally (Jarvik, 1942). This double series of bones is also present in *Eusthenopteron*, *Eusthenodon* (osteolepiforms), panderichthyids, *Holoptychius* (porolepiforms), *Powichthys* and lungfishes. It is very probable that this double series also occurs as a primitive condition in other sarcopterygians which usually show cosmine-covered snouts and which otherwise obscures the underlying bone pattern. For instance, a double series has been reported in specimens of *Osteolepis* (Westoll, 1936), *Porolepis* and *Glyptolepis* (Jarvik, 1972).

By contrast, most primitive actinopterygians do not show a double series. In *Mimia* and *Moythomasia* (Gardiner, 1984a), and *Polypterus* (Allis, 1922) there is only a parietonasal series; although the fossil forms mentioned do show an enlarged dermosphenotic which has grown forward lateral to the frontal and sometimes occupying the territory of the supraorbital(s) in sarcopterygians. A distinct supraorbital is developed within actinopterygians at the hierarchical level of Chondrostei in the phylogeny put forward by Gardiner (1984a: fig. 146).

It is not readily apparent whether the presence (sarcopterygians) or absence (primitive actinopterygians) of a supraorbito-tectal series should be regarded as a derived condition. It is of some significance that *Chierolepis*, judged to be the most primitive actinopterygian (Gardiner, 1984a), has been restored with two anamestic bones lying adjacent to the nasal (Pearson and Westoll, 1979: fig. 20). These bones, termed the supraorbital and preorbital, can only be questionably identified in available material, which always shows the delicate snout bones to be crushed (a preorbital has not been seen in *C. canadensis* Whiteaves). Because of this I prefer to regard the supraorbito-tectal series as primitively absent in actinopterygians. The skull roof pattern of placoderms is admittedly difficult to compare with osteichthyans. However, it is to be noted that there is no anamestic series lateral to the main canal-bearing bones (preobitals and postnasals) and in front of the eye. From this it follows that there is some reason to believe that a supraorbito-tectal series is a derived character.

The sarcopterygian supraorbito-tectal series is primitively anamestic (actinopterygians, *Eusthenopteron*, *Osteolepis*, *Holoptychius*, panderichthyids and *Ichthyostega*) and separation of tectals from supraorbitals is somewhat arbitrary. When Jarvik (1942) introduced the term tectals he was referring to those elements of the series in *Latimeria* which lie above the nasal capsule and this remains a reasonable criterion of identification. Tubules may penetrate into the supraorbitals in some (Säve-Söderbergh, 1933; Jarvik, 1944, 1972). This means that the main supraorbital canal is confined to the frontal–nasal series. In coelacanths the supraorbital sensory canal runs in the sutures separating the supraorbito-tectal series from the parieto-nasal series. In *Powichthys* the canal runs from the nasals into the supraorbitals while in lungfishes it runs through parts of the supraorbito-tectal series (L, M, N, P bones).

Rosen *et al.* (1981: 257) suggested that association of the sensory canal with the supraorbitals is a synapomorphy of coelacanths, lungfishes and tetrapods. This suggested grouping might be modified in two ways. First, *Powichthys* must be included. Second, it is to be admitted that the tetrapod condition is not clear cut. In *Ichthyostega* the canal is restored as running through the parietonasal series close to the sutural contact with the supraorbitals (Säve-Söderberg, 1933: fig. 15). However, in several other 'labyrinthodonts' the canal seems to wander from one series to the other (Säve-Söderberg, 1935). It is therefore possible that the association of the canal and the bone in tetrapods is weak and that this is a possible phenomenon to be taken into consideration when trying to follow bone–sensory canal patterns in tetrapods. This idea of a weak association might be supported by the facts that the canals are usually found in shallow open grooves and that they do not show any constant relationship with centres of radiation. Thus, an attempt to try and identify one or another pattern in primitive tetrapods with respect to the path of the supraorbital sensory canal above the eye might not be appropriate.

The infraorbital sensory canal of primitive actinopterygians passes anteriorly from the lachrymal (infraorbital) into the premaxilla and traverses the rostral as the ethmoid commissure. It takes a course through the centre of ossification of the premaxilla and from here it may pass through the ossification centre of the median rostral. The rostral is a characteristically large and broad shield-like bone, always forming a significant part of the snout (Gardiner, 1984a: 270) and may in rare forms separate the premaxilla and bear teeth (Gardiner, 1984a: fig. 48).

In more derived actinopterygians, the rostral is accompanied by a phylogenetic fragmentation of the premaxilla to form separate elements – premaxilla, antorbital, lateral dermethmoids ('pholidophorids'). The fragmentation patterns have been documented by Gardiner (1963) and Patterson (1975) and because they concern more derived actinopterygians their importance to understanding the relationships between the osteichthyan groups is limited to an example of polarity of character change (viz. fragmentation).

The condition of the sensory canals of the snout and nostrils in sarcopterygian fishes is rather different from that in primitive actinopterygians. In 'osteolepiforms', panderichthyids, porolepiforms and coelacanths the supraorbital canal, contained within the nasals, runs forward and joins the ethmoid commissure anterior (dorsal) to the external narial opening. The nasal bone has no association with the nostril(s). Instead the nostrils are located between the tectal and premaxilla and/or a lateral rostral. The infraorbital canal runs forward ventral to the external nostril(s) into the premaxilla and crosses the midline of the snout as the ethmoid commissure within a median rostral. There is little doubt that the sarcopterygian median rostral corresponds to the single rostral of primitive actinopterygians but in the former it is always very small and the sensory canal may or may not be closely associated with the centre of ossification.

In primitive coelacanths, *Powichthys* (Jessen, 1980), *Youngolepis* (Chang, 1982) and *Diabolepis* (Chang and Yu, 1984), the infraorbital canal/ethmoid commissure follows a sutural course between the premaxilla and tectals (coelacanths) or between the premaxilla and an otherwise consolidated shield. And in more derived coelacanths, where the premaxilla has fragmented to form several rostral ossicles, the ethmoid commissure runs between ossifications. The sutural course is regarded as a derived condition within bony fishes because in placoderms, actinopterygians, 'osteolepiforms', porolepiforms and over most of the head of rhizodontiforms the sensory canals are related to centres of ossifications.

The differing relationships between the nostrils and the sensory canals of actinopterygians and sarcopterygians must be judged against conditions in other fish groups and tetrapods. In chondrichthyans both the supraorbital and infraorbital canals are elaborately developed (Garman, 1888; Allis, 1923) but they usually join behind the two nostrils. This is a pattern fundamentally different from that in bony fishes. The infraorbital canal of elasmobranchs runs forward ventral to the nostrils and curves medially to run close to its antimere but does not apparently join to form an ethmoid commissure (Allis, 1923). In batoids such as *Raja* an ethmoid commissure is developed (Allis, 1934). The pattern in placoderms is poorly known because canals are rarely preserved on the snout. In *Coccosteus* (Miles and Westoll, 1968: fig. 24) the supraorbital sensory canal is restored as running forward through the

postnasal, lateral (posterior) to the nasal opening. The infraorbital canal runs through the suborbital to converge upon the supraorbital canal.

Both canals run off the bones and clearly extended within the soft tissue of the snout. Whether they joined and whether there was an ethmoid commissure is not certainly known, but both conditions are restored by Stensiö (1963: fig. 9) as lying within restored labial folds. Antiarchs have a similar pattern, although the supraorbital canal is very reduced and often absent, and sometimes the ethmoid commissure is incomplete within the premedian plate.

In *Wuttagoonaspis*, a placoderm with very clearly marked sensory line grooves, the supraorbital and infraorbital lines meet with the infraorbital canal curving around the ventral surface of the snout. But it is uncertain whether there is a complete ethmoid commissure.

Irrespective of which phylogeny of placoderms we might accept (Denison, 1978; Goujet, 1984), the patchy information suggests that the primitive condition for this group involves a union between supraorbital and infraorbital canals with the supraorbital canal passing behind the nostrils, and the development of an ethmoid commissure. The relationship between nostrils and canal is therefore similar to that in elasmobranchs.

The pattern of the sensory canals of the snout in lungfishes has been the subject of some discussion (Jarvik, 1968; Rosen *et al.*, 1981), particularly with respect to the relationship with the nostrils. In lungfishes the supraorbital and infraorbital canals lie dorsal to both nostrils. This is undoubtedly a derived condition within adult verebrates, and similar to the pattern seen in chimaeriforms (Allis, 1934). Jarvik (1968: 232) ranks this as an homology. The pattern in Recent lungfishes is different to the extent that an ethmoid commissure is absent and the infraorbital canal ends just above the anterior nostril. And this is also true of fossil lungfishes (*Chirodipterus* and *Dipnorhynchus*) where conditions are known. In chimaeriforms the infraorbital canal continues anteriorly to form an ethmoid commissure. In comparison with other fish groups, therefore, lungfishes are derived in lacking an ethmoid commissure with the incurrent nostril marking the termination of the infraorbital canals. This is a pattern similar to that seen in Recent amphibians (Platt, 1896; Escher, 1925) in which sensory lines are developed (urodeles and anuran tadpoles). Coelacanths are similar to lungfishes to the extent that the infraorbital sensory canal is interrrupted at the level of the anterior nostril.

Extrascapular series

The extrascapulars form a transverse series of bones carrying the supratemporal commissure and lying behind the skull roof. As separate elements they are restricted to placoderms and bony fishes. Jarvik's theory of the homology of skull roofing bones includes the idea that the extrascapulars, with the contained supratemporal commissure, have fused with the posterior skull bones of tetrapods.

Primitive actinopterygians, porolepiforms, onychodonts, 'osteolepiforms' and coelacanths show clearly that the extrascapulars are quite distinct from skull roofing bones, an idea emphasized by Allis (1899: 64). The adjacent skull roofing bones – parietals and supratemporals – are firmly attached to the underlying neurocranium by variously developed descending processes or flanges which may also support the extrascapulars. The extrascapulars themselves, however, never bear such processes – notwithstanding Jollie's (1975) idea that the supraoccipital is phylogenetically derived from such a process.

In placoderms extrascapulars are known only in a few arthrodires (*Dickosteus*, *Coccosteus*, Miles and Westoll, 1963: fig. 1 and *Sigaspis* Goujet 1973). Where they do occur they lie behind and free from the nuchal and

paranuchal which bear descending flanges. Furthermore, the extrascapulars rest upon markedly bevelled edges on the nuchal and paranuchal, and these edges have led to the prediction of the presence of extrascapulars in many other arthrodires (Stensiö, 1963), an association which has, however, been questioned by Goujet (1984: 98).

The extrascapular series in 'osteolepiforms', porolepiforms, rhizodontiforms, onychodonts, coelacanths and lungfishes consists of a median extrascapular flanked by one or more laterals. In contrast, the series in actinopterygians is usually paired around the midline (Allis, 1922; Nielsen, 1942; Gardiner, 1984a). *Acipenser* is exceptional among actinopterygians in showing a large median extrascapular, but because this arises from an ontogenetically paired ossification (Sewertzoff 1926: 479), and because the Recent sister group – *Polyodon* (Gardiner, 1984b) – shows only paired extrascapulars, then I consider this to be a derived condition within actinopterygians. Two other instances of median extrascapulars have been reported in teleosts: Patterson (1970: 177) suggested that the supraoccipital of clupeomorphs is the result of phylogenetic fusion with a median extrascapular; and Taverne (1973: fig. 1) records a separate median extrascapular in *Mastacembelus congicus* Boulenger. I regard both of these teleostean examples as independently derived because of the sporadic distribution of this character in any of the many proposed classifications of teleostean fishes.

Which of the two conditions – the possession of a median extrascapular or an exclusively paired series – is to be regarded as the derived condition for bony fishes is a moot point. Goodrich (1925: fig. 1) considered the possession of a median extrascapular to be primitive. As mentioned above, extrascapulars are known in a few placoderms, but they are restricted to a few arthrodires and it is by no means certain that they are primitive for the group. Where they are known some are clearly paired (in *Coccosteus* and *Dickosteus*), whereas *Sigaspis* shows a median element which, however, is markedly bilobed (Goujet, 1973: fig. 3), suggesting a paired origin. On this, admittedly weak, evidence I would propose that the median extrascapular is a derived character of sarcopterygians with similar conditions in a few actinopterygians being homoplasious conditions.

One final aspect of the extrascapular pattern of coelacanths needs to be mentioned. In some coelacanths such as *Laugia*, *Coelacanthus granulatus*, *Rhabdoderma* and possibly *Diplocercides*, the extrascapulars overlap one another. Although the overlap is never great it is always laterad. In other coelacanths extrascapulars do not overlap each other. The laterad overlap, far better developed, is also seen in porolepiforms but Jarvik (1944: 36) noted that in *Eusthenopteron* and other 'osteolepiforms' (Jarvik, 1972: 144) the median extrascapular is overlapped by the lateral extrascapulars. Such is probably also the case in rhizodontiforms. In actinopterygians, lungfishes, and onychodonts (Dr S. M. Andrews, pers. comm.), there is no overlap between extrascapulars. At present, therefore, we can only say that coelacanths and porolepiforms show laterad overlap relationships whereas 'osteolepiforms', *Panderichthys* (Vorobjeva, 1977: fig. 2) and probably rhizodontiforms show mediad overlap relationships.

The main points of the discussion of extrascapular variation may be summarized in the following: extrascapulars are a synapomorphy of osteichthyans (and possibly placoderms) – primitively they are not part of the skull roof; incorporation of extrascapulars into the skull roof is a derived feature which may have arisen by a forward shift (*Acipenser*, *Mawsonia*) or by fusion with a parietal or supraoccipital (some teleosts); a pattern of a median plus one lateral extrascapular is a synapomorphy of sarcopterygians, although it may occur secondarily in actinopterygians (e.g. *Acipenser*); porolepiforms and coela-

canths show laterad overlap relationships between extrascapulars while 'osteolepiforms' and *Panderichthys* show mediad overlap.

As a postscript to this discussion of extrascapulars, and for the sake of completeness, conditions in lungfishes and a few problematic Lower Devonian sarcopterygians may be noted. Extrascapulars have not been described for *Powichthys*, *Youngolepis* or *Diabolepis*, yet in all these forms such a series was presumably present because the otic canal runs off the back of the skull roof within the supratemporal without giving rise to the supratemporal commissure (Jessen, 1980; Chang, 1982; Chang and Yu, 1984).

Although Jessen did not find articulated extrascapulars with *Powichthys*, he did describe an isolated lateral extrascapular (1980: 183, pl. 1, fig. 4) suspected to belong to *Powichthys*. This bone appears to show an area overlapped by a median extrascapular as in porolepiforms and coelacanths.

In primitive lungfishes such as *Uranolophus*, *Dipnorhynchus* and *Speonesydrion*, no extrascapulars have been found (Campbell and Barwick, 1982). The only trace of the supratemporal commissure is a short path running near the posterior edge of the right 'I' bone of *Uranolophus* (Denison, 1968: fig. 6) and the edge of the left 'I' bone in *Speonesydrion* (Campbell and Barwick, 1984a: fig. 5, who incidentally deny the previous commissure reconstruction of Lehmann and Westoll, 1952, for *Dipnorhynchus lehmanni*). In these forms the 'I' bones meet in the midline behind the median 'B' bone. Thus, if a normal canal arrangement existed in these fishes, the triple junction of canals and most of the commissure must have been located behind the skull roof in unknown extrascapulars. In the majority of Devonian forms the path of the commissure is more clearly marked. In *Dipterus* (White, 1965), *Chirodipterus australis* (Miles, 1977), *Scaumenacia* and *Fleurantia* (Graham-Smith and Westoll, 1937; Westoll, 1949) a 'Z' bone contains the triple junction and the supratemporal commissure passes through 'I' and through a median extrascapular (bone 'A'). And like most lungfishes the 'I' bones are separated by a median 'B' bone. This pattern has also been restored for the Carboniferous *Conchopoma* (Schultze, 1975). In some specimens of both *Chirodipterus australis* and *Dipterus* the supratemporal commissure passes directly from 'Z' to 'A' without penetrating bone 'I' (White, 1965: fig. 24; Miles, 1977: fig. 186). These variants therefore display the presumed primitive sarcopterygian condition, where 'Z' and 'A' are the lateral and median extrascapulars. However, the extrascapulars barely touch one another and do not show any mutual overlap relationship.

The association of the supratemporal commissure with bone 'I' is considered a derived condition which may be synapomorphy of most dipnoans, excepting *Diabolepis*, notwithstanding the exceptional examples mentioned above. It may also be an instance of a phylogenetic 'movement' of a canal, because there is no indication of the former existence of an extra canal-bearing extrascapular which may have fused with 'I'. The canal passes through the 'I' bone, which bears the usual middle and posterior pit lines associated with a parietal and often bears posterior flanges. Furthermore, the canal does not run through the centre of ossification, even though it is deeply embedded within the bone. These are indications that the association between the commissure and the 'I' bone (parietal) is weak and probably secondary. (A similar condition may be seen in some arthrodires where the occipital commissure crosses the nuchal with a groove running well away from the centre of ossification: Goujet, 1975, fig. 1c; Goujet, 1984: fig. 93.)

It is generally believed that the commissure also became associated with the median bone 'B' in *Sagenodus*, *Uronemus* and *Ctenodus* (Westoll, 1949). The evidence for this is not convincing but is perhaps most clear cut in *Tellerodus* (Teller 1891: pl. 1) and *Ptychoceratodus* (Schultze, 1981), where the canal appears to run quite separately from the centre of ossification. In any event, the association of the commissure with bone 'B' is a derived condition within lungfishes; and this may be associated with the loss of the median 'A' bone.

Another derived condition within lungfishes appears to be the incorporation of bone '2' (equated here with a lateral extrascapular) into the skull roof (Miles 1977: 305, point 10). This is analogous to the incorporation of extrascapulars into the skull roof of *Mawsonia* and *Acipenser*.

Cheek and operculo-gular bones

The osteichthyan cheek is primitively completely covered by at least five bones. Canal-bearing bones are the lachrymal, jugal, postorbital/dermosphenotic which carry the infraorbital canal beneath the eye, and the preoperculum which carries the preopercular canal from the cheek to the mandible. Additionally, there is an anamestic quadratojugal which carries a small division of the cheek pit line. Sarcopterygians are distinct in that the cheek has an additional bone, the squamosal, which carries a horizontal portion of the preopercular canal (usually called the jugal canal) forward from the preoperculum to join the infraorbital canal within the jugal bone (or between the jugal and postorbital in coelacanths). This pattern of sensory canals is found in chondrichthyans and acanthodians and probably represents the general gnathostome pattern. In actinopterygians there is no squamosal but the preoperculum is much larger and reaches far dorsally to contact the skull roof and carries the preopercular canal to contact the otic canal directly. Actinopterygians also show a dermal bone, the dermohyal, attached to the underlying head of the hyomandibular. This appears to have no homologue in sarcopterygians.

Some sarcopterygians may show additional cheek bones. For instance, porolepiforms show small anamestic bones beneath the squamosal. These subsquamosals often carry pit lines. Porolepiforms also show an extra bone which carries the preopercular sensory canal between the cheek and the lower jaw. This bone lies wholly behind the palate and is the preoperculosubmandibular.

In the dorsal part of the sarcopterygian cheek there may be a small extra bone wedged between the postorbital and the skull roof. This is the pre- or postspiracular. No sarcopterygian has both and therefore it is possible that they are the same bone with different assumed positions to the spiracular opening, which can rarely be seen in sarcoptergians. The only fish with more than one spiracular bone is *Polypterus*, which has a short series of five or six spiraculars lying between the skull roof and the cheek. This is an autapomorphic condition.

Although there are generally more bones in the sarcopterygian cheek than the actinopterygian cheek, coelacanths have fewer bones than most. Here the lachrymal and jugal are replaced by a single bone (lachrymojugal, p. 95) and there is no quadratojugal (this may also be absent from onychodonts). A special note must be made of lungfishes. Primitive lungfishes have a cheek made up by many bones. Usually, these bones are described by numbers but there are clearly bones which may correspond to all those of other sarcopterygians. In addition to the usual lachrymal (bone 7 or 6), jugal (5), postorbital (4), squamosal (8), preoperculum (9), quadratojugal (10) and subsquamosal (11), there may also be a preoperculosubmandibular (9b–f in *Griphognathus*, or the un-numbered bones in *Chirodipterus*: Miles, 1977). A homologue of the spiracular may be absent but bone '13' of *Dipnorhynchus* occupies the same relative position.

There is, in addition to the constituent bones, great variation in the relative sizes of the individual bones and the positional relationships of one bone to another, and these

have been used as indicators of phylogenetic relationships. Particular points which have been used are the facts that in some sarcopterygians the jugal and quadratojugal bones contact or do not contact one another (Panchen and Smithson, 1987) or in others the preoperculum contacts the postorbital (Schultze, 1994). The relative size of the jugal and its positional relationship with the eye (Schultze, 1994), or the shape of the preoperculum and whether it is orientated horizontally or vertically (Long, 1985), have also been regarded as phylogenetically significant characters. There is, however, great variation in all of these attributes and it is difficult to be precise about which particular pattern some taxa show.

The opercular gular series is usually formed by a large operculum and smaller suboperculum with the former overlapping the latter. Together they cover the lateral face of the gill chamber. The operculum is usually a rounded bone while the suboperculum is usually crescent-shaped. Exceptionally, in porolepiforms the operculum and suboperculum are proportionately much smaller and long and narrow (Jarvik, 1972), imparting a rather characteristic shape to the entire opercular series. In tetrapods the operculum is lost with the suboperculum and more ventral bones of the operculo-gular series.

The branchiostegal rays, which are attached to the ceratohyal, are very numerous in actinopterygians and run as a series between the mandibles. In sarcopterygians there are primitively fewer branchiostegals. Porolepiforms show about six and they are confined to a level behind the jaw. In most other sarcopterygians there is a single branchiostegal ray which appears to be in series with the submandibulars. Coelacanths, *Polypterus* among actinopterygians, and most tetrapods show no branchiostegals.

The submandibular series lies ventral to the infradentaries and it is usually developed as a series of small equidimensional bones, some of which may bear pit lines. In lungfishes they are very broad bones; this may be a lungfish specialization. Gulars are primitively present in osteichthyans. Many show a median gular in addition to the two lateral gulars. The median gular appears to have a very homoplasious systematic distribution: it is present in most non-teleost actinopterygians (except *Polypterus*), some lungfishes, some porolepiforms as well as most 'osteolepiforms'. Lateral gulars are usually seen in fish-like osteichthyans. They are usually large and may overlap one another, although the degree of overlap appears to vary considerably between and amongst commonly recognized groups (e.g. porolepiforms).

Neurocranium

The neurocranium of sarcopterygian fishes is relatively conservative and different from that of actinopterygians (Patterson, 1975; Gardiner, 1984a). The dominant feature of the sarcopterygian fish neurocranium is the presence of the intracranial joint which is present in most representatives (see below). The ventral fissure, which is the ventral end of the intracranial joint, represents the adult retention of the embryonic division between the trabecular and parachordal regions of the skull. This is also present in primitive actinopterygians (Gardiner, 1984a) but in sarcopterygian fishes the fissure is usually very large. The fissure is spanned by the unconstricted notochord which reaches forward to end anteriorly in a notochordal pit (Fig. 6.2, n.p) and the level of the intracranial joint. Additionally, the ventral part of the oticooccipital part of the braincase remains largely unossified as the basicranial fenestra (Fig. 6.2, b.fen). This fenestra is occupied by ventral arcual plates (catazygals: Fig. 6.1(A), a.Cat, p.Cat) which, in *Latimeria*, are embedded within the connective tissue surrounding the notochord. Jarvik (1972) considers these plates to be evidence of cranial vertebrae. In the evolution of actinopterygians, the ventral fissure closed up by the backward growth of

the parasphenoid. The ventral fissure also became occluded in early tetrapods, but the process is not clear because so few early tetrapod braincases are known in sufficient detail. Among sarcopterygians, *Youngolepis* is unusual in that the basicranial fenestra and catazygals are absent (in this it resembles actinopterygians) but the notochord remains unconstricted and passes through the posterior half of the skull.

The lungfish neurocranium is highly apomorphic and the ventral fissure and basicranial fenestra is obliterated, catazygals are absent and the palate is fused with the neurocranium. This has resulted in a drastic restructuring of the neurocranium, making it difficult to compare point-for-point with the neurocranium of other sarcopterygians.

At the anterior end of the neurocranium there are differences between sarcopterygian fishes in the proportions of the snout and the manner in which the palate articulates with the neurocranium. These differences, which have been documented in detail by Jarvik (1942, 1972, 1980), together with his study of the tongue (Jarvik, 1963), led him to formulate his theory of the diphyletic origin of tetrapods. Jarvik (1972: 160) lists some 15 differences between the snout anatomy of *Porolepis* and *Eusthenopteron*. Many of these differences concern the reconstructed shapes of the nasal sacs, positions of the nasal soft-tissue tubes and the nerve supply. Unfortunately, very few other sarcopterygian fishes are known in such detail and it is therefore difficult to evaluate the significance of all of these differences. However, some of the more obvious and superficial skeletal differences can be seen and checked in many taxa.

In porolepiforms the snout is broad and the nasal cavities are widely spaced and separated by paired internasal pits which lie medial to the separated vomers. In osteolepiforms the nasal cavities lie close together (or if far apart they are separated by a solid internasal wall), there are no internasal pits and the vomers lie close together and come into contact with one another.

The neurocranium of sarcopterygian fishes is usually completely ossified in one or two units, depending on whether there is an endoskeletal intracranial joint present or not. Both coelacanths and lungfishes show reduction of this complete ossification during their separate histories (for coelacanths see p. 237). In most sarcopterygian fishes there is a pronounced suprapterygoid process at the posterodorsal corner of the orbit which receives the ascending process of the metapterygoid. This lies above the basal articulation commonly found in teleostomes. Another distinctive feature of the sarcopterygian neurocranium is the development of the otic shelf which reaches forward from the lateral commissure. It is particularly well developed in coelacanths (Chapter 6).

The braincase of most plesiomorphic sarcopterygians is divided by the intracranial joint into ethmosphenoid and otico-occipital portions each covered by dermal bones. The function of the joint is presumably to increase the gape of the mouth rapidly during opening phases (Thomson, 1967). There has been considerable debate about whether the joint is homoguous throughout all groups (Bjerring, 1973, 1978; Miles, 1977; Wiley, 1979; Jarvik, 1980; Chang, 1982; Gardiner, 1984a; Ahlberg. 1991). The joint is composed of a dermal joint separating parietals from postparietals. The endoskeletal part of the joint is more complicated, with an articulation between front and rear halves above and below the cranial cavity. These were called supracerebral and infracerebral divisions by Bjerring (1973). In some sarcopterygians such as *Powichthys*, *Glyptolepis* and *Holoptychius*, as well as in all coelacanths, there is a complex joint developed just below the cranial cavity whereby the otic shelf articulates with the processus connectens of the basisphenoid, and the basisphenoid articulates with one of the arcual plates which separate the cranial cavity dorsally

from the notochordal canal beneath. The entire joint is best developed in coelacanths (Chapter 6). In other sarcopterygians (e.g. *Eusthenopteron*) the endoskeletal part of the joint remains, but the articulations between the anterior and posterior portions are weakly developed. In yet others (e.g. *Panderichthys*) the endoskeletal part remains, but the dermal roof is fused up (Vorobjeva and Schultze, 1991) such that any functional advantage of the joint must have been lost.

The ventral part of the joint is present also in actinopterygians as the ventral fissure and this portion is a very ancient feature of the vertebrate skull. The more dorsal division – the supracerebral division – which divides the neurocranium at or near the levels of the exit of profundus, trigeminal and facial nerves (see below), would therefore be regarded as derived within osteichthyans and a primitive feature of sarcopterygians. This interpretation is complicated by the facts that in some sarcopterygian fishes and tetrapods the two halves of the neurocranium are fused at this level (as in actinopterygians), leading to the possibility that a fully developed intracranial joint may have developed on more than one occasion. Bjerring (1973) justified this view by suggesting that because of the different positions of the nerve exits between coelacanths (p. 168) and other sarcopterygians with intracranial joints (i.e. porolepiforms and many osteolepiforms), then the joint was developed independently in coelacanths. However, Ahlberg (1991) showed that there are at least four variations in the topographic relationships between the joint and nerve exits, and this implies that little significance can be attached to these subtle variations and that the dorsal part of the intracranial joint is homologous throughout sarcopterygians. It is probable, therefore, that the entire intracranial joint of sarcopterygians is homologous and that it has been closed on more than one occasion (Ahlberg *et al.*, 1996).

One of the main reasons why 19th century biologists considered lungfishes to be close relatives of tetrapods was the fact that lung-fishes have internal nostrils (choanae). Rosen *et al.* (1981) surveyed the history of studies concerning the presence of a choana in lung-fishes, as well as the restoration of the choana in extinct rhipidistian fishes. Their conclusion was that the opening into the roof of the mouth in a primitive lungfish such as *Griphognathus* occupies the same topological relationship to surrounding bones (premaxilla, vomers, dermopalatine and maxilla) as does the opening in a primitive tetrapod such as *Ichthyostega*, an animal agreed by all to have a true choana. Therefore, Rosen *et al.* argued that the choana in a lungfish and a tetrapod must be regarded as a putative synapomorphy between lungfishes and tetrapods. Campbell and Barwick (1984b) attempted to show that the bones surrounding the internal nostril of *Griphognathus* are neomorphic and therefore the enclosed opening was not a choana. Their view depends on the assertion that *Griphognathus* (and presumably *Holodipterus* – another lungfish with 'supernumerary' palatal bones) is not a primitive lungfish. However, ideas on the basal phylogeny of lungfishes (Miles, 1977; Schultze and Marshall, 1993) do not clarify this point. Furthermore, *Diabolepis* also shows an opening into the roof of the mouth partly surrounded by vomer and premaxilla (dermopalatine and maxilla remain unknown). In my view there is no reason to judge these bones in primitive lungfishes as non-homologous on grounds of topological relationship.

Rosen *et al.* (1981) went further to suggest that while there was no doubt about the presence of a choana in lungfishes, there was doubt about the restoration of a choana in *Eusthenopteron* ('osteolepiforms') and porolepiforms. For *Eusthenopteron* they argued that (1) the continuity of the infraorbital sensory canal between the presumed choana and the external nostril does not accord with our understanding of the ontogeny of the nostrils/choana in tetrapods, and (2) the

presumed choana is the site of a moveable joint between the braincase and the palate + maxilla and is therefore unlikely to have been a passageway for water or air. For porolepiforms they argued (1) that in these fishes the restored choana is a tiny triangular opening, totally unlike the large oval or keyhole opening of 'osteolepiforms' and tetrapods and (2) that there are two external nostrils providing necessary explanation for incurrent and excurrent openings into and out of the nasal cavity. If the choana is a modification of the fish posterior nostril (reviews: Rosen *et al.*, 1981; Panchen, 1967) then porolepiforms cannot have a choana. However, others (Jarvik, 1942) have suggested that the tetrapod choana (and the opening in 'osteolepiforms' and porolepiforms) is a new development and that the original fish posterior nostril has been modified to form the nasolachrymal duct. For either scenario the porolepiform fishes, with two external nostrils and a presumed choana, are problematical. Because studies of the comparative anatomy and development of the choana of modern animals have not revealed the history of the choana, it is doubtful if fossils will fare any better. Clearly, our interpretations of fossil forms are laden with assumptions about development.

In recent years more articulated 'osteolepiforms' have been found (especially *Gogonasus* and *Medoevia*) and there is a large oval palatal opening associated with a single external nostril. I would accept the fact that this could be a choana which has precisely the same relationships to surrounding bones as in primitive lungfishes and tetrapods. Accepting the additional assumption that the choana is a modification of the fish posterior nostril, then porolepiforms do not have choanae.

Palatoquadrate and hyoid arch

The palatoquadrate of sarcopterygians is primitively long, reaching back to flank the side wall of the otico-occipital part of the braincase and extending posteroventrally to form a very oblique jaw suspension. The palatoquadrate is formed by three endoskeletal ossifications: (a) the autopalatine, which articulates with the ethmoid region of the neurocranium; (b) the metapterygoid, which articulates with the neurocranium at the basal articulation and with the suprapterygoid process; and (c) the quadrate, which articulates with the articular (and a separate retroarticular in derived coelacanths). Lungfishes and Recent amphibians are apomorphic in lacking the autopalatine, probably as a result of the palate having become fused to the neurocranium. The dermal bones associated with the palatoquadrate consist of a large pterygoid (distinctively shaped in coelacanths), a dermopalatine and one or more ectopterygoids lying near the outer edge of the pterygoid. Additionally, actinopterygians show dermal metapterygoids which lie medial to the pterygoid. All of these bones primitively bear tiny villiform teeth. In some actinopterygians (e.g. *Pteronisculus* Nielsen, 1942) and at least one 'osteolepiform' (Ahlberg and Johanson, in press), there may be additional tooth plates lying between the pterygoid and the parasphenoid. The pterygoid bones usually lay separate from one another in both actinopterygians and primitive sarcopterygians. A clearly derived condition is for the pterygoids of either side to meet one another in the midline and completely or partially exclude the parasphenoid from the roof of the mouth.

Special mention must be made about the palate of lungfishes, which is usually thought to be quite different from that of other fishes (Schultze and Cloutier, 1991). In most lungfishes the pterygoids are the only bones remaining. These meet in the midline and carry the specialized grinding tooth plates so characteristic of most lungfishes. However, some supposedly primitive lungfishes such as *Griphognathus* and *Holodipterus* show tooth-bearing bones anterior and ante-rolateral to the pterygoids (Miles 1977). Rosen *et al.* (1981) homologized these bones with the 'normal' osteichthyan vomers, dermopalatines and ectopterygoids. However, Campbell and Barwick (1984b) consider that these extra bones in *Griphognathus* are neomorphic (see also arguments about the premaxillae and maxillae, p. 274). Identification of these bones has relevance to the identification of a choana because part of the topological criteria of the internal nostril is that it is surrounded by a particular suite of bones including the dermopalatine, vomer, premaxilla and maxilla. The idea that the bones in *Griphognathus* and *Holodipterus* are neomorphic for these lungfishes is not based on phylogenetic reasoning but rather that these fishes have a specialized method of feeding. If it should be shown that these taxa, carrying the possible homologues of the vomers, dermopalatines and ectopterygoids, are very primitive lungfishes, then the theory of homology with the generalized osteichthyan palate would be strengthened. I see no reason why these bones in lungfishes and other osteichthyans should not be regarded as homologous.

The palate is supported against the braincase by the hyomandibular, which is primitively long and lies along and tightly applied to the full length of the rear edge of the palate. In coelacanths, lungfishes and tetrapods this relationship is modified because the hyomandibular (stapes of tetrapods) is reduced in size and lies at an angle to the palate. The hyomandibular may also become decoupled from its primitive association with the ceratohyal.

Upper jaw

The upper jaw consists of premaxilla and maxilla which are sutured to snout and infraorbital bones in most primitive osteichthyans and most workers accept this as the plesiomorphic condition. Von Wahlert (1968) is one of the few workers to suggest that the absence of a maxilla in coelacanths and modern lungfishes is primary and this is his chief reason for suggesting that *Latimeria* is the most plesiomorphic osteichthyan. Others (Jarvik, 1968; Campbell and Barwick, 1984b; Schultze, 1987, 1991) have suggested that the bones surrounding the upper jaw of lungfishes are not comparable with any in other osteichthyans and are instead autapomorphies for lungfishes. The arguments have relevance for the identification of a choana because the tetrapod choana is recognized, in part, on the relationships of the internal opening into the mouth to the surrounding dermal bones. Rosen *et al.* (1981: 181, figs 7, 8) argued that both a premaxilla and a maxilla were present in early lungfishes and cited evidence from the Devonian lungfishes *Griphognathus* and *Ganorhynchus*, in which there are marginal teeth lying anteriorly and posteriorly to the anterior (external) nostril which bit outside the opposing lower jaw. These are the positional relationships expected of maxillae and premaxillae. The 'maxilla' of *Griphognathus* is called the subnasal ridge and the lateral nasal tooth plate by Miles (1977). Identification of both a premaxilla and maxilla were questioned by Campbell and Barwick (1984b), who suggested that these bones in *Griphognathus* did not carry true teeth and are variable in their development from specimen to specimen. Their argument was complicated and relied on 'knowing' the primitive lungfish condition, which they took to be something like *Uranolophus*, which has a bone mosaic in this region of the skull. To some extent the arguments have now been superseded by the discovery of *Diabolepis*, which most workers (not Campbell and Barwick) accept as a plesiomorphic lungfish. *Diabolepis* shows well-developed premaxillae (Chang and Yu, 1984) which lie separate from the otherwise consolidated snout and bear well-developed teeth. There is therefore no need to deny the existence of premaxillae in lungfishes although, as Schultze and Cloutier (1991)

remarked they may have 'moved' from marginal position to a palatal position. Maxillae are unknown in *Diabolepis* because all specimens are broken at this point. This might suggest that if they were present then they would articulate with the snout, as in most fishes, rather than be integrated into a solid snout as are the bones identified as maxillae in *Griphognathus*. I conclude that it is reasonable to accept that lungfishes have premaxillae. Identification of maxillae in *Griphognathus* amongst lungfishes is not so easy although, in my opinion, there is no convincing reason to deny homology. One of the features of the premaxilla of lungfishes is that it is strongly inturned so that it is only obvious in a ventral view of the snout.

Lower jaw

Structure of the lower jaw has been used on many occasions to suggest relationships among sarcopterygians (Westoll, 1943; Jarvik, 1954, 1963; Holmes, 1985; Schultze, 1987, 1994). The most frequent comparisons made concern the similarity, or near identity (Westoll, 1943; Schultze, 1987) between the lower jaws of osteolepiforms, porolepiforms and primitive tetrapods. Particular similarities that are often cited concern the shared possession of a long dentary, four infradentaries, three coronoids and a prearticular. During the last few years, the lower jaws of many more sarcopterygians and primitive actinopterygians have been described in detail (Miles, 1977; Jessen, 1980; Gardiner, 1984a; Campbell and Barwick, 1988; Clack, 1988; Ahlberg, 1991, 1995; Chang 1991b, 1995; Young *et al.*, 1992; Lebedev and Clack, 1993; Ahlberg *et al.*, 1994; Fox *et al.,* 1995; Lebedev, 1995; Ahlberg and Clack, unpublished data). This increased knowledge has identified many more potential characters as well as questioning the 'near identity' of osteolepiform jaws with those of tetrapods. A re-evaluation of lower jaw structure is therefore desirable.

The primary lower jaw is formed by Meckel's cartilage which usually ossifies perichondrally and endochondrally. The primitive gnathostome pattern consists of anterior (mentomeckelian) and posterior (articular) ossifications (Nelson, 1973; Gardiner, 1984a) and these sometimes join to form a complete meckelian bone. In most tetrapods the anterior end of Meckel's cartilage/bone is very reduced (Clack, 1988) and this is probably correlated with a particular pattern of dermal bones at the anterior end of the jaw of tetrapods. In fishes, there is usually a vacuity – the precoronoid fossa (anterior mandibular fossa of Fox *et al.*, 1995) – lined with the mentomeckelian or Meckel's cartilage (see below). In most tetrapods the dermal bones cover the mentomeckelian entirely, leaving no vacuity. In lungfishes the meckelian bone is exposed laterally as well due to development of a labial pit (see below). In most fishes the symphysis is formed, in part, by the median contact, or fusion (lungfishes) between the mentomeckelian of either side. Further posteriorly, Meckel's cartilage/bone is primitively exposed on the medial face of the jaw between the prearticular and the infradentaries. The exposed portion is the site of origin of the intermandibularis in *Polypterus*, *Latimeria*, lungfishes and larval salamanders (Edgeworth, 1935) and this is presumably the primitive condition. In some sarcopterygians (e.g. *Eusthenopteron*, Jarvik, 1980: fig. 132) the prearticular sutures with the infradentaries and the meckelian bone is not exposed, meaning that the intermandibular muscles insert onto dermal bone.

The posterior end of Meckel's cartilage usually ossifies as a single articular ossification. In actinopterygians, other than the most primitive, and in derived coelacanths (p. 238), there are separate articular and retroarticular ossifications, each contributing to the formation of the glenoid. The subdivision of the single articular is here regarded as a derived condition (Nelson, 1973). Within the taxon sampling used for this systematic analysis separate articular and retroarticular ossifications are seen only in *Latimeria*. A retroarticular process (whether or not this is formed by a separate ossification) extending posteriorly to the glenoid is also a derived feature within gnathostomes, because it is not found in chondrichthyans, placoderms or acanthodians. The glenoid cavity is primitively double, with concavities either lying side by side or slightly offset relative to one another. In some lungfishes (e.g. *Rhynchodipterus* and *Griphognathus*) the glenoid is narrow and single, and this has been thought to have some functional significance if correlated with the denticulated dentition (Campbell and Barwick, 1983, 1987, 1990). Lungfishes also show a well-developed preglenoid process developed from the articular and this is flanked laterally and medially by dorsal extensions of the neighbouring dermal bones, surangular and prearticular. In *Neoceratodus* the preglenoid process receives a posterior division of the adductor mandibulae.

The dermal bones surrounding the lateral face of the meckelian cartilage/bones consist of the dentary and a variable number of infradentaries (splenial, postsplenial, angular and surangular). On the medial surface the coronoid series (including symphysial plates) and the prearticular are exposed. The dermal bones covering the lateral face of the mandible of lungfishes are sometimes regarded as non-homologous with those of other osteichthyans (Jarvik, 1967; Schultze and Campbell, 1987; Schultze, 1994) and they are either given an osteichthyan terminology in quotes ('angular', 'splenial' etc.) or a neutral terminology (Jarvik, 1967). Schultze and Campbell (1987) agree with Jarvik that the canal system within the lower jaw bones of a lungfish is so different from that of other osteichthyans that the canal-bearing bones cannot be homologous. While it is true that the canal system in lungfishes is different (see below), the dermal bones of a primitive lungfish are similar in number and mutual relationships to those of primitive sarcopterygians, such that topographic homologues can be recognized easily. It is accepted here that the outer dermal bones are homologous with those of other osteichthyans (Watson and Gill, 1923; Rosen *et al.*, 1981; Gardiner, 1984a) and that the same names can be applied. Theories suggesting that they are non-homologous lead to no alternative phylogenetic hypothesis (Miles, 1977). With respect to the inner dermal bones then all workers, except Jarvik (1967), accept that the principal tooth-bearing bone is homologous with the osteichthyan prearticular.

The dentary is the tooth-bearing element on the lateral face of the jaw. It usually forms the majority (actinopterygians) or the entire oral margin, reaching back as far as the glenoid. Exceptions are coelacanths and some lungfishes where one or both of these taxa are often described as having a short dentary (Forey *et al.*, 1992; Schultze, 1994; Cloutier and Ahlberg, 1996). Here the teeth are restricted to the anterior third of the jaw. There is usually little difficulty in recognizing the different states. However, it may be of more importance to recognize the fact that in most osteichthyans the dentary reaches back to reach the level of the adductor fossa, whereas in coelacanths and some lungfishes it fails to do so. Described in this way, there is at least one lungfish (*Diabolepis* – Chang, 1995) which has a long dentary. The distribution of a long tooth-bearing dentary suggests that this is the primitive osteichthyan condition, although teeth may be absent in several taxa (some coelacanths, some lungfishes, many teleosts). A restricted condition is seen in some sarcopterygians where the extreme anterior end of the dentary lacks teeth which are otherwise well developed more posteriorly (Jarvik, 1972; Ahlberg, 1991). A further modification of the dentary is seen in many primitive sarcopterygians where the anterior end of the inner margin of the dentary is developed as a medial lamina which

supports an adsymphysial plate (anterior-most coronoid).

There has been some debate as to whether a dentary is present in lungfishes. In Devonian lungfishes there is a large, usually median bone covering the oral border of the symphysis. It is often cosmine covered and thus clearly of dermal origin. Those who deny homology between the dermal bones of lungfishes and other osteichthyans accept this as a neomorph peculiar to lungfishes. However, this bone also carries true teeth (*Diabolepis*, Smith and Chang, 1990) or, as is sometimes described, denticles (*Uranolophus, Griphognathus*) and bites inside similar toothed/denticulated areas on the upper jaw. It therefore has all the topographic relationships of a dentary. Marginal dentitions in the lower jaws are quite common in Devonian lungfishes such as *Dipnorhynchus kurikae* (Campbell and Barwick, 1985), *Andreyevichthys epitomus* (Krupina 1992) and *Ichnomylax halli* (Long *et al.*, 1994), but unfortunately the occlusal relationships with upper jaw elements are not known in these forms. The dentary may also be present as a transitory structure in the ontogeny of *Neoceratodus* where a patch of teeth lie anterior to the prearticular tooth plate of either side (Kemp, 1995: fig. 1). However, this tooth plate may also be a coronoid (Gardiner, 1984a). In this work it is accepted that a dentary is present in lungfishes and can be compared with that of other osteichthyans.

Beneath the dentary there is a maximum of four infradentaries, which from front to back are called splenial, postspenial, angular and surangular. In sarcopterygians each of the infradentaries carries part of the mandibular sensory canal. Coelacanths and some lungfishes have two infradentaries, usually called splenial and angular (sometimes postsplenial and angular or compound names – Stensiö, 1947; Jarvik, 1967). In both of these groups the posterior element forms most of the outer surface of the jaw. However, it is unlikely that this pattern is homologous because primitive lungfishes have the usual complement of at least four infradentaries. The terminology of the coelacanth infradentaries is given on p. 129. Some primitive lungfishes such as *Uranolophus* show a mosaic of small bones wedged between the anterior infradentary and the dentary. This is probably a derived condition.

A very different pattern of bones was described for the lungfish *Melanognathus* by Jarvik (1967), who considered that the primitive lungfish jaw was covered laterally by two parallel series of infradentaries, a lateral series carrying a long oral canal (a lungfish synapomorphy) and a medial series carrying the usual mandibular canal. However, the reconstruction provided by Jarvik was challenged by Thomson and Campbell (1971: fig. 24), who recognized the usual single series pierced by two parallel canals (oral and mandibular). Primitive lungfishes therefore fall into line with most other sarcopterygians in the number of infradentaries.

In actinopterygians the dentary is the dominant dermal bone, forming most of the lateral surface. Posterior to the dentary lies a single infradentary, which is called the angular but is topographically in the same position as the surangular of sarcopterygians and carries the oral pit line. The mandibular sensory canal passes through the ossification centres of both the angular and dentary. Some primitive actinopterygians show an additional bone dorsal to the angular (surangular) and posterior to the dentary and which is called the supra-angular (Gardiner, 1984a). However, this bone has almost certainly been acquired within the actinopterygian lineage (Patterson, 1982) and has no homologue within sarcopterygians. In sum, the lateral face of the lower jaw of actinopterygians is very different from that of sarcopterygians, but it is not clear as to which pattern should be regarded as plesiomorphic because acanthodians and placoderms lack any comparable dermal bones.

The dermal bones on the medial side of Meckel's cartilage are the coronoids and the prearticular. The prearticular is the largest of the bones; it usually forms the medial border of the adductor fossa and extends forward beneath one or more of the coronoids. The prearticular is thought to form the lower tooth plate in lungfishes, although there have been some opinions that coronoids may be fused in as well (Jarvik, 1967; Smith and Chang, 1990). However, embryology of dental development in *Neoceratodus* shows no evidence of fusion of separate elements (Kemp, 1995). In most lungfishes the prearticular tooth plates of either side are sutured in the midline, a presumed synapomorphy. Morphology of the coronoid series has often been regarded as important in theories of sarcopterygian relationships, where it is suggested that the condition of three coronoids is a character of rhipidistians and tetrapods (Ahlberg, 1991; Vorobjeva and Schultze, 1991; Schultze, 1994; Cloutier and Ahlberg, 1996). This stands in contrast to the condition in primitive actinopterygians and in coelacanths where there are four or five (more derived actinopterygians have a variable number, Gardiner, 1984a). However, it needs to be stated that there is often an additional coronoid in 'rhipidistians' which is called the adsymphysial or parasymphysial plate: that this is the serial homologue of more posterior elements is not disputed. Thus, all osteichthyans primitively have four coronoids and the number would seem insignificant for this systematic problem. The putative derived condition within some sarcopterygians is the specialization of the anteriormost coronoid, which is perhaps most extreme in porolepiforms and onychodonts where a lamina on the dentary is also present and there are highly modified symphysial teeth. However, it is not clear that the so-called adsymphysial plate of many osteolepiforms or primitive tetrapods is any more specialized than the anteriormost coronoid of primitive actinopterygians, although it is possible that the morphology and histology of the teeth borne upon this plate may be specialized in some species. Some lungfishes (e.g. *Chirodipterus*) show the presence of a small median symphysial plate lodged between the anterior ends of the prearticular tooth plates. In *Neoceratodus* a single transitory tooth occupies a similar position (Jarvik, 1967; Kemp, 1995). This is a derived condition within lungfishes. The coronoids of many sarcopterygian fishes show intercoronoid fenestrae which are depressions floored by meckelian bone (or cartilage in life). Such fenestrae are absent from the coronoid series of actinopterygians and some primitive tetrapods. The latter also show the presence of one or two small, forwardly directed foramina close to the suture between the anterior two coronoids (Ahlberg *et al.* 1994, mesial and lateral parasymphisial foramina). Both Clack (1988) and Ahlberg *et al.* (1994) note that in some tetrapods, the anterior infradentary (presplenial) wraps round the ventral margin of the lower jaw to suture with the coronoid series and this reflects a retreat of the prearticular.

The dentition within the lower jaw varies considerably amongst primitive osteichthyans. Teeth may be in either single or multiple rows (a shagreen) along the length of the dentary and/or coronoid series. And some of the teeth on the dentary and/or the coronoids may be very much enlarged in comparison with their neighbours. These are often called fangs. These large teeth are nearly always associated with a replacement socket, a condition which is described as a 'fang pair' (Cloutier and Ahlberg, 1996).

The sensory canal within the lower jaw is developed as a mandibular canal which passes through the infradentary series (sarcopterygians) or the angular and dentary (actinopterygians). Lungfishes have an additional canal, the oral canal, which runs above and parallel to the mandibular canal. A similar canal is found in holocephalans and this was regarded as a special similarity of phylogenetic significance by Jarvik (1967). However, the oral canal is seen in a wide variety of lower vertebrates (some Recent urodeles, *Chlamydoselache*, holocephalans – Stensiö, 1947; some acanthodians – Watson, 1937; as well as some temnospondyl amphibians – Nilsson, 1943, where it is represented by grooves). This systematic distribution led Miles (1977) to consider that it is probably a primitive vertebrate character and that the reduction of this canal (to absence or to a pit line) is derived, and this seems a reasonable interpretation. In many other fishes, the oral canal may be represented by a small anterodorsal branch from the mandibular canal which arises at the level of the jaw articulation. This is present in young stages of *Amia* and in *Lepisosteus* (Hammerberg, 1937) where it is replaced by a pit line in adults. Jarvik (1944: fig. 11c; 1972: pl 31.3) describes a similar short canal in *Eusthenopteron* and *Holoptychius*. In restorations of *Eusthenopteron* (Jarvik, 1944: Fig. 11a), there is an associated pit line lying over the canal, implying that the canal and the pit line cannot be homologues. In all other instances there is either a canal or a pit line. And it may be of significance that in at least one specimen of *Thursius pholidotus* Traquair (Jarvik, 1948: fig. 68E) the oral canal leads into a pit line. Clearly, the observations of *Eusthenopteron* need to be checked in more specimens. Here it is accepted that the small pit line at the posterior end of the jaw represents a pattern which may be a remnant of the oral canal. Another pattern is seen in some sarcopterygians (e.g. Jarvik, 1948) where the pit line reaches well forward to at least the second infradentary and often bends ventrally to lie close to the mandibular sensory canal.

One final feature of the sensory system needs to be mentioned. Ahlberg (1991) noted the presence of large infradentary foramina in the porolepiforms *Holoptychius*, *Laccognathus*, *Glyptolepis* (Ahlberg, 1989b – some species) and *Youngolepis*. Because these foramina are absent from *Powichthys* and *Porolepis*, Ahlberg considered this character as slightly problematic, requiring caution if considered as a potential homology between these taxa. This view has been justified because it is now known that *Gogonasus* (a typical osteolepid) shows prominent infradentary foramina (Fox *et al.*, 1995: fig. 52).

Shoulder girdle and pectoral fin/limb

Shoulder girdle

Characters of the shoulder girdle and pectoral fin have been used on several occasions to recognize sarcopterygian groups (Jarvik 1944; Andrews and Westoll, 1970a,b; Rosen *et al.*, 1981; Long, 1985; Panchen and Smithson, 1987; Schultze, 1987; Ahlberg, 1989a,b, 1991). Many authors have emphasized the morphological similarities between the shoulder girdle and limb construction of osteolepiforms (*Eusthenopteron*) and tetrapods (Gregory, 1911; Watson, 1913; Westoll, 1943, 1958; Jarvik, 1965; Thomson, 1968, 1972; Andrews and Westoll, 1970a; Rackoff, 1980; Holmes, 1985; Panchen and Smithson, 1987) with the intention of demonstrating that the prerequisites for walking on land are present in these fishes. However, some earlier authors made close anatomical comparisons between the limbs of Recent lungfishes and urodeles (Braus, 1901; Holmgren, 1933, 1949a,b; Rosen *et al.*, 1981). It is necessary therefore to evaluate this work and to consider characters potentially useful for the classification of sarcopterygians.

The shoulder girdle consists of the dermal (exoskeletal) girdle and the endoskeletal scapulocoracoid which bears the fin articulation. The chondrichthyan girdle is exclusively endoskeletal and is homologous with that in other gnathostomes. In all other fishes, a dermal pectoral girdle is developed to a greater or lesser degree and might be thought of as a potential synapomorphy of placoderms + teleostomes. However, there are considerable differences in basic construction, sufficient in fact to suggest that the dermal girdle may have arisen on three separate occasions.

In acanthodians the dermal bones associated with the scapulocoracoid are restricted to non-overlapping plates which lie beneath the fin insertion. These dermal plates are presumed to have been lost in derived acanthodians. In placoderms there are large overlapping plates which completely encircle the body and superficially may be matched with those in osteichthyans (Stensiö, 1959; Forey, 1980; Jarvik, 1980). Thus, the anterior median ventral, anteroventral and anterior lateral of placoderms lie in the same positions as the interclavicle, clavicle (ventral cleithrum) and cleithrum (dorsal cleithrum) of osteichthyans. The dorsal part of the dermal girdle of placoderms is formed by the anterior dorsolateral, which carries the lateral line and is therefore equivalent to the post-temporal according to Jarvik (1980: fig. 290). Differences from the girdle of osteichthyans include the presence of a median dorsal (a synapomorphy of placoderms); there does not appear to be a separate bone corresponding to the osteichthyan supracleithrum and the scapulocoracoid is inserted to the medial surface of two bones (anterolateral and anterior ventrolateral) rather than exclusively on the cleithrum. For these reasons of topographic difference, it is questionable whether we could be justified in using a common terminology.

The dermal girdle of osteichthyans shows a dorsoventrally overlapping series above the scapulocoracoid (post-temporal, supracleithrum and cleithrum) and a clavicle and median interclavicle ventrally. The clavicle overlaps both the cleithrum and the interclavicle. The post-temporal and supracleithrum carry the lateral line and the former connects the girdle with the skull.

Sarcopterygian fishes show an additional element, the anocleithrum, situated between the supracleithrum and the cleithrum and overlapped by both those bones. Rosen *et al.* (1981) suggested that it was not primitively part of the osteichthyan girdle but that it may correspond to the scale-like dorsalmost postcleithrum of actinopterygians which has become incorporated into the sarcopterygian girdle. Jarvik (1944) took the opposite view and suggested that the anocleithrum is a primitive osteichthyan feature lost or modified in actinopterygians and *Polypterus*. There seems no way to polarize this character (but see arguments in Rosen *et al.*, 1981). Like the post-temporal and supracleithrum, the anocleithrum is absent from the girdle of more derived sarcopterygians and this reflects the decoupling of the pectoral girdle from the skull. However, the size, shape and relationship of the anocleithrum to the surrounding bones and scales vary within sarcopterygians. In some, such as osteolepiforms, the anocleithrum lies superficially and carries ornament and is exposed in lateral view. In others, such as porolepiforms, lungfishes and coelacanths, the anocleithrum is unornamented and lies beneath the surface and is sometimes completely obscured in lateral view by the supracleithrum and body scales (e.g. *Glyptolepis*, Ahlberg, 1989a). If similarity between the anocleithrum and the actinopterygian postcleithrum is accepted, then the superficial ornamented anocleithrum may be judged as a plesiomorphic condition for sarcopterygians – a view accepted here. However, Long (1985) suggests that the ornamented anocleithrum is a synapomorphy of osteolepiforms. His argument is based on functional grounds and assumes that there was point of flexure within the pectoral girdle between the supracleithrum/cleithrum in actinopterygians, or the anocleithrum/ cleithrum in sarcopterygians. In osteolepiforms, he argues, this primitive flexibility has been eliminated, or greatly reduced, by the incorporation of a firmly sutured, ornamented anocleithrum. This idea contrasts with that of Andrews and Westoll (1970a) who argue for flexibility at this point of the girdle in the osteolepiform *Eusthenopteron*. Functional scenarios are difficult to substantiate in extinct fishes and Long's argument is

not accepted here. Instead, it is of interest to note that other elements of the pectoral girdle (interclavicle and clavicle) appear to have lost ornament and to have 'sunk' beneath the surface in cladistically more derived members of actinopterygians on the one hand and tetrapods on the other. A similar fate is assumed here for the anocleithrum.

A superficial and ornamented interclavicle is primitively present throughout osteichthyans, although the subsequent fate may vary. It may be wholly subdermal and covered by the overlying clavicles (e.g. *Acipenser* among actinopterygians, some coelacanths) or it may even become (phylogenetically) 'fused' with the urohyal (pholidophorid actinopterygians, Patterson, 1977b). In most bony fishes the interclavicle is restricted to a position between or anterior to the clavicles, but in *Ichthyostega* and some other early tetrapods it bears a posterior process, sometimes called the parasternal process (Brough and Brough, 1967), which extends backwards between the clavicles and which may reflect a change in the insertion for the pectoralis muscles (Romer, 1924). The only sarcopterygian fish known to show a posteriorly extended interclavicle is *Panderichthys* (Vorobjeva and Schultze, 1991: 81), and it has been scored as such in Table 10.1 (below), but this observation needs to be checked in more material.

In all the taxa under consideration here, the cleithrum and the clavicle are the dominant dermal elements of the girdle. In primitive actinopterygians and sarcopterygians the clavicle bears an ascending process, of varying length, which lies along the anterior edge of the cleithrum and often wraps around onto the inner face of that bone. In the Devonian lungfish *Chirodipterus* and, to a lesser extent in *Griphognathus*, the ascending process is expanded to form an inturned branchial lamina (Campbell and Barwick, 1987). The cleithrum and the clavicle usually lie superficially and are uniformly ornamented, but Andrews and Westoll (1970b) noted that in the rhizodonts *Strepsodus* and

Screbinodus, the posterior edge of the clei-thrum above the fin insertion was inturned and unornamented as a depressed lamina (they suggested this was the phylogenetic homologue of the scapular blade of tetrapods – but see Patterson, 1977b: 103). A similar lamina is present in some Devonian lung-fishes (Campbell and Barwick, 1987) and the rugose surfaces presumably meant that trunk musculature attached here. In most osteichth-yans the clavicle overlaps the cleithrum in a simple suture, but Andrews and Westoll identified a complex reverse overlap relation-ship between these bones in the rhizodonts *Strepsodus*, *Sauripterus* and *Screbinodus*. In the dorsal half of the girdle the clavicle overlies the cleithrum; in the ventral half the clei-thrum overlaps the clavicle.

Rosen *et al.* (1981) and Long (1985) used the size ratio of clavicle to cleithrum and the relative position of the fin insertion as char-acters suggesting that in more derived sarcopteryians, the fin inserted higher on the body than in actinopterygians. Because the scapulocoracoid, bearing the glenoid, is attached to the cleithrum, then the position of the fin insertion is also a function of the relative size of the two bones. Certainly, when comparing a primitive actinopterygian, such as *Mimia*, with a rhizodont such as ?*Strepsodus* (Andrews, 1971), the differences in fin insertion and size ratios of the clei-thrum and clavicle are striking. However, Schultze (1987) criticized the usefulness of these observations for classifying sarcopter-ygians because of the extreme variability of this character. Furthermore, the position of the fin insertion relative to the height of the girdle may also be a function of the curva-ture of the lower half of the body. Given the difficulties of compensating for body curva-ture in reconstructing fossil vertebrates, the position of the fin insertion might be expres-sed as specifying the location of the glenoid as nearer to the dorsal or to the ventral margin of the cleithrum.

The scapulocoracoid is ossified in one piece in all sarcopterygian taxa considered here. Gardiner (1984a) reviewed the ossifica-tion centres in the endoskeletal pectoral girdle in gnathostomes pointing out that in acanthodians, derived actinopterygians, Recent tetrapods and many fossil tetrapods there are two, sometimes three separate bones (names such as scapular, suprascapu-lar, coracoid, procoracoid and mesocoracoid are used). He concluded that separate ossifi-cation centres characterize the early growth stages of members of all vertebrate groups and that retention of these centres as separate bones into the adult stage is derived and attained on several occasions. He also poin-ted out that a tripodal girdle is widespread among osteichthyans: that is, the scapulocor-acoid is attached to the cleithrum by three buttresses which delimit a supraglenoid canal above and a supracoracoid canal below the glenoid. These canals allow nerves and blood vessels to pass out to the dorsal and ventral fin muscles. Because of the widespread distribution of the tripodal girdle, Ahlberg (1989a) regarded the imperforate scapulocor-acoid without separate buttresses as derived and I agree with this assessment. Even in forms with an imperforate scapulocoracoid the single contact surface may retain its basi-cally tripartite shape (*Glyptolepis* – Ahlberg, 1989a,b). Long (1989) added another argu-ment by suggesting that while the tripodal scapulocoracoid is a primitive osteichthyan feature, the development of three *equal-sized* buttresses delimiting large canals is shared only by rhizodonts, osteolepiforms (only *Eusthenopteron* known) and tetrapods. The sizes of the buttresses and the canals vary considerably amongst the few sarcopter-ygians known in this respect and this makes it difficult to characterize taxa as conforming to these descriptors.

The shape of the glenoid fossa, and by implication the shape of the head of the humerus, has been noted on many occasions. Usually, authors have stressed the similarity in shape between the glenoid of *Eusthe-*

nopteron and *Ichthyostega* (Andrews and Westoll, 1970a; Jarvik, 1980; Panchen and Smithson, 1987; Long, 1989) describing each as concave and screw-shaped. The precise shape of the glenoid is very difficult to determine in many fossil sarcopterygians because the end of the glenoid often passes into an unfinished cartilage-covered surface (Rosen *et al.*, 1981: fig. 40; Ahlberg, 1989a). However, it is usually possible to determine if it was convex or concave and this may be of greater functional importance than the precise surface shape. A few sarcopterygians (*Glyptolepis* – Ahlberg, 1989a; *Youngolepis* – Chang, 1991b) have an unusual strap-shaped glenoid which is very different from the rounded glenoid of other sarcopterygians. Ahlberg (1989a) proposed that this type of glenoid, together with a flattened head of the humerus as seen in *Glyptolepis* (not known for *Youngolepis*), constrains the fin to movement largely in a single plane. This is in contrast to the possibility of rotational movements shown by both lungfish and coelacanth pectoral fins and assumed to have been present in most fossil sarcopterygians. Parenthetically, it must be noted here that the glenoid of actinopterygians is strap-shaped but here there are at least four radials attached and consequently this is regarded as a fundamentally different type of glenoid.

Pectoral fin

Anatomy of the pectoral (and pelvic) fin has been one of the most intensely discussed aspects of the fish–amphibian transition. Yet, it has also been one of the most confused studies of sarcopterygian anatomy and relationships (Rosen, *et al.*, 1981; Holmes, 1985; Panchen and Smithson, 1987; Ahlberg, 1989a,b). This has come about for several reasons. There have been difficulties, or disagreements, in identifying topographic homologues of separate bones between fish and tetrapod limbs. Thus, there are disagreements (Watson, 1913; Holmgren, 1933; Jarvik, 1980) about which carpel bones and which digit represent the primary axis of the tetrapod limb, or whether tetrapod digits are neomorphs, or whether they represent mesomeres and pre- and/or postaxial radials inherited from the fish endoskeleton. In addition, there have been discussions about whether there has been phylogenetic rotation of the pectoral fin (Romer and Byrne, 1931; Rosen *et al.*, 1981) to bring the preaxial side of the fish fin into a dorsal position, or whether some sarcopterygian fishes have complex elbow and knee joints, so making them more highly favoured evolutionary precursors of tetrapods than others (Rackoff 1980). Lastly, there has been a tendency to make very limited comparisons (e.g. *Eusthenopteron* with *Ichthyostega* or *Eryops*, *Sauripterus* with *Eryops*) and, although this has depended on availability of fossil material, it has had the effect of channelling evolutionary scenarios into very restricted paths. A good history of the earlier ideas on the fish–tetrapod limb transition is provided by Andrews and Westoll (1970a). Throughout all these discussions, the study of pattern and inferred processes have become closely intertwined and for these reasons Rosen *et al.* (1981) and Ahlberg (1989a) have attempted to confine discussion to pattern and to examine as broad a range of taxa as possible.

The number of fossil sarcopterygian fishes in which even the proximal few segments of the fin endoskeleton are known with any degree of completeness is very limited and includes the following: the osteolepiform *Sterropterygion* (Rackoff, 1980) *Eusthenopteron* (Jarvik, 1965, 1980; Andrews and Westoll, 1970a), the rhizodonts *Sauripterus* (Andrews and Westoll, 1970b), and *Barameda* (Long, 1989), the porolepiform *Glyptolepis* (Ahlberg, 1989a,b), and the coelacanths *Laugia* (Stensiö, 1932) and *Panderichthys* (Worobjeva, 1975). In addition, the basal element and sometimes the second elements of the series are known in the lungfish *Chirodipterus* and the osteolepiform *Beelarongia* (Long, 1987). The tetrapod limbs with which comparisons are usually made are those of *Eryops* and *Proterogyrinus* (the forelimb of *Ichthyostega* is incompletely known) but there is now reasonably complete evidence available for *Acanthostega*, which in some respects is more fish-like than previously known tetrapods (Coates and Clack, 1991).

Rosen *et al.* (1981) review the structure of the paired fins in all jawed fishes and provide an assessment of primitive conditions which help to stabilize polarity decisions of variation within the sarcopterygian fins. They conclude that the endoskeleton of the primitive elasmobranch, placoderm and actinopterygian fin articulates with the scapulocoracoid by means of a segmented metapterygium which supports the trailing edge of the fin and that these are a variable number of more anteriorly placed radials. The metapterygium is the main axis of the fin and is made up by three to five segments, and all or most of the radials are in a preaxial position: that is, positioned towards the leading edge of the fin. The radials, in turn, are segmented along their length. In both elasmobranchs and actinopterygians, there is a reduction in the number of preaxial radials articulating directly with the scapulocoracoid, and in most actinopterygians there is an enlarged and fenestrated radial – the propterygium (Jessen 1972; Rosen *et al.*, 1981; Gardiner, 1984a) – supporting the leading edge of the fin. The reduction in the number of radials attaching directly to the scapulocoracoid appears to have happened independently in elasmobranchs and in actinopterygians. One final feature to be noted is that the metapterygium carries its own radials and that there is, usually, a one-to-one correspondence between metapterygial segments (mesomeres) and supported radials. The sarcopterygian fin differs from this primitive gnathostome fin and is derived in that the metapterygium is the sole element articulating with the shoulder girdle.

Accepting a higher-level phylogeny which groups actinopterygians as the sister group to sarcopterygians, and elasmobranchs as the sister group to this combined group, a number of conclusions follow about the primitive sarcopterygian fin. A single-axis fin consisting of a metapterygium with few mesomeres (3–5), radials supported on the preaxial side, with a one-to-one correspondence between radial and mesomere and, probably, segmented radials, are features plesiomorphic for sarcopterygian fins. This assessment of the primitive sarcopterygian fin agrees with that of Ahlberg (1989a) in most respects and this allows some of the characters listed below to be polarized.

However, this assessment differs radically from that suggested by Schultze (1987), who proposed that a biserial long archipterygium, with equally developed pre- and postaxial radials, such as is developed in pleuracanth sharks, is primitive. The conclusions which flow from this assumption are that the long 'archipterygium' of lungfishes, and porolepiforms which are now known to be of this type (Ahlberg 1989a), are primitive for sarcopterygians. And the short uniserial fin is synapomorphous for osteolepiforms, rhizodonts and tetrapods (Vorobjeva and Schultze, 1991). Schultze's assumption is not accepted here for three reasons: first, it is unlikely that the pleuracanth fin is primitive for elasmobranchs; second, the fin of lungfishes is not symmetrical because, unlike the preaxial radials, the postaxial radials do not show a strict numerical relationship with their mesomere supports (they also develop in a different way: Shubin and Alberch, 1986); third, actinopterygians never show a symmetrical, archipterygial fin (contra Schultze, 1987: 64).

The anatomy of the tetrapod limb with respect to the conditions mentioned above is not entirely clear and the ambiguities have been discussed by Ahlberg (1989a). Estimation of the length and number of mesomeres composing the metapterygial axis of the tetrapod limb is unclear because it is not agreed whether the digits are neomorph structures (Holmgren, 1933; Westoll, 1943; Shubin and Alberch, 1986) or whether they are homologous with fish mesomere segments of a metapterygial axis (irrespective of which digit the axis is drawn through) or segmented pre- and postaxial radials (Jarvik, 1980; Rosen et al., 1981; Shubin and Alberch, 1986). In the absence of condradictory evidence it does not seem unreasonable, and indeed it is more parsimonious, to regard one or other of the carpal and phalange series as representing the original metapterygial axis. The minimum number of segments in any proximal–distal series is six (*Acanthostega* – Coates and Clack, 1990; *Proterogyrinus* – Panchen and Smithson, 1987). I would suggest therefore that the tetrapod limb is derived with respect to the short fin of 3–5 mesomeres.

Whether the tetrapod limb is to be regarded as symmetrical, with equally developed pre- and postaxial radials, or whether it is asymmetrical with only preaxial or only postaxial radials depends on where the axis of the fin is drawn (Westoll 1943; Jarvik 1980; Rosen et al., 1981) as well as on theories of fin rotation (Romer and Byrne, 1931; Rosen et al., 1981). There are difficulties with identifying both the axis and rotation and it seems premature to speculate without considerably more evidence. I agree with Ahlberg (1989a: table 1) that it is difficult to speculate about fin/limb rotation in tetrapods, but I would also question whether the tetrapod limb could be described as biserial (that is, whether there are both pre- and post axial radials). In the lungfish *Neoceratodus* alone, the pectoral fin is known to rotate during development (Semon, 1898), leading to the original preaxial side becoming postaxial. And the adult asymmetrical condition of the pectoral fin of *Latimeria* strongly suggests that this fin is also rotated, but whether this happens in development is unknown.

Most sarcopterygian fins (osteolepiforms,

rhizodonts and tetrapods) agree in having a large basal mesomere (humerus) followed by a paired segment consisting of an axial mesomere lying alongside a preaxial radial element attached to the humerus. Usually, the shorter of these is called the ulna and is thought to represent the axial mesomere (depending on where the axis of the fin is drawn), and the longer is the radius. In osteolepiforms and rhizodonts the radius is very much longer than the ulna, whereas in tetrapods they are of subequal length. And in *Sauripterus* and *Rhizodus* this radial is characteristically blade-like. The ulna, in turn, articulates with the ulnare (the fin axis according to most authors) and a radial (sometimes called the intermedium). The exceptions to this pattern are seen in coelacanths, porolepiforms and lungfishes which, at first sight, appear very different.

In coelacanths the basal mesomere, or humerus, is succeeded by a single large axial mesomere, and the radial associated with the humerus, as with the second and succeeding mesomeres, is very much smaller and reduced to a nubbin of cartilage straddling the joint between the first and second mesomeres. These highly reduced radials, some of which are segmented and others not, may be a synapomorphy of coelacanths. The chief difference between the coelacanth fin and that of osteolepiforms, rhizodonts, and perhaps tetrapods, is that the fin is topographically reversed, leading Rosen et al. (1981) to suggest that the preaxial side is rotated to a postaxial position.

In *Neoceratodus* the fin is also rotated and the humerus of the adult appears to be without an associated preaxial radial (Ahlberg, 1989a). However, Rosen et al. (1981: fig. 31) pointed out that the embryo *Neoceratodus* has a separate preaxial radial articulated with the humerus in the same way as the tetrapod ulna articulates with the humerus. This fuses with the second mesomere in the adult. Thus, I consider that in lungfishes there is a humeral preaxial radial (contra Ahlberg,

1989a) and that fusion of this element with the mesomere is an autapomorphy of *Neoceratodus*. In *Glyptolepis* there are no radials (pre- or postaxials) attached to the first two mesomeres and the fin is not rotated (Ahlberg 1989a).

Postaxial radials are present in coelacanths, lungfishes, *Glyptolepis* and possibly rhizodonts. They are here regarded as being absent from osteolepiforms but comment is necessary because Westoll (1943) and Jarvik (1980) regard the well-developed processes on the more distal mesomeres to be postaxial radials. I agree with Ahlberg (1989a), who points out that these processes are always fused with the mesomeres and never articulated as radials. Although there is some variation in the processes on the mesomeres in *Eusthenopteron*, the third mesomere (ulnare) of osteolepiforms usually bears a very large process (pisiforme, Jarvik 1980). This may be a synapomorphy of osteolepiforms. The problem of tetrapod postaxial radials has been alluded to above; Ahlberg (1989a) considers they are probably present.

A great deal of attention has been paid to the presence and nature of the processes present on mesomeres of the metapterygium. In particular, attention is usually focused on the processes associated with the humerus, with very detailed comparisons made between those of osteolepiforms, rhizodonts and tetrapods (Andrews and Westoll, 1970a,b; Jarvik, 1980; Rackoff, 1980; Panchen and Smithson, 1987; Long, 1989). Thus, in addition to the dorsal, or entepicondyle process, separate or conjoined ventral deltoid, supinator and ectepicondyle processes have been identified. Humeral processes are a sarcopterygian character. They are usually confined to the humerus but, as Ahlberg points out, *Glyptolepis* and coelacanths show dorsal and ventral processes on all mesomeres. It is uncertain whether this should be regarded as primitive or derived within sarcopterygians. The entepicondyle process is a constant feature of sarcopter-

ygians but osteolepiforms, rhizodonts and tetrapods show that this process is fenestrated for the ulnar nerve. Beyond this, close comparisons of the details of the ventral processes (deltoid, supinator and ectepicondylar) are very difficult to make considering the great diversity of their expression (e.g. *Neoceratodus*, Rosen *et al.*, 1981; *Barameda*, Long, 1989; *Ichthyostega*, Jarvik, 1980; *Acanthostega*, Coates and Clack, 1990; *Proterogyrinus*, Panchen and Smithson, 1987; *Eusthenopteron*, Jarvik, 1980; *Sterropterygion*, Rackoff, 1980; *Beelarongia*, Long, 1987).

One aspect of the sarcopterygian endoskeletal pectoral girdle that has been considered of significance concerns the nature of the joint surfaces along the metapterygial axis. The head of the humerus of osteolepiforms, rhizodonts and tetrapods is convex, contrasting with the concave surfaces in lungfishes, coelacanths and the flat surfaces in *Glyptolepis*, *Youngolepis* and *Panderichthys* (Vorobjeva and Schultze, 1991). The degree of convexity varies considerably from nearly flat in *Acanthostega* to a prominent ball in rhizodonts and *Eryops*, and no doubt this reflects some unknown functional variation. The joint surfaces between the humerus and the radius and ulna are usually flat but they are essentially ball-and-socket-shaped in *Neoceratodus*. Much has been made of the fact that the facets on the distal end of the humerus receiving the radius and ulna lie in slightly different planes, thus implying that there may have been a complex rotational elbow joint (Andrews and Westoll, 1970a; Rackoff, 1980) in some sarcopterygian fishes. But this may not have been the case in primitive tetrapods such as *Acanthostega*. In any event, if the ulna and radius are compared with the second axial mesomere and the preaxial radial attached to the humerus, such a difference in the plane of attachment would be expected as a primitive feature of the sarcopterygian fin.

The overall shape of the fin varies within sarcopterygians. Many are highly asymmetrical and have long fin rays attached, others such as rhizodont fins have very short fin rays confined as a fringe at the fin margin. A very different kind of overall fin shape is shown by lungfishes and porolepiforms, where the fin is long and leaf-like with a near-equal development of pre- and postaxial margins.

Pelvic girdle and fin

The pelvic girdle and fin is known in very few fossil sarcopterygian fishes and far less understood. As a consequence very few comparative remarks can be made. Rosen *et al.* (1981) outlined a possible history of the girdle and the fin. They pointed out that actinopterygians are unique in that the pelvic girdle is formed by the phylogenetic homologue of metapterygial elements which have replaced or become fused with (Stensiö, 1921, 1925) the original pelvic bone. This means that the actinopterygian pelvic girdle is not homologous with the girdle of chondrichthyans, placoderms or sarcopterygians (acanthodians remain unknown). Panchen and Smithson (1987) suggested that the pelvic girdle of lungfishes, osteolepiforms and tetrapods showed a similar history to that of actinopterygians. This latter idea was discussed and rejected by Ahlberg (1989a) and his arguments are accepted here, leaving the incorporation of the metapterygium to the pelvic girdle as a synapomorphy of actinopterygians (Patterson, 1982). Thus, actinopterygians are unsuitable as an outgroup taxon to polarize characters of the pelvic girdle and aspects of the fin structure (e.g. number of mesomeres).

With the exception of the actinopterygians, the history of the pelvic girdle amongst fishes has been very conservative. The primitive gnathostome condition of the pelvic girdle includes posteriorly diverging pelvic bones/cartilages which abut one another or are connected via connective tissue anteriorly (Ahlberg, 1989a). The posterior end of each pelvic bone/cartilage is expanded as the acetabulum, alongside which there is a lateral expansion or process and, usually, a medial expansion or process. These lateral and medial processes are variously developed in fishes and there appears to be little pattern of development congruent with any recognized pattern of relationships. The fossil lungfishes *Chirodipterus* and *Griphognathus* (Young *et al.*, 1989) as well as the Recent lungfishes have an additional anterolateral process which may be a lungfish synapomorphy (but some urodeles may also show these processes – e.g. *Proteus* Goodrich, 1930: fig. 213). Recent lungfishes may be further derived in showing the pelvic cartilage extended forwards as a long median process (again, it is interesting to note that urodeles and anurans show a median prepubic cartilage).

In tetrapod terminology (usually applied in descriptions of *Eusthenopteron*, Andrews and Westoll, 1970a; Jarvik, 1980) the anterior ramus of the pelvic bone is equated with the tetrapod pubis, the lateral process with the ilium and the medial process with the ischium. Three separate ossifications within the pelvic girdle is a character found only in more derived tetrapods. Unlike the nature of the ossification pattern of the endoskeletal shoulder girdle, there is no living sarcopterygian in which this can be followed, and it is impossible to say whether the pelvic girdle may have ossified from one, two or three centres. Tetrapods are also unique in the hypertrophy of the lateral process to contact the vertebral column through the sacroiliac connection. It remains to be determined at what hierarchical level this took place: it is present in *Ichthyostega* (Jarvik, 1980).

There are two further differences between the pelves of fishes and primitive tetrapods. In tetrapods the two halves of the girdle are suturally united along a broad contact in the ventral midline and the acetabulum is located anteriorly, resulting in a very short pubic region. These features are, no doubt, associated with the demands of weight bearing and force transmission as well as reorientation of the pelvic limb relative to the girdle. The Devonian lungfishes *Chirodipterus* and *Griphognathus* approach the tetrapod condition in that the pelvic bones contact each other in the ventral midline over a long sutural contact. Rosen *et al.* (1981) suggested that lungfishes and tetrapods are unique in showing fusion between left and right pelvic girdles. They made their comparisons between Recent lungfishes and urodeles. The discovery of the girdles of Devonian lungfishes (Young *et al.*, 1989) and the fact that the primitive tetrapod girdle is formed of paired ossifications suggest that the similarity between the Recent taxa is homoplasious.

With respect to the pelvic limb then we meet the same problems as in the pectoral fin of identifying possible rotation, the path of the metapterygial axis and hence the number of elements of which it is composed, whether postaxial radials are present in tetrapods as they are in some fishes. Additionally, while the pelvic limb endoskeleton is known in *Eusthenopteron* (Andrews and Westoll, 1970a; Jarvik, 1980), it is only very incompletely known in the osteolepiform *Sterropterygion*, the lungfish *Chirodipterus* and the porolepiform *Glyptolepis* (Ahlberg 1989a). Thus, the basis for comparison amongst fossil taxa is very limited indeed. Rosen *et al.* (1981) suggested that coelacanths, lungfishes and tetrapods were synapomorphous in showing structural similarity of the endoskeleton between the pectoral and pelvic fins. Those authors were criticized by Panchen and Smithson (1987) and by Ahlberg (1989a) for the ambiguity of this character. Ahlberg went further by suggesting that under three criteria of similarity (number of axial mesomeres, presence of similar processus and arrangement of radials), only the coelacanth complies with the statement of similarity (and even this may not be true with respect to the condition of the preaxial radials). Rosen *et al.*'s intention was to point out that in coelacanths, lungfishes and tetrapods the fore- and hindlimbs are equally developed whereas in osteolepiforms and actinopterygians the hindlimb is considerably shorter than the forelimb. However, it remains true that only *Latimeria* shows exactly the same number of mesomeres in both limbs (Ahlberg, 1989a).

As with the forelimb the primitive sarcopterygian condition is taken to be a uniserial fin in which the metapterygium is formed by few (3–4) mesomeres, each associated with a segmented preaxial radial. This assumption demands that the fins of *Glyptolepis* (porolepiforms) and lungfishes be considered derived. In *Latimeria* and *Neoceratodus* there is no preaxial radial associated with the basal mesomere and the second bears two preaxial radials, suggesting the possibility that the first preaxial radial has shifted to an association with the second mesomere. This may be a potential synapomorphy of coelacanths and lungfishes, but the anatomy of fossil lungfishes and more porolepiforms needs to be described before this can be verified. Only lungfishes appear to have equally developed pre- and postaxial radials. In both coelacanths and porolepiforms, the pelvic fin postaxial radials stop far short of the base of the fin axis.

10.6 CHARACTER LIST AND CHARACTER CODES

The following list of characters has been used in the morphological data matrix (Table 10.1). Assessments of the polarity are discussed in the preceding text. Many of these characters have been recognized previously and used in previous analyses (particularly those of Schultze, 1994, and Cloutier and Ahlberg, 1996), although not all codings will be the same from one analysis to another (e.g. disputes about the presence of a choana, or whether a fin is rotated).

1. Skull arched with lateral orbits (0), flat with eyes close together (1).

2. Tectals absent (0), present (1).
3. Number of supraorbitals, none (0), one or two (1), many (2).
4. Parietal–supraorbital contact, absent (0), present (1).
5. Supraorbital canal not associated with the supraorbital series (0), associated with the supraorbital series (1). Those taxa which do not have supraorbitals will be scored '0' for this character. Also, 'association' with the supraorbital series does not discriminate whether the canal follows a sutural course between the parietals and supraorbitals or whether it runs through the ossification centres of the supraorbitals.
6. Pattern of median pair of skull roof bones: two large pairs of bones (0), three large pairs (1), at least two pairs with anterior mosaic (2).
7. 'B' bone absent from skull roof (0) present (1).
8. Nasal bone associated with nostril (0), not associated (1).
9. Nasals separated in midline (0), meeting in midline (1).
10. Pineal opening absent (0), present (1).
11. Extratemporal absent (0), present (1).
12. Intertemporal present (0), absent (1).
13. Median extrascapular absent (0), present (1).
14. Overlap relations of extrascapulars: mediad (0), laterad (1), abutting (2).
15. Lateral rostral (naroidal) absent (0), present (1).
16. Rostral tubuli absent (0), present (1).
17. Sclerotic ring with five or fewer ossicles (0), many ossicles (1).
18. Premaxilla not inturned (0), inturned (1).
19. Maxilla present (0), absent (1).
20. Dorsal part of endoskeletal intracranial joint absent (0), present (1).
21. Two external nares present (0), one external nares present (1).
22. Anterior (incurrent) naris located away from jaw margin (0), close to or at jaw margin (1).

23. Choana absent (0), present (1).
24. Vomers separated in midline (0), joined in midline (1). Those taxa which have a median vomer are coded '1'. In *Polypterus* there is no vomer in the adult but paired vomers are present in the embryo (Gardiner, 1984a: 295).
25. Posterior process on vomer absent (0), present (1).
26. Pterygoids of either side not meeting in midline (0), meeting in midline (1).
27. Internasal pits absent (0), present (1).
28. Pterygoid elongated (0), triangular (1).
29. Fossa autopalatina absent (0), present (1).
30. Unconstricted notochord running through base of neurocranium absent (0), present (1).
31. Hyomandibular long and resting against hind edge of palate (0), short and inclined at an angle to palate (1).
32. Hyomandibular linked to ceratohyal (0), decoupled from ceratohyal (1).
33. Separate lachrymal and jugal bones (0), 'fused' lachrymal and jugal (1).
34. Dermohyal absent (0), present (1).
35. Squamosal absent (0), present (1).
36. Preopercular small (0), large and carrying preopercular canal dorsally to contact the otic canal (1).
37. Subsquamosal absent (0), present (1).
38. Preoperculosubmandibular absent (0), present (1).
39. Pre/postspiracular absent (0), present (1).
40. Quadratojugal absent (0), present (1).
41. Operculum absent (0), present (1).
42. Suboperculum absent (0), present (1).
43. Branchiostegals absent (0), present (1).
44. Submandibulars absent (0), present (1).
45. Median gular absent (0), present (1).
46. Lateral gulars absent (0), present (1).
47. Supraorbital and infraorbital canals joining one another anteriorly (0), remaining separate from one another (1).
48. Supraorbital canal passing between incurrent and excurrent nares (0),

passing wholly in front of (or dorsal to) both nares (1).

49. Ethmoid commissure passing through centre of premaxilla (0), taking a sutural course above premaxilla (1).

50. Supraorbital sensory canal failing to join the otic canal (0), joining the otic canal (1).

51. Supraorbital sensory canals remaining separate from one another anteriorly (0), joining one another (1).

52. Infraorbital sensory canal straight (0), describing a medial loop anteriorly (1).

53. Infraorbital sensory canal running uninterrupted beneath the anterior nostril (0), canal interrupted (or terminates) at level of anterior nostril (1).

54. Articular glenoid double (0), single (1).

55. Meckelian bone/cartilage exposed anteriorly (0), not exposed (1). This character is synonymous with the presence of a structure called the precoronoid fossa in sarcopterygians; but a more descriptive term is used here to cover the situation in actinopterygians which is similar to that in primitive sarcopterygians. In the lungfish *Griphognathus* the meckelian bone is not exposed in the medial surface of the jaw because the extensively denticulated prearticulars cover the lingual surface of the symphysis. However, the meckelian bone is exposed in the inner wall of the labial pit (Miles, 1977).

56. Retroarticular process absent (0), present (1). In some tetrapods a retroarticular process is sometimes described (Clack, 1988). This is, however, little more than the elevated posterior edge of the glenoid facet. A process quite distinct from the glenoid is not present in primitive tetrapods.

57. Preglenoid process absent (0), present (1).

58. Meckelian bone exposed between the prearticular and infradentary series (0), not exposed (1). This character refers to the fact that in most fishes the prearticular fails to contact the infradentaries. In fossils this is most clearly evident when there is a deep groove beneath the prearticular.

59. Meckelian bones of either side remaining separate from one another (0), Meckelian bones of either side fused in midline (1). In *Diabolepis*, only separate rami have been found. However, the symphysial region appears to be roughly broken and this might suggest that the meckelian bone of either side has fused together.

60. Labial pit absent (0), present (1).

61. Prearticulars or coronoids separated or sutured anteriorly (0), fused in the midline (1). The condition of the symphysis in *Diabolepis* is not entirely clear. The mutual edges of the separate rami may be broken (Chang, 1995), suggesting fusion but this needs to be confirmed.

62. Long dentary, reaching the adductor fossa (0), short dentary, failing to reach the adductor fossa (1).

63. Dentary tooth row extending to symphysis (0), anterior toothless area on dentary (1).

64. Anterior end of dentary simple (0), with medial lamina supporting an adsymphysial plate (1).

65. Number of infradentaries, one (0), two (1), four (2).

66. Modified parasymphysial/adsymphysial plate absent (0), present (1).

67. Dentary fangs absent (0), present (1).

68. Coronoid fangs absent (0), present (1).

69. Intercoronoid fossae absent (0), present (1).

70. Dentary teeth forming multiple rows (including a shagreen) or single row.

71. Medial suture between anterior infradentary and coronoid series absent (0), present (1).

72. Oral canal developed (0), represented as long pit line (1), represented as short pit line (2).

73. Parasymphyseal foramen (foramina) absent (0), present (1).
74. Coronoid dentition developed as a shagreen (0), teeth in single row (1).
75. Groups of sensory pores present on infradentaries (1).
76. Anocleithrum absent (0), present (1).
77. Anocleithrum ornament absent (0), present (1).
78. Anocleithral process absent (0), present (1).
79. Post-temporal absent (0), present (1).
80. Supracleithrum absent (0), present (1).
81. Interclavicle absent (0), present (1). An interclavicle has not been described for *Cladarosymblema* but the presence of this bone is inferred from the obvious overlap surfaces on the clavicles. Unfortunately nothing can be said about the form of the interclavicle. Denison (1968) described an interclavicle in *Uranolophus* but Campbell and Barwick (1988) redescribed the overlap surfaces on this 'interclavicle' and suggested that it should be interpreted as a median ventral scale, a view which is accepted here.
82. Interclavicle ornamented (0), unornamented (1).
83. Interclavicle extent: lying between or anterior to clavicles (0), exending as a process posteriorly between clavicles (1).
84. Clavicle and cleithrum with simple overlap (0), reverse overlap (1).
85. Glenoid closer to the ventral margin of cleithrum (0), nearer to the dorsal margin (1).
86. Scapulocoracoid perforated (0), imperforate (1). The scapulocoracoid is very badly preserved in *Cladarosymblema* but there are clearly separate areas of attachment, so it is likely that better material will confirm a perforated scapulocoracoid. In *Uranolophus* the scapulocoracoid is unknown but the very clear separated areas of attachment scars (Campbell and Barwick, 1988)

suggest the presence of a perforated scapulocoracoid.
87. Glenoid articulating with several radials (0), with single mesomere (1). This character dependently expresses the condition of the head of the humerus because there is no known condition of convex (or concave) glenoid and humerus head in the same species.
88. Scapulocoracoid glenoid surface circular/ovoid (0), strap-shaped (1). In *Acanthostega* the shape of the scapulocoracoid is unknown. But Coates and Clack (1990) describe the head of the humerus as being convex and this implies a concave glenoid.
89. Glenoid surface flat (0), concave (1), convex (2). This character reflects the shape of the head of the humerus which is commented on by several authors. It also stands for the shape of the acetabulum and head of the femur, which in all taxa known have the same surface relationships between the girdle and the proximal mesomere in the fore- and hindlimb.
90. Metapterygial axis of pectoral fin consisting of 3–4 mesomeres (0), 6–8 meso-meres (1), 15 or more mesomeres (2).
91. Radials segmented (0), unsegmented (1).
92. Postaxial radials in pectoral and pelvic fins absent (0), present (1). As pointed out above, the pelvic fins of coelacanths and porolepiforms are different from those of lungfishes in that in the latter, the postaxial radials extend as far down the fin as the preaxial radials. In porolepiforms and coelacanths, distribution of pre- and postaxial radials is highly asymmetrical.
93. Fin unrotated (0), rotated (1).
94. Prominent postaxial process on third mesomere absent (0), present (1).
95. Dorsal humeral process (entepicondyle) absent (0), present (1).

96. Entepicondylar foramen absent (0), present (1).
97. Elaborate ventral humeral processes absent (0), present (1).
98. Dorsal and ventral processes of distal mesomeres absent (0), present (1).
99. Fin rays segmented near to base (0), unsegmented bases (1).
100. Fins outwardly short and asymmetrical (0), long and leaf-like (1).
101. Cosmine absent (0), without horizontal canals (1), with horizontal canals (1).
102. Ganoine absent (0), present (1).
103. Plicidentine absent (0), present and dendrodont (1), present and polyplocodont (2).
104. Supraneurals present (0), restricted to anterior trunk segments or absent (1).
105. Second dorsal fin and anal fins with equal number of radials (0), second dorsal with more radials (1).
106. Unbranched radials in second dorsal fin (0), with branched posterior radials (1).
107. Dorsal fin(s) present (0), absent (1).
108. Anal fin present (0), absent (1).

Any analysis would ideally include all species relevant to a systematic problem. Usually this is not possible because of computer limitations. And in this particular case which involves many incompletely known fossils, there are a large number of question marks introduced for many taxa, resulting in a plethora of cladograms, many of which will be spurious trees (p. 230). Some initial selection must be made and this introduces one level of bias. It turns out that taxon selection is particularly important in this case – a phenomenon also found by Cloutier and Ahlberg (1995).

Here a sample of taxa representing most of the commonly recognized higher taxa is made. The only exception is the exclusion of onychodonts about which there is very little published information. For this data set (Table 10.1), well over 60% of the codings would be question marks and potentially disruptive to any analysis. The taxa sampled consist of two actinopterygians (*Mimia* and *Polypterus*), three coelacanths (*Miguashaia*, *Rhabdoderma* and *Latimeria*), three porolepiforms (*Porolepis*, *Glyptolepis* and *Holoptychius*), three lungfishes showing different degrees of ossification and/or dentition (*Griphognathus*, *Chirodipterus* and *Neoceratodus*), three genera the relationships of which have been discussed in recent years (*Powichthys*, *Youngolepis* and *Diabolepis*), three traditionally recognized tetrapods (*Ichthyostega*, *Acanthostega* and *Crassigyrinus*), *Panderichthys*, a rhizodont (*Barameda*) and three 'osteolepiforms' taken to represent different kinds usually assigned to different families (*Cladarosymblema*, *Osteolepis* and *Eusthenopteron*).

10.7 RESULTS OF CLADISTIC ANALYSIS

A PAUP analysis (PAUP 3.1.1, Swofford, 1993) was carried out using data listed in Table 10.1. The actinopterygians *Mimia* and *Polypterus* were used as the outgroup, constrained to be a monophyletic. This analysis omitted *Crassigyrinus* (see below) and resulted in eight most parsimonius cladograms, the strict consensus of which is shown in Fig. 10.4. The cladogram was fully resolved (i.e. completely bifurcating) as far as the node embracing traditionally recognized 'osteolepiform' taxa (*Eusthenopteron*, *Cladarosembleyma*, *Osteolepis* and *Panderichthys*) and the clade (*Panderichthys* + tetrapods). The eight alternative solutions involving these taxa arose as a result of conflicting data, and further characters or other criteria (e.g. stratigraphic position) are needed to attempt resolution. These cladograms are extremely 'sensitive' to taxon sampling. For instance, the topology of the cladograms changes completely if *Osteolepis* is removed, such that coelacanths become the sister group of the remaining 'osteolepiforms' + tetrapods. Alternatively, if *Crassigyrinus* is included, then resolution is completely destroyed beyond the recognition of the separate groupings of

Table 10.1 Data matrix for 21 sarcopterygian taxa plus two actinopterygians. The traditional systematic assignments are as follows (see pp. 259–260) with the age given in square brackets (D, Devonian; C, Carboniferous; R, Recent): *Mimia* [D] and *Polypterus* [R] are actinopterygians; *Miguashaia* [D], *Rhabdoderma* [C] and *Latimeria* [R] are coelacanths; *Barameda* [C] is a rhizodont; *Cladarosymblema* [C], *Osteolepis* [D] and *Eusthenopteron* [D] are osteolepiforms; *Panderichthys* [D] is a panderichthyid; *Porolepis* [D], *Glyptolepis* [D], and *Holoptychius* [D] are porolepiforms; *Uranolophus* [D], *Chirodipterus* [D], *Griphognathus* [D] and *Neoceratodus* [R] are lungfishes; *Ichthyostega* [D], *Crassigyrinus* [C] and *Acanthostega* [D] are traditionally recognized tetrapods; *Youngolepis* [D], *Powichthys* [D] and *Diabolepis* [D] are taxa which have been referred to one or another of the other groups or left as taxa of uncertain relationships. The 108 characters used here are described and numbered in Chapter 10. The 'N' coding is used for illogical scores: that is, character state cannot be coded because the structure does not exist in that particular taxon. For computational purposes 'N' and '?' are treated equally and as question marks

CHARACTERS

```
                                                                                                    111111111
              1111111111222222222233333333334444444444555555555566666666667777777777888888888899999999990000000000
TAXA          12345678901234567890123456789012345678901234567890123456789012345678901234567890123456789012345678

Mimia           0020000010002000000000000000000001010001110110000000000000000000020000NN11000000N0202000000100NN00
Polypterus      0020000000102000N00000000000000007210100NN1100110010000100000000010000NN11000N000N100700000000NN00
Miguashaia      01211100011010122??2?0100020010021?????010101000010010200111001100?7011001110211021011010000010000
Rhabdoderma     01211101001101101010010010?10100010010100010010100011101010?00?70110011002???????????7?000001????
Latimeria       01211010001210N011010000011101010100010011101000?1??????01000007011102111021011010000010000
Barameda        0110000101010?010?710022?100??????1??11010002011101100117??????7010?????1????7?10072001?001?????
Cladarosymblema 0???????01072??21?1000020010?2007001110011?????????0???2?1??7?00??????7??7111102??2?????01?????
Youngolepis     0?????7?01072??2100001001172?7?71111101100?7000000017017010012???7?11107???7??71101?????7101????
Powichthys      0?702207120117012700000100170?710111011?7011117??7???2?1021021027??7???7??7111027?????7101????
Porolepis       0110020017111700100700010110000010111111101010000007000000720111210?7112101??712201001002?1022??00
Glyptolepis     0110200001111011010000011110001011111110101010000000000001121011112?7111220100101010020110100
Holoptychius    01210200011110101000000011070100001011111111110101010001201121011021011107711?7772???710722?700
Diabolepis      0?707717?000?71170?110702?1000101??7??????2???????71070111102010711?71111??202007?07?2?10022?700
Uranolophus     0110121?7000121710011101020011000011107001177111?202000000N07?27?77722007222??2002?7210072??7?7
Chirodipterus   0201217?000121710111101?0101010001107??7720100711110020007200N01111177001022??7701002771000710071100
Griphognathus   0201217?000121717011101001?2011N0171011011111001100100072000001110072001111000N111021010102001100
Neoceratodus    000?0N1??001NN00NN1011111011101?770010?07200010010111111101000000200002??7?2121011110072?00202217?00
Osteolepis      01100010110101001011100010000000011111110101000200110000000201001117000010011110000010000
Eusthenopteron  01110010010101011110000102?70010100010000111110101000002011000700101111100000101100111000
Panderichthys   11110211110100NN?700011110100010100011111?012100017?112010017111102007?0?2011NN11
Ichthyostega    1111211011001NN?70001111011011000010100010101?200012110000NN?2100?71?10721022?00727211NN011
Crassigyrinus   11NNN107110ONN?00017110007110000?2?00210000000002?7010000000201012110N2NN?710107710?2211ONN001NN11
Acanthostega    11101011100NN?0100011101000?11001000100001000001000?0110000201001211020NN?71???21?11107?110?110NN001NN11
```

the coelacanths, *Diabolepis* + lungfishes, and the tetrapods (*Ichthyostega*, *Crassigyrinus*, *Acanthostega*).

Most of the nodes in this consensus cladogram are poorly supported (see legend to Fig. 10.4). Those that are best supported are those leading to coelacanths (node 15), *Powichthys* + lungfishes + porolepiforms (node 3), *Youngolepis* + porolepiforms (node 9), *Diabolepis* + lungfishes (node 5) and *Panderichthys* + tetrapods (node 13). Nodes 2 and 12 are very poorly supported, yet for the answer to the initial question "is *Latimeria* a missing link?", these are the most crucial. For instance, if either the lungfish + porolepiform clade (node 3) or any of the 'osteolepiform' taxa were replaced by coelacanths, there may well be a case for considering *Latimeria* as a missing link. Considering this consensus cladogram to represent eight alternative phylogenetic hypotheses, then node 2 is particularly important. At present, the evidence suggesting that coelacanths are the most plesiomorphic sarcopterygians is supported by a single synapomorphy, suggesting that most sarcopterygians are more derived in the elaboration of the ventral processes on the humerus. However, there are several other characters at this node which may be regarded as synapomorphies if we are willing to make assumptions about subsequent evolutionary loss and/or modification. For instance, all sarcopterygian fishes except coelacanths show submandibulars (absent in tetrapods), four infradentary bones in the lower jaw (two in the lungfishes *Griphognathus* and *Neoceratodus*) and plicidentine (absent in lungfishes and most tetrapods). These can be regarded as characters primitively absent in coelacanths but present in sarcopterygians minus coelacanths. I would consider the plesiomorphic status of coelacanths to be further supported when the details from soft anatomy and physiology are added (see p. 249 above). However, the same line of argument does not hold for characters at node 12, which is that linking 'osteolepi-

forms' with tetrapods because, in this case, most of the characters must be regarded as parallelisms with other sarcopterygians (characters 21, 24, 49, 81). And we have no possibility of adding soft characters to this problem.

The consensus cladogram here agrees to some extent with the recent analysis of Cloutier and Ahlberg (1995), reproduced here as Figure 10.3(M). Both place coelacanths as the most plesiomorphic sarcopterygians; both recognize a large clade including lungfishes and porolepiforms, including *Powichthys* and *Youngolepis* – although the relationships of these last two taxa are differently resolved; both recognize a grouping of 'osteolepiforms' and tetrapods. Furthermore both recognize that the relationships among 'osteolepiforms' are problematical. Cloutier and Ahlberg (1995) went further in their study to carry out a bootstrap analysis in an attempt to evaluate some statistical support for the internal nodes on their tree. The result of that analysis, in as far as it went (computer limitations restricted the bootstrap to 100 replications), showed a consensus placing coelacanths closer to tetrapods than either is to lungfishes. Cloutier and Ahlberg (1995) carried out other 'experiments' in using hypothetical taxa (created by taking the optimization of characters at internal nodes in the initial analysis to 'build' hypothetical animals). Although this is further divorced from real animals this reduces considerably the number of taxa and allows more thorough computation. As a result the coelacanths were resolved as plesiomorphic sarcopterygians but there were still problems with the relationships of 'osteolepiforms'. Another strategy to look for stability is to optimize the data set against other phylogenies which have been suggested (Fig. 10.3(A–L)). This exercise does not test the validity of the original hypotheses but it does give some idea of the high degree of homoplasy within the current data set. When this is done there is very little difference in the

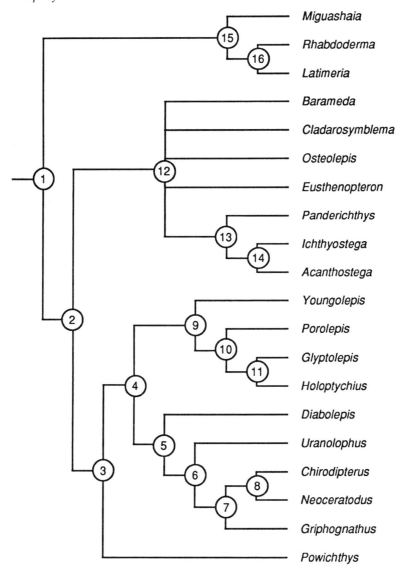

Fig. 10.4 Result of analysing 108 characters for 22 osteichthyan taxa coded for in Table 10.1 (*Crassigyrinus* omitted) with the PAUP 3.1.1 parsimony program (Swofford, 1993). The actinopterygians *Mimia* and *Polypterus* were constrained as the monophyletic outgroup, all multistate characters were run unordered under the heuristic search option (TBR). The figure shows the strict consensus cladogram resulting from eight equally parsimonious solutions (length = 226, consistency index = 0.505, retention index = 0.704). Characters were optimized to one of the trees using the ACCTRAN option and those relating to the numbered nodes are as follows (those starred are resolved as synapomorphies):

Node   1. 2(0→1), 3(0→1), 6(0→1), 8(0→1), 13(0→1)*, 14(2→1), 15(0→1), 17(0→1)*, 20(0→1), 30(0→1), 34(1→0)*, 35(0→1)*, 36(1→0)*, 47(1→0), 48(0→1)*, 49(0→1), 65(0→1), 68(0→1), 76(0→1), 87(0→1)*, 9295(0→1), 102(1→0)*, 104(0→1).

Node   2. 10(0→1), 23(0→1), 43(0→1), 44(0→1), 45(0→1), 65(1→2), 91(1→0), 97(0→1)*, 103(0→1).

Node 3. 4(1→0), 6(1→2), 8(1→0), 16(0→1), 27(0→1)*, 90(0→2), 100(0→1)*, 101(0→1), 104(1→0), 105(0→1)*, 106(0→1)*.

Node 4. 10(1→0), 18(0→1), 20(1→0), 29(0→1)*, 50(1→0), 72(2→0).

Node 5. 5(0→1), 7(0→1), 14(1→2), 15(1→0), 21(0→1), 22(0→1), 30(1→0), 31(0→1), 32(0→1), 51(0→1), 52(0→1)*, 53(0→1), 54(0→1), 57(0→1)*, 59(0→1)*, 60(0→1)*, 61(0→1)*, 68(1→0), 70(1→0), 93(0→1), 103(1→0).

Node 6. 24(0→1), 26(0→1), 47(0→1), 62(0→1).

Node 7. 2(1→0), 3(1→2), 43(1→0), 65(2→1), 101(1→0).

Node 8. 54(1→0), 58(0→1)*, 77(0→1).

Node 9. 11(0→1), 23(1→0), 37(0→1)*, 38(0→1)*, 39(0→1), 64(0→1), 66(0→1), 69(0→1), 72(0→1), 85(0→1), 86(0→1), 88(0→1)*, 98(0→1).

Node 10. 12(0→1), 18(1→0), 20(0→1), 49(1→0), 50(0→1), 63(0→1), 74(0→1), 103(1→2).

Node 11. 4(0→1), 16(1→0), 72(1→2), 101(1→0).

Node 12. 14(1→0), 21(0→1), 24(0→1), 49(1→0), 81(0→1), 89(2→1)*, 92(1→0).

Node 13. 1(0→1)*, 22(0→1), 32(0→1), 47(0→1), 67(0→1), 74(0→1), 83(0→1)*, 107(0→1)*, 108(0→1)*.

Node 14. 9(0→1), 20(1→0), 26(0→1), 31(0→1), 41(1→0)*, 43(1→0), 44(1→0), 45(1→0), 46(1→0), 50(1→0), 55(0→1)*, 68(1→0), 71(0→1)*, 73(0→1)*, 76(1→0), 79(1→0), 80(1→0), 90(0→1), 96(0→1).

Node 15. 3(1→2), 5(0→1), 19(0→1), 22(0→1), 28(0→1)*, 31(0→1), 33(0→1)*, 39(0→1), 40(1→0)*, 53(0→1), 56(0→1), 62(0→1), 77(0→1), 79(1→0), 80(1→0), 82(0→1)*, 85(0→1), 86(0→1), 90(0→1), 93(0→1), 98(0→1).

Node 16. 12(0→1), 70(1→0).

Accepting this as a summary of eight equally likely evolutionary trees the following Linnaean classification can be given which incorporates the plesion convention suggested by Patterson and Rosen (1977):

Subclass Sarcopterygii Romer 1955
  Infraclass Actinistia Cope 1871
       Mignashaia
       Rhabdoderma
       Latimeria
  Infraclass Choanata Säve-Söderberg 1934
    Series Dipnoi Müller 1844
      plesion *Powichthys*
      Order Porolepiformes Jarvik 1942
        Youngolepis
        Porolepis
        Glyptolepis
        Holoptychius
      Order Dipnoiformes
      plesion *Diabolepis*
      plesion *Uranolophus*
      plesion *Griphognathus*
      plesion *Chirodipterus*
      Neoceratodus
    Series Tetrapoda Rosen *et al.* 1981
      incertae sedis *Barameda*
      incertae sedis *Cladarosymblema*
      incertae sedis *Osteolepis*
      incertae sedis *Eusthenopteron*
      incertae sedis Eutetrapoda
      Panderichthys
       Acanthostega
      Ichthyostega

lengths of the trees, underlining the fact that consideration of these skeletal characters may not give a strong phylogenetic signal. Schultze (1994) also came to a similar conclusion. He used a large morphological data set (213 characters) and arrived at a single cladogram – in this case resolving lungfishes and coelacanths as living sister-groups (Fig. 10.3(K)). However, two cladograms were a single step longer: one of these gave the result arrived at by Schultze (1987: Fig. 10.3(H)), suggesting that amongst the living taxa coelacanths and tetrapods are sister groups; the other suggesting a hypothesis close to that of Cloutier and Ahlberg (1985). Thus, within a single step (306–307 steps), all three resolutions amongst the Recent taxa were obtained.

I conclude from all of these studies that, to date, a consideration of the many Devonian fossil taxa which are so often studied does not provide unequivocal resolution of the relationships of primitive sarcopterygians. However, from this rather indecisive result we can learn something. The most robust part of all equally parsimonious solutions places porolepiforms with lungfishes. This grouping always occurs and leads to consequences for the coelacanth. It suggests that two features shared by coelacanths and lungfishes (the short lower jaw with an abbreviated dentary and the apparent rotation of the pectoral fin) are independently derived. Also consideration of fossil lungfishes (as well as porolepiforms) shows that other 'similarities' between modern lungfishes and *Latimeria*, such as absence of branchiostegals, submandibulars and maxillae, are homoplasious. Lungfishes are known to have lost these structures in their own history.

From the three lines of evidence presented above – skeletal + fossil, soft anatomy and molecular – it is the soft anatomy which gives the strongest phylogenetic signal: lungfishes are the Recent sister group to tetrapods. The evidence from consideration of palaeontological data provides some weak support but, more importantly, does not contradict this view. Evidence from molecular data is very mixed and, at present, difficult to evaluate. Phylogenetic conclusions arrived at with different molecular data sets seem very susceptible to taxon sampling (as with some of the palaeontological studies). However, one of the two most favoured molecular solutions places lungfishes with tetrapods. Therefore, on balance, we must conclude that *Latimeria* is the most primitive of living sarcopterygians and it cannot be considered a 'missing link' between fishes and tetrapods.

11

TAXONOMY

11.1 INTRODUCTION

In this chapter, diagnoses are given for all taxa recognized in this work, as well as a list of taxa doubtfully referred to the coelacanths. Full diagnoses for each taxon are given rather than diagnoses relevant only to their position on the phylogenetic tree (Fig. 9.7) and as a consequence there is some repetition of information. This approach will enable the reader to gain knowledge about a particular taxon without having to read a great deal of preliminary text as well as make such diagnoses more resistant to changes in coelacanth classification.

In the following diagnoses there are several measurements and ratios cited. Figure 11.1 provides a key to these measurements with abbreviations used and their meanings described in the legend. The abbreviation pt, & cpt, refers to part and counterpart of individual specimens.

The list here is divided to four categories (Sections 11.3–11.6). The first lists those species that can be placed within a phylogenetic tree. The second category lists those considered to be valid species but about which we know so little that their relationships cannot be accurately specified: however, they can be placed as incertae sedis at various hierarchical levels in a classification. These taxa are useful for calculating species diversity but are useless for estimations of morphological or clade evolution. The third category lists those coelacanth taxa which cannot be demonstrated to be valid species or of which the species status can be

questioned. The last category lists those taxa simply identified as 'coelacanth', nomen nuda or those taxa which have been referred to coelacanths but which either are not determinable as coelacanths or can be referred to non-coelacanth taxa. In each category the taxa are arranged alphabetically.

Where possible the location and specimen number of the holotype is given, together with details of other specimens that have been examined in this study.

11.2 LINNAEAN CLASSIFICATION OF COELACANTHS

A Linnaean classification fully reflecting the ideas of phylogenetic relationships is difficult to construct without introducing paraphyletic taxa and/or many empty ranks. Part of this difficulty arises because there is a single Recent species. Another aspect is that the basal part of the tree is highly pectinate, theoretically requiring many suprageneric ranks. It seems desirable to try and reflect some of the structure of the tree. For instance, the two major monophyletic clades of Mesozoic coelacanths should be distinguished. These are those taxa embraced by *Holophagus* to *Latimeria* (group 1) and *Chinlea* to *Garnbergia* (group 2) in Fig. 9.7, as well as the sister-group relationships between *Laugia* + *Coccoderma* and *Spermatodus* + *Sassenia*. Within each of these groups taxa can be sequenced denoting a pectinate phylogeny. However, applying the Hennigian principle that sister groups be given equal rank, we soon run into problems. We could give

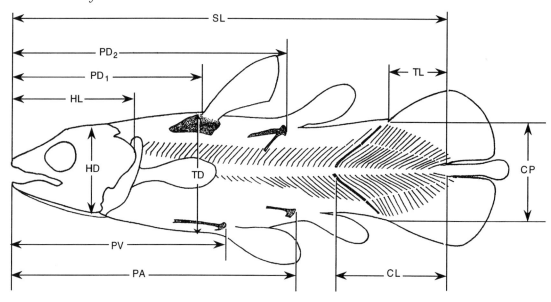

Fig. 11.1 Outline diagram to show the measurements used in the species diagnoses. Also included here are abbreviations used for counts of fin rays and neural arches. SL, standard length: distance from snout to the base of the supplementary caudal fin (taken as the distal tip of the neural spine supporting the last fin ray in the principal caudal fin lobe). PD_1, predorsal$_1$ distance: distance from the snout to the base of D_1, taken as the point of insertion of the anteriormost fin ray to the supporting basal plate. PD_2, predorsal$_2$ distance: distance from the snout to the distal tip of the basal plate supporting D_2 (this is used because it is rare to be able to recognize a clear outline of the lobe suporting this fin or the endoskeleton). ID, interdorsal distance: distance between the point of insertion of the anteriormost fin ray of D_1 to the supporting basal plate and the distal tip of the basal plate supporting D_2. HL, head length: snout to posterior limit of the operculum. HSL, distance from the posterior limit of the operculum to the base of the supplementary caudal fin (see SL). HD_1, posterior limit of operculum to the point of insertion of the anteriormost fin ray of D_1 to the supporting basal plate. HD_2, posterior limit of operculum to the distal tip of the basal plate supporting D_2. HCP, posterior limit of operculum to the level of insertion of the first principal caudal fin ray within the upper lobe. HD, depth of head: measured from the highest point of the postparietal shield to the ventral edge of the retroarticular. TD, maximum depth of the body. CP, depth of caudal peduncle: measured as the depth at the level of insertion of the first principal caudal fin ray within the upper lobe. PV, prepelvic distance: measured from the level of the snout to the posterior tip of the pelvic girdle. PA, preanal distance: measured from the level of the snout to the posterior tip of the anal basal plate. D_1, number of rays in the first dorsal fin. D_2, number of rays in the second dorsal fin. P, number rays in the pectoral fin. V, number of rays in the pelvic fin. A, number of rays in the anal fin. C, number of rays in the principal dorsal/ventral lobes of the caudal fin. AbdV, number of abdominal neural arches. CbdV, number of abdominal neural arches counted posteriorly to the last neural arch supporting rays of principal caudal lobes (i.e. number of neural arches in supplementary lobe omitted).

family status to group 1 and group 2 and link both into a suprafamily rank (e.g. suborder). This would mean that the sister-group genus *Whiteia* would be ranked as a suborder and, following this logic, the genus *Coela-*

canthus would be a more inclusive rank (say, an order). We would soon run out of ranks. To avoid this problem, Patterson and Rosen (1977) introduced the neutral rank plesion for fossil taxa and suggested that these be ordered according to their cladistic appearance on the tree, the convention being that the Recent representative(s) appeared last. In this particular case every taxon except *Latimeria* would be a plesion. This strategy would be possible here; we could still recognize *Laugia* + *Coccoderma* and a family Laugiidae and this could be placed as a plesion to cladistically more derived taxa. But it also is instructive to recognize at least one node on the tree (that numbered 11 in Fig. 9.7) where some significant changes took place in coelacanth history. To do this I have decided to use the order Coelacanthiformes which does not have a similarly rank-named sister group but it would be the family Laugiidae.

There are quite a few genera which can only be placed as *incertae sedis* at various levels in the classification. These are inserted at the most inclusive hierarchical level allowed by the known data for those taxa. For example, *Lualabaea* is known to have the dentary specialization of *Axelrodichthys* and *Mawsonia* and therefore it may be placed as incertae sedis before listing those genera. It may also be possible to place other coelacanths at this level but only on phenetic resemblances which are mentioned in the diagnosis. In other instances there may be very little useful information. There are several genera (e.g. *Alcoveria* and *Axelia*) which are known to have either a fragmented neurocranium and/or denticle-covered fin rays. This would place them somewhere between nodes 11 and 12 (Fig. 9.7), at the level of *Whiteia*.

In an attempt at a Linnaean classification, the following conventions are used: taxa are ranked sequentially such that the succeeding taxon is cladistically more derived than that preceding; supragenetic categories are indented and the sequencing format followed again; taxa incertae sedis are placed at the most inclusive point. The rank plesion is not used here, only because all except one taxon would be a plesion leading to unnecessary repetition.

The suprageneric classification is as follows.

Infraclass Actinistia Cope 1871
    Family Sassenidae nov.
    Family Laugiidae Berg 1940
    Order Coelacanthiformes
        Suborder Latimeroidei Schultze 1993
            Family Mawsoniidae Schultze 1993
            Family Latimeriidae Berg 1940

Diagnoses for these higher categories are as follows.

Infraclass Actinistia Cope 1871

Diagnosis

Sarcopterygian fishes in which the head shows a well-developed intracranial joint, rostral organ, two external nostrils, a single bone (lachrymojugal) beneath the eye, upright jaw suspension with a triangular palate, tandem jaw articulation between the quadrate and articular and between the symplectic and retroarticular, lower jaw in which there is a short dentary, two infradentaries of which the angular is by far the largest with a characteristic dorsal expansion, posteriormost coronoid much expanded and separated from anterior coronoids; maxilla, submandibulars and branchiostegals absent, urohyal subdermal: in the postcranial skeleton shoulder girdle free from skull and showing an extracleithrum, first dorsal fin sail-like and lacking radials, second dorsal and anal fin endoskeleton mirror images of each other and similar to the paired fin endoskeletons, caudal fin with a single series of radials distal to neural and haemal spines, scales circular and deeply overlapping lacking ganoine or cosmine, ornamented with enamel-capped ridges, tubercles or denticles.

Family Sasseniidae nov.

Diagnosis

Actinistians with long cheek region and quadrate/articular jaw joint placed relatively far back, dermal intracranial joint strongly interdigitate, lateral rostral without pronounced ventral process, posterior openings of the rostral organ closely juxtaposed separated only by a narrow bridge of bone, supratemporal forming a very small part of postparietal shield, ornament of small rounded tubercles closely packed, sensory canal openings very small.

Family Laugiidae Berg

Diagnosis
Actinistians with pelvic fins lying anterior to the level of the dorsal fin, pelvic bones of either side expanded and joined in ventral midline over their entire length, cheek bones reduced, sensory canals over head with marked increase in pore size towards snout and jaw symphysis.

Order Coelacanthiformes

Diagnosis
Actinistians in which the otico-occipital portion of the neurocranium is fragmented to separate prootic, opisthotic, basioccipital and supraoccipital bones, with a complex suture between the prootic and basioccipital, loss of vestibular fontanelle and buccohypophysial canal.

Suborder Latimeroidei Schultze

Diagnosis
Coelacanthiformes in which the postparietal shield is markedly shorter than pareito-nasal shield, showing a postparietal descending process, preorbital and sclerotic ossicles usually absent, usually more than five extrascapulars, retroarticular and articular separate ossifications, denticles present upon rays of first dorsal and caudal fins.

Family Mawsoniidae Schultze

Diagnosis
Latimeroidei in which long, ossified ribs are present in most, ornament tends to be rugose, spiracular and suboperculum usually absent, reduction or loss of supratemporal descending process.

Family Latimeriidae Berg

Diagnosis
Latimeroidei in which the parasphenoid shows anteriorly placed ascending processes, restricted parasphenoid dentition, reduction of bone lining the temporal excavation, anterior branches of the supratemporal commissure, lateral line scales with complex secondary tubes and multiple openings.

The full classification is as follows.

Infraclass Actinistia
Miguashaia
incertae sedis *Euporosteus, Chagrinia*
Lochmocercus
Allenypterus
Hadronector
Polyosteorhynchus
Caridosuctor
Rhabdoderma
Family Sasseniidae
    Spermatodus
    Sassenia
Family Laugiidae
    incertae sedis *Piveteauia*
    Laugia
    Coccoderma
Order Coelacanthiformes
    Coelacanthus
    incertae sedis *Alcoveria, Axelia, Graphiurichthys,*
        Mylacanthus, Sinocoelacanthus, Ticinepomis, Wimania
    Whiteia
  Suborder Latimeroidei
    Family Mawsoniidae
      Garnbergia
      Libys
      incertae sedis *Changxingia, Heptanema, Indocoelacanthus*

Diplurus
Chinlea
incertae sedis *Lualabaea, Mega-locoelacanthus, Moenkopia*
Mawsonia
Axelrodichthys
Family Latimeriidae
incertae sedis *Macropoma willemoesii*
Holophagus
Undina
incertae sedis *Macropomoides*
Macropoma
Latimeria

11.3 TAXA THAT CAN BE PLACED WITHIN A PHYLOGENETIC TREE (Category 1)

Genus *Allenypterus* Melton 1969

Diagnosis (emended)

Distinctively shaped small coelacanth with a deep compressed body and head, and with a long filamentous tail. Dorsal profile is rounded while the ventral profile is straight. Notochord characteristically arched above the abdominal cavity. Head short and deeper than long. The skull roof has a steep profile and, as a consequence, the jaw articulation lies behind the level of the operculum. The eye is large, exceeding one-third of the head length and is surrounded by many sclerotic ossicles. The dermal bones of the skull roof are ornamented with coarse, flattened vermiform ridges. The parietal shield is short, equal to approximately 45% of the parietonasal shield. There may be three or five extrascapulars; but in either case the lateral element is very broad. The parietonasal shield is also broad and includes a series of quadrangular tectals and supraorbitals, and there is also an ornamented preorbital pierced by the posterior openings of the rostral organ which lie close to one another. The snout is made up of star-shaped rostrals separated by large pores associated with the sensory canals. Above the eye the supraorbital sensory canal opens

through medium-size pores located along the sutures. The cheek is completely covered with closely fitting bones, all of which are very thin and unornamented. The postorbital is very narrow and the squamosal is deeper than wide, but the preoperculum is a very large triangular element filling most of the lower part of the cheek. Both operculum and suboperculum are narrow. In the lower jaw the angular is deep while the splenial is very small. Both of these elements are ornamented with longitudinal ridges. The dentary is substantially larger than the splenial and is unornamented: it is of unusual shape, resembling a dorsally concave rod. The mandibular sensory canal runs the entire length of the angular and splenial: there is no dentary pore. The oral pit line curves forward from the centre of the angular to run off the anterior end of the bone. In the shoulder girdle both the cleithrum and extracleithrum are very slender and ornamented with prominent ridges. The first dorsal is long based while the second dorsal, which is located well forward within the anterior half of the body, is very small as is the anal fin. The distinctive caudal fin is highly asymmetrical. The dorsal lobe begins anterior to the midlength of the fish and continues as a low fringe supported upon 50 radials. There is a greater than one-to-one correspondance between fin ray and supporting radials. The ventral lobe is restricted to a few tiny rays posteriorly. However, the ventral radial series, which contains about 40 elements, continues considerably in advance of the fin ray series. The supplementary lobe continues with little differentiation from the main lobes. It carries a few slender fin rays but these disappear distally. The scales are small and ornamented with closely set longitudinal ridges which, on some scales, converge posteriorly. There is a mid-ventral row of six to nine enlarged ridge scales extending from the shoulder girdle to the level of the pelvic fin insertion.

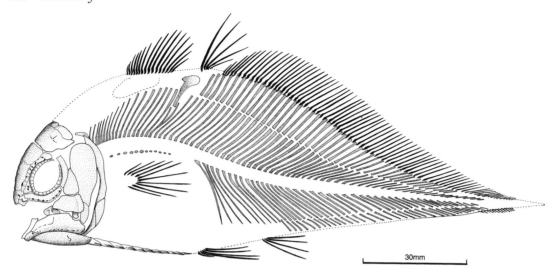

Fig. 11.2 *Allenypterus montanus* Melton. Restoration of fish. Details of pelvic girdle, anal fin support and tip of tail are not sufficiently well known to allow restoration. Based on FMNH PF10016 (pt & cpt), PF10025, PF10027, PF10939, PF10940, PF10942, PF10943a,b, PF13511, PF13512, PF14223a,b.

Type and only species: *A. montanus* Melton

Allenypterus montanus Melton
Figs 3.5, 4.6, 8.2(H), 11.2.
1969 *Allenypterus montanus* Melton: 199, figs
    4–8; pls 1 and 2.
1985 *Allenypterus montanus* Melton; Lund
    and Lund: 39, figs 57, 59, 60–66.

Diagnosis (emended)
Species reaching 160 mm SL, HL/SL = 17%, $PD_1/SL = 22\%$, $PD_2/SL = 36\%$, TL/SL = 60%, CL/SL = 70%, TD/SL = 36%, CP/SL = 35%, PV/SL = 36%; $D_1 = 15$, $D_2 = 6$, C = 72/15, P = 9, V = 6, A = 6; AbdV = 22, CauV = 52.

Holotype
UMON 2555, a complete fish in pt and cpt, from the Lower Carboniferous (Namurian A), Heath Formation, Bear Gulch Limestone Member, Montana, USA.

Material
FMNH PF10016 (pt and cpt), PF10025, PF10027, PF10939, PF10940, PF10942,

PF10943a,b, PF13511, PF13512, PF14223a,b, all from the Lower Carboniferous (Namurian A), Heath Formation, Bear Gulch Limestone Member, Montana, USA.

Genus *Axelrodichthys* Maisey 1986b

1991 *Axelrodichthys* Maisey; Maisey: 303.

Diagnosis (emended)
Monotypic genus of large-headed coelacanths. The head is extended as a short shallow snout: the eye is relatively small, equal to about 18% of the total head length and is located well forward within the anterior half of the skull; sclerotic ring absent. Postparietal shield short and broad, with the lateral margins parallel-sided in dorsal view. Extrascapulars incorporated to the postparietal shield where a median element persists. Descending process upon the supratemporal is very weakly developed. The parietonasal shield is at least 2.5 times as long as the postparietal shield and is narrow and concave in lateral profile. Parie-

tal and nasals elongated; lateral rostral very narrow, barely larger than the contained sensory canal; preorbital absent. Openings of the otic and supraorbital sensory canals are rare except within the tectals and the rostrals. Palate long and shallow, particularly anteriorly; posterior vertical limb narrow and much of the pterygoid is formed of very thin bone; autopalatine is very small. Cheek incompletely covered with bones which do not meet one another; postorbital and squamosal large, robust and irregularly shaped, lachrymojugal large and deep. Preoperculum small and thin and a suboperculum is absent. Most of postorbital lying adjacent to parietonasal shield with infraorbital sensory canal passing through the centre of the postorbital. Lower jaw in which the articular and retroarticular are separate, principal coronoid very large, saddle-shaped, located well forward within the anterior half of the jaw and sutured with the angular. Most of the dermal bones of the skull are thick and ornamented with fine rugae. Fin rays in all fins very closely articulated, particularly in the pectoral fin. Anterior caudal rays and D_1 rays ornamented with a few scattered denticles. The

scales bear an ornament of fine longitudinal ridges. The ossified swim bladder is very large reaching well back to the level of the basal of the anal fin; there is some evidence that it was divided into anterior and posterior chambers.

Type and only species: *A. araripensis* Maisey 1986b

Axelrodichthys araripensis
Figs 4.17, 5.10, 7.1(F), 11.3, 11.7.
1986b *Axelrodichthys araripensis* Maisey: 13.
1991 *Axelrodichthys araripensis* Maisey; Maisey: 303, un-numbered figures appearing on pp. 303–308, 310.

Diagnosis
Species reaching about 1 metre SL. HL/SL = 25%, $PD_1/SL = 35\%$, TL/SL = 18%, CL/SL = 32%, TD/SL = 25%, CP/SL = 20–25%, $D_1 = 9–10$, $D_2 = 10$, C = 15/15, P = 12–15, V = 17–18. Gular plate is relatively short, less than 75% of the jaw length. Scales ornamented with short, irregularly spaced horizontal ridges, those scales towards rear of body show a large conspicuous central ridge.

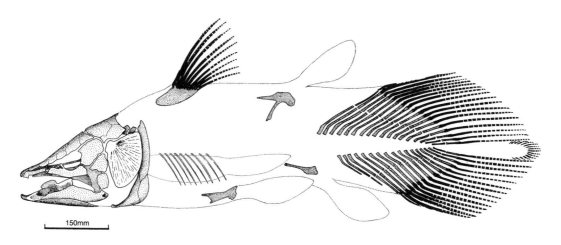

150mm

Fig. 11.3 *Axelrodichthys araripensis* Maisey. Restoration of skeleton, after Maisey (1991) with added information from FMNH PF11856.

Holotype
AMNH 1759, a complete fish from the Lower Cretaceous (Apto-Albian), Lower Romualdo Member, Santana Formation, Ceara, Brazil.

Material
FMNH PF 10726, PF 11856, PF 12840, all from the Lower Cretaceous (Apto-Albian), Lower Romualdo Member, Santana Formation, Ceara, Brazil.

Genus *Caridosuctor* Lund and Lund 1984

Diagnosis (emended)
A monotypic genus very similar to *Rhabdoderma*. The body is slender with a small head with a convex profile. Postparietal and parietonasal shields are of about equal length. The postparietal shield shows long postparietals and there are five scale-like extrascapulars lying along a straight transverse posterior margin. In the parietonasal shield the premaxillae are large, perforated by the anterior opening to the rostral organ and each bears three or four very stout teeth. The cheek bones are overlapping and completely cover the cheek. As in *Rhabdoderma* the postorbital, squamosal and preoperculum are relatively broad. The operculum is also broad dorsally but tapers rapidly ventrally. The lower jaw is shallow anteriorly although there is a rounded coronoid expansion. The principal coronoid is triangular. There are large pointed teeth upon the anterior coronoids and the dentary lacks a hook-shaped process. The shoulder girdle shows a broad cleithrum and a large extracleithrum. The caudal fin is asymmetrical and relatively elongate with a long supplementary lobe. The basal support for D_1 is kidney-shaped with the ventral margin emarginated to receive the tips of the adjacent neural arches. Pelvic bone with several anterior processes and a strongly digitate medial process.

Type and only species: C. populosum Lund and Lund

Caridosuctor populosum
Figs 3.3(C), 5.3(A), 11.15(A).
1984 *Caridosuctor populosum* Lund and Lund: 238, fig. 1.
1985 *Caridosuctor populosum* Lund and Lund; Lund and Lund: 17, figs 16–34.

Diagnosis (emended)
Species reaching 220 mm SL, HL/SL = 22%, PD_1/SL = 38%, PD_2/SL = 58%, TL/SL = 20%, CL/SL = 33%, TD/SL = 26%, CP/SL = 16%, PV/SL = 44%; D_1 = 12, D_2 = 14, C = 19/16, P = 19, A = 14; AbdV = 38, CauV = 18.

Holotype
UMON 6021, a complete fish from the Lower Carboniferous (Namurian A), Heath Formation, Bear Gulch Limestone Member, Montana, USA.

Material
FMNH PF12920, Lower Carboniferous (Namurian A), Heath Formation, Bear Gulch Limestone Member, Montana, USA.

Genus *Chinlea* Schaeffer 1967

Diagnosis (emended)
A monotypic genus reaching 70 cm SL. The head is large and relatively shallow, being about half as deep as long. The postparietal shield is short and broad, less than half the length of the parietonasal shield. The snout is elongated, being at least one-third of the total head length. The bones of the parietonasal shield are rectangular with sutures between successive paired elements orientated transversely. In lateral view the dorsal profile of the parietalonasal shield is markedly concave. Five extrascapulars closely sutured to the parietals. The lateral rostral is long, reflecting the long snout. Supraorbito-tectal series with about 12 elements. The post-

orbital and preoperculum are broad and the lachrymojugal is deep with a prominent anterodorsal excavation for the posterior opening from the rostral organ. Preorbital absent. A spiracular bone and suboperculum were probably absent. The operculum is triangular with a rounded dorsal margin. Teeth on the premaxilla, ectopterygoid, dermopalatine and coronoid 4 are relatively large, while other teeth are small and villiform. Basisphenoid with short, stout antotic processes which are triangular in dorsal view. The ornament is rugose over most of the skull bones and the sensory canals open through small pores. The scales are ornamented with narrow longitudinal ridges which vary in length over a single scale. The vertebral column shows long ribs. The basal of the first dorsal fin is a large triangular plate whereas that of the second dorsal fin is the typical bifurcated shape. Rays of the first dorsal fin are substantially stouter than those of all other fins and there are no ornament denticles. The supplementary lobe of the caudal fin is well developed.

Type and only species *C. sorenseni* Schaeffer

Chinlea sorenseni
1928 *Macropoma* sp. Warthin: 17, pl. 1, figs 2, 3.
1961 un-named coelacanth Schaeffer and Gregory: 10, figs 4, 5.
1967 *Chinlea sorenseni* Schaeffer: 323, figs 13, 14; pls 26–28.
1987 *Chinlea sorenseni* Schaeffer; Elliott: 47, figs 1–3.

Diagnosis
As for genus, only species.

Holotype
AMNH 5652, nearly complete fish, lacking posterior half of caudal skeleton. Upper Triassic (Carnian), Chinle Formation of Little Valley, San Juan County, Utah.

Material
AMNH 3201, 5653–5660, 5704; UM 9360, 38320; YPM 3928, TTCM 527, MNA V5470 (cast of specimen in private collection). Upper Triassic (Carnian), Chinle Formation of Utah, Colorado and New Mexico and the Tecovas Formation of Texas.

Genus *Coccoderma* Quenstedt 1858
(emended Reis 1888)

1858 *Kokkoderma* Quenstedt: 810.

Diagnosis (emended)
A genus of elongate, slender fishes with relatively small head, and pelvic fins located well anterior to the level of the anterior dorsal fin. The postparietal shield is nearly as long as the parietonasal shield and is relatively slender, particularly at the level of the intracranial joint. There are five extrascapulars which are closely sutured to one another and to the postparietals and the supratemporals. The tripartite division of the otic canal, supratemporal commissure and lateral line lies within the separate lateral extrascapular. The snout is composed of a mosaic of star-shaped ossicles which together form a solid snout. Interorbital septum ossified; sclerotic ring contains many ossicles. The palate shows a very shallow anterior limb of the pterygoid. The cheek is incompletely covered with elements which are well separated from one another: squamosal and preoperculum are somewhat reduced, the latter to a vertically orientated tube enclosing the preopercular canal. Preorbital and suboperculum are absent. In the lower jaw the angular shows a prominent coronoid expansion which bears a shallow notch in the posterior margin. The coronoid is quadrangular and considerably deeper than wide. The sensory canals of the head open through many small pores which tend to increase in size considerably at the anterior end of the jaw and over the snout. The shoulder girdle is broad with a large

extracleithrum closely sutured to the cleithrum and there is a broad plate-like anocleithrum. The pelvic girdle is large with the two pelvic bones ossified such that they meet each other in the ventral midline over most of their length. The posterolateral angles are produced dorsally as hook-shaped processes. The pelvic fin is large, located close to the pectoral girdle and the rays are expanded and they are closely articulated throughout their distal halves. The rays of the anal fin are also slightly expanded. The caudal fin is rather square-cut in outline and the supplementary caudal fin is large. The fin rays within the principal lobes and the supplementary fin are expanded and closely articulated in their distal halves. The scales are ornamented with discrete longitudinal ridges. Lateral line scales bear a very wide sensory tube opening to the surface through several small pores. Ornamentation upon the head bones is confined to tiny spine-like denticles sparsely distributed on the angular and the gular. Denticles are also present covering most of the pelvic fin rays, the anterior rays of both D_1 and the caudal fin and, as a very sparse covering, upon the anal and the supplementary caudal fin rays where there appears to be a denticle associated with each segment.

Remarks

Coccoderma resembles *Laugia* in the slender form of the body, shape and proportions of the parietonasal shield and extrascapulars, reduced cheek covering, shape of the coronoid and the shoulder girdle and in the anterior location of the pelvic fins.

Type and only species *C. seuvicum* Quenstedt

Coccoderma suevicum

Figs 3.10, 4.11, 4.12, 5.7.

1858 *Kokkoderma seuvica* Quenstedt: 810, pl. 100, fig. 14.

1874 *Coelacanthus harlemensis* Winkler: 101, pl. 4.

1881 *Coelacanthus harlemensis* Winkler; Vetter: 13, pl. 2, fig. 4.

1888 *Coelacanthus harlemensis* Winkler; Reis: 5.

1888 *Coccoderma suevicum* Quenstedt; Reis: 51.

1888 *Coccoderma nudum* Reis: 58, pl. 3, figs 14, 15, 16, pl. 4 figs 9, 18, pl. 5, fig. 1.

1891 *Coccoderma seuvicum* Quenstedt; Woodward: 415.

1906 *Coccoderma seuvicum* Quenstedt; Heineke: 11, pls 6, 7.

1921 *Coccoderma seuvicum* Quenstedt; Stensiö: 116, fig. 53.

1954a *Coccoderma seuvicum* Quenstedt; White: fig. 6.

1992 *Coccoderma seuvicum* Quenstedt; Lambers: 10, figs 1–12.

Diagnosis (emended)

Species reaching 300 mm SL, HL/SL = 22–25%, PD_1/SL = 40%, PD_2/SL = 60–63%, TL/SL = 14%, TD/SL = 25%, CP/SL = 17–20%; D_1 = 9–11, D_2 = 12–14, C = 21/19, A = 13–16; AbdV = 50–54, CauV = 24–26.

Holotype

UT, an articulated quadrate, metapterygoid and partial pterygoid from the Upper Jurassic (Tithonian) of Nusplingen, Bavaria, Germany.

Material

BMNH P.8356, BSM 1870.XIV.23 (holotype of *C. nudum*), 1870.XIV.505, 1870.XIV.506, JME 1956.2, SOS.2187, SOS.2193, SOS.2194, TM T.13280 (holotype of *Coelacanthus harlemensis*). All from the Upper Jurassic (Tithonian) of Bavaria, Germany.

Remarks

C. nudum Reis was erected for a small individual which in so many respects is similar to the type species and it has been placed in synonymy here. The chief differences originally recognized are the almost total absence of ornament, the relatively shorter and unexpanded fin rays of the pelvic fin and the more oblique outline of the caudal fin, all of which may be juvenile characters.

Coccoderma sp. Fabre *et al.* (1982) has been mentioned from the Lower Cretaceous (Berriasien), Var, France. The skull and position of the pectoral fins agree well with *C. suevicum* but it is insufficiently described to be sure of specific identity.

Genus *Coelacanthus* Agassiz 1839

Diagnosis (emended)
A genus of elongate coelacanths in which the body is relatively shallow and the head is small and the supplementary caudal fin is well developed. The skull roof is narrow throughout. The parietonasal and postparietal shields are of equal length and most of the bones are thin. The parietonasal shield is composed of two pairs of parietals (the posterior substantially the larger), at least two pairs of nasals and a snout which consists of small, loosely associated ossicles, the pattern of which remains unknown. The premaxillae are small and delicate, and there may have been more than one pair on each side. The postparietal is very large in relation to the supratemporal, the latter bears a prominent descending process. There are three extrascapulars lying between the supratemporals; they are, however, small and the posterior margin of the skull roof is not markedly embayed. A well-developed sclerotic ring of at least 30 ossicles is present. The cheek consists of a robust lachrymojugal which is associated with small stud-shaped ossicles which must have lain free in the skin (similar to those in *Spermatodus*), a narrow postorbital and a squamosal and preoperculum which are developed as tubes around the sensory canal. No preorbital has been found, but this region of the skull is not well known. The lower jaw is long, equal in length to the skull roof. The angular is perfectly smooth and does not show evidence of a pit line. Both dentary and splenial are very narrow bones and the dentary lacks a hook-shaped process. The principal coronoid is quadrangular and the anterior three coronoids are rounded tooth plates each bearing at least 40 conical teeth. The fourth coronoid and the dermopalatine/ectopterygoids bear prominent conical teeth. The basisphenoid is narrow, the antotic processes are weakly developed and the sphenoid condyles are close together. The shoulder girdle shows a large extracleithrum and a broad anocleithrum. The pelvic bone is also broad with two prominent anterior processes with a web of thin bone between them. The basal of D_1 is oval while the basal supporting the anal fin is uniquely deep and plate-like. The arrangement of the fin rays in D_2 suggests a well-developed lobe while the anal fin is very asymmetrical, suggesting that the lobe was weakly developed. The paired fins are rather weakly developed. An ossified air bladder is preserved.

Type species: *C. granulatus* Agassiz

Coelacanthus granulatus
Figs 3.7, 5.4, 11.4.
1835 'Fossil fish' Sedgwick: 118, pl. 9, fig. 3, pl. 11.
1839 *Coelacanthus granulatus* Agassiz: pl. 62.
1842b *Coelacanthus hassiae* Münster: 49.
1844 *Coelacanthus granulosus* Agassiz: 172.
1850 *Coelacanthus granulatus* Agassiz; Egerton: 235.
1850 *Coelacanthus caudalis* Egerton: 236, pl.XXVII, fig. 2.
1866 *Coelacanthus caudalis* Egerton; Huxley: 14, pl. 5, fig. 5.
1869 *Coelacanthus macrocephalus* Willemoes-Suhm: 74, pl. 11, fig. 2.
1869 *Coelacanthus hassiae* Willemoes-Suhm: 76, pl. 10, fig. 1, pl. 11, fig. 1.
1888 *Coelacanthus macrocephalus* Willemoes-Suhm; Reis: 68.

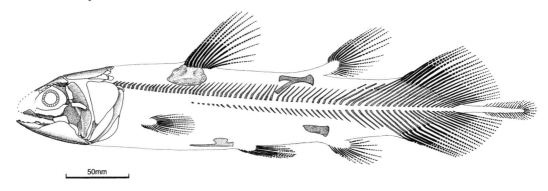

Fig. 11.4 *Coelacanthus granulatus* Agassiz. Restoration of entire fish: proportions based on a small individual (HM G50.66).

1888   *Coelacanthus hassiae* Willemoes-Suhm; Reis: 69, pl. 3, fig. 22, pl. 4, figs 7, 12, 15, 16, 19.

1891   *Coelacanthus granulatus* Agassiz; Woodward: 400.

1932   *Coelacanthus granulatus* Agassiz; Stensiö: 40.

1935   *Coelacanthus granulatus* Agassiz; Moy-Thomas and Westoll: 447, figs 1–12.

1948   *Coelacanthus granulatus* Agassiz; Schaeffer: fig. 5.

1952b *Coelacanthus granulatus* Agassiz; Schaeffer: fig. 15c.

1978   *Coelacanthus granulatus* Agassiz; Schaumberg: 196, figs 1–21.

Diagnosis (emended)

Elongate species reaching 70 cm SL, HL/SL = 20%, PD_1/SL = 36%, PD_2/SL = 60%, TL/SL = 12%, CL/SL = 28%, TD/SL = 17%, CP/SL = 12%, PV/SL = 45%; D_1 = 12, D_2 = 19–20, C = 19–20/18, P = 17–20; AbdV = 58, CauV = 19. Cranial bones smooth, except for some granular ornament upon the lachrymojugal. All fin rays smooth, unsegmented for proximal one-third to half of their length. Scales ornamented with many closely spaced tubercles of regular size.

Holotype

BMNH P.3338, caudal region, from the Upper Permian (Gaudaloupian), Marl Slate, Durham and Northumberland, England.

Material

BMNH P.554, P.555/P.3335 (pt and cpt), P.3339, P.3339a, P.3340; HM G26.52, G26.53, G26.54, G50.66, G50.68, SM D.435, from the Upper Permian (Gaudaloupian), Marl Slate, Durham and Northumberland, England. BMNH 40372, 43429, P.753, P.754, P.754/P.3342b, P.3342/P43427, P.38586, P.43426: casts of specimens B and C of Schaumberg (1978): MGUH VP2345a–b, MGUH VP2344a–b all from the Upper Permian (Gaudaloupian), Kuperschiefer of Germany.

Remarks

Coelacanthus is the type genus of the family Coelacanthidae. Historically many species have been referred to the genus but most have now been transferred to other genera. The genus still has several named species the status of which remains doubtful (see sections 11.4 and 11.5).

C. arcuatus Hibbard 1933	? to *Rhabdoderma elegans*
C. elongatus Huxley 1866	to *Rhabdoderma elegans*
C. elegans Newberry 1856	to *Rhabdoderma elegans*
C. evolutus Beltan 1979	to *Whiteia woodwardi*
C. exiguus Eastman 1902	? to *Rhabdoderma* sp.
C. granulostriatus Moy-Thomas 1935a	to *Rhabdoderma tingleyense*
C. guttatus Woodward 1912	to ? *Sassenia* sp.
C. harlemensis Winkler 1874	to *Coccoderma suevicum*
C. huxleyi Traquair 1881	to *Rhabdoderma huxleyi*
C. kayseri Woodward 1898b	to *Diplocercides kayseri*
C. kohleri Münster 1842a	to *Undina penicillata*
C. lepturus Agassiz 1844	to *Rhabdoderma elegans*
C. minutus Willemoes-Suhm 1869	to *Undina cirinensis*
C. mucronatus Pruvost 1914	to *Rhabdoderma tingleyense*
C. newarki Bryant 1934	to *Diplurus newarki*
C. newelli Hibbard 1933	? to *Rhabdoderma elegans*
C. ornatus Newberry 1856	to *Rhabdoderma elegans*
C. phillipsi Agassiz 1844	? to *Rhabdoderma tingleyense*
C. picenus Bassani 1896	to *Undina picenus*
C. robustus Newberry 1856	to *Rhabdoderma elegans*
C. striolaris Münster 1842a	to *Undina penicillata*
C. summiti Wellburn 1903	to *Rhabdoderma elegans*
C. watsoni Aldinger 1931	to *Rhabdoderma elegans*
C. sp. Aldinger 1931	to *Diplocercides* sp.

Genus *Diplocercides* Stensiö 1922a

1937 *Nesides* Stensiö: 44.

Diagnosis (emended)

A genus of primitive coelacanths known mostly by cranial material, suggesting small fishes with a head about 5 cm long and which is relatively shallow. Eye surrounded by many small sclerotic ossicles. Each half of the braincase ossified as a single unit. Basisphenoid region shows a prominent basipterygoid process articulating with a cup-shaped depression upon the metapterygoid. The parasphenoid is broad throughout and carries a dentition of small villiform teeth over most of its length. Skull roof is relatively flat with little transverse curvature. The postparietal shield shows small supratemporals and three extrascapulars, the lateral elements being broad and containing the junction between the lateral line, the supratemporal commissure and the otic canal. The postparietals are marked with prominent, long posterior pit-lines as well as middle and anterior pit lines. Descending processes upon the postparietals and the supratemporals are absent. Parietonasal shield shows very small supraorbitals and several internasals. The palate is deep over most of its length with the pterygoid rising gradually posteriorly without the prominent angle found in the palate of most coelacanths. Cheek bones completely covering the cheek and closely abutting one another, although overlap is minimal. The squamosal reaches dorsally to contact the skull roof. The preoperculum is the largest element in the cheek. The lower jaw is long and low throughout with the angular only weakly arched dorsally. Articular and retroarticular fused. The dentary is without a hook-shaped process and is entirely ornamented upon the lateral surface and there is no dentary pore.

Anterior coronoid series of simple tooth plates; principal coronoid relatively small. Teeth fused to the dentary. Dentition upon the coronoids, prearticular and pterygoids is formed of tiny conical teeth which are aligned in definite rows along the ventral edge of the prearticular and the dorsal edge of the pterygoid. Ornamentation upon the head consists of tubercles and ridges which tend to be aligned antero–posteriorly. Shoulder girdle is broad compared with that in most coelacanths and the cleithrum, clavicle and extracleithrum are ornamented with prominent ridges running the length of the bones. Sensory canals open to the surface through many tiny pores; the oral pit line is long and runs from the angular onto the dentary to end on the splenial. The post-

cranial skeleton is poorly known: in the caudal fin the fin rays outnumber the radial supports by approximately two to one. The scales are longer than deep and ornamented with longitudinal ridges.

Type species: *D. kayseri* (v. Koenen 1895)

Diplocercides kayseri
Figs 3.4, 4.5, 5.2, 6.2, 7.1(C,D), 11.5.

1895 *Holoptychius kayseri* v. Koenen: 28, pl. 2, fig. 2.

1898b *Coelacanthus* ? *kayseri* (v. Koenen); Woodward: 529, fig. 1.

1922a *Diplocercides kayseri* (v. Koenen); Stensiö: 169, figs 1–8; pl. 3, figs 1–4, pls 4, 5.

1922b *Diplocercides kayseri* (v. Koenen); Stensiö: 1259, figs 5, 6.

1932 *Diplocercides kayseri* (v. Koenen); Stensiö: 17, figs 8, 11, 14.

1936 *Diplocercides kayseri* (v. Koenen); Holmgren and Stensiö: fig. 266.

1937 *Diplocercides kayseri* (v. Koenen); Stensiö: 36, figs 1, 4, 7, 11, 16–19; pls 1–6; pl. 7, figs 1, 3; pl. 8, fig. 2; pl. 10, fig. 1; pl. 11, figs 1, 2.

1937 *Nesides schmidti* Stensiö: 44, figs 2, 3, 5, 8, 22, 23; pl. 9; pl. 10, figs 2–4.

1942 *Diplocercides kayseri* (v. Koenen); Jarvik: 554, pl. 16, figs 2, 3.

1942 *Nesides schmidti* Stensiö; Jarvik: 554, figs 75b, 76, 77a, 78, pl. 16, fig. 4.

1947 *Nesides schmidti* Stensiö; Stensiö: 85, fig. 23A.

1947 *Diplocercides kayseri* (v. Koenen); Stensiö: 85, fig. 23B.

1967 *Nesides schmidti* Stensiö; Bjerring: 225, figs 4–6.

1971 *Nesides schmidti* Stensiö; Bjerring: fig. 5a.

1972 *Nesides schmidti* Stensiö; Bjerring: 58, figs 1–5, 6a.

1973 *Nesides schmidti* Stensiö; Bjerring: figs 1b, 4c, 9.

1977 *Nesides schmidti* Stensiö; Bjerring: figs 23–25, 28.

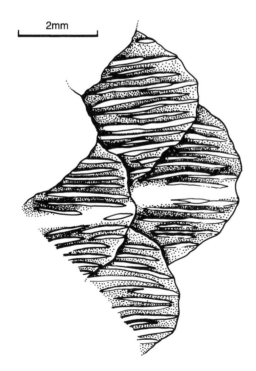

2mm

Fig. 11.5 *Diplocercides kayseri* (von Koenen). Camera lucida drawing of cast of scales from beneath the first dorsal fin as preserved in specimen 'a' of Stensiö (1937).

1978   *Nesides schmidti* Stensiö; Bjerring: fig. 2c.
1980   *Nesides schmidti* Stensiö; Jarvik: figs 206b, 208a,b, 212, 215, 216, 217a, 218, 221.
1980   *Diplocercides kayseri* (v. Koenen); Jarvik: fig. 224c,d.
1985   *Nesides schmidti* Stensiö; Bjerring: 234, fig. 5a.
1985   *Nesides schmidti* Stensiö; Bjerring: figs 3, 5, 6b.
1986   *Nesides schmidti* Stensiö; Bjerring: fig. 5.
1991a   *Nesides schmidti* Stensiö; Cloutier: fig. 2.

Diagnosis (emended)
Diplocercides in which the lachrymojugal is expanded posteroventrally. Suboperculum substantially smaller than preoperculum/quadratojugal. Many tiny teeth fused to the dentary in several rows. Ornament pattern consists of small rounded tubercles juxtaposed to each other upon the skull roof and cheek plates, tending to ridges upon the operculum. Lower jaw and gular plates ornamented with long ridges. Scales ornamented with closely-spaced longitudinal ridges, some of which run the entire length of the exposed portion.

Holotype
Gö 470–1: slightly disarticulated head from Frasnian of Gerolstein, Germany.

Material
Holotype. Specimens a, c, d (belonging to UB) of Stensiö (1937) and specimen b (belonging to UP) from the Upper Devonian (Frasnian) of Ense, near Wildungen, Germany. Cast (of Gö 470-1) holotype of *Nesides schmidti* from the Frasnian of Braunau, Wildungen.

Remarks
The genus *Nesides* was erected by Stensiö (1937) with the type species *N. schmidti* for a single specimen originally described as *Diplo-*

cercides kayseri. The specimen was serially ground and forms the basis for our knowledge of the primitive coelacanth braincase. In nearly all respects it is closely similar to *Diplocercides kayseri*. The stated differences in the shapes of the angular and the coronoid, the ventral margin of the operculum do not stand critical examination. The dentary teeth are more rounded and the ornament of the scales is intermediate between that of *D. kayseri* and *D. jaekeli*. There does not appear to be sufficient evidence to retain this as a separate genus and species but *Nesides* is a name used in the literature to refer specifically to the wax model remaining after the serial grinding of the holotype.

D. heiligenstockiensis (Jessen 1966)

1966   *Nesides ? heiligenstockiensis* Jessen: 376, fig. 15a–f, i, l; pl. 21, figs 2–4, pl. 22.
1973   *Nesides ? heiligenstockiensis* Jessen; Jessen: 170, figs 2b, 3, pls 24, 25.
1991a   *Nesides heiligenstockiensis* Jessen; Cloutier: fig. 3.

Diagnosis (emended)
Diplocercides reaching an estimated 13 cm SL. D_2 13, A 14. The lachrymojugal is narrow and tube-like throughout. Suboperculum equal in size to the preoperculum/quadratojugal. About 15 teeth fused to the dentary in a single row. Ornament pattern consists of small rounded tubercles juxtaposed to each other upon the skull roof and cheek plates. Operculum bears large tubercles separated from each other. Lower jaw and gular plates ornamented with long ridges. Anteriormost fin rays of at least the lower lobe of the tail bear fine denticles. Scales ornamented with closely spaced longitudinal ridges some of which run the entire length of the exposed portion.

Holotype
NRM P.53777, a disarticulated head and scales of the ventral side of the body from

the Upper Devonian (Frasnian) of Bergisch-Gladbach, West Germany.

Other known material
NRM P7775 and P4878, from the Upper Devonian (Frasnian) of Bergisch-Gladbach, Germany. Only the holotype was examined in this study.

D. jaekeli Stensiö 1922

1922a *Diplocercides jaekeli* Stensiö: 195, pl. 3, fig. 5.
1937   *Diplocercides jaekeli* Stensiö; Stensiö: 40, figs 20, 21; pl. 7, fig. 2; pl. 8, fig. 1.

Diagnosis (emended)
Diplocercides in which the lachrymojugal is expanded posteroventrally. Ornament upon the postorbital, supraorbitals, operculum and lachrymojugal consists of short ridges. The scales are ornamented with sparse, small longitudinal ridges and a group of tiny tubercles near the centre.

Holotype
UB, a disarticulated head and anterior scales from the Upper Devonian (Frasnian) of Ense, Wildungen, Germany. Only the holotype is known.

D. davisi (Moy-Thomas 1937)

1883 *Coelacanthus* sp. Davis: 524, pl. 43, figs 7, 9, 11, 12, (not figs 8, 10).
1903 *Coelacanthus abdenensis* Traquair: name only.
1937 *Rhabdoderma davisi* Moy-Thomas: 410, pl. 4, fig. f.
1937 *Rhabdoderma abdenense* Moy-Thomas: 410, pl. 4, figs a–e.
1981 *Diplocercides davisi* (Moy-Thomas); Forey: 217, fig. 218.

Diagnosis
A poorly defined species based on isolated bones which bear an ornament of sharp-crested ridges.

Syntypes
BMNH P.3346 and P.3350, an operculum and angular: from the Lower Carboniferous (Viséan P_1) Carboniferous Limestone, County Armargh, Ireland.

Material
Syntypes plus BMNH P.3348, P.3351, P.4194a–d, Lower Carboniferous (Viséan P_1), Carboniferous Limestone, County Armargh, Ireland. RSM 1905.10.22–34 (isolated head bones forming the syntype series of *Coelacanthus abdenensis*) and BMNH P.11726, all from the Lower Carboniferous, Viséan P_1, Abden Bone Bed, Calciferous Sandstone Series, Fife, Scotland.

Diplocercides sp.

1974 *Diplocercides* sp. Janvier: 25, fig. 6a; pl. 2, fig. 1; pl. 4, figs 1, 2.
1977 *Diplocercides* Janvier: 285, fig. 5.
1979 *Diplocercides* sp. Janvier and Martin: 498, fig. 1a, b; pl. 1.

Material known
Museum of the Geological Survey of Iran M.G.S.I. 16 and Iran National Museum of Natural History M.M.T.T. DF0012; two specimens of the posterior end of the lower jaw, both from the Upper Devonian (Frasnian) of Central Iran. Not examined in this study.

Genus *Diplurus* Newberry 1878

[*Osteopleurus* Schaeffer, 1941]

Diagnosis (emended)
A genus of slender coelacanths in which the shallow head occupies less than one quarter of the SL. The postparietal shield is short and broad with parallel lateral margins and is less than half the length of the parietonasal shield. The postparietals and the supratemporals end at the same transverse level and there are seven small free extrascapulars, barely larger than the sensory canal they

carry. The parietonasal shield is also broad and composed chiefly of three pairs of large bones – two parietals and one nasal – preceded by markedly smaller nasals and rostrals. The posterior parietals are distinctly larger than the anterior parietals. The premaxillae are represented as narrow splints bearing very small teeth. The supraorbito-tectal series contains approximately nine bones of which one is particularly large and lies in front of the orbit to contact the lachrymojugal. There is no preorbital. The lateral rostral is very shallow and the anteroventral process is poorly developed. The parasphenoid is broad throughout and bears teeth over most of its oral surface. The basisphenoid is short with very divergent antotic processes, a shallow dorsum sella and weakly developed sphenoid condyles which are placed very close together. The otic portion of the braincase is poorly known but appears to be composed of well-separated elements as in *Whiteia* and more derived coelacanths. In keeping with the shallow head the palatoquadrate is shallow with the pterygoid nearly twice as long as deep. The quadrate condyle is placed relatively far forwards and the jaw articulation lies beneath the centre of the orbit. The eye is large, resulting in the cheek bones, particularly the postorbital, being very narrow. The lachrymojugal shows a deep excavation anteriorly and this probably contained the posterior openings of the rostral organ. There are no sclerotic ossicles. The lower jaw is of distinctive shape with a parallel-sided angular and a dentary and splenial which are angled sharply downwards anteriorly. The teeth upon the dentary are extremely small. The majority of the skull bones are without ornament. A few ridge-like denticles occur on the cheek bones and the operculum. The sensory canals of the head generally open through large pores. Those of the supraorbital canal lie between successive supraorbitals. There are one or two pores piercing the postparietal, indicating the presence of a medial branch of the otic sensory canal. The post-

orbital, squamosal and preoperculum each have one or two large pores. The cleithrum and clavicle are inturned to form a prominent postbranchial lamina, the extracleithrum is small. The first dorsal fin support is a broad triangular plate and the other supports are of typical coelacanth form. The anterior rays of the first dorsal fin and those of both the upper and lower lobes of the caudal fin bear denticles. The supplementary caudal fin is long, equal to about the length of the principal lobes. There are long ossified ribs and these extend nearly to the mid-ventral line. The scales are ornamented with prominent ridges which are aligned longitudinally. The ornament varies between species but also between scales from different parts of the body; those scales lying beneath and between the two dorsal fins are the most densely ornamented.

Type species: *D. longicaudatus* Newberry

1857 *Rhabdiolepis elegans* Emmons: 191 *nomen nudum.*
1878 *Diplurus longicaudatus* Newberry: 127.
1889 *Diplurus longicaudatus* Newberry; Newberry: 70.
1932 *Diplurus longicaudatus* Newberry; Stensiö: 40.
1948 *Diplurus longicaudatus* Newberry; Schaeffer: 4, figs 1–4.
1952a *Diplurus longicaudatus* Newberry; Schaeffer: 53, pl. 15, pl. 16, fig. 2.
1954 *Diplurus longicaudatus* Newberry; Schaeffer: 53.
1959 *Rhabdiolepis longicaudatus* (Newberry); Bock: 17, fig. 3, pl. 1, fig. 1, pl. 2, figs 2–4, pl. 3, figs 1–4.

Diagnosis (emended)
Species reaching about 700 mm. Operculum entirely covered with ornament. $D_1 = 11$, $C = 16/16$, $P = 18$, $V = 20$, $A = 20$; a total of about 50 neural arches. Mid-flank scales show 25–30 slender ridges which are all of regular size.

Holotype
AMNH 630, a complete fish from the Lower Jurassic (Sinemurian), New Jersey, USA.

Material
AMNH 1529, AMNH 1531–3, AMNH 1536–7, AMNH 4800, from various formations of the Lower Jurassic (Hettangian–Sinemurian) of Connecticut, New Jersey, Virginia, USA.

D. newarki (Bryant)

Figs 4.16(A), 7.6(E), 8.2(L), 11.6, 11.7(B).
1934   *Coelacanthus newarki* Bryant: 323.
1941   *Osteopleurus newarki* (Bryant); Schaeffer: 1; figs 1–4, 5f, 6E, 7E, 8E, 9E.
1943   *Osteopleurus milleri* Shainin: 272.
1943   *Osteopleurus milleri grantonensis* Shainin: 274.
1952a *Diplurus newarki* (Bryant); Schaeffer: 54; figs 1–12A–C; pls 5–14, 16, fig. 1.
1959   *Osteopleurus newarki* (Bryant); Bock: 24; pl. 1, figs 2–4; pl. 2, fig. 5; pl. 3, fig. 6.

Diagnosis (emended)
Species reaching 200 mm SL. Operculum with ornament confined to the dorsal half. Mid-flank scales show approximately 12 ridges of which the centrally placed ones are larger and more robust than the peripheral ridges (Fig. 11.7(B)). HL/SL = 25%, PD$_1$/SL = 38%, PD$_2$/SL = 63%, TL/SL = 12%, CL/SL = 28%, PV/SL = 50%; D$_1$ = 8–9, C = 14/13, P = c.13, V = 15–17; AbdV = 25–27, CauV = 15–16.

Holotype
PUGM 13695, partial skeleton, Upper Triassic (Carnian), Lockatong Formation, North Wales, Pennsylvania, USA.

Material
BMNH P.41631, P.62103–8, Princeton University 14943a, 14944, 14958b, YPM WS-687, all from Upper Triassic (Carnian), New Jersey, USA. The species is also known from contemporaneous localities in North Carolina and Pennsylvania (Olsen *et al.*, 1982).

Genus *Garnbergia* Martin and Wenz 1984

Diagnosis (emended)
Monotypic genus in which the head is about 10 cm long and shows a relatively long snout and proportionately small orbit, the former being about one-third and the latter about

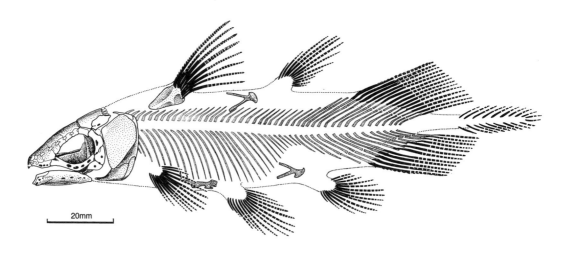

Fig. 11.6 *Diplurus newarki* (Bryant). Restoration of skeleton: slightly modified after Schaeffer (1952a).

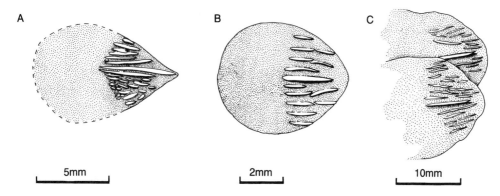

Fig. 11.7 Scales of some genera in which the scale ornament is characterized by long central ridges and short lateral tubercles (see also *Macropoma* – Fig. 11.13). (A) *Lualubaea henryi* Saint-Seine – from Saint-Seine (1955). (B) *Diplurus newarki* (Bryant) – after Schaeffer (1952a). (C) *Axelrodichthys araripensis* Maisey – FMNH PF11856.

one-fifth of the head length. The parietonasal shield has a marked concave lateral profile and the snout is rather shallow: postparietal shield short and broad, equal to about half the length of the parietonasal shield. The supraorbital series includes one elongate element at the anterior limit of the orbit and this has a broad contact with the deep lachrymojugal. Preorbital and sclerotic ossicles absent. Cheek narrow with a particularly narrow postorbital; spiracular and subopercular bones probably absent. Lower jaw relatively shallow throughout, especially the angular. Basal plate of D_1 is triangular and the fin rays are smooth. Scales ornamented with many fine longitudinal ridges (up to 40 per scale).

Type and only species: *G. ommata* Martin and Wenz, 1984

1984 *Garnbergia ommata* Martin and Wenz: 2, figs 1, 2.

Diagnosis
See Martin and Wenz (1984).

Holotype
SMNS 51035, a head and partial trunk showing D_1, from the Middle Triassic (Ladinian) of Baden-Wurttemberg, West Germany. Specimen not examined in this study.

Remarks
Garnbergia shows similarity to *Chinlea* in the shapes and proportions of the snout, orbit, jugal and in the shallow jaw. It is also apparently similar to *Coelacanthus lunzensis* Teller (1891), a species poorly known from remains of the trunk and the tail and it is possible that further, more complete material may show that they should be regarded as synonyms.

Genus *Hadronector* Lund and Lund 1984

Diagnosis (emended)
A monotypic genus of short stocky coelacanths reaching about 110 mm SL. The head is deep: within the parietonasal shield there are at least three internasals separating the nasals of either side; there is a preorbital and a premaxilla which bears many small teeth; the postparietal shield shows a straight posterior margin and is bordered by a small median and large lateral extrascapulars, the latter carrying the junction between the lateral line, otic canal and the supratemporal commissural sensory canal. The cheek and

opercular bones form a tightly interlocking and complete covering to the cheek: the postorbital is narrow, the squamosal is also narrow and reaches far dorsally behind the postorbital; there is a small spiracular; both the preoperculum and suboperculum are large and equidimensional while the operculum is narrow and deep. The lower jaw is shallow throughout with a narrow dentary and splenial. Ornament consists of closely packed ovoid tubercles upon the parietonasal shield which tend to closely packed ridges upon the postparietal shield, cheek, opercular and lower jaw bones; the ridges have a predominantly longitudinal orientation. In the postcranial skeleton the two dorsal fins are placed close to each other in the centre of the back and D_1 is relatively long-based. The contour of the tail is rather square-cut. The fin rays of all fins, but especially those of the D_2, P, V, A, are particularly fine with long lepidotrichia segments. All fin rays are unornamented. Scales are ornamented with a dense covering of longitudinal ridges, which tend to be of equal length.

Type and only species: H. donbairdi

Hadronector donbairdi
Figs 3.3(B), 4.7.
1984 *Hadronector donbairdi* Lund and Lund:
      240, fig. 2.
1985 *Hadronector donbairdi* Lund and Lund;
      Lund and Lund: 25, figs 35–45.
1991a *Hadronector donbairdi* Lund and Lund;
      Cloutier: fig. 4.

Diagnosis
Species reaching 110 mm SL; HL/SL = 24–28%, PD_1/SL = 36%, PD_2/SL = 52%, TL/SL = 19%, CL/SL = 36–40%, TD/SL = 40%, CP/SL = 25%; D_1 = 10, D_2 = 10, C = 21/23, P = 11, V = 10–12, A = 10; AbdV = 25 (estimated – the number of vertebrae beneath the operculum is not known), CauV = 22.

Holotype
UMON 3635, a complete fish in pt and cpt, from the Lower Carboniferous (Namurian A), Heath Formation, Bear Gulch Limestone Member, Montana, USA. Specimen not examined in this work.

Material
Latex casts of specimens CM 27308A, CM 30711A, CM 30712A, from the Lower Carboniferous (Namurian A), Heath Formation, Bear Gulch Limestone Member, Montana, USA.

Genus *Holophagus* Egerton 1861

Diagnosis
Genus of coelacanths very similar to *Undina* but differing primarily in details of the lower jaw, preoperculum, scale ornament and the fin rays. The body is relatively stout and the head exceeds 25% of the SL; the posterior profile of the tail is rather square-cut and the tail is proportionately small. In the head the skull roof shows a short postparietal shield, which is deeply embayed posteriorly, flared posterolaterally and strongly convex in the transverse plane. The parietonasal shield contains large, quadrangular supraorbitals; these and the postparietals and supratemporals are perforated with many small sensory pores similar to conditions in *Macropoma*. The braincase appears to be developed similar to that of *Macropoma* but remains incompletely known. The parasphenoid is markedly expanded anteriorly and teeth are confined to the anterior half. The pterygoid has a deep anterior wing and a narrow, upright posterior limb; the pterygoid dentition is arranged in parallel rows close to the dorsal margin. The lower jaw consists of a low angular, without a marked coronoid expansion and is constricted midway along its length. The dentary is relatively long (up to half the length of the jaw) and is splint-like anteriorly. The mandibular sensory canal opens through a regular series of very small pores

and the dentary pore is large; the oral pit line occurs beneath the principal coronoid; a subopperculum branch was developed. The principal coronoid is saddle-shaped, similar to that in *Undina*. The cheek is covered by a jugal, postorbital, squamosal and preoperculum; all these bones are large but very thin. The preoperculum is deep and triangular (unlike the narrow bone in *Undina*) and fails to overlap the angular. A preorbital is absent. The anterior four or five neural arches are much enlarged. There are short ribs present in the anterior half of the abdominal region. The basal of D_1 is large and triangular with a pronounced strengthening ridge. Each pelvic bone bears two anterior prongs and, in large specimens, the space between is filled with thin bone. The rays of all the fins are expanded and an ornament of many denticles is present upon the rays of D_1 and the caudal fin. All fin rays are very closely segmented and the sutures between the segments are often sigmoid in shape. The scales bear an ornament of very fine elongate ridges and tubercles which are closely spaced. A large ossified gas bladder is present.

Type species: *H. gulo* Egerton 1861

Holophagus gulo
Figs 3.18, 5.12(A,B), 6.9, 8.2(M), 11.8, 11.9(A).
1861 *Holophagus gulo* Egerton: 19.
1866 *Holophagus gulo* Egerton; Huxley: 26, pl. 6.
1868 *Holophagus gulo* Egerton; Egerton: 502.
1872 *Holophagus gulo* Egerton; Huxley: 36, pl. 10.
1891 *Undina gulo* (Egerton); Woodward: 411, fig. 53.
1941 *Undina gulo* (Egerton); Schaeffer: fig. 9f.
1960 *Holophagus gulo* Egerton; Gardiner: 327, figs 55–58, pl. 42.

Diagnosis (emended)
Species reaching 650 m SL, HL/SL = 26–30%, PD_1/SL = 42%, PD_2/SL = 60%, TL/SL = 10–12%, CL/SL = 25%, TD/SL = 30%, CP/SL = 15–19%, PV/SL = 50–53%; D_1 = 10–11, D_2 = 20, C = 18/17, P = 23 V = 18, A = 17; AbdV = 44, CauV = 21. Scales ornamented with many tiny short ridges (Fig. 11.9A), head ornamented with a few sparse tubercles

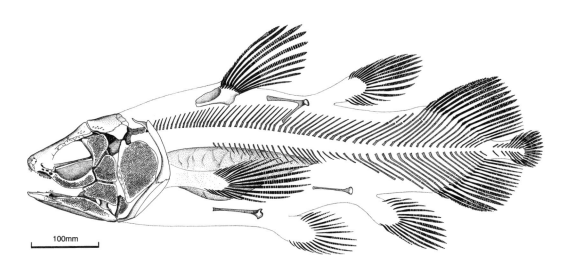

Fig. 11.8 *Holophagus gulo* Egerton. Restoration of fish. Based on BMNH P.7795, P.2022, P.3344.

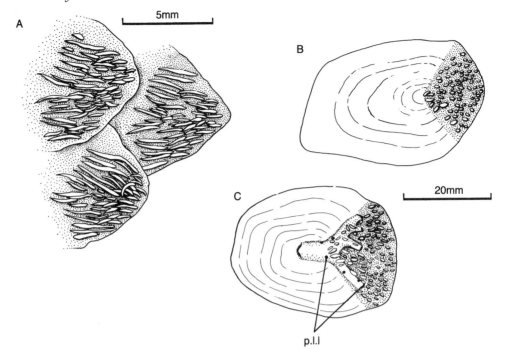

Fig. 11.9 (A) Scales of *Holophagus gulo* Egerton. Camera lucida drawing of scales immediately anterior to first dorsal fin, BMNH P.7795 (reversed). (B) *Latimeria chalumnae* Smith, scale from holotype of *Malania anjouauae* Smith BMNH P.34360. C. *Latimeria chalumnae* Smith, lateral line scale from the level of the pectoral fin, redrawn from Millot and Anthony (1965).

on the roof and many small, closely set tubercles on the cheek, jaw and the gular plates. Measurements could only be taken from three specimens.

Holotype
GSM 28832, a body minus head from the Lower Jurassic (Sinemurian), Dorset, England.

Material
BMNH P.875, P.2022, P.2022a, P.3344, P.7795, all from Lower Jurassic (Sinnemurian), Dorset, England.

Genus *Latimeria* Smith 1939a

Latimeria Smith 1939a: 455.
Latimeria Smith 1939c: 6.
Malania Smith 1953: 99.

Diagnosis (emended)
Large plump-bodied coelacanth growing to 1.8 m and up to 95 kg (adult females considerably larger than males); steel blue marked with white blotches, without countershading. Cranial roof broad posteriorly and strongly arched in transverse plane. Postparietal shield short, less than half the length of the parieto-nasal shield; snout covered with tiny ossicles mostly lying separate from one another, premaxillae represented as thin tooth-bearing splints, posterior parietals and postparietals with raised areas overlying ossification centres, nine extrascapulars reduced to a chain of small ossicles lying free within the posterior embayment of the posterior skull roof margin. Cheek covered by lachrymojugal, postorbital, squamosal, preoperculum and suboperculum – most of these bones lying separate from one

another. Preorbital and sclerotic ossicles absent. Posterior openings of the rostral organ lying anterodorsal to lachrymal. Neurocranium fragmented to lateral ethmoids and basisphenoid in the anterior moiety and prootic, basioccipital, opisthotic and supraoccipital in posterior moiety. Parasphenoid with anterior ascending processes and toothed area confined to anterior half. Dermopalatines and coronoid 4 with fangs. Cephalic sensory canals profusely branched and opening through many tiny pores in the skin, the subopercular branch is particularly well developed and the supratemporal commissure shows anterior branches. Rays of D_1 and principal caudal lobes stout and ornamented with many tiny denticles. Supplementary caudal lobe barely extending beyond principal lobes. Lateral line scales contain complex branched tubes which open via several pores within each scale (Fig. 11.9(C)). Air bladder without ossified walls, filled with oil; kidney lying along ventral body wall.

Type and only species: *Latimeria chalumnae* Smith

Latimeria chalumnae
Figs 2.1–2.11, 3.1, 3.2, 4.1–4.3, 5.1, 6.1, 7.1(A,B), 7.3–7.5, 7.6(A,B), 7.6(I), 8.1, 8.3(A,B), 8.4(A,B), 11.9(B,C).
1939a *Latimeria chalumnae* Smith: 455.
1939c *Latimeria chalumnae* Smith: 10, figs 1–19, pls 1–44.
1953 *Malania anjouanae* Smith: 99, fig. 1.

Diagnosis (emended)
Species reaching 1.7 m SL, HL/SL = 24–26%, PD_1/SL = 40%, PD_2/SL = 63–65%, TL/SL = 15%, CL/SL = 26–28%, TD/SL = 27%, CP/SL = 20–22%, PV/SL = 43–48%; D_1 = 8, D_2 = 29–31, C = 22–25/21–22, P = 30–32, V = 33, A = 29–32; AbdV = 68, CauV = 25–26 (the vertebral counts are taken as far as the SL to make comparison with counts in fossil species: the supplementary caudal lobe contains an extra 20–25 neural arches).

Holotype
Complete individual mounted as a dry specimen in the East London Natural History Museum, Republic of South Africa, caught at 72–100 m, off Chalumnae River, Indian Ocean. The holotype was not examined.

Material
Two unregistered specimens in BMNH, RSM 1969.76, two embryo specimens AMNH 32949SW, BMNH 1976:7:1:16, all from the Comores Islands, Indian Ocean. A full inventory to 1991 of known captures is given by Bruton and Coutividis (1991).

Remarks
Latimeria lives at 150–500 m (prefers 180–210 m), living below the 18°C thermocline on steep sand-free submarine slopes. Most are found around the Comores Islands but there have been at least three finds along the south east coast of RSA and the west coast of Madagascar (Chapter 2). There is a great deal of literature about *Latimeria* and this has been listed by Bruton *et al.* (1991).

Genus *Laugia* Stensiö 1932

Diagnosis (emended)
A genus of relatively small coelacanths in which the body is slender and the head is short. The skull roof is narrow throughout and the postparietal shield is as long as the parietonasal shield. The joint margins of the two shields are strongly interdigitate and closely matching, such that there is very little space left when the joint is in the abducted position. The parietonasal shield contains a single pair of parietals flanked by several very narrow supraorbitals. The snout is blunt and consolidated, perforated by large sensory canal pores. Each premaxilla bears five or six small conical teeth. The postparietal shield is formed largely by the postparietals: the supratemporals are relatively small and elongate, showing a long suture with the postparietals. The descending process of the

supratemporal is well developed but there is no postparietal descending process. There are five large extrascapulars, closely sutured with each other and with the postparietals and the supratemporals. The neurocranium is extensively ossified. The ethmosphenoid moiety shows large lateral ethmoids which must have covered much of the lateral surface as well as the floor of the nasal capsules. The basisphenoid is continued forward as an ossified interorbital septum. The antotic processes are narrow and divergent. The otico-occipital moiety is completely ossified but has distinct prootic, basioccipital and complex opisthotic/exoccipital ossification centres. There is no separate supraoccipital. There is a large vestibular fontanelle notched posteriorly for the glossopharyngeal nerve. The cheek is incompletely covered by three bones which lay free from one another. The postorbital is a narrow bone which is individually variable in shape. The jugal is a slender tube-like bone and the squamosal is large and triangular. There is a well-developed sclerotic ring of at least 30 ossicles. The operculum is relatively small, triangular with a slightly emarginated posterior margin. The lower jaw is long and shallow; the principal coronoid is rectangular and deeper than long; the quadrate/articular joint is located well posteriorly beneath the posterior level of the skull roof. The vertical limb of the palatoquadrate is broad while the anterior limb is long and shallow reflecting the proportions of the head. The basibranchial is particularly large and bears many paired tooth plates which decrease in size away from the midline. The sensory canals open over most of the head through many small pores arranged in linear fashion. Over some areas such as the supraorbital, otic and mandibular canals the pores are often grouped, which in life probably reflected underlying neuromast distribution. Ornament upon the head is confined to sparse tubercles on the postparietal shield and tubercles on the cheek plates and the dorsal half of the operculum. The first six or seven neural arches are short and broad and overlap one another distally forming a short consolidated structure. Cleithrum with expanded dorsal end. The endoskeleton of all fins is ossified and in all except the radials supporting the caudal fin there is both endo- and perichondral bone. The pattern of the endoskeleton is very similar to that seen in *Latimeria*. The pectoral fin endoskeleton has a very prominent proximal axial mesomere. The pelvic girdle and fin endoskeleton is also well ossified but shows considerable variation in position. Generally, in small individuals the pelvic girdle is located beneath the fin support of D_1. In larger individuals it is usually found further forward, sometimes contacting the pectoral girdle. The pectoral fin is small and the fin rays are very narrow. Some individuals show a marked expansion and lengthening of the inner pelvic rays and when this occurs these fin rays are always very closely articulated. Such expanded fin rays are confined to those individuals with anteriorly positioned girdles. None of the fin rays is ornamented. The caudal fin is asymmetrical with a longer dorsal lobe. The caudal outline tapers gradually. A long supplementary lobe extends some way beyond the principal lobes. Within the vertebral column the abdominal region shows short ribs which partly overlie an ossified air bladder. The scales are ornamented with flattened longitudinal ridges, usually up to 10–12 per scale. The lateral line scales bear a large unbranched tube carrying the lateral line which opens within each scale through a large terminal pore.

Type and only species: *L. groenlandica*

Laugia groenlandica

Figs 3.8, 3.9, 4.10, 4.12(A), 5.5, 5.6, 6.6, 6.7, 7.6(C), 8.2(J), 8.3(C,D), 8.4(C,d), 11.10.

1932 *Laugia groenlandica* Stensiö: 48, figs 18–27, pls 1–5, pl. 4, figs 1, 2, pls 7, 8, pl. 9, fig 2.

1976 *Laugia groenlandica* Stensiö; Bendix-Almgreen: 561, fig. 460M, 461.

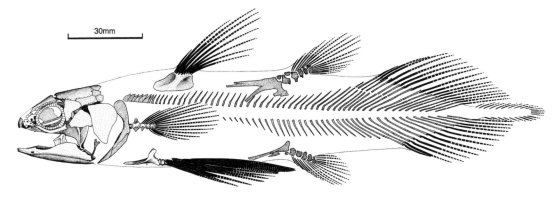

Fig. 11.10 *Laugia groenlandica* Stensiö. Restoration of entire fish. Based on many specimens in Mineralogical Museum, University of Copenhagen.

Diagnosis

Species reaching 180 mm SL; HL/SL = 20%, PD_1/SL = 30–35%, PD_2/SL = 50%, TL/SL = 26%, CL/SL = 30%, TD/SL = 20–22%, CP/SL = 15%; D_1 = 9, D_2 = 16, C = 17/13–14, P = 16, V = 15, A = 17; AbdV = 42, CauV = 18.

Holotype

MGUH VP3170a–b, a complete fish from the Lower Triassic (Scythian), Wordie Creek Formation, Cape Stosch, East Greenland.

Material

MGUH VP.1017, VP.1018, VP.2306a–b, VP.2307, VP.2308a–b, VP.2309, VP.2311, VP.2312a–b, VP.2313a–c, VP.2315a–b, VP.2316a–b, VP.2317, VP.2338–40, VP.2342, VP.3171a–b, VP.3173, VP.3174, VP3175a–b, VP.3253. Lower Triassic (Scythian), Wordie Creek Formation, Cape Stosch, East Greenland.

Laugia? sp. Stensiö

Remarks

Stensiö (1932: 73) described an un-named species of *Laugia* based on an isolated second dorsal fin support which lacked the posteroventral process. This appears to be an imperfectly preserved specimen and I consider there are no grounds for separating this from the type species.

Genus *Libys* Munster 1842a

Diagnosis

A genus of medium-sized, relatively deep-bodied coelacanths. The head is nearly as deep as long. The postparietal shield is less than half the length of the parietonasal shield and is much expanded posterolaterally where the supratemporals are particularly large. The palate is deep and tapers rapidly anteriorly and the symplectic is also long, in keeping with the rather deep head. The cheek is covered with large thin bones which are, however, well separated from one another. The postorbital and jugal are broad and the preoperculum is long and strap-like. The operculum is substantially deeper than broad. Sclerotic ossicles, suboperculum and a preorbital are absent. In the lower jaw the angular shows a prominent dorsal expansion and the principal coronoid is developed posterodorsally as a rounded finger-like process. The most obvious specialization of *Libys* are the sensory canals which open to the surface through very large pores. The teeth covering the palate and the lower jaw are very small; most are rounded and bear delicate radiating ridges. Ornamentation is absent from the skull bones. The shoulder girdle shows a very narrow cleithrum, clavicle and extracleithrum, whereas the anocleithrum is expanded. The pelvic fin is located well behind the level of D_1 and is supported by very narrow pelvic bones. The rays of D_1 and the caudal fin are ornamented with many prominent denticles. Fin rays of pectoral, ventral, D_2 and anal are expanded and very closely articulated close to their bases. The scales are covered with a sparse ornament of short ridges and the lateral line scales carry a large sensory tube which opens through several secondary tubules. An ossified air bladder is present.

Type and only species: *L. polypterus* Münster

Libys polypterus
Fig. 3.17.
1842a *Libys polypterus* Münster: 45.
1866 *Coelacanthus (Undina) kohleri* Huxley: 42 (errore).
1887 *Libys polypterus* Münster; Zittel: 174.
1887 *Libys superbus* Zittel: 175, fig. 179.
1888 *Libys polypterus* Münster; Reis: 37, 50, pl. 3, figs 1–11.
1888 *Libys superbus* Zittel; Reis: 41, 50, pl. 2, figs 1–4, pl. 4, fig. 17.
1892 *Libys polypterus* Münster; Reis: pl. 1, fig. 7.
1898a *Libys superbus* Zittel; Woodward: 414.
1992 *Libys superbus* Zittel; Lambers: 28, figs 1, 2, 3, 4a, 6, pls 1, 2.
1992 *Libys polypterus* Münster; Lambers: 37, fig. 7.

Diagnosis
Species reaching 600 mm SL. HL/SL = 25–28%, PD_1/SL = 40%, PD_2/SL = 55%, TL/SL = 20%, CL/SL = 35%, TD/SL = 35–40%, CP/SL = 25%, PV/SL = 45%; D_1 = 10, D_2 = 15–20, C = 21/19, P = 16, V = 19, A = 18–20; AbdV = 47, CauV = 23. Pectoral fin is relatively long, reaching back to posterior level of pelvic fin.

Holotype
BSM 1870.XIV.502, head only from the Upper Jurassic (Tithonian), Bavaria, Germany.

Material
BSM AS.I.801a,b, 1870.XIV.501a,b, 1870.XIV.503, 1870.XIV.504, 1870.XIV.507a,b, BMNH P.3337, JME SOS.2189, SOS.2202.

Remarks
Two species of *Libys* have been recognized: the type species, known only by the head and anterior part of the vertebral column, and *L. superbus*, known by several complete specimens. They are here placed in synonymy with the type species taking priority. Reis (1888) found difficulty in separating the two species. Woodward (1891) maintained the two species, distinguishing them on the relative width of the gular plates (narrow in the type species, broader in *L. superbus*). A review of further specimens than those used by Woodward suggests this distinction is not valid. Otherwise the head of *L. polypterus* is similar to that of *L. superbus* from which the body proportions are measured.

Genus *Lochmocercus* Lund and Lund 1984

Diagnosis (emended)
A monotypic genus which is very incompletely known. Relatively deep bodied with a small, rounded caudal fin. Complete development of the cheek and opercular bones, the cheek bones fitting closely together. Small preorbital present. The lower jaw shows a simple dentary without a hook-shaped process. The premaxillae and anterior coronoids bear large, pointed teeth. Sclerotic ring well developed. Basal supporting D_1 is kidney shaped with emarginated ventral margin matching positions of neural spines. All fin rays smooth. Caudal fins rays are rather short and outnumber the supporting radials by two-to-one.

Type and only species: *L. aciculodontus*

1984   *Lochmocercus aciculodontus* Lund and Lund: 243, fig. 5.
1985   *Lochmocercus aciculodontus* Lund and Lund; Lund and Lund: 45, figs 67–70.

Diagnosis
Reaching at least 90 mm SL, $D_1 = 11$, $C = 25/25$.

Holotype
UMON 6044, a poorly preserved complete fish from the Lower Carboniferous (Namurian A), Heath Formation, Bear Gulch Limestone Member, Montana, USA. Specimen not examined in this work.

Material
Casts of CM 27406, 30715, 35201, from the Lower Carboniferous (Namurian A), Heath Formation, Bear Gulch Limestone Member, Montana, U.A.

Genus *Macropoma* Agassiz 1835

Diagnosis (emended)
A genus of medium-sized coelacanths in which the body is relatively deep and well-rounded. The head is comparatively small. The postparietal shield is broadly triangular and strongly vaulted; it is wider than long and has a deep posterior embayment receiving seven extrascapulars. Postparietal and supratemporal descending processes well developed. The parietonasal shield is narrow, widening slightly in front of the eye. In the type species the premaxillae and rostrals are consolidated to give a solid hemispherical tip to the snout which is ornamented with prominent denticles. The snout of remaining species remains unknown. In lateral view the parietonasal shield has a concave profile. The pareito-nasal series has three nasals followed by two pairs of parietals, the posterior being substantially the larger. There are three tectals and seven to nine supraorbitals. The cheek is completely covered by a jugal, post-orbital, squamosal and preoperculum but there is no overlap between these bones. The jugal is expanded anteriorly and grooved for the passage of the two posterior ducts from the rostral organ. The anterodorsal corner of the postorbital is also deeply excavated and this probably received a tough connective tissue attachment with the skull roof. Spiracular, subopercular and preorbital bones are absent. In the lower jaw there are distinct articular and retroarticular ossifications: the principal coronoid is produced poster-odorsally as a rounded process. The parasphenoid is narrow over most of its length but expands anteriorly as a pair of ascending wings; the parasphenoid teeth are confined to the anterior half. The teeth lining the mouth are very small and villiform but a few upon the dermopalatines and at least one upon the premaxilla are larger and pointed. Most of the teeth show fine ridges which converge to the tips. The sensory canals open to the surface through many small pores; along the supraorbito-tectal series and in the lateral rostral these may be accompanied by larger pores. The supratemporal commissure sends long anterior branches into the post-parietals. The pectoral girdle is narrow throughout and the ventral end of the cleithrum is almost hemispherical in cross section, resulting in a deep concavity poster-iorly which probably received the scapulo-coracoid cartilage in life. The clavicle is relatively small and is distinctively angled ventrally. The extracleithrum is very narrow. The anocleithrum is large and flattened with a well-developed anterior process. Fin rays of D_1 and the anterior rays of the principal caudal lobes are ornamented with denticles. Details of the supplementary fin are unknown. The first five or six neural arches are short and very broad and closely applied to one another. The ossified air bladder is particularly large and reaches back to the level of the anal fin insertion. The lateral line scales show enlarged and branched sensory tubes with dorsal and ventral openings.

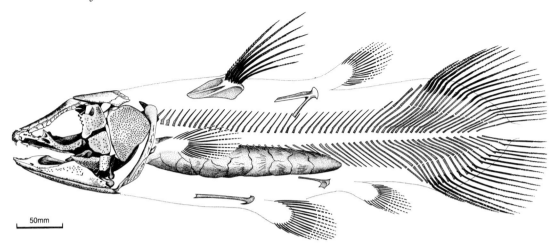

50mm

Fig. 11.11 *Macropoma lewesiensis* (Mantell). Restoration of fish: details of the supplementary fin are insufficiently known to allow restoration. Based on many specimens in The Natural History Museum, London.

Type species: *M. lewesiensis* (Mantell)

Macropoma mantelli

Figs 3.19(A), 3.20, 3.21(B), 4.18, 4.19, 6.10, 6.12, 7.2, 7.6(H), 7.7, 11.11, 11.12(A,B).

1822 *Amia ? lewesiensis* Mantell: 239, pls 37, 38.

1835 *Macropoma mantelli* Agassiz: 35.

1849 *Macropoma mantelli* Agassiz; Williamson: 462, pl. 42, figs 25, 26, pl. 43, figs 27–30.

1850 *Macropoma mantelli* Agassiz; Dixon: 368, pl. 34, fig. 2.

1866 *Macropoma mantelli* Agassiz; Huxley: 27, pls 7, 8.

1888 *Macropoma mantelli* Agassiz; Woodward: 284.

1891 *Macropoma mantelli* Agassiz; Woodward: 416, pl. 14, fig. 3.

1907 *Macropoma mantelli* Agassiz; Woodward: 136, pl. 8, figs 7, 8.

1909 *Macropoma mantelli* Agassiz; Woodward: 172, figs 49, 50, pl. 35, figs 9, 10, pls 36, 37, pl. 38, figs 1–5.

1921 *Macropoma mantelli* Agassiz; Watson: 321, figs 1–5.

1932 *Macropoma mantelli* Agassiz; Stensiö: 26, figs 12, 15.

1937 *Macropoma mantelli* Agassiz; Stensiö: fig. 15.

Diagnosis

Species reaching 600 mm SL. Head bones ornamented with prominent tubercles which are particularly large on the snout, $HL/SL = 25\%$, $PD_1/SL = 40\%$, $PD_2/SL = 60\%$, $TL/SL = 16\%$, $CP/SL = 16–18\%$, $PV/SL = 45\%$; $D_1 = 7$, $D_2 = 17$, $C = 18–20/15–18$, $P = 16$, $V = 18$, $A = 16$; $AbdV = 38$, $CauV = 22$. Scales show many irregularly-sized denticles (Fig. 11.12(A,B)); those scales in the posterior half of the body show one to three particularly large central denticles which are hollow.

Holotype

BMNH 4219, a complete fish from the Upper Cretaceous (Cenomanian) of Sussex, England.

Material

BMNH 115, 4213, 4217, 4221, 4223–4228, 4232–4234, 4236–4238, 4241–4248, 4251–4253, 4256, 4258, 4259–4262, 4264, 4269, 4270, 4289,

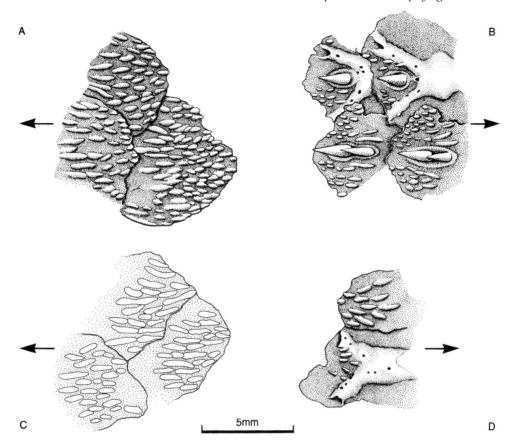

Fig. 11.12 *Macropoma* camera lucida drawings. *M. lewesiensis* (Mantell): (A) scale from the flank immediately beneath D_1 (BMNH 4269); (B) lateral line and neighbouring scales from the caudal peduncle (BMNH 49832). *M. precursor* Woodward: (C) scales from the flank immediately beneath D_1 (BMNH P.11003); (D) lateral line and neighbouring scale from the caudal peduncle (BMNH P.10810). Arrows indicate anterior end of scales.

4298, 4910, 6288, 9837, 25870, 25871, 25782, 25789, 25923a, 25944, 28388, 39070, 39086, 41642, 41669, 43851, 49094, 49096, 49110, 49832–49836, 49887, P.742, P.742a, P.2051, P.3352, P.4547, P.4548, P.4637, P.4638, P.5407, P.6454, P.6455, P.7654, P.10224, P.10755, P.10918, P.10919, P.11003, P.11272, P.12886, P.12887, P.33540, P.33373, P.42027, P.42028, P.63999 – P.64002. SM B.9118–B.9124, GN.250, GN.252–GN.254, GN.256, all from the Upper Cretaceous (Albian–Turonian), England.

Remarks

Fragmentary specimens of scales and head bones from the Cenomanian of Portugal were referred to *Macropoma* and likened to this and following species by Jonet (1981). I have not seen the material described but the illustrations given by Jonet (1981) do not suggest identity with *Macropoma*. The scales appear to be covered with many equally-sized denticles.

Macropoma precursor Woodward

Macropoma precursor
Figs 3.19(B), 6.10, 6.13, 11.12(C,D).
1909 *Macropoma precursor* Woodward: 181,
    pl. 38, figs 8–10.

Diagnosis
Species reaching 450 mm SL. Cranial bones smooth, without denticles. Head shorter and relatively deeper than in type species. Scales covered with denticles of uniform size.

Holotype
BMNH 35700, head from the Upper Cretaceous (Cenomanian) of Kent, England.

Material
BMNH 47239, 47240, 49828, P.3353, P.6453, P.10810, P.10816, P.10917, P.11002, P.12884, P.12885, P.16979, all from the Cretaceous (Albian–Cenomanian), SE England.

Macropoma speciosum Reuss

1857 *Macropoma speciosum* Reuss: 33, pls 1,
    2.
1986 *Macropoma speciosum* Reuss; Tima: pls
    1–4.

Diagnosis
Species approximately 450 mm SL distinguished from the type species in having eight rays in D_1, up to 24 rays in the principal lobes of the caudal fin and scales bearing two or three large denticles which are flanked by denticles relatively larger than those in the type species.

Holotype
NMP Oc.114, a complete specimen from the Upper Cretaceous (Turonian), The Czech Republic. Specimen was not examined.

Material
BMNH P.9007, from the Upper Cretaceous (Turonian), The Czech Republic. Many more specimens of this species have been described by Tima (1986).

Genus *Mawsonia* Woodward

1907 *Mawsonia* Woodward: 134.

Diagnosis (emended)
A genus known mostly from cranial remains. Some species must have exceeded 2 m SL. The head is large and deep with most of the dermal bones being very thick, ornamented with coarse rugosities and with prominent ridges upon the operculum, angular and gular. The postparietal shield is short and broad and is parallel-sided in its posterior half. There are two extrascapulars firmly sutured into the postparietal shield and resembling an additional pair of postparietals. A median extrascapular is absent. The parietonasal shield is narrow and generally more than twice as long as the postparietal shield. There are two pairs of elongate parietals and the snout is composed of a mosaic of star-shaped ossicles. The lateral rostral is particularly slender anteriorly, extending well in front of the eye and angled dorsally at the anterior tip. A preorbital and sclerotic ossicles are absent. The tectal and supraorbital series is as wide as the parietonasal series and consists of relatively few elements (usually six). A descending process upon the supratemporal is absent, a condition which in *Mawsonia* is regarded as secondarily derived within coelacanths. Within the braincase the lateral ethmoid has a very pronounced posterodorsal process which contacts the undersurface of the skull. The basisphenoid is stout with prominent antotic processes which are parallel-sided and in this respect resembles the basisphenoid described as *Moenkopia*. The palatoquadrate shows a very shallow anterior limb of the pterygoid, with a small autopalatine. The cheek is covered by a large postorbital bearing a prominent anterior process, a shallow and elongate lachrymojugal and quadrangular squamosal. The coronoid is a

prominent element markedly saddle-shaped in lateral view with the posterior limb considerably the larger. The teeth covering the palatoquadrate and the coronoids and the prearticular are each marked with very fine striations which radiate from the apex. The sensory canals generally open to the surface through many small pores, those upon the roofing bones and the cheek are often obscured by the coarse ornamentation. The exceptions are the pores in the mandible which are large. The infraorbital canal runs through the centre of the postorbital. The basal plate of D_1 is a large triangular plate with a very thickened articular face. All fin rays are slender and are very closely segemented; ornament appears to be absent from the fin rays.

Type species: *M. gigas* Woodward

Mawsonia gigas
Fig. 5.11(A,B).
1907   *Mawsonia gigas* Woodward: 134, pl. 7, pl. 8, figs 1–6.
1908   *Mawsonia minor* Woodward: 358, pl. 42, figs 1–3.
1982   *Mawsonia gigas* Woodward; Carvalho: 522, pls 2–8.

Diagnosis
Large species in which the angular measures up to 400 mm. Ornament of coarse rugae which are arranged in strong longitudinal ridges on the parietals, postparietals, and angular. Operculum ornamented with ridges which radiate from centre of ossification. D_1 with at least eight rays. Caudal with about 20 rays in upper and lower principal lobes.

Holotype
BMNH P.10355, partial head from the Lower Cretaceous (?Aptian), Bahia, Brazil.

Paratypes
BMNH P.10356, P.10357 from the Lower Cretaceous (?Aptian), Bahia, Brazil.

Material
BMNH P.9621, P.10060, P.10355, P.10358, P.10359, P.10362, P.10363, P.10365, P.10366, P.10367, P.10567 (holotype of *M. minor*), P.10569, P.10605, P.19148, P.33370, P.33372, P.33375, all from the Lower Cretaceous (Neocomian or Aptian), Bahia, Brazil.

Mawsonia cf. gigas

1986b   *Mawsonia* cf. *gigas* Woodward; Maisey: 3, figs 1–11.
1991   *Mawsonia* cf. *gigas* Woodward; Maisey: 317, un-numbered figures on pp. 316, 318, 320–1.

Remarks
Maisey (1986b, 1991) described a large coelacanth from the Albo-Aptian Santana Formation, Ceara, Brazil, which he compared with *M. gigas*. This form is equally as large as the type species and shows the longitudinal ridged ornament upon the skull roof bones. I agree with Maisey in comparing this form to *M. gigas*. Campos and Wenz (1982) mention two forms from Ceara under the names *Mawsonia* sp. and 'forme B'. The first is said to be large, comparable in size with *M. gigas* but to differ in having a narrower snout. Unfortunately the snout is unknown in type material of *M. gigas* and true comparisons cannot be made. 'Forme B' is described for a small body which shows denticulations on the D_1 fin rays, a condition similar to *Axelrodichthys araripensis*, while at least a small specimen of *M. gigas* shows no such denticles. Because I have not seen the material described by Campos and Wenz I cannot comment further.

Mawsonia lavocati Tabaste

Mawsonia lavocati
Fig. 5.11(D).
1963   *Mawsonia lavocati* Tabaste: 466, pl. 13, figs 4–6.
1981   *Mawsonia lavocati* Tabaste; Wenz: 3, figs 1–4, pls 1, 2.

Diagnosis
See Tabaste (1963: 466).

Holotype
MNHN MRS.78 posterior end of left angular from the Albian of Morocco.

Remarks
This species is as large as the type species but said to be distinguished by the form of the ornament which is more vermiform. Wenz described further fragmentary cranial material from the Albian of Algeria.

Mawsonia libyca Weiler

Mawsonia libyca
Fig. 5.11(E).
1935 *Mawsonia libyca* Weiler: 11, fig. 1, pl. 1, figs 5–10, 12, 17–29, 31–34, 42–46, 50–52, pl. 2, figs 4, 9, 27, 35, 36, pl. 3, figs 1–6, 11, 13, 18.

Diagnosis
See Weiler (1935: 11).

Remarks
Species similar in size to the type species. The form of the ornament on the postparietal shield and the angular is reminiscent of that in the type species and it is possible that *M. libyca* should be regarded as a synonym. Unfortunately all type material has been destroyed (Wenz 1981). *M. libyca* comes from Baharija, Egypt and was originally dated as Cenomanian. However, Tabaste (1973) suggests this should be considered as Albian.

Mawsonia tegamensis Wenz

Mawsonia tegamensis
Fig. 5.11(C).
1975 *Mawsonia tegamensis* Wenz: 177, figs 1, 2, 6, pls 1–5.

Diagnosis
See Wenz (1975: 188).

Holotype
MNHN GDF.401, a complete head from the Lower Cretaceous (Aptian) of Niger.

Paratypes
MNHN   GDF.402–405,   GDF.407–410, GDF.412, GDF.417–422, cranial fragments plus pelvic bone, all from the Lower Cretaceous (Aptian) of Niger.

Remarks
This is a small species (angular about 50 mm in length) with a relatively wide parietonasal shield and uniform ornament, rarely developed as long ridges.

Mawsonia ubangiensis Casier

1961 *Mawsonia ubangiana* Casier: 23, figs 4b, 5b, 6, 8b, 9b, pls 2, 3, figs 1, 2.
1969 *Mawsonia ubangiensis* Casier: 16, pl. 2, fig. 2.

Holotype
MRAC RG13.604, a right half of a postparietal shield from the Lower Cretaceous (Neocomian) of Zaire. Specimen not examined in this work.

Material
Framents of an angular and quadrate from the type locality have been referred to this species (Casier, 1969).

Remarks
In my opinion the original descriptions do not clearly allow separation of this species from *M. gigas*, but since I have not examined original material it is retained here.

Mawsonia sp. Wenz 1981

Remarks
Wenz (1981: 10, pl. 2, figs c–e) described a single basisphenoid which is said to differ from the basisphenoid of other species of *Mawsonia* in its more slender form. It may

Fig. 11.13 *Miguashaia bureaui* Schultze. Restoration of skeleton, from Cloutier (1996) with permission.

represent a new species but more comparative material of other species is needed before this suggestion can be confirmed.

Genus *Miguashaia* Schultze 1973

Diagnosis (emended)

Plump-bodied coelacanth with heterocercal tail and second dorsal and anal fins placed well posteriorly. Postparietal shield constituting at least half of total skull roof length and flanked by separate intertemporal and supratemporal bones, separated from parieto-nasal shield by a narrow joint margin. Supraorbital and otic sensory canal running through centres of canal-bearing bones. Middle and posterior pit lines placed within posterior half of postparietal. Three extrascapulars present (a large median plus paired laterals). Orbit small and postorbital portion of the cheek very broad with a large squamosal which expands to reach the skull roof and a large preoperculum: suboperculum large and lying directly beneath the operculum. In the lower jaw, both dentary and splenial are very shallow, the former carries about 11 teeth which are fused to the dentary. In postcranial skeleton, first dorsal is broad-based and second dorsal does not show the typical lobe present in other coelacanths. Anal fin has small lobe in large specimens. Caudal fin highly asymmetrical with small epichordal and large hypochordal lobes. Number of caudal fin rays greatly exceeds the number of radials. Many fin rays in the second dorsal, anal and caudal fins are branched. Ornament on the exposed surface of the scales is very variable: usually developed as small tubercles passing into ridges or to an almost continuous enamel covering.

Type and only species: *M. bureaui* Schultze

Miguashaia bureaui

Figs 3.3(A), 4.4, 8.2(G), 11.13.

1973 *Miguashaia bureaui* Schultze: 189, figs 1–3, pl. 31, fig. 2, pl. 32, fig. 2, pl. 33, fig. 7.

1991a *Miguashaia bureaui* Schultze; Cloutier: 379.

1991b *Miguashaia bureaui* Schultze; Cloutier: 25.

1996 *Miguashaia bureaui* Schultze; Cloutier: figs 1–8, fig. 9(A), 10(A), 11(A,B), figs 12–17.

Diagnosis (emended)

Miguashaia reaching an estimated 450 mm SL, HL/SL = 25%, PD_1/SL = 32–36%, PD_2/SL = 65%, $D_1 = 18$, $D_2 = 19$–27, caudal fin heterocercal with many very short epichordal rays and 31–37 hypochordal rays.

Holotype

ULQ Escuminac 120 a–b, entire juvenile specimen in pt and cpt, from Upper Devonian (Frasnian), Escuminac Formation, Migausha, Quebec, Canada.

Material

BMNH P.58691a,b, BMNH P.58692a,b, PMNH P.58693a,b, BMNH P.62794, MHNM 06-41, MHNM 06-264, MHNM 06-274, MHNM 06-494, MHNM 06-641, all from Upper Devonian (Frasnian), Escuminac Formation, Migausha, Quebec, Canada.

Genus *Polyosteorhynchus* Lund and Lund 1984

Diagnosis (emended)

Monotypic genus of small coelacanths (<180 mm SL) with a relatively deep body equal to one-third SL. The head is nearly as deep as long. The postparietal shield, which is two-thirds the length of the parietonasal shield, is flat with a straight transverse posterior margin. There are three or five large scale-like extrascapulars. Within the parietal shield there is a large preorbital and the premaxilla bears three or four large conical teeth. The cheek is completely covered by large, overlapping bones: the postorbital is rather narrow while the preoperculum is the largest and is trapezoidal in shape. The operculum is narrow and deep and is accompanied by a suboperculum ventrally and spiracular dorsally. There is a well-developed sclerotic ring. The lower jaw is shallow and the dentary is without a hooked-shaped process. Ornamentation upon the skull roof and cheek bones consists of closely packed ovoid tubercles while on the angular it tends to

ridges. The caudal fin is relatively short and rounded. An ossified swim bladder is present.

Type and only species: *P. simplex* Lund and Lund

1984　*Polyosteorhynchus simplex* Lund and Lund: 241, fig. 3.
1985　*Polyosteorhynchus simplex* Lund and Lund; Lund and Lund: 32, 46–56.

Diagnosis

See Lund and Lund (1985: 32).

Holotype

UMON 6043, a complete fish from the Lower Carboniferous (Namurian A), Heath Formation, Bear Gulch Limestone Member, Montana, USA. Specimen was not examined here.

Material

No material of this species was examined.

Genus *Rhabdoderma* Reis 1888

(*Coelacanthus* in part)

Diagnosis (emended)

A genus of small to medium-sized coelacanths. The type species is the only species known in any detail. Body slender throughout and the head is small with a convex dorsal profile. The parietonasal and postparietal shields are of approximately equal length. At least two pairs of nasals and two pairs of parietals with the anterior pair slightly the smaller: supraorbital series consists of three or four elements of which one is usually elongated. The hind margin of the parietonasal shield is straight. There are five scale-like extrascapulars and the lateral extrascapular contains the junction between the lateral line and the supratemporal commissure. The braincase is extensively ossified: anterior moiety is composed of lateral ethmoids and a large basisphenoid which encloses the optic foramen: otico-occipital portion composed of a single ossification occupying the prootic, opisthotic and basioccipital regions. There is a descending process of the supratemporal but a postparietal descending process is absent. The parasphenoid is flat and bears tiny teeth over most of its length. The preorbital is large; it enters the orbital margin and is pierced by two well-spaced openings from the rostral organ. The premaxilla completely encloses the anterior opening of the rostral organ while the ethmoid commissure runs along its dorsal margin. Each premaxilla bears four or five prominent teeth which are considerably larger than those upon the dentary tooth plates. The cheek is completely covered by closely fitting bones and there is considerable overlap between the postorbital and the squamosal and between the squamosal and the preoperculum. There is a small spiracular dorsally and a suboperculum ventrally, the latter is extensively overlain by the preoperculum. The operculum has a broadly rounded posterior margin. The two limbs of the pterygoid meet in a narrow angle. The lower jaw contains a triangular coronoid and the gular plates extend behind the jaw rami. The sensory canals open onto the parieto-nasal shield through large pores but elsewhere upon the postparietal shield and the cheek and lower jaw there are many small pores. Additionally, there may be one or two pores upon the postparietal shield immediately in front of the pit line and above the jugal canal within the squamosal. The pit lines are very well marked on the parietal, squamosal, preoperculum and the angular. The posterior pit line upon the parietal is particularly long, nearly reaching the supratemporal. Ornament is variable between individuals of a species, and between species, but it generally consists of rounded tubercles or elongate ridges which often follow the margins of the cheek, opercular and gular bones. In the postcranial skeleton the basal

supporting the first dorsal is kidney-shaped with the ventral margin contoured to interdigitate with three or four neural spines. The pelvic girdle shows an anterior division strengthened by one to three ridges and a posteromedial process, denticulated along the contact margin with its antimere. Ossified ribs are absent. Ossified air bladder present. The caudal fin shows a prominent supplementary lobe extending well beyond the principal lobes of the tail. All fin rays are devoid of ornament. The scales are ornamented with ridges/tubercles which converge posteriorly. The lateral line scales are not clearly distinguished from the surrounding scales, each is marked only by a single pore leading to the lateral line canal.

Type species: *R. elegans* (Newberry)

Rhabdoderma elegans
Figs 3.3(D), 3.6, 4.8, 4.9, 5.3(C), 6.4, 6.5, 7.1(E), 8.2(I), 11.14, 11.15(B).
1844 *Coelacanthus lepturus* Agassiz: *nomen nudum.*
1844 *Holopygus binneyi* Agassiz: *nomen nudum.*
1856 *Coelacanthus elegans* Newberry: 98.
1856 *Coelacanthus robustus* Newberry: 98.
1856 *Coelacanthus ornatus* Newberry: 98.
1866 *Coelacanthus lepturus* Huxley: p. 16; pl. 2, figs 1–4; pl. 3, figs 1–3; pl. 4, figs 1–6.
1866 *Coelacanthus elongatus* Huxley: 23, pl. 5, figs 6, 7.
1866 *Coelacanthus elegans* Newberry; Huxley: 20; pl. 5, figs 1–4.
1872a *Coelacanthus lepturus* Agassiz; Hancock and Atthey: 256, pl. 17, fig. 4.
1872b *Coelacanthus lepturus* Agassiz; Hancock and Atthey: 416, pl. 15, fig. 4.
1873 *Conchiopsis filiferus* Cope: 342.
1873a *Coelacanthus robustus* Newberry; Newberry: 341, pl. 40, fig. 2.
1873a *Coelacanthus elegans* Newberry; Newberry: 339; pl. 40.
1873b *Coelacanthus elegans* Newberry: Newberry: 425, fig.1.

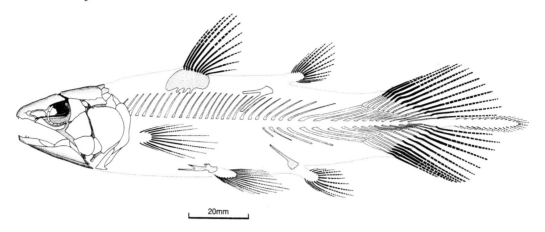

Fig. 11.14 *Rhabdoderma elegans* (Newberry). Restoration of entire fish. From Forey (1981).

1873 *Conchiopsis anguliferus* Cope: 342.

1876 *Coelacanthus lepturus* Agassiz; Davis: 339.

1888 *Rhabdoderma lepturus* Reis: 5.

1888 *Rhabdoderma elegans* Reis: 5.

1888 *Rhabdoderma robustum* Reis: 5.

1889 *Coelacanthus elegans* Newberry; Newberry: 213.

1890 *Coelacanthus lepturus* Agassiz; Ward: 168, pl. 5, figs 1, 3.

1891 *Coelacanthus elegans* Newberry; Woodward: 403, pl. 14, fig. 2.

1891 *Coelacanthus robustus* Newberry; Woodward: 406.

1891 *Coelacanthus elongatus* Huxley; Woodward: 406.

1895 *Coelacanthus elegans* Newberry; Dean: 64.

1898 *Coelacanthus elegans* Newberry; Wellburn: 426, pl. 41, fig. 12.

1900 *Coelacanthus robustus* Newberry; Hay: 115.

1901 *Coelacanthus elegans* Newberry; Wellburn: 167.

1902b *Coelacanthus* sp. Wellburn: 474, fig. 1.

1903 *Coelacanthus summitti* Wellburn: 71.

1908 *Coelacanthus elegans* Newberry; Hussakof: 54.

1908 *Coelacanthus ornatus* Newberry; Hussakof: 54, pl. 4, fig. 2.

1914 *Coelacanthus elegans* Newberry; Pruvost: 928.

1927 *Coelacanthus elegans* Newberry; Chabakov: 301, pl. 1.

1931 *Coelacanthus elegans* Newberry; Wehrli: 119, pl. 23, fig. 13.

1931 *Coelacanthus watsoni* Aldinger: 187, figs 1–6, pl. 6.

1935a *Coelacanthus elegans* Newberry; Moy-Thomas: 39.

1935a *Coelacanthus robustus* Newberry; Moy-Thomas: 40.

1936 *Coelacanthus elegans* Newberry; Wood: 486.

1937 *Rhabdoderma elegans* (Newberry); Moy-Thomas: 399, figs 1, 3–7, 9–12.

1981 *Rhabdoderma elegans* (Newberry); Forey: 214, figs 1–9, 10a,b.

1985 *Rhabdoderma elegans* (Newberry); Lund and Lund: 9, figs 2–12.

1985 *Rhabdoderma lepturus* (Agassiz); Lund and Lund: 15, fig. 13.

1991 *Rhabdoderma elegans* (Newberry); Forey: figs 3, 7c.

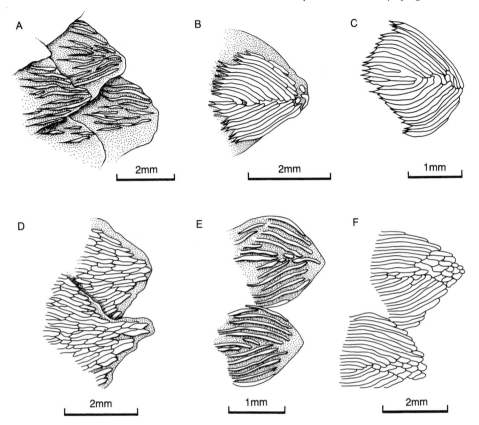

Fig. 11.15 Camera lucida drawings of scales of *Caridosuctor* and *Rhabdoderma*, all taken from close to the first dorsal fin. (A) *Caridosuctor populosum* Lund and Lund (FM PF.12920). (B) *Rhabdoderma elegans* (Newberry) (BMNH 36477). (C) *R. ardrossense* Moy-Thomas (BMNH P.19244). (D) *R. tinglyensis* (Davis) (BMNH P.1188). (E) *R. huxleyi* (Traquair) (BMNH P.4080a). (F) *R. madagascariensis* (Woodward) (BMNH P.10768).

Diagnosis

Reaching 400 mm SL but most specimens considerably less (usually < 250 mm SL), D_1 = 10, D_2 = 14 or 15, C = 12–13/12–13, A $\geqslant$ 13, P = 11, V = 14–16, AbdV = 30, CauV = 15; approximately 55 vertical scale rows anterior to supplementary caudal fin which extends posteriorly beyond the main caudal lobes. Ornament upon the skull roof consists of closely packed ovoid tubercles. Cheek and opercular bones and the angular show elongate tubercles surrounded by ridges. Gular plates show parallel ridges converging anteriorly. Scale ornament formed by adjoining long ridges which converge posteriorly (Fig. 11.15(B)), the more marginal sometimes embracing the centrally placed ridges. Hind edge of scale occasionally drawn out.

Syntypes

AMNH SO3, 2024, nearly complete fishes from the Upper Carboniferous (Westphalian D) of Linton, Ohio, USA. Specimen not examined in this work.

Material

BMNH 21464, 21952, 30572, 36477, 37956, 41197, 42382, 48055, P.748, P.751, P.1187a,

P.3330–33, P.5177, P.5379, P.6101, P.6286, P.6614, P.6663, P.6663a, P.6664–6671, P.7731, P.7765, P.7767, P.7905–14, P.7916–18, P.7920, P.7924, P.8433, P.8470, P.10473–4, P.11240–1, P.12924, P.24508–9, P.24809–10, P.30090, P.30091, P.30581, P.40393, P.42382, P.45638, P.45814–5, P.57928, P.57930, P.57931, P.57936, P.57938, P.57941, P.57945, P.57946, P.57957, P.57971, P.57981, P.58689, SMC E.168, E.169, E.170, E.171, E.172, E.4812, all from Upper Carboniferous (Namurian–Westphalian) localities in the British Isles. BMNH P.9607 Westphalian of Ireland. BMNH P.334, P.579, P.580, P.581, P.746, P.747, P.749, P.48452–3, P.57944, P.58690; AMNH 493(G), 494(G), 495(G), 503, 600, 900, 1154(G), 2024, 8562; CM 43902, 43919, 43920, 43924, 43927, 43932, 43938, 43941, 43949, 43951, 43953, 43955, 43958, 43959, 43965, 43966, 43978, 43990, 43991, 43994, 44004, 44006, 44033, 44034, 44059, 44087, 44094 44101, all from the Upper Carboniferous (Westphalian D) of Linton, Ohio, USA. Also known from northern France, Belgium, Holland, West Germany and the Stephanian of the Ukraine.

R. tinglyense (Davis 1884)

Rhabdoderma tingleyense
Fig. 11.15(D).
1884 *Coelacanthus tinglyensis* Davis: 427, pls 44–49.
1888 *Rhabdoderma tingleyense* (Davis); Reis: 72.
1891 *Coelacanthus tinglyensis* Davis; Woodward: 402.
1901 *Coelacanthus tinglyensis* Davis; Wellburn: 162.
1931 *Coelacanthus mucronatus* Pruvost; Wehrli, 119, pl. 23, fig. 14.
1935a *Coelacanthus corrugatus* Moy-Thomas: 43, pl. 5, figs 1–4.
1935a *Coelacanthus granulostriatus* Moy-Thomas: 44, pl. 6.
1936 *Coelacanthus mucronatus* Pruvost; Wood: 486.

Diagnosis (emended)
Reaching 600 mm SL, $D_1 = 15$, C = 12–13/12–13. Ornament upon the scales consists mainly of closely packed elongate tubercles which break up into shorter ovoid tubercles towards the hind edge of the scale, which is usually drawn out into a short 'tail'.

Syntypes
BMNH P.7733–48, Upper Carboniferous (Westphalian) of Yorkshire, British Isles.

Material
BMNH 21423, P.1187, P.1188, P.3412a, P.7766, P.15322, P.27367, P.57925–27, P.57929, P.57932–35, P.57937, P.57939, P.57943, P.57947, P.57948, P.57951–56, all from the Upper Carboniferous (Westphalian) of the British Isles. Also known from northern France, Belgium, Holland and West Germany.

R. ardrossense Moy-Thomas 1937

Rhabdoderma ardrossense
Fig. 11.15(C).

Diagnosis
Small species of about 100 mm SL, with a long supplementary lobe of the caudal fin. Dermal bones of skull ornamented with a few coarse ridges. Scales show closely packed ridges similar to those of *R. elegans*.

Holotype
BMNH P.19244 and cpt RSM 1958.1.3994, Lower Carboniferous (Visean P_1), Calciferous Sandstone Series, Fifeshire, Scotland.

Material
BMNH P.22005–6 Lower Carboniferous (Visean P_1), Calciferous Sandstone Series, Fifeshire, Scotland.

R. huxleyi (Traquair 1881)

Diagnosis
Small species reaching approximately 130 mm SL, with a long supplementary lobe in

the caudal fin; D_1 = 9–10, C = 16–17/15–16. Ornament of the head restricted to the angular and to the gular plates where it consists of a few widely-spaced sharp ridges. Scales similarly ornamented with sparse ridges.

Syntypes
BGS (Edinburgh) 4693, M 2149c, M 2297c, M 2150; from the Lower Carboniferous (Visean B), Calciferous Sandstone, Glencartholm Volcanic Group, Dumfriesshire, Scotland.

Material
BMNH P.4079a (pt and cpt), P.4079b (pt and cpt), P.4079c (pt and cpt), P.4079c (pt and cpt), P.4080a, P.4080b, P.4080c, P.4080d, P.4080e, P.5983, P.115290, P.20140–1(pt and cpt), RSM 1885.54.6, 1885.54.7, 1885.54.8a, 1885.54.8b, 1885.54.9, 1885.54.10, 1891.53.4, 1891.53.5, 1891.53.49, 1891.53.50, 1966.48.22, 1978.43.1 all from the Lower Carboniferous (Visean B), Calciferous Sandstone, Glencartholm Volcanic Group, Dumfriesshire, Scotland.

Remarks
There are a few minor differences in skull proportion and the shape of the pelvic plate from other species which led Lund and Lund (1985) to refer this species to a monospecific genus *Dumfregia*. The differences are not considered significant enough to warrant the erection of a new genus and, more particularly, there is no evidence suggesting that this species is more closely related to species of other genera.

R. (?) alderingi *Moy-Thomas (1937)*

This is a species known only by the holotype (UCL P. 124, 125 pt and cpt) which shows a crushed head and part of the trunk. Ornament very similar to that of *R. huxleyi* and perhaps should be regarded as synonymous. Upper Carboniferous (Namurian A), Clwyd, Wales.

Rhabdoderma madagascariensis (Woodward)

Rhabdoderma madagascariensis
Fig. 11.15(F).
1910 *Coelacanthus madagascariensis* Woodward: 5, pl. 1, fig. 5.
1924 *Coelacanthus madagascariensis* Woodward; Priem: pl. 2, fig. 4 (not figs 1–3).
1935b *Coelacanthus madagascariensis* Woodward; Moy-Thomas: 223, figs 9, 11.
1981 *Rhabdoderma madagascariensis* (Woodward); Forey: 221, fig. 13.

Diagnosis
See Woodward (1910: 5).

Holotype
BMNH P.10768 pt and cpt, a complete fish lacking only the supplementary lobe from the Lower Triassic (Scythian) of Madagascar.

Material
MNHN 1972–7, from the Lower Triassic (Scythian) of Madagascar.

Remarks
Poorly known species but differing from *R. elegans* in the deeper body, wider spaced dorsal fins, and the ornament of closely packed regular tubercles on the cranial bones (Forey 1981). The ornament on the scales (Fig. 11.15(F)) is most like that of *R. elegans* except that the apical tubercles are rounded and close-set.

Rhabdoderma (?) newelli (Hibbard)

1933 *Coelacanthus newelli* Hibbard: 280, pl. 27, figs 2, 3.
1933 *Coelacanthus arcuatus* Hibbard: 282, pl. 26, fig. 8, pl. 27, fig. 1.
1963 *Synaptotylus newelli* (Hibbard); Echols: 479, figs 1–7.

Diagnosis
See Echols (1963).

Holotype
KUVP 786F from the Upper Carboniferous (Stephanian) of Kansas, USA. Specimen not examined in this work.

Remarks
This species was re-evaluated by Forey (1981). It appears to be distinct in showing reduced ornament on the cranial bones and the very short, stout pelvic bones.

Rhabdoderma exiguum (Eastman)

1902 *Coelacanthus exiguus* Eastman: 536, fig. 2.
1972 *Rhabdoderma exiguum* (Eastman); Schultze: 90, figs 1, 2.
1980 *Rhabdoderma exiguum* (Eastman); Schultze: 101, fig. 2.

Diagnosis
See Eastman (1902).

Material
FM PF.725, PF.726, PF.3199, PF.3660, PF.3779, PF.5494, PF.5521, PF.5757, PF.5760, PF.5906, PF.6270, PF.6825, PF.7338, PF.7528, PF.7529, PF.8342, PF.8551, PF.8574, PF.8663, PF.8666–8670, PF.8673, PF.8724, PF.8858, PF.8859, PF.8888, PF.8945, PF.9953–9955, PF.12835, PF.12836, FM UC.9110, UC.14389, UC.14391, all from the Upper Carboniferous (Westphalian C) of Illinois, USA.

Remarks
This is a problematic species. It was erected by Eastman (1902) on the basis of 10 small specimens (now in Peabody Museum, Yale and Museum of Comparative Zoology, Harvard – not examined in this study) from Upper Carboniferous of Mazon Creek, Illinois. Eastman distinguished the species on the large head (over 25% of total length). Subsequently a very large number of specimens have been collected. By far the majority are of small (<60 mm) obviously young individuals and many still have the yolk sacs

attached (Schultze, 1972, 1980). There are a few large individuals (200 mm SL) associated in the same beds and it is assumed that they belong to the same species. This seems reasonable to the extent that the fin ray counts are similar (D_1 = 12–13, C = 15–16/16) in most of the small specimens and the large supplementary caudal lobe is proportionately very long (more so in the small specimens than in the large). However, there is very little else by which they can be compared. Assuming this association is correct then the larger specimens show that this species is probably a member of the genus *Rhabdoderma*. The scale ornament consists of posteriorly converging adpressed ridges, the shape of the pterygoid is very much like that of *Rhabdoderma*, as is the triangular principal coronoid. One of the smaller specimens (FM PF.8888) shows the presence of four large premaxillary teeth, as seen in *R. elegans* (Fig. 3.6). Specific distinction from *R. elegans* is evidenced by higher D_1 and caudal fin ray counts and by considerably more vertebrae (55–58). These last counts can only be made on the small specimens. Lastly, it should be noted that two small specimens examined here (FMNH PF.5660 and PF.8664) show much higher vertebral counts (64 and 70) as well as higher caudal fin ray counts (22/20 and 20/18). It is therefore possible that there may be more than one species represented in the Mazon Creek fauna. In summary, I would maintain the specific distinction of *R. exiguum* based on fin ray and vertebral counts. The large head size which impressed Eastman may be a juvenile character.

Genus *Sassenia* Stensiö

Diagnosis (emended)
Medium-sized coelacanths known mostly by cranial remains. The skull roof consists of a narrow postparietal shield and a parietonasal shield of approximately equal length. The preorbital is large and the two posterior

openings of the rostral organ are situated very close together. The lateral rostral and the jugal are both deep bones. The cheek bones are large and abut one another to completely cover the cheek. The postorbital is truncated at the anterodorsal angle. The operculum is large and rounded and is as broad as deep. The lower jaw is relatively long with the jaw articulation placed well back beneath the level of the posterior margin of the skull roof. The angular is generally shallow but is expanded as a broad and rounded coronoid expansion. The principal coronoid is deeper than wide with a markedly thickened head. The braincase is extensively ossified: the ethmosphenoid portion shows an ossified interorbital septum as well as the usual basisphenoid and lateral ethmoids. The otico-occipital portion is ossified as a prootic, reaching up to the parietals and a large occipital ossification enclosing the posterior semicircular canals, foramen magnum and the notochordal canal. The ornament upon the dermal bones of the head and the scales consists of small to ovoid, flat-topped tubercles of regular size and arranged in a uniform pattern. The ornament is rather similar to that of *Spermatodus* but there is no evidence of one generation of tubercles overlying another. The sensory canals open to the surface through very small pores.

Remarks

Sassenia is very similar to *Spermatodus* in the shapes of the cheek and opercular bones, the proportions of the various components of the skull roof and the lower jaw, the rather completely ossified braincase and in the pattern of ornament.

Type species: *S. tuberculata* Stensiö

Figs 35–38, pl. 10.
1921   *Sassenia tuberculata* Stensiö: 85, figs 35–38, pl.10.

Diagnosis

Head reaching about 60 mm maximum depth, preoperculum triangular, scales ornamented with minute tubercles of unequal size, closely adpressed to one another.

Syntypes

UP P.224, P.225, P.226, Lower Triassic (Scythian), Sassendalen Group, Sticky Keep Formation, West Spitzbergen.

Remarks

The type species is very poorly known and most of the generic diagnosis is based on the following species.

S. groenlandica sp. nov.

Sassenia groenlandica
Figs 3.11, 4.13, 5.8(B), 6.8, 11.16, 11.17.
1936   *Sassenia* sp. Nielsen: 14.
1976   coelacanthid Bang: pl. 1A.
1976   *Sassenia* sp. Bendix-Almgreen: 561.

Diagnosis

Head reaching about 80 mm long and 60 mm deep, preoperculum pentagonal, scales ornamented with short ridges near focus passing to rounded and equally sized tubercles distally which are closely adpressed to one another and relatively larger than those in the type species.

Holotype

MGUH VP.2326 (Fig. 11.16), head and shoulder girdle plus anterior scales, from the Lower Triassic (Scythian), Wordie Creek Formation, Cape Stosch, East Greenland.

Material

MGUH VP.2325a–b, VP.2327a–b, VP.2337, VP.3258, all from the Lower Triassic (Scythian), Wordie Creek Formation, Cape Stosch, East Greenland.

Fig. 11.16 Holotype of *Sassenia groenlandica* sp. nov. MGUH VP.2326 showing the head and shoulder girdle plus anterior scales in left lateral view: Lower Triassic (Scythian), Wordie Creek Formation, Cape Stosch, East Greenland. Scale bar in mm.

Genus *Spermatodus* Cope 1894

Diagnosis (emended)

A monotypic genus known only by cranial remains and a few isolated scales and cleithra. Head and opercular apparatus reaching approximately 150 mm in length. The postparietal shield is narrow and flat and the supratemporal forms less than 15% of the total surface area. The extrascapular series appears to be represented as many small ossicles. The postparietal shield is longer than the parietonasal shield. In the parietonasal shield there are four nasals followed by anterior and posterior parietals; the latter are very unequal in size with the anterior pair considerably less than half the length of the posterior pair. The intracranial joint margins are deeply notched and the parietal descending process is very broad. There are four tectals and at least three supraorbitals of which the second is considerably larger

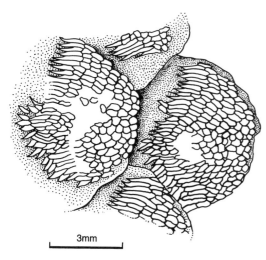

Fig. 11.17 *Sassenia groenlandica* n. sp. Camera lucida drawing of scales from beneath the first dorsal fin as preserved in MGUH VP2327a.

than the adjacent frontal. The preorbital is large, ovoid and pierced by closely-spaced posterior openings from the rostral organ. The premaxilla is large and robust, containing the anterior opening to the rostral organ. The area between the premaxilla and the anterior tectal is filled with many small, irregularly shaped ossicles. The braincase shows a very large basisphenoid which is produced posteriorly as two well-developed sphenoid condyles. The parasphenoid bears teeth over the anterior two-thirds. The otico-occipital portion of the braincase is well ossified and there is a complete undivided occipital portion. A descending process of the supratemporal forms a strong union between the neurocranium and the overlying skull roof but it is not certain if there was a postparietal descending process. The cheek is very broad such that the quadrate/articular joint lies well behind the orbit, on a vertical level with the posterior edge of the postparietals. Jugal, postorbital, squamosal and preoperculum are all large bones which form a complete covering to the cheek and

which show mutual overlap areas. The squamosal is particularly large and contacts the postparietal shield. The operculum is large with a rounded posterior margin. In the lower jaw the principal coronoid is deeper than wide; it is broad dorsally where it has a thickened head; the posterior margin is straight, in contrast to the rounded anterior margin. The premaxilla bears many rounded teeth and the dentary carries four large tooth plates, each bearing about 30 rounded teeth. Other teeth within the mouth are villiform. All teeth bear many minute striations radiating from the apex. Ornament is present over the exposed surfaces of most of the dermal bones but is absent from the premaxilla, dentary and splenial and is sparse upon the angular. The ornament consists of closely spaced and irregularly sized hemispherical denticles which are often seen to partly overlie each other. Each denticle has a cap of dentine and enamel. There are also many small polygonal ossicles present behind the parietals, on the snout and particularly along the posterodorsal margin of the orbit. Some of the large ossicles bear the characteristic ornament. The scales bear a similar dense covering of rounded denticles which abut one another. The sensory canals appear to be quite narrow, in contrast to those in most other coelacanths and they open to the surface by way of many small pores.

Type and only species: *S. pustulosus* Cope

Spermatodus pustulosus
Figs 3.12, 3.13, 3.14, 5.8(A).
1894 *Spermatodus pustulosus* Cope: 438, fig. 4.
1911 *Spermatodus pustulosus* Cope; Hussakof: 172, pl. 32.
1939 *Spermatodus pustulosus* Cope; Westoll: 3, figs 1–5.

Diagnosis
Same as genus, only species.

Holotype
AMNH 7245 partially disarticulated head from the Lower Permian of Texas (precise locality unknown but probably from Admiral Formation, Wichita Group, Texas, USA.

Material
AMNH 4612; UM 10308, 11047, 11658, 11660, 16163, 16071 (these specimens are each collections of fragments from many individuals); MCZ 8428, 8830, 8847, 8910, 8949, 13739, 13740, 13755, 13757, 13758. All specimens from the Lower Permian (Admiral Formation, Wichita Group) of Archer County, Texas, USA.

Genus *Undina* Münster 1834

Diagnosis (emended)
A genus of relatively plump-bodied coelacanths in which the head is relatively large (25% SL), the posterior profile of the tail is square-cut and the supplementary caudal lobe is prominent. The head is deep: the parietonasal shield is substantially longer than the postparietal shield. The postparietal shield is broad, flared posteriorly and markedly convex and is deeply embayed posteriorly, the embayment containing several free extrascapulars. The skull roof bones are closely sutured to one another but they are smooth with very little ornamentation. The supraorbitals and supratemporals are perforated by many small pores. The braincase must have been largely cartilaginous, developed to the same degree as in *Macropoma* except that an ascending process of the prootic was probably absent. The parasphenoid is slender posteriorly but it broadens considerably within the anterior half where ascendings wings are developed. Teeth are confined to the anterior half. The pterygoid has a deep anterior limb and a very narrow, upright posterior limb. The pterygoid teeth are arranged in rows close to the dorsal margin. In the lower jaw the angular is very shallow throughout its length and the principal coronoid is saddle-shaped. The posterior member of the anterior coronoid series bears enlarged and pointed teeth which are oposed by similar teeth upon the dermopalatine. The mandibular sensory canal opens through a regular series of tiny pores which runs the entire length of the splenial and the angular; oral pit line located well posteriorly close to jaw articulation. The cheek consists of a broad jugal, large postorbital and squamosal bones and a short, shallow but elongate preoperculum. A spiracular bone is absent. The cheek, operculum, angular, splenial and gular plates are ornamented with dense tubercular ornament. There is a narrow smooth area on the angular overlapped by the preoperculum. The gill arches bear small tooth plates alongside pointed teeth. The shoulder girdle is very narrow and the anocleithrum is sigmoid. The pelvic fin rays are usually expanded but most of the rays of other fins are usually narrow. The rays of D_1 and the caudal rays are ornamented with a dense covering of small denticles. The scales are ornamented with elongate tubercles and ridges. Lateral line scales bear a complex arrangement of secondary tubules. A large ossified air bladder is present.

Type species: *U. penicillata* Münster 1834

Undina penicillata
Figs 5.12(C), 7.6(G), 8.5.
1834　*Undina penicillata* Münster: 539.
1842a　*Coelacanthus striolaris* Münster: 40.
1842a　*Coelacanthus kohleri* Münster; Münster: 40.
1842b　*Coelacanthus striolaris* Münster; Münster: 57, pl. 2, figs 1, 3, 5, 6, 8–10, 12, 14, 16.
1842b　*Coelacanthus kohleri* Münster; Münster: 59, pl. 2, figs 2, 4, 7, 11, 13, 15, 17.
1844　*Undina striolaris* (Münster); Agassiz: 171.
1844　*Undina kohleri* (Münster); Agassiz: 171.
1863　*Undina penicillata* Münster; Wagner: 696.

1869   *Coelacanthus penicillatus* Willemoes-Suhm: 80, pl. 10, figs 2, 3, pl. 11, fig. 3.

1887   *Undina penicillata* Münster; Zittel: 175, fig. 177.

1887   *Undina acutidens* Zittel: 175, 177b.

1888   *Undina penicillata* Münster; Reis: 10, 36, pl. 1, figs 2–6, 8–24.

1891   *Undina penicillata* Münster; Woodward: 410.

1906   *Undina acutidens* Zittel; Heineke: 163, fig. 1, pl. 29, fig. 1.

1921   *Undina penicillata* Münster; Stensiö: 116, fig. 52.

1927   *Undina penicillata* Münster; Watson: 435, pls 1, 2.

1930   *Undina acutidens* Zittel; Aldinger: 22, figs 1–10.

1931   *Undina penicillata* Münster; Abel: 99, figs 1–5.

1933   *Undina penicillata* Münster; Broili: 7, pl. 1, fig. 1.

1935   *Undina penicillata* Münster; Gross: 43, fig. 17.

Diagnosis (emended)
Species reaching 400 mm SL, HL/SL = 25–27%, PD_1/SL = 45%, PD_2/SL = 65%, TL/SL = 15–18%, CL/SL = 31–35%, TD/SL = 33%, CP/SL = 20–23%, PV/SL = 45%; D_1 = 8, D_2 = 15–17 P = 19, V = 21, C = 18/16, A = 14, AbdV = 50, CauV = 20–22. The head is nearly as deep as long. Scales covered with a uniform covering of short ridges. Anterior rays of both caudal lobes are much enlarged and ornamented with a dense covering of denticles.

Holotype
BSM AS.I.820, a complete fish from the Upper Jurassic (Tithonian), Bavaria, Germany.

Material
BSM 1870.XIV.22, 1870.XIV.508–516, 517, 1889.XV.22, 1899.I.10, 1899.I.40, JME SOS.2195, SOS.2198, SOS.2201, SOS.2206, BMNH 37032, P.5543, P.10884, all from the Upper Jurassic (Tithonian), Bavaria, Germany.

Undina cirinensis Thiollière 1854

1854   *Undina cirinensis* Thiollière: 10, pl. 1.

1863   *Coelacanthus minutus* Wagner: 697.

1869   *Coelacanthus cirinensis* (Thiollière); Willemoes-Suhm: 79, pl. 11, fig. 4.

1891   *Undina cirinensis* Thiollière; Woodward: 411.

1949   *Undina cirinensis* Thiollière; Saint-Seine: 77, figs 34–37, pls 5, 6.

Diagnosis
See Saint-Seine (1949: 77).

Remarks
No material of this species was examined. It is much smaller than the type species and the scales show a posterior fringe of tiny pointed denticles.

Undina purbeckensis Woodward 1916

1916   *Undina purbeckensis* Woodward: 22, pl. 4, fig. 1.

Diagnosis
See Woodward (1916: 22).

Holotype
BMNH P.11925, nearly complete fish (much of head lacking) from the Upper Jurassic (Purbeckian), Dorset, England.

Remarks
Only the holotype is known: this is poorly preserved, and it is not certain that this is a species distinct from *U. penicillata*. The ornament of the scales is finer and less dense than that of the type species but in all other respects appears to be similar.

Genus *Whiteia* Moy-Thomas 1935

[*Coelacanthus* in part of Priem, 1924]

Diagnosis (emended)
A genus of small coelacanths in which the

body is relatively slender and the head relatively large with a slightly elongated snout with the preorbital length greater than one-third the length of the skull roof. The parietal shield is short and broad and is approximately half the length of the parietonasal shield. Both postparietal and the supratemporal show prominent descending processes buttressing the skull roof to the neurocranium. There are five extrascapulars. Parietonasal shield relatively narrow with three equidimensional nasals followed by two frontals of equal size. There are four or five tectals and five supraorbitals: the anteriormost supraorbital is particularly large and nearly meets the jugal to exclude the preorbital from the orbital margin. The preorbital is ovoid and includes two widely spaced posterior openings of the rostral organ. Premaxilla shallow and includes the anterior opening to the rostral organ. The cheek is completely covered by a jugal, postorbital, squamosal and preoperculum which abut one another but do not overlap. Additionally, there is a small spiracular dorsally and a suboperculum ventrally. The lachrymojugal is large and distinctly angled anteriorly as it runs forward to contact the lateral rostral. The operculum is rounded dorsally and posteriorly but is pointed ventrally. The lower jaw is shallow with a saddle-shaped coronoid. The gular plates do not extend behind the jaw rami. The basisphenoid is short and broad, with narrow antotic processes: parasphenoid broad and covered with tiny teeth which extend posteriorly as far as the level of the basisphenoid. The otico-occipital portion of the braincase consists of, at least, separate prootic and basioccipital portions, developed as in *Latimeria*. The teeth upon the premaxilla, palatine, ectopterygoids and the dentary tooth plates are small and villiform. The sensory canals are developed as in other coelacanths; postparietals contain both a median branch of the otic canal and a middle pit-line; the junction of the otic

sensory canal and the supratemporal commissure lies within the supratemporal. As the infraorbital canal passes through the anterior half of the lachrymojugal it opens through two large, dorsally directed pores. Well-marked pit lines are also present upon the squamosal, preoperculum and the angular. The sensory canals generally open through relatively few pores, those upon the angular and splenial are elongated. Ornament upon the skull consists of tubercles which are absent, or very sparse upon the skull roofing bones, dentary and splenial, and are variously developed upon the cheek bones, opercular bones and the angular. The extracleithrum is broad. The endoskeletal fin supports are very poorly known but the D_1 support is a simple triangular plate with a single anteroventrally directed strengthening ridge. The fin rays of D_1 and the anterior rays of the principal caudal lobes show a double row of pointed denticles on either side. Upper and lower lobes of the caudal fin symmetrical. The scales are well preserved in specimens of all species implying that they were relatively thick. The lateral line scales are prominent and each contains a large lateral line canal which sends off one dorsal and, usually, one ventral branch.

Type species: *W. woodwardi* Moy-Thomas

Whiteia woodwardi
Figs 3.15, 3.16(A), 4.14, 4.15, 5.9(A–C), 7.6(D), 8.29K), 11.18, 11.19(C).
1924 *Coelacanthus madagascariensis* Woodward; Priem: 7, pl. 2, figs 1–3 (not fig. 4).
1935b *Whiteia woodwardi* Moy-Thomas: 216, figs 3–6.
1952 *Whiteia woodwardi* Moy-Thomas; Lehman: 17, figs 3a, 4–7, 9, 10c, pl. 1, 2, 3a, 4d, 5a,c.e.
1968 *Whiteia woodwardi* Moy-Thomas; Beltan: 113, figs 42, 43, pls 46–49.
1979 *Coelacanthus evolutus* Beltan: 453, pl. 9.

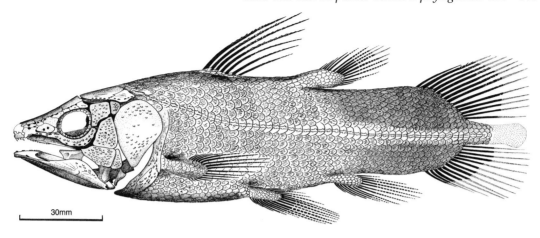

Fig. 11.18 *Whiteia woodwardi* Moy-Thomas. Restoration of entire fish, supplementary caudal lobe not well known. Based on BMNH P.17200–1, P.17169, P. 17208–9, P.16636–7.

Diagnosis (emended)
Species reaching 160 mm SL, HL/SL = 33%, PD_1/SL = 40%, PD_2/SL = 62%, TL/SL = 12%, TD/SL = 27%, CP/SL = 18%, PV/SL = 50%; D_1 = 7–8, D_2 = 15, P = 16, V = 18, A = 13, C = 15/14 (measurements and counts are available on very few specimens and some are taken from single specimens). About 45 scales along lateral line from skull to base of supplementary fin. Several openings from the supraorbital canal per supraorbital. Preorbital length equal to 25% or more of head length, orbit very small at 20% head length. Ornament consists of sparse, slightly elongated tubercles on the head and many ridges which remain separate from one another on the scales (Fig. 11.19(C)).

Holotype
BMNH P.19500–1, pt and cpt of head plus anterior part of body, Lower Triassic (Scythian), Middle Sakemena Group, northern Madagascar.

Material
BMNH P.16236, P.16636–7, P.17167, P.17168, P.17169, P.17202, P.17204–5, P.17206–7,

P.17208–9, P.17210–11, P.17212–3, P.19500–1, P.19624–5. Lower Triassic (Scythian), Middle Sakemena Group, northern Madagascar.

W. tuberculata Moy-Thomas

Whitiea tuberculata
Figs 5.9(F), 11.19(B).
1935b *Whitiea tuberculata* Moy-Thomas: 220, figs 7, 8.
1952 *Whiteia tuberculata* Moy-Thomas; Lehman: 32, figs 14, 15, pl. 4, pl. 5b.

Diagnosis (emended)
Small species, approximately 115 mm SL, D_1 = 7. Cheek bones narrower than in the type species. Pores opening from the supraorbital and mandibular sensory canals are very large, with only a single pore between successive supraorbitals. Ornament upon the head, shoulder girdle and scales consists of small, ovoid and widely-spaced tubercles.

Holotype
BMNH P.17214–5, pt and cpt of entire fish, Lower Triassic (Scythian), Middle Sakemena Group, northern Madagascar.

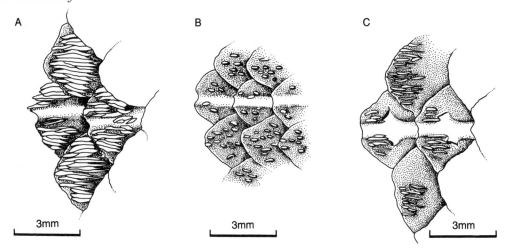

Fig. 11.19 Scales of *Whiteia*. Camera lucida drawings of scales from beneath the first dorsal fin. (A) *W. neilseni* n. sp. (MGUH VP2335a – reversed). (B) *W. tuberculata* Moy-Thomas (BMNH P.17215). (C) *W. woodwardi* Moy-Thomas (BMNH P.17201).

Material
BMNH P.17203, Lower Triassic (Scythian), northern Madagascar.

Whiteia nielseni n. sp.

Whiteia nielseni
Figs 3.16(B), 5.9(D,E), 11.19(A).
1936   undetermined Coelacanthid Nielsen: figs 8–10, 14, 16.
1936   *Whiteia* sp. Nielsen: figs 11–12.
1976   Coelacanthid; Bang: pl. 1A.

Diagnosis
A species reaching an estimated 240 mm SL. $D_1 = 9$. Parietonasal shield has a concave lateral profile. Lachrymojugal with pronounced ventral angle beneath the eye. Preorbital length equal to about 30% head length. Preorbital very long, equal at least to 15% of head length and at least twice as long as high. Supraorbital sensory canal opens between successive supraorbitals through single small pores. Ornament upon the cheek, operculum and the central portion of the angular consists of closely packed and ovoid/elongate tubercles. Ornament upon the scales consists of many elongate ridges which are closely adpressed to one another (Fig. 11.19(A)).

Holotype
MGUH VP.133, a complete head figured by Nielsen (1936: figs 11, 12) from the Lower Triassic (Scythian), Wordie Creek Formation, Cape Stosch, East Greenland.

Material
MGUH VP.986a,b, VP.989, VP.2323a–b, VP.2328–30, VP.2331a–d, VP.2332–4, VP.2335a,b, 92–94, 96,97, 126–130, 133, 136. Lower Triassic (Scythian), Wordie Creek Formation, Cape Stosch, East Greenland.

Etymology
Species name after Dr Eigel Nielsen, a pioneer vertebrate palaeontologist who collected many vertebrates from the Devonian and Triassic of Greenland.

Whiteia africanus (Broom)

1905 *Coelacanthus africanus* Broom: 339.
1909 *Coelacanthus africanus* Broom; Broom: 253, pl. 12, fig. 3.
1931 *Coelacanthus africanus* Broom; Brough: 238, pl. 1, fig. 1.
1975 *Coelacanthus africanus* Broom; Jubb and Gardiner: 411.

Diagnosis
See Broom (1905: 339).

Holotype
SAM-6027, a caudal region from the Lower Triassic of Republic of South Africa.

Material
BMNH P.11210, P.16033, P.16034, all from the Lower Triassic of Orange Free State, Republic of South Africa.

Remarks
This species has never been satisfactorily distinguished from other coelacanths and it may well be conspecific with *Whiteia woodwardi*. The overall shape of the head with a slightly elongate snout, the shapes of the angular and low, saddle-shaped principal coronoid, the relatively large preoperculum, low fin ray counts in both the D_1 (7 or 8) and the caudal (approximately 16 rays in each lobe) all agree with conditions in *Whiteia*. The sparse tubercular ornament upon the operculum and the scale ornament is more like *W. woodwardi* than other species. The chief difference is that there do not appear to be any denticles on the rays of either D_1 or the principal caudal rays.

Whiteia sp. Gardiner 1966

Remarks
Gardiner (1966) and Schaeffer and Mangus (1976: 551, fig. 19) both mention specimens from the Lower Triassic (Scythian) Sulphur Mountain Formation, British Columbia that almost certainly can be referred to the genus *Whiteia*. This material remains to be described. It is probably conspecific with *Coelacanthus banffensis* Lambe (Gardiner, 1966: 93), from the Lower Triassic (Scythian) Spray River Formation, Banff, Alberta; a species known only by the holotype which consists only of a pectoral fin and scales.

11.4 COELACANTHS RECOGNIZED AS BEING VALID GENERA AND SPECIES BUT WHICH CANNOT BE PLACED ACCURATELY WITHIN A PHYLOGENETIC TREE (Category 2)

Genus *Alcoveria* Beltan 1972

Diagnosis
Monotypic genus – see species.

Type and only species: *A. brevis* Beltan

1972 *Alcoveria brevis* Beltan: 285, fig. 1, pls 1–3.
1984 *Alcoveria brevis* Beltan; Beltan: 117, pl. 1.

Diagnosis
See Beltan (1972).

Holotype
MSB M.140a,b, a complete fish in part and counterpart, from the Middle Triassic (Ladinian), Taragana Province, Spain. A silicone rubber cast was examined.

Material
MSB M123a,b; M20a,b, Middle Triassic Ladinian), Taragana Province, Spain.

Remarks
This is a monotypic genus of small coelacanths (< 180 mm SL) which is very poorly known. Both the head and the body are relatively deep: the maximum depth of the body is equal to nearly one-third SL. The head is proportionately large, exceeding one-third of the SL, and being

nearly as deep as long. Little detailed information is available concerning the head but significant points are the large orbit apparently lacking sclerotic ossicles, and the very narrow postorbital and squamosal. The operculum is also narrow and acutely pointed ventrally. In the postcranial skeleton the pelvic fins are supported by very narrow pelvic bones and they originate at a level midway between the dorsal fins. The caudal fin is relatively small and the supplementary fin remains unknown. The anterior rays of D_1 and both caudal lobes are covered with many small denticles. Details of ornamentation and squamation remain unknown.

Genus *Axelia Stensiö* 1921

1921   *Axelia* Stensiö; Stensiö: 43.

Diagnosis (emended)
Large coelacanths in which the head is broad with a bluntly rounded snout and a very large, deep operculum. Parietonasal shield nearly as broad as the postparietal shield and only slightly longer. Single pair of parietals, flanked by large, quadrangular supraorbitals. Postparietal shield markedly embayed posteriorly. Parasphenoid very broad and bears teeth over nearly its entire palatal surface. These teeth are pebble-like and irregular in size which, with similar teeth upon the pterygoid and basibranchial tooth plates, form a crushing dentition. The palatoquadrate has a broad posterior limb and a shallow anterior limb reflecting a general shallowness of the head. Cheek plates highly reduced, represented only as narrow tubes: jugal and preoperculum remain unknown and may well have been absent. The sclerotic ring is well developed. Angular is deep in profile, almost triangular in shape. The sensory canals within the parietonasal and the angular open through very large pores.

Type species: A. robusta *Stensiö*

1921   *Axelia robusta* Stensiö: 90, figs 39–47, pls 11–15, pl. 16, figs 1–5.
1932   *Axelia robusta* Stensiö; Stensiö: fig. 15b.

Diagnosis
Species in which the head reaches 200 mm long. Ornament upon the cranial bones consists of very fine ridges and small tubercles which are arranged longitudinally and are densely spaced. Within the postcranial skeleton the anterior caudal fin rays are ornamented with small denticles. Segmentation in the anterior caudal fin rays and the pectoral fin is confined to the distal half of each ray. The scales are marked with strong longitudinal ridges.

Syntypes
UP P.187, P.189, P.190, P.193, P.195 from the Lower Triassic (Scythian), Sassendalen Group, Sticky Keep Formation, West Spitzbergen.

A. elegans Stensiö 1921

1921   *Axelia elegans* Stensiö: 106, figs 48, 49, pls 16, figs 6, 7, pl. 17, figs 1, 2.

Diagnosis
Species much smaller than type species with head length about 100 mm. Ornament on head bones of elongate, widely spaced, sharp edged tubercles.

Syntypes
UP P.220–Triassic (Scythian), Sassendalen Group, Sticky Keep Formation, West Spitzbergen.

Gnus *Chagrinia* Schaeffer 1962

Diagnosis
See Schaeffer (1962: 8).

Type and only species: *C. enodis*

1962   *Chagrinia enodis* Schaeffer: 8, figs 1–4.
1996   *Chagrinia enodis* Schaeffer; Cloutier: 235, fig. 9(D).

Diagnosis
As for genus, only species.

Holotype
CM 7997, partial head, trunk and tail from the Upper Devonian (Fammenian), Chagrin Shale, Ohio, USA

Material
CM 8066, Upper Devonian (Fammenian), Chagrin Shale, Ohio, USA.

Remarks
This is a monotypic genus based on specimens which are poorly preserved and shows few distinguishing features. The body is very slender with a particularly narrow caudal peduncle. The tail is symmetrical and, like that of *Diplocercides*, *Allenypterus* and *Lochmocercus*, the fin rays appear to outnumber the endochondral supports. The scales appear to lack ornament of any kind, and this is the basis for distinguishing this as a distinct genus and species. However, lack of ornament may be a preservational artefact. No material of this species was examined here.

Genus *Changxingia* Wang and Liu 1981

Diagnosis
See Wang and Liu (1981: 305).

Type and only spcies: *C. aspratilis* Wang and Liu

1981   *Changxingia aspratilis* Wang and Liu: 305, figs 1–2, pls 1–2, pl. 3, fig. 4.

Diagnosis
See Wang and Liu (1981: 305).

Holotype
IVPP V.6133.1 from the Upper Permian, Zhejiang Province, China.

Material
No material of this species was examined.

Remarks
This monotypic genus is poorly known. The body is about 300 mm SL and appears to be short and deep with a rounded posterior profile of the tail. Denticles are present on the D_1 fin rays and anterior fin rays of the principal caudal lobes. The scales are marked with several parallel elongate ridges running rostro-caudally. Ossified ribs are said to be present (Wang and Liu, 1981).

Coccoderma bavaricum Reis 1888

1888   *Coccoderma bavaricum* Reis: 60, pl. 5, fig. 2.
1891   *Coccoderma bavaricum* Reis; Woodward: 415.

Diagnosis
See Reis (1888: 60).

Holotype and Remarks
This species is based on the rear half of a head and the anterior part of the trunk (BSM 1870.XIV.24a,b) from the Upper Jurassic (Tithonian), Bavaria, Germany. The shapes of the angular, coronoid, the narrow pelvic bones and the opening of the sensory canals through many small openings arranged regularly within the angular are all features different from *Coccoderma* but closely similar to those of *Undina*.

Coccoderma gigas Reis 1888

1888   *Coccoderma gigas* Reis: 57, pl. 3, figs 17–19.
1891   *Coccoderma gigas* Reis; Woodward: 415.

Remarks

This species is based on a specimen (BSM 1870.XIV.25) showing the two lower jaws. These display a long low angular and principal coronoid totally unlike that of the type species, *Coccoderma suevicum* (page 143). The species comes from the Upper Jurassic (Tithonian), Bavaria, Germany.

Coccoderma substriolatum (Huxley 1866)

1866   *Macropoma substriolatum* Huxley: 39, pl. 9, 10.
1888   *Coccoderma substriolatum* (Huxley); Reis: 51.
1891   *Coccoderma substriolatum* (Huxley); Woodward: 415.

Holotype

SMC J27415a,b, a specimen showing a partial head and anterior portion of trunk and pectoral fin from the Upper Jurassic (Kimmeridgian), Cambridgeshire, England.

Remarks

This is a recognizably distinct species based on the form of the scale ornament which consists of a dense covering of regularly sized, small ridges which diverge towards the posterior margin of the scale. Huxley (1866) made comparsions between this form and *Macropoma* but the shape of posterior margin of the postparietal shield, the post-orbital and the slightly expanded pectoral fin rays are more like *Holophagus*.

Coelacanthus lunzensis Teller

1891   *Coelacanthus lunzensis* Teller: 3 (name only).
1900   *Coelacanthus lunzensis* Teller; Reis: 25, pls 9, 10.

Remarks

This species is known only by a single specimen (Geological Survey Museum, Vienna, in part and partial counterpart described most completely by Reis, 1900). The head is poorly represented by fragments of the jaw only. Few features are of diagnostic value. The scales appear to show a regular ornament of elongate ridges. Martin and Wenz (1984) suggest that it is possible that this species and *Garnbergia ommata* are synonymous. Both have a similar shape to the basal plates of the dorsal fins and the fin rays of D_1 are without denticles. These are features found in other coelacanths and are weak evidence of relationship and, for now, the species is kept separate. This species has not been examined in this study. Upper Triassic (Carnian), Lunz Sandstone, Austria.

Coelacanthus welleri Eastman

Fig. 11.20.
1908   *Coelacanthus welleri* Eastman: 358, fig.1.

Diagnosis

See Eastman (1908: 358).

Holotype

FMNH UC.21631, partial head and body without fins. Upper Devonian (Fammenian), Maple Mill Shale, Kinderhook Limestone Series, Burlington, Iowa, USA.

Remarks

Distinctive features were given in the original description as spinous ornament on operculum and cheek plates, fine longitudinal ridges of ornament on scales, and the prominent simple unbranched lateral line tube passing through about 40 scales to the midpoint of the main caudal lobes. None of these features is unique and in most it compares with species of *Diplocercides* such as *D. jaekeli*. Unfortunately little detail can be seen on the only known specimen but it is possible to determine some structure in the cheek which shows a large squamosal and preoperculum, a markedly angled jugal canal and a ventral tube-like portion to the post-orbital, all of which features are seen in *D. jaekeli*.

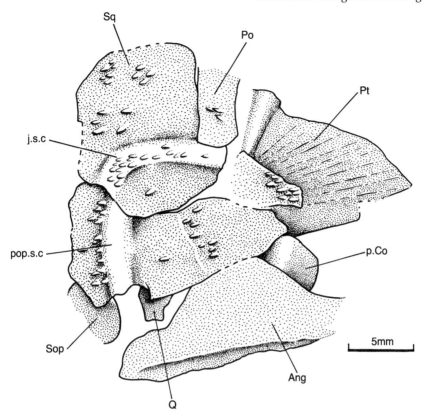

Fig. 11.20 *Coelacanthus welleri* Eastman. Camera lucida drawing of the holotype (FMNH UC.21631) to show the similarity to *Diplocercides kayseri* in the angular path taken by the jugal sensory canal through the squamosal (cf. Fig. 4.5).

Genus *Euporosteus* Jaekel 1927

Diagnosis (emended)
Monotypic genus based on a single specimen representing the completely ossified ethmosphenoid portion of a braincase about 2 cm long. Most of the dermal bone is missing, but some sutural impressions remain on the underlying neurocranium suggesting that there were two or three internasals. The nasal region is large occupying at least half of the length of the ethmosphenoid. The endoskeletal posterior openings for the rostral organ are well spaced. Both suprapterygoid and antotic processes are well developed but remain distinct from one another, separated by a deep notch. The parasphenoid is very broad; it extends well forward beneath the nasal capsules and bears teeth arranged in whorls over the entire surface. Impressions for the vomers suggests that these elements were separated in the midline. The supraorbital canal is very large and probably opened to the surface through a regular series of about 15 large ovoid pores.

Type and only species: *E. eifeliensis* Jaeckel

Euporosteus eifeliensis
Fig. 6.3.
1927   *Euporosteus eifeliensis* Jaeckel: 916, figs 23, 49.

1937   *Euporosteus eifeliensis* Jaeckel; Stensiö:
        48, figs 9, 10; pl. 11, figs 3, 4; pl. 12.
1942   *Euporosteus eifeliensis* Jaeckel; Jarvik:
        555, figs 73, 74a, 75a, 77b,c, 79; pl. 14,
        15, 16, fig. 1.

Diagnosis
As for genus, only species.

Holotype
UB, an ethmosphenoid portion of the braincase lacking most of the dermal bones – from
the Middle Devonian (Givetian) of Gerolstein, Germany.

Genus *Graphiurichthys* White and Moy-Thomas

1866   *Graphiurus* Kner: (name preoccupied).
1937   *Graphiurichthys* White and Moy-
        Thomas: 286.

Diagnosis
Monotypic genus of small (< 15 cm), plump
coelacanths with large head and relatively
small tail. This genus appears to be distinct
in that all fin rays, including those of the
paired fins are short and expanded (the last
feature seen elsewhere in *Libys*). The caudal
fin consists of relatively few rays (about 10)
in both the dorsal and the ventral lobes. The
pelvic fins are located behind the level of D_1.
The head bones and the scales are ornamented with narrow, elongate tubercles.

Type and only species: *G. callopterus* (Kner)

1866   *Graphiurus callopterus* Kner: 155, pl. 1,
        figs 1–3.

Holotype
Complete fish from the Upper Triassic
(Carnian), Raibl, Austria.

Genus *Hainbergia* Schweizer

1966   *Hainbergia* Schweizer: 225.

Diagnosis
See Schweizer (1966: 225).

Remarks
A very poorly known coelacanth represented
only by the type specimen. The head remains
virtually unknown but it appears that the
parietal shield is short, less than half the
length of the parietonasal shield. The anterior
dorsal fin is composed of smooth fin rays
and the scales show an ornament of many
parallel ridges. These last two features are
similar to those in the contemporaneous
Coelacanthus lunzensis Teller.

Type and only species: *H. granulata*
Schweizer

1966   *Hainbergia granulata* Schweizer: 225.

Diagnosis
As for genus, only species.

Holotype
Gö 538-1, complete but poorly preserved
individual from the Middle Triassic (Ladinian), Gottingen, West Germany. Specimen
was not examined here.

Genus *Heptanema* Belloti 1857

1932   *Heptanema* Stensiö.

Diagnosis
A monotypic genus showing a slender body
form of about 250 mm SL. The head is relatively long and shallow and shows a short
parietal shield equal to approximately half of
the parietonasal shield. The details of the
individual head bones remain unknown and
it is therefore difficult to compare this coelacanth with others. D_1 is very robust with
about seven large rays articulated only in the
distal half. The anterior ray at least bears a
double row of prominent denticles. Denticles
are also present over the anterior few principal rays of the caudal lobes. The scales are

ornamented with a prominent median ridge, flanked by one or two smaller ridges. This type of ornamentation is very similar to that seen in *Diplurus*.

Type and only species: *H. paradoxum* Belloti 1857

1857 *Heptanema paradoxum* Belloti: 435.
1910 *Heptanema paradoxum* Belloti; De-Alessandri: 39, pl. 1, fig. 11.

Diagnosis
As for genus, only species.

Genus *Indocoelacanthus* Jain 1974

Diagnosis (emended)
Monotypic genus of large coelacanths known only by incomplete cranial remains, scales and scattered postcranial material. The head is estimated to be about 18 cm in length. The cheek is distinctive in showing a very large rectangular postorbital, a small triangular squamosal located wholly behind the postorbital and a smaller rounded preoperculum resulting in a cheek which shows some similarities to that of *Axelrodichthys*. In the lower jaw the dentary bears a large hook-shaped process, the angular is relatively shallow and there is a large saddle-shaped principal coronoid. The cranial bones are ornamented with coarse rugae and tuberculations rather similar to that seen in *Mawsonia*. The palatoquadrate is deep with the dorsal margin rising gradually and without a notch. Within the postcranial skeleton the pelvic bones are distinctively broad and have several short anteriorly divergent finger-like processes. The scales are ornamented with many fine ridges which run parallel to one another.

Type species: *I. robustus* Jain

1974 *Indocoelacanthus robustus* Jain: 49, figs 2–7, pls 1 and 2.
1980 *Indocoelacanthus robustus* Jain; Jain: 107, fig. 6.5.

1980 "Kota bird" Jain: 109, fig. 6.6.

Diagnosis
As for genus, only species.

Holotype
ISI P.40, fragmented head in pt and cpt, from the Lower Jurassic, Kota Formation, Kota, India.

Material
Only photographs of the holotype were examined.

Genus *Lualabaea* Saint-Seine 1955

Diagnosis
Coelacanth in which the postparietal shield is very short and broad with the anterolateral edges flared, supratemporals large; three large extrascapulars, a median plus laterals which completely fill the deep posterior embayment of the skull roof. Parietonasal shield long (about twice the length of the postparietal shield) and narrow. Sclerotic ossicles absent and the cheek bones appear to be reduced such that in life they would have lain well separated from one another. The operculum is relatively large and shaped as a quarter-circle with the circumference facing posteroventrally. Palate is rather shallow with a long anterior limb. Angular is low with only a weak coronoid expansion developed anteriorly as a small prominence: dentary bears a large hook-shaped process and is swollen laterally: principal coronoid is low and saddle-shaped. Ornamentation upon the skull roof consists of coarse ridges. Denticles are present upon at least the anterior two rays of D_1 and the anterior rays of both caudal lobes. Rays of D_2 and the caudal are very slender. Scales are relatively small, ovoid and ornamented with a prominent central ridge which runs the entire length of the exposed portion and flanked by two or three thinner longitudinal ridges.

Type species: *Lualabaea lerichei* Saint-Seine 1955

1955 *Lualabaea lerichei* Saint-Seine: 7, fig. 2, pls 1–3.

Diagnosis
Species reaching 350 mm SL. Scales ornamented with central large ridge and a few lateral ridges, the latter tending to be of equal size.

Holotype
Head and parts of the body and pectoral fin MRAC R.G.10.046 from the Upper Jurassic (? Kimmeridgian), Lualabaea Series, Maosaosa, Zaire.

Material
No material was examined.

Remarks
In many respects, *Lualabaea lerichei* is very similar to *Mawsonia* and *Axelrodichthys*. The proportions of the parietonasal and post-parietal shields of the skull roof are very similar, as is the shape of the postparietal shield with a prominent angle along the lateral margin. The operculum is also of very similar shape, in all three genera having the shape of a quarter circle. The palate is shallow with a long anterior limb of the pterygoid. The dentary shows a lateral swelling and the principal coronoid is low and saddle-shaped. The ornament is developed as coarse rugosity in the bone. At present no autapomorphy of *Lualabaea* is known.

L. henryi Saint-Seine 1955

1955 *Lualabaea henryi* Saint-Seine: 14, figs 3–4.

Diagnosis
See Saint-Seine (1955).

Syntypes
Isolated scales and fin rays, MRAC R.G.9458 and R.G.94560 from the Upper Jurassic (? Kimmeridgian), Lualabaea Series, Maosaosa, Zaire.

Material
No material was examined.

Remarks
Species named for isolated scales and denticulated fin rays which are larger than those of the type species. The scales show more than three lateral ridges which tend to be of irregular size. It is very possible that this species simply represents mature fragments of the type species but more material is needed.

Macropoma willemoesii Vetter 1881

1881 *Macropoma willemoesii* Vetter: 1, pl. 1, fig. 1.
1888 *Heptanema willemoesii* (Vetter); Reis: 64, pl.3, figs 20, 21.
1891 *Heptanema willemoesii* (Vetter); Woodward: 416.

Holotype
A nearly complete specimen in the Geological Museum, Dresden from the Upper Jurassic (Tithonian), Bavaria, Germany. This specimen was not examined.

Material
JME SOS.2190, SOS.2192, SOS.2203, SOS.2204, all from the Upper Jurassic (Tithonian), Bavaria, Germany.

Remarks
This species, which grows to about 500 mm SL is distinctive in showing very coarse tuberclar ornament on the cranial bones and the scales. The proportions of the head and the shape of the preoperculum are similar to both *Macropoma* and *Undina*. The body is particularly deep immediately behind the

head, and the supplementary caudal lobe appears to be wholly placed between the principal caudal lobes. The lachrymojugal is expanded anteriorly as in coelacanths belonging to the family Latimeriidae. Unlike *Macropoma lewesiensis* there is a separate premaxilla. The scales in the middle of the flank are ornamented with 5–8 large ridges, each bearing fine striations, and in a few scales from beneath the D_1 the central ridge is considerably larger than those situated peripherally. The lateral line tubes branch at least once before opening. This species is probably closely related to either *Undina* or *Macropoma* but is insufficiently known to allow positive determination.

Genus *Macropomoides* Woodward 1942

Diagnosis
A monotypic genus. The body is relatively deep and reaches about 300 mm SL. The head bones are without ornament; a preorbital is absent; the lachrymojugal is narrow beneath the eye and barely larger than the enclosed sensory canal; postorbital is deep, expanded dorsally with a narrow ventral limb; the squamosal is very small and both the spiracular and the preoperculum may be absent. The premaxilla carries a few stout teeth. The operculum is rounded posterodorsally with a very oblique ventral margin. Sensory canals open by a few large pores on the parietonasal shield; the angular and splenial each have four large sensory pores. Teeth upon the parasphenoid are restricted to the anterior third of the bone. The principal coronoid has a distinct waist and a longitudinally expanded head. The gular plates are twice as long as broad. The anocleithrum is forked dorsally with a narrow dorsal limb and a broad anterodorsal limb. Short ribs are developed throughout the posterior half of the abdominal region. The caudal fin has a rounded posterior margin which encloses the supplementary lobe. Pointed denticles are present on at least the first three rays of D_1

and the leading rays of the principal caudal lobes. The pelvic bone is a simple rod with a proximal lateral expansion and the D_1 support has a prominent anteroventrally directed thickened ridge. The scales are ornamented with many closely spaced denticles which, like the denticles on the fins, bear many fine striations. Those scales beneath and behind the level of D_1 show a prominent central denticle (the only denticle present in small specimens).

Type and only species: *M. orientalis* Woodward

M. orientalis Woodward
Fig. 11.21.
1942 *Macropomoides orientalis* Woodward: 560, fig. 4.
1975 *Macropomoides orientalis* Woodward; Gaudant: 959, fig. 1.
1975 Coelacanth 'B' Gaudant: 959, figs 1, 3.
1975 Coelacanth 'C' Gaudant: 959, figs 1, 2.

Diagnosis
Species reaching 30 cm estimated SL, HL/SL = 26%, PD_1/SL = 35%, TL/SL = 18%, CL/SL = 32%, TD/SL = 30%, CP/SL = 20%, PV/SL = 43%, D_1 = 8, D_2 = 14, C = 20/19, AbdV = 55, CauV = 20. Proportional measurements based on two small specimens: meristic counts based on four small specimens.

Holotype
AUB 108935, specimen showing D_1 and D_2 plus associated squamation, from the Upper Cretaceous (Middle Cenomanian), Hajula, Lebanon.

Material
Holotype plus MNHN 1939-18-205d/g (pt and cpt), MNHN HDJ-73-20, HDJ-73-21, HDJ-73-22; AMNH 10398; P.62532; JME GPKR 22a,b (pt and cpt), from Upper Cretaceous (Middle Cenomanian), Lebanon. Most specimens are from Hajula; P.62532 is from Hakel; AMNH 10398 and JME GPKR 22a,b from either Hakel or Hajula.

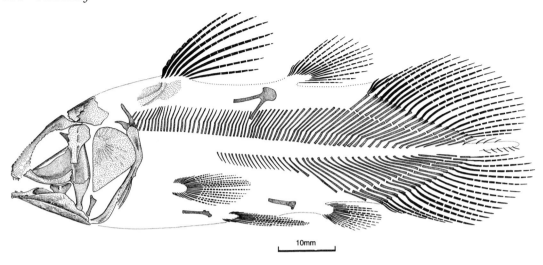

Fig. 11.21 *Macropomoides orientalis* Woodward. Based on MNHN 1939-18-205d–g, MNHN HDJ-73-20, HDJ-73-21, HDJ-73-22; AMNH 10398; BMNH P.62532; JME GPKR 22a,b.

Remarks

The generic and specific diagnoses are based primarily on small specimens and it is possible that when more complete large individuals are found they will have to be revised. Some features such as the lack of ornament and the weak development of the squamosal may be juvenile features. However, there remains the possibility that the small and large individuals represent different taxa. The restoration provided in Fig. 11.21 is based on small individuals. Some remarks are therefore necessary.

Woodward (1942) erected the species solely on the holotype which, based on proportions of the complete smaller specimens, must have been a fish of about 30 cm. Another large, but disarticulated tail (MNHN HDJ-73-22) is known, again probably from a comparable-sized fish. The remaining specimens, although articulated, are much smaller (< 10 cm SL) and there are differences in certain features between them and the holotype. The scales of the small individuals are more delicately orna-mented or lack ornament completely, and the rays of D_1 lack expanded tips (a feature considered diagnostic by Woodward). A number of features suggest that the small specimens are juvenile: the endochondral fin supports and pelvic girdle are lightly ossified; the scales are very thin and carry little ornamentation (see below). All these features are seen in young *Rhabdoderma exiguum* and in *Coccoderma nudum* which also are thought to be a juvenile forms.

Gaudant (1975) was also unsure whether the small specimens were juvenile *M. orientalis* or separate species. She further recognized two forms of small individuals which she called 'coelacanthe B' (two specimens) and 'coelacanthe C' (one specimen). The differences between them were alleged to be the slightly more anterior placement of the pelvic bone relative to the pectoral lobe and the dorsal fin in 'B' than in 'C' and the fact that form 'B' showed scales with a small central spine-like denticle (in large specimens the scale ornament consists of a spine-like central denticle flanked by many small denticles).

Further specimens showing intermediate fin placements, plus the fact that either the dorsal fin or the pelvic plate have been distorted in the specimens suggest that subtle differences in pelvic position do not justify a separation into two forms. The presence of a central spine-like denticle in some specimens but not others may suggest different growth stages because in *Latimeria* the central denticles (which are not enlarged) arise first (Smith, 1979).

I therefore suggest that coelacanthe 'B' and 'C' are the same, and in all probability should be referred to *Macropomoides orientalis*.

Genus *Megalocoelacanthus* Schwimmer et al. 1994

Diagnosis
See Schwimmer *et al.* (1994).

Type and only species: *Megalocoelacanthus dobiei*

Diagnosis
See Schwimmer *et al.* (1994).

Syntypes
Columbus College, Columbus, Georgia CCK 88-2-1,19; a collection of isolated cranial bones from the Upper Cretaceous (early Campanian), Barbour County, Alabama. Specimens were not examined.

Material
Other fragmentary material referred to this species comes from various localities in Alabama, Georgia and New Jersey, ranging in age from late Santonian to mid-Maastrichtian (Schwimmer *et al.* 1994). None has been examined in this study.

Remarks
This genus is known only by isolated cranial remains (basisphenoid, palate, operculum, ceratohyal, ceratobranchials, gular plate and lower jaw) and shoulder girdle (cleithrum).

It is said to be distinguishable from other coelacanths by its giant size, shape of the pterygoid and principal coronoid, and by the absence of marginal dentition. In some respects the estimated size depends upon model chosen: for instance, the known elements are smaller relative to total length in *Latimeria* than they are in *Axelrodichthys*. The size of the lower jaws and gular plate of *Megalocoelacanthus* as given in Schwimmer *et al.* (1994: fig. 2) are directly comparable with the same elements in the holotype of *Mawsonia gigas*, and assuming similar overall shape, the fishes would have been about the same size (estimated at about 3 metres, Maisey, 1991). So, *Megalocoelacanthus* is certainly considerably larger than *Latimeria* but may not have been the largest coelcanth known. It is, however, the youngest well-documented fossil coelacanth at 75 million years.

The relationships of *Megalocoelacanthus* are uncertain. The palate is very similar to that of *Macropoma*, *Libys* and *Latimeria* in showing a deep but narrow vertical limb and a rounded process along the ventral profile of the pterygoid, immediately anterior to the quadrate. The shape of the lower jaw, showing the insertion of the principal coronoid lying well anteriorly, is similar to that of *Axelrodichthys* and *Mawsonia*. There is also a large lateral and posteriorly directed foramen near the posterior end of the jaw which suggests that a posteriorly directed subopercular branch of the sensory canal was developed as in *Holophagus*, *Macropoma* and *Latimeria*. The absence of marginal dentition may not be significant because the dentary tooth plates of more derived coelacanths are separate from the supporting bone and are usually not preserved.

Genus *Moenkopia* Schaeffer and Gregory 1961

Diagnosis
See Schaeffer and Gregory (1961: 3)

Remarks

A genus based upon distinctive features of basisphenoid morphology; otherwise known by fragments of palatoquadrate, ceratohyal and cleithrum associated in the same deposit. The basisphenoid is large, similar in size to that of adult *Latimeria*. It is distinctively long and squat, the sphenoid condyles are widely spaced, the dorsum sellae is wide anteroposteriorly and the pituitary fossa is very narrow. The antotic processes project laterally only slightly beyond the level of the sphenoid condyles and they are parallel sided and rectangular in posterodorsal aspect, a feature similar to that of the basisphenoid of *Mawsonia* and *Axelrodichthys*. The lateral wall of the basisphenoid is without foramina implying that the oculomotor, abducens and profundus nerves left the cranial cavity anterior to the basisphenoid.

Type and only species: *M. wellesi* Schaeffer and Gregory 1961

1961 *Moenkopia wellesi* Schaeffer and Gregory: 5, figs 1, 2, 3k.

Diagnosis
As for genus only species.

Holotype
UCMP 36193, a basisphenoid from the Middle Triassic (Anisian), Moenkopi Formation, Arizona.

Genus *Mylacanthus* Stensiö 1921

1921 *Mylacanthus* Stensiö: 107.
1932 *Mylacanthus* Stensiö; Stensiö: 44.

Diagnosis
See Stensiö (1932).

Remarks
This is a very poorly defined genus which appears similar to *Axelia* in having a crushing dentition of pebble-like teeth and a deep triangle-shaped angular, penetrated by very large sensory canal pores. The genus is distinguished by Stensiö (1921, 1932) from *Axelia* on the basis that the hind margin of the operculum is modified (see below) and that the vertical limb of the palatoquadrate is narrow (versus broad). It is possible that both of these differences are preservational artefacts, leaving no features by which to distinguish this genus from *Axelia*. Two species are recognized, based on differences in the margin of the operculum. The genus and the two species are listed here but it is very likely that, with more material, these will be placed in synonymy with species of *Axelia*.

Type species: *M. lobatus* Stensiö

1921 *Mylacanthus lobatus* Stensiö: 108, fig. 50, pl. 18, figs 1, 4.

Diagnosis
See Stensiö (1921).

Syntypes
UP P.216–218, from the Lower Triassic (Scythian), Sassendalen Group, Sticky Keep Formation, West Spitzbergen.

M. spinosus Stensiö

1921 *Mylacanthus spinosus* Stensiö: 110, pl. 19, fig. 3, pl. 20, fig. 2.

Diagnosis
See Stensiö (1921).

Syntypes
UP P.687, P.703, from the Lower Triassic (Scythian), Sassendalen Group, Sticky Keep Formation, West Spitzbergen.

Remarks
The two species of *Mylacanthus* recognized by Stensiö (1921) are distinguished from each

other only by the shape of the posteroventral margin of the operculum. In *M. lobatus* the margin is lobed while in *M. spinosus* it is jagged. It is likely that both of these distinguishing features are preservational artefacts.

Genus *Piveteauia* Lehman 1952

Diagnosis
See Lehman (1952).

Remarks
Incompletely known monotypic genus. Slender coelacanth in which the pelvic fins are inserted anterior to the level of D_1. The caudal fin is asymmetrical with the dorsal lobe longer and beginning in advance of the ventral lobe. The scales are ornamented with short, ovoid tubercles which are arranged in parallel fashion and form a dense covering. The tubercles on more posterior scales become larger.

Type and only species: *P. madagascariensis* Lehman

1952   *Piveteauia madagascariensis* Lehman: 36, figs 16, 17, pl. 3, fig. b, c.

Diagnosis
See genus, only species.

Holotype
MHNP 1946.1.6, nearly complete fish, Lower Triassic (Scythian), Middle Sakemena Group, northern Madagascar.

Material
No material other than holotype was examined.

Genus *Rhipis* Saint-Seine

1950   *Rhipis* Saint-Seine: 11.

Diagnosis
See Saint-Seine (1950: 11).

Type species: *R. moorseli* Saint-Seine

1950   *Rhipis moorseli* Saint-Seine: 11, fig. 1, pl. 1 a,b,c.
1965   *Rhipis moorseli* Saint-Seine; Casier: 43, pl. 16, figs 2, 4, 6.

Diagnosis
See Saint-Seine (1950: 11),

Syntypes
MRAC 2892 and 3091, isolated scales from the Upper Jurassic, Kimbau, Zaire.

Remarks
Based on scales which are circular, about 20 mm in diameter and ornamented with thickened ridges or rows of tuberculations which fan out towards the posterior margin. No material was examined in this study.

R. tuberculata Casier 1965
1965   *Rhipis tuberculatus* Casier: 44, pl. 16, figs 1, 3.

Diagnosis
See Casier (1965).

Holotype
MRAC 11.141, a negative imprint of a scale from Upper Jurassic, Kinko, Zaire. No material was examined in this study.

Genus *Scleracanthus* Stensiö 1921

1921   *Scleracanthus* Stensiö: 111.
1932   *Scleracanthus* Stensiö; Stensiö: 45.

Diagnosis
See Stensiö (1932).

Type and only species: *S. asper* Stensiö

1921   *Scleracanthus asper* Stensiö: 111, pl. 17, fig. 3, pl. 18, fig. 2, pl. 19, figs 1, 2, pl. 20, fig. 1.

Diagnosis
As for genus, see Stensiö (1932).

Syntypes
UP P.239–P.242, P.709, from the Lower Triassic (Scythian), Sassendalen Group, Sticky Keep Formation, West Spitzbergen.

Remarks
This is a very poorly defined monotypic genus which is very similar to *Axelia*, and may be synonymous with one of the included species. Only fragmentary cranial remains, scales and isolated fins are known and these are relatively larger than those of either species of *Axelia*. Like *Axelia*, the pterygoid bears pebble-like teeth and the lower jaw is relatively short with a deep angular pierced by few, large, sensory pores. The genus was originally distinguished on the very fine, ridged sculpture upon the scales and the unusual feature of D_1 in which the fin rays are very stout with segmentation confined to their distal half and ornamented by many tiny denticles. If the D_1 of *Axelia* spp. were better known, this distinction may no longer be valid.

Genus *Sinocoelacanthus* Liu

1964   *Sinocoelacanthus* Liu: 211.

Diagnosis
See Liu (1964: 211).

Type and only species: S. fengshanensis *Liu*

1964   *Sinocoelacanthus fengshanensis* Liu: 211, pl. 1.

Diagnosis
See Liu (1964: 211).

Holotype
IVPP V.2895 from the Lower Triassic of Kwangsi Province, China.

Material
No material of this species was examined.

Remarks
Monotypic genus known by a single specimen showing only an incomplete tail. The supplementary lobe is comparatively small, barely extending beyond the main outline of the tail and one of the principal lobes appears longer than the other. There are more than 20 rays in each of the principal lobes and the fin rays are ornamented with ridge-like denticles. This is a very poorly defined coelacanth but the high number of rays may set it apart from many others.

Genus *Ticinepomis* Rieppel 1980

Diagnosis
Monotypic genus of small coelacanths. The head is relatively shallow and the postparietal shield is short, less than half the length of the parietonasal shield. Postorbital and probably also the squamosal are narrow bones, while the operculum and preoperculum are both relatively large and rounded. Premaxilla bears four stout, conical teeth. The jaw articulation occurs well forward beneath the eye and the dentary shows a pronounced ventral angle midway along its length. The palatoquadrate is deep throughout with the pterygoid showing a deep anterior limb. Ornament upon the operculum, preoperculum and the roofing bones consists of rounded tubercles. The shoulder girdle shows a cleithrum which is expanded dorsally. All of the fin rays of the median fins are expanded to some degree, particularly those of D_1. Denticles are borne upon the anterior rays of both D_1 and the caudal fin. The scales are densely covered with many elongate, sometimes pointed tubercles.

Type and only species: T. peyeri *Rieppel*

Ticinepomis peyeri
Fig. 4.16(B).
1980   *Ticinepomis peyeri* Rieppel: 923, figs 1–8.

Diagnosis
Species reaching at least 180 mm SL: $D_1 = 8$, $D_2 = 22–23$, C = 18/18, P = > 17; AbdV = 33, CauV = 18.

Holotype
ZM T3925, nearly complete fish from the Middle Triassic (Ladinian), Grenzbitumen Horizon, Monte San Giorgo, Kanton Tessin, Switzerland.

Material
No material of this species was examined.

Remarks
This species appears very similar to *Urocomus picenus* Costa (known from the Norian of Salerno, Italy) and may be synonymous (Rieppel, 1980) but appears to show less regular ornament of the scales.

Trachymetopon Genus Hennig 1951

Diagnosis
See Hennig (1951).

Type and only species: *T. liassicum* Hennig

Trachymetopon liassicum
1951   *Trachymetopon liassicum* Hennig: 69, pls 6–8

Holotype
UT 19050, Complete fish 1.6 m long, from the Lower Jurassic (Toarcian) of Westphalin, Germany. This specimen was not examined in this study.

Material
SMNS 150 91, Lower Lias (Sinemurian) of Westphalia, Germany.

Remarks
This is a large coelacanth which may have reached about 1.7 m (about the same size as *Latimeria*) with a proportionately large head from the Lower Jurassic of Germany. One specimen of this species was examined in

this work. From the description and the illustrations given by Hennig (1951), some interesting similarities with *Mawsonia* and *Axelrodichthys* can be seen. The parietonasal shield appears very long and narrow and ornamented with coarse rugosities, some of which are aligned longitudinally, the pterygoid is very long and shallow while the metapterygoid is broad, and the angular rises to a sharp pointed angle, the basibranchial dentition consists of a pair of very elongate tooth plates (Hennig, 1951: pl. 7, figs 3, 4). One obvious difference from *Mawsonia* and *Axelrodichthys* is the shape of the antotic process of the basisphenoid which in *Trachymetopon* appears strongly divergent rather than parallel-sided.

Undina (?) *barroviensis* Woodward 1890

1890   *Undina* (?) *barroviensis* Woodward: 436, pl. 16, fig. 5.
1891   *Undina* (?) *barroviensis* Woodward; Woodward: 413.

Diagnosis
See Woodward (1890: 436).

Holotype
BMNH 21335 and P.3343, a nearly complete fish in pt and cpt, from the Lower Jurassic (Lower Lias), Leicestershire, England.

Material
Warwickshire Museum G.562, a poorly preserved complete fish from Lower Jurassic (Lower Lias), Warwickshire, England.

Remarks
Poorly defined species of uncertain affinities known by small (< 16 cm) and very imperfect specimens. Almost nothing is known of the head but it appears as if the supraorbital sensory pores are relatively large. The fin rays are not expanded and ornament is absent except for a few denticles on the anterior rays of D_1. The scale ornament is distinc-

tive, consisting of a few stout tubercles per scale. The lightly ossified skeleton may suggest this to be a juvenile form. The D_1 support is similar in shape to that of *Holophagus gulo*. Further, the number of neural arches that can be counted from the D_1 support to the middle of the caudal lobe is exactly the same as in *H. gulo*.

Undina grandis Eastman 1914

1914 *Undina grandis* Eastman: 358, pl. 48, fig. 2.

Diagnosis
See Eastman (1914: 358).

Remarks
Eastman erected this species for a single specimen (CM 4748) of a large coelacanth tail from the Upper Jurassic (Tithonian) of Cerin, France. He distinguished it from the coeval *U. cirinensis* Thiollière by size, and from all other species of *Undina* by the fact that the tail is asymmetrical (a stated 25 rays in the upper lobe and 19 in the lower). The drawing of the specimen shows only 18 dorsal caudal rays and the ventral lobe is clearly incomplete anteriorly. Thus, it is unlikely that we can assume that the fin is asymmetrical. The fin rays of D_2, anal and the caudal fin are all expanded (a feature seen in *Holophagus gulo*) and the illustration (Eastman, 1914: pl. 48, fig. 2) shows an unusual feature in that the neural arches and spines are considerably thicker than the matching haemal arches and spines. At present the status of this species cannot be evaluated.

Genus *Wimania* Stensiö 1921

1918 *Leioderma* Stensiö: 121 (name preoccupied).
1921 *Wimania* sp. Stensiö: 79, pl. 9, figs 2, 3.

Diagnosis (emended)
Large coelacanths, head approximately 180 mm long, shallow. Postparietal shield short and strongly embayed posteriorly, less than half the length of the parietonasal shield. Parietonasal shield as broad as the postparietal shield, with a single pair of large parietals bordered by very narrow supraorbitals. Neurocranium ossified as many separate ossifications including lateral ethmoids, basisphenoid, and prootic with posterior wings. The parasphenoid widens markedly at the anterior end and bears tiny teeth over its anterior two-thirds. The palatoquadrate is elongate and shallow anteriorly and broad posteriorly, the dorsal angle between the anterior and vertical limbs is marked by a notch. Lachrymojugal, postorbital, squamosal and preopercular cheek bones are present; although large, they do not overlap each other. There is a well-developed sclerotic ring made up of approximately 20 quadrangular ossicles. In the lower jaw, coronoid 4 bears enlarged and pointed teeth.

Type species: *W. sinuosa* Stensiö 1921

1918 *Leioderma sinuata* Stensiö: 121, fig. 2.
1921 *Wimania sinuosa* Stensiö: 53, figs 19–30, pl. 4–7, pl. 8, fig. 1.
1932 *Wimania sinuosa* Stensiö; Stensiö: fig. 15a.

Diagnosis
Species in which the scales are ornamented with many tiny elongate ridges which are arranged irregularly.

Syntypes
UP P.257, P.255, P.677, Lower Triassic (Scythian), Sassendalen Group, Sticky Keep Formation, West Spitzbergen.

Remarks
Although Stensiö (1921) did not designate a holotype it is clear that UP P.257 is the specimen on which most of the description of this species is based. The remaining two specimens consist only of isolated scales and fin rays.

W. (?) multistriata Stensiö 1921

1921 *Wimania (?) multistriata* Stensiö: 81, figs 31–34, pl. 8, figs 2–7, pl. 9, fig. 1.

Diagnosis
See Stensiö (1921).

Syntypes
UP P.288, P.244–P.250, P.252, P.254, P.674. Lower Triassic (Scythian), Sassendalen Group, Sticky Keep Formation, West Spitzbergen.

Remarks
This is a poorly diagnosed species which Stensiö referred to the genus because of similarity of the scale ornamentation. It is a very much larger species than the type species, being estimated to reach the size of *Mawsonia gigas* but known only by very fragmentary cranial remains, fin rays and scales.

Wimania sp. Stensiö 1921

Remarks
This form is founded on two specimens (UP P.253, P.256) showing parts of the caudal skeleton and scales. It is associated with the type species because of similarity in scale ormanent. The tail shows about 20 fin rays in both dorsal and ventral lobes. This is a high count and may suggest specific status. Unfortunately the tail of the type species is unknown and it is possible that *Wimania* sp. should more properly be referred to *W. sinuosa*.

Genus *Youngichthys* Wang and Liu

1981 *Youngichthys* Wang and Liu: 306.

Diagnosis
See Wang and Liu (1981: 306).

Type and only species:. *Y. xinghuainsis* Wang and Liu

1981 *Youngichthys xinghuansis* Wang and Liu: 306, fig. 3, pl. 3, figs 1–3.

Diagnosis
See Wang and Liu (1981: 306).

Holotype
IVPP V.53315, a complete fish from the Upper Permian, Zhejiang Province, China.

Material
No material of this species was examined.

Remarks
This monotypic genus is poorly known. Unlike the coeval *Changxingia*, this form is said to lack ossified ribs but otherwise appears to be similar.

11.5 COELACANTHS THAT CANNOT BE RECOGNIZED AS DISTINCT GENERA OR SPECIES (Category 3)

Undina picena (Costa 1862)

1862 *Urocomus picenus* Costa: 23.
1896 *Undina picenus* (Costa); Bassani: 179, pl. 11, fig. 1, pl. 15, figs 56–63.
Upper Triassic (Norian), Italy.

Remarks
This coelacanth was named for a fragment of caudal fin. A second specimen was described as *Undina picenus* by Bassani (1896). This is a nearly complete fish but badly preserved. The operculum is rounded and shows regular rounded tubercles. The scales are covered with closely spaced, regular-sized tubercles. Rieppel (1980) considers it to be closely similar to *Ticinepomis*.

Coelacanthopsis curta Traquair 1901

1901 *Coelacanthopsis curta* Traquair: 113.
1905 *Coelacanthopsis curta* Traquair; 84, pl. 5, fig. 4.

Remarks
This is a form described from the Lower Carboniferous (Visean) of Scotland. The holotype (RSM 1950.38.78/79 pt and cpt) is clearly distorted and the shortness of the body, which is supposed to be the distinguishing feature, is a preservational artefact. Moy-Thomas (1937: 14) considered this form to be undefinable.

Coelacanthus dendrites Gardiner 1973

1973 *Coelacanthus dendrites* Gardiner: 33, fig. 1.

Remarks
A species recognized on scales (BMNH P.10510) from the Lower Permian of Somkele, Natal, RSA. These scales show an ornament pattern very similar to that seen on scales of *Rhabdoderma*. Our current knowledge does not justify this as a distinct species.

Coelacanthus stensioei Aldinger 1931

1931 *Coelacanthus stensioi* Aldinger: 191, figs 7–14.

Remarks
This form is named for a single specimen of head fragments from the Namurian E_1 of Germany. The pterygoid is very deep and covered with teeth arranged in whorls. In my view there is nothing specifically distinct about this specimen. It may prove to belong with *Rhabdoderma elegans*.

Coelacanthus sp. Chabakov 1927

Remarks
Based on a coelacanth scale from the Upper Carboniferous (Stepahanian) of the Ukraine.

Coelacanthus sp. Fletcher 1884

Remarks
This is a record of a single specimen of a gular plate from the Upper Carboniferous of Nova Scotia, Canada.

cf. *Rhabdoderma* Forey and Young 1985

Remarks
This coelacanth is represented by a fragmentary head from the Lower Carboniferous (Dinantian), Calciferous Sandstone Series, Scotland. It shows some resemblance to *R. huxleyi* in the proportions of the cheek bones but cf. *Rhabdoderma* displays ornament which is absent from *R. huxleyi*.

? Actinistia gen et sp. indet. Janvier *et al.* 1984

Remarks
This is based on a collection of coelacanth scales from the Lower Carboniferous (Tournasian) Köprülü Shales, Turkey.

? Actinistia gen et sp. indet. Schultze and Möller 1973

Remarks
Based on scales from the Middle Triassic (Anisian) Muschelkalk of Germany.

Coelacanth of Patton and Tailleur 1964

Remarks
Based on a specimen of a partial coelacanth trunk from the Lower Triassic (Scythian) Shublick Formation, Alaska, USA.

Coelacanth of Anderson *et al.* 1994

Remarks
This record is based on 10 specimens of a very small coelacanth (< 6 cm) recognized by the shape and position of the fins as well as the distinctively coelacanth-shaped urohyal. This coelacanth remains to

be described. It has a fully symmetrical caudal fin.

Material
Specimens in the Albany Museum, Grahamstown from the Upper Devonian (Fammenian), Witport Formation, Witteberg Group, Eastern Cape Province, Republic of South Africa.

'Coelacanth' Lelièvre and Janvier 1988

Remarks
Some fragmentary coelacanth remains have been described from the marine Upper Devonian (Fammenian) from two localities in Morocco. These remains consist of part of an ethmosphenoid, a postparietal, supratemporal, a urohyal and a scale. It is possible that they represent a single taxon but there are some slight differences in ornament between the dermal bones preserved and there is insufficient evidence to identify a named taxon. The partial ethmosphenoid shows similarities with other Devonian coelacanths (*Euporosteus* and *Diplocercides* – see p. 171). The supratemporal shows a smooth undersurface and there is no descending process. The scale ornament is very similar to *Diplocercides kayseri* in showing parallel, sharp-crested ridges.

Undina leridae Sauvage

1903 *Undina leridae* Sauvage: 471, pl. 1, fig. 2.
1988 ? '*Holophagus leridae*' (Sauvage); Sanz *et al.*: 622, pl. 2, fig. 7.

Remarks
This species is based on a poorly preserved specimen showing the posterior part of the trunk and the caudal skeleton. The tail appears to show a relatively low number of caudal rays (about 10) in each of the lobes.

Sassenia (?) *guttata* (Woodward) 1912

1912 *Coelacanthus guttatus* Woodward: 291.

Remarks
Lower Triassic (Scythian), Sassendalen Group, Sticky Keep Formation, West Spitzbergen. The single specimen on which this species is based is much smaller than the other species of *Sassenia* and it consists of the posterior half of the head and most of the trunk and is insufficient to refer it to the genus unequivocally. Both the scale ornament of several short ridges and the sparse ornament upon the cranial bones, together with the shape of the operculum are more like those of *Whiteia*.

Coelacanth Wenz 1979

Remarks
'Basisphénoide 1' from the Middle Jurassic (Callovian), northern France.

Coelacanth Wenz 1979

Remarks
'Carré-entopterygoide 1' from the Upper Jurassic (Kimmeridgian), northern France.

Coelacanthe indéterminé Martin and Wenz 1984

Remarks
Isolated scales from Middle Triassic (Ladinian) Muschelkalk of Germany.

Coelacanth cf. *Undina* Forey et al. 1985

Remarks
Partial crushed head (BMNH P.61627) showing spinous dentition on the ceratobranchials similar to that in *Undina*. The single specimen is from the Upper Jurassic (Tithonian) Akkuyu Formation of Turkey.

Coelacanth indet. Schultze and Chorn 1989

Remarks
Isolated scales and an articular from the Upper Carboniferous (Stephanian) Bern Limestone Formation of Kansas, USA.

Coelacanth indet. Martin 1981

Remarks
Isolated angular from the Upper Triassic of Morocco.

Coelacanth indet. Zidek 1975

Remarks
Scales and fragments of a trunk from the Upper Carboniferous (Stephanian) Wild Cow Formation of New Mexico, USA. More material was described by Schultze (1992). The pelvic bone shows some similarity with that of *Rhabdoderma* in being very broad and having a strong medial process which shows interdigitations with the pelvic bone of the opposite side.

Coelacanth indet. Dehm 1956

Remarks
Partial trunk from the Middle Triassic of France.

Coelacanth indet. Gall et al. 1974

Remarks
Complete but very poorly preserved skeleton from the Lower Triassic (Scythian) of Buntsandstein, France.

Coelacanth indet. Berger 1832

Remarks
This is a partial head from the Upper Triassic Keuper of Coberg, Germany. It is one of the earliest coelacanth remains to be described but was not recognized as such, simply being referred to as a fossil fish.

11.6 FOSSIL FISHES DESCRIBED AS COELACANTHS BUT NOMEN NUDA, NOT POSITIVELY RECOGNIZABLE AS COELACANTHS OR KNOWN TO BE OTHER FISHES (Category 4)

Genus *Bogdanovia* Obrucheva 1955

Type and only species:. *B. orientalis* Obrucheva

1955 *Bogdanovia orientalis* Obrucheva.

Diagnosis
See Obrucheva (1955).

Remarks
This species is known from isolated head bones from the Upper Devonian (Frasnian) of Central Kazakstan. According to Cloutier (pers. comm.) this species is a rhizodontiform.

Bunoderma haini de Saez 1940

Remarks
Scattered remains present on core samples from the Upper Jurassic of Argentina. This has been identified as a teleost (Cione and Pereira, 1990).

Coelacanthus abdenensis Traquair 1903

Remarks
Lower Carboniferous (Visean), Scotland, nomen nudum.

Coelacanthus distans, C. hindi, C. spinatus, C. tuberculatus, C. woodwardi Wellburn 1902

Remarks
All from the Upper Carboniferous of England, nomen nuda. Specimens now lost.

Coelacanthus giganteus Winkler 1880

Remarks
Triassic of Germany. Nomen nudum.

Coelacanthus minor Agassiz 1844

Remarks
Middle Triassic of France. Nomen nudum.

Coelacanthus munsteri Agassiz 1844

Remarks
Upper Carboniferous, Lebach, Germany. Nomen nudum and the specimen is a lung-fish, probably *Conchopoma gadiforme* Kner.

Dictyonosteus arcticus Stensiö 1918

Remarks
This taxon is founded on a single specimen of an ethmosphenoid portion of a sarcopterygian braincase from the Middle Devonian of West Sptizbergen. The single specimen has been redescribed by Stensiö (1921) and Jarvik (1942), but there is no clear evidence that this is a coelacanth.

Rhabdoderma (?) aegyptiaca Heide 1955

Remarks
This taxon named for a scale from the Lower Carboniferous of Egypt. It is a rhizodont scale, showing the typical ornament and the well-developed central boss on the inner surface.

Celacantideo Richter 1985

Remarks
These remains from the Lower Permian (Irati Formation) of Brazil are not determinable as coelacanth.

Coelacanth of Ørvig 1986

Remarks
This piece of bone from the Palaeocene of Sweden was described as having a dentine tubercle overlying basal cellular bone. It was recognized as possibly being coelacanth on the basis of comparsions with a section published by Weiler (1935) of *Mawsonia libyca*. If correct this would be the youngest fossil coelacanth known. However, there is nothing diagnostic allowing its identification as a coelacanth.

Coelacanth remains (C) of Gardiner 1966

Remarks
This record is from the Upper Devonian of Alberta, Canada. These remains are those of a dipnoan.

Coelacanth of Dziewa 1980

Remarks
This coelacanth was recognized on a portion of bone from the Lower Triassic Knocklofty Formation of Tasmania, Australia. The bone is not determinable.

Coelacanthidae gen. non det. Woodward 1895

'Coelacanth' Schaeffer 1941

Remarks
This taxon was described from a pectoral fin skeleton from the Jurassic Talbragar Beds, New South Wales, Australia. The fin skeleton is that of an actinopterygian in showing several radials articulating with the shoulder girdle. It bears some resemblance to a polypterid fin.

Coelacanthinien gen. non det. Casier 1961

Remarks
Small pieces of ornamented dermal head bones from the Lower Cretaceous of Zaire. These remains are not determinable.

12

CONCLUSIONS

Throughout this monograph an attempt has been made to update and summarize information about coelacanth fishes. Inevitably the questions of most general interest concern the significance of the single living species, *Latimeria chalumnae*, to our understanding of the origin of tetrapods as well as what it may reveal about the importance of long-lived, low-diversity lineages. In other words we need to reassess the reputation of *Latimeria* as a 'living fossil' by returning to the questions posed in Chapter 1, namely: is *Latimeria* a missing link between fishes and tetrapods?; is *Latimeria* anatomically close to the earliest coelacanths?; does *Latimeria* signify a long, unrepresented fossil record?; and is it an organism with an unusually restricted geographic range?

The systematic position of coelacanths can no longer provide justification for the claim that *Latimeria* is a 'missing link' between fishes and tetrapods and we need to re-evaluate its contribution to knowledge about how vertebrates moved to land. *Latimeria* is a bony fish, it shares with most a complex covering of dermal bones over the braincase, cheek, palate, lower jaw and shoulder girdle which conforms to a common pattern; it shares a similar pattern of gill arch elements and the inclusion of an interhyal within the hyoid arch as well as lepidotrichia supporting the fins, the presence of endochondral bone and a swim/air bladder (although this is reduced and filled with oil). This said, it also retains a large number of ancient vertebrate characters including an unconstricted notochord, a relatively simple organization of

the brain, weakly differentiated endocrine organs and perhaps the most primitive of all adult vertebrate hearts in which the auricle, ventricle and conus arteriosus divisions are laid out in a simple line (Millot *et al.*, 1978), rather than being reflexed over and under one another. Some of these features had led some workers to ally *Latimeria* with cartilaginous fishes. However, in the light of the bony fish characters these similarities are best explained by assuming that in some respects *Latimeria* is a paedomorphic animal.

Amongst bony fishes *Latimeria* is a sarcopterygian showing the single-axis pectoral and pelvic fin structure which characterizes this group, as well as the presence of enamel and of pulmonary and posterior vena cava veins. It also shares with other primitive sarcopterygians an intracranial joint, a characteristic pattern of cheek bones and enclosed sensory canals and a lower jaw in which the tooth-bearing dentary is separate from the sensory canal bones. All or some of these latter features are subsequently lost during the evolution of tetrapods and sarcopterygian fishes.

Studies of the skeletal anatomy and soft anatomical features suggest that *Latimeria* is the most primitive of sarcopterygian fishes. The other living group of sarcopterygian fishes – lungfishes – share many more characters with tetrapods chiefly concerned with the potential to breathe air and involving modifications of the lungs, the associated blood supply and the heart.

The idea that *Latimeria* and other coelacanths are more distantly related to tetrapods than are the lungfishes is only weakly supported by evidence from molecular data and from consideration of skeletons of the great variety of Palaeozoic sarcopterygian fishes. Analysis of molecular data leads to results that are highly conflicting and the different solutions often depend on particular taxon sampling. To some extent this is also true with fossil data. At best we can say that neither class of evidence contradicts strongly the view that coelacanths are the most primitive sarcopterygians. Amongst the fossil groups, then, there is good evidence that the porolepiforms and lungfishes are sister groups. However, the relationships of various 'osteolepiform' taxa, both to one another and to other sarcopterygians remain poorly resolved in the sense that any hypothesis put forward carries the burden of much homoplasy. Some, such as *Panderichthys*, do appear closely related to tetrapods.

Accepting the rather primitive position of coelacanths, we may ask what can *Latimeria* tell us about the origin of tetrapods? There are probably two main areas, locomotion and 'hearing'. Although *Latimeria* does not walk underwater, the paired limb coordination is like that of a tetrapod (and at least one species of lungfish). There have been reports of enlargements of the spinal cord above the pectoral and pelvic fins (Millot and Anthony, 1965) of *Latimeria*. Such enlargements are not present in lungfishes examined so far (Northcutt, 1987). Their presence may well be associated with such coordination and need to be examined in more detail together with the relationship between the spinal cord and the cerebellum. The second area concerns the presence in *Latimeria* and tetrapods of a basilar papilla. In tetrapods this is associated with hearing in air. The function of the basilar papilla in *Latimeria* is as yet unknown, but it would be of interest to investigate.

A unique feature of *Latimeria* among living vertebrates is the possession of the intracranial joint. Because this is present in many fossil sarcopterygians there was great hope

that at last we would understand the function of this unusual structure. In *Latimeria* the joint appears to be a mechanism whereby the gape can suddenly be increased during mouth opening and this is associated with a forward movement of the lower jaw. If we have interpreted the mechanics correctly then the opening depends upon the palate being securely tied only to the ethmosphenoid portion of the braincase and the position of the jaw joint lying well forward close to a vertical level beneath the intracranial joint. Neither of these facts is universal in fossil joint-bearing sarcopterygians. In most, the jaw joint lies well behind the level of the intracranial joint. And in *Eusthenopteron* at least, there is an additional connection between the palate and the posterior portion of the braincase (Jarvik, 1954) meaning that the palate effectively forms a brace between the two halves of the braincase. Furthermore, other sarcopterygians do not have the reduced hyomandibular and symplectic connection between hyoid arch and lower jaw which are important elements in jaw opening. Whether we accept the function of the joint in *Latimeria* as the primitive function within sarcopterygians or, more likely, as a modification of some pre-existing function, the amount of information it can give us about the intracranial joint in general is limited.

Coelacanths are a monophyletic group of sarcopterygians distinguished by several unique characters: (1) a median rostral organ opening by three paired openings; (2) an upright jaw suspension which includes a triangular pterygoid; (3) a tandem jaw articulation; (4) a characteristically shaped lower jaw in which the dentary is very short, the angular is the largest external bone with a large dorsal expansion, and the posteriormost coronoid is large and separated from small anterior coronoids; (5) a single bone (lachrymojugal) beneath the eye; (6) a shoulder girdle containing an extracleithrum bone; (7) a first dorsal fin which is located well forward and lacks radials; (8) the endoskeleton of the second dorsal and the endoskeleton of the anal fin are mirror images of each other; (9) a caudal fin which is supported by a single series of radials distal to the neural and haemal spines. All coelacanths are very unusual but not unique amongst sarcopterygians in lacking a maxilla, branchiostegal rays and submandibulars and in having a subdermal urohyal. All coelacanths have thin, circular and deeply overlapping scales.

As a group they are known as far back as the late Middle Devonian (Givetian). About 80 valid species can be recognized (although this is certainly a lower estimate to be increased when better material is known for some of the doubtful species) in a fossil record which extends 305 million years to the mid-Maastrichtian (75 million years ago). The absence of a Tertiary fossil record is difficult to explain and the explanation given by White (1954a) remains the most likely. He suggested that at the end of the Cretaceous the dwindling coelacanth lineage occupied deeper seas, beyond the continental shelf, from which we have little palaeontological record. However, neither *Latimeria* nor any of its immediate fossil relatives show any obvious structural modifications usually associated with deep-sea dwelling (gross reduction of the skeleton, thinning of the bones, drastic increase in the size of the jaws and/or stomach). One fact may suggest that coelacanths moved to deeper water in the Cenozoic. Most fossil coelacanths, including *Macropoma*, the sister taxon to *Latimeria*, show a large ossified air bladder. Assuming this was, in fact, filled with air during life, then we may suggest these were shallow-water dwellers. The air bladder of *Latimeria* is reduced in size compared with that of most fossil coelacanths and is filled with oil (Millot and Anthony, 1958) and this fact would agree with the adoption of a deeper-water existence. Thomson (1991) added the idea that coelacanths became geographically limited to the western Indian Ocean during the Tertiary. We have no evidence of Tertiary deep-sea fish-bearing sediments from this part of the world. The very existence of *Latimeria* should warn against relying too heavily on such negative evidence to construct a scenario of geographic and habitat restriction. It may be the best we have at present but its fragility will be exposed by a single find elsewhere of either a fossil or living coelacanth. Reassuringly, if a Tertiary coelacanth is found it would be readily identifiable.

The peak of coelacanth diversity occurred in the Lower Triassic but there is a significant peak again in the Upper Jurassic. Throughout their history, coelacanths have mostly been inhabitants of the sea. The notable times where freshwater coelacanths dominate (Upper Carboniferous and Lower Cretaceous) is probably no more than a general reflection of the pattern of the fossil fish record.

The phylogeny of coelacanths shows a regular pectinate lineage of genera throughout the Palaeozoic followed by a dichotomy into two main lineages in the Mesozoic, one of these latter becoming extinct with some of the largest of all coelacanths (*Mawsonia*, *Megalocoelacanthus*).

Morphological evolution of coelacanths has been marked by an overall regressive development of the skeleton involving a 'fragmentation' of neurocranial ossification, reduction of the cheek bone cover, loss of sclerotic ossicles and increase in the numbers of extrascapulars, this last may be thought of as a case of fragmentation. In the early stages of coelacanth evolution the most obvious changes were: (1) 'fragmentation' of the anterior half of the neurocranium into lateral ethmoid and basisphenoid separate ossifications; (2) a probable reduction in brain size so that the brain becomes restricted to the cranial cavity entirely behind the intracranial joint; (3) development of ventral processes from the undersurface of the skull roof to brace the underlying neurocranium; (4) loss of the primitive basipterygoid articulation between the braincase and the palate; (5) acquistion of a symmetrical caudal fin with two principal lobes between which the notochord extends as a small terminal supplementary lobe; and (6) development of a one-to-one correspondence between the caudal fin rays and the radial supports.

In the early Mesozoic the dominant changes concerned: (7) 'fragmentation' of the posterior half of the neurocranium into prootic, basioccipital, opisthotic and supraoccipital ossifications; (8) general reduction in cheek cover (details vary in separate genera); (9) development of denticulate ornament upon the rays of the first dorsal and principal caudal fin rays; (10) the appearance of a distinctively hook-shaped dentary; and (11) an elaboration of the sensory canal system, resulting in multiple openings of the lateral line canal within the lateral line scales and, in many, the development of a subopercular branch of the mandibular canal (in *Latimeria* this branch ramifies through a prominent subopercular flap of skin).

In describing a 'living fossil' as a sole survivor of a once abundant group *Latimeria* is a good example (although we may question whether coelacanths were ever abundant). However, in describing a 'living fossil' as a living example of the primitive member of a lineage, then *Latimeria* does not agree. It is neither a primitive coelacanth, nor a primitive sarcopterygian. There have been numerous, albeit small, morphological changes affecting the lineage to *Latimeria*.

Certainly, coelacanth morphological evolution has been conservative. There are examples of other fish taxa where some 80 species may be said to show less variation (e.g. species of cyprinids such as *Barbus*), but very

few where such limited variation extends over 300 million years (perhaps hagfishes and lampreys might be examples although practically nothing is known of their fossil record).

The claim that the coelacanth lineage shows rapid initial evolution followed by stasis needs to be re-evaluated, or at least redefined. Along the lineage from the most primitive coelacanth leading to *Latimeria*, there was an initial increase in the rate of morphological evolution (character change) which gradually, rather than dramatically, declined after the Carboniferous. At no time did stasis set in. Plotted against cladogenetic event rather than absolute time then there is no such clear pattern and there are three major peaks corresponding with cladogenetic events occurring in Upper Devonian, Upper Permian–Lower Triassic and Upper Jurassic. Finally, if we measure morphological disparity and cladogenesis over the entire coelacanth tree through time, there appears to be a gradual increase in the rate of character change.

One final thought: *Latimeria* turned out to be a superb test case for measuring the palaeontologist's ability to restore fossils. For nearly 100 years, numerous vertebrate palaeontologists had been restoring coelacanth fossils, piecing together the skeleton from flattened or fragmentary fossils in the absence of a Recent template. When the modern species turned up, nearly all their restorations were confirmed, the chief exception being the confusion between nostrils and the hitherto unknown rostral organ. This is powerful endorsement for the science of comparative anatomy and vertebrate palaeontology.

APPENDIX A ABBREVIATIONS USED IN THE FIGURES

A.b — basal plate of anal fin
a.Cat — anterior catazygal
Acl — anocleithrum
a.c.v — anterior cerebral vein
add.lig — adductor ligament
add.md — adductor mandibulae muscle
add.op — adductor opercular muscle
Ana — anazygal
Ang — angular
ant.pr — antotic process
a.o.io.s.c — anterior openings from infraorbital sensory canal
a.p.l — anterior pit line
a.ros — anterior opening for the rostral organ
Art — articular
art.ant.pr — surface for articulation with the antotic process
art.bpt.pr — articulation for basipterygoid process
art.Eb1 — articulation for first epibranchial
art.Ih — articulatory surface for interhyal
Art–Rart — articular–retroarticular
art.Uhy — articulation for urohyal
a.stt.com — anterior branch(es) of supratemporal commissure
Aup — autopalatine
a.w.Par — ascending wing of parasphenoid
ax.mes. 1– — axial mesomeres (numbered)

Bb — basibranchial
b.fen — basicranial fenestra

b.hyp.c — buccohypophysial canal
Boc — basioccipital
bpt.pr — basipterygoid process
Bsph — basisphenoid
bucc.can — buccal canal

Cat — catazygal
Cb — ceratobranchial
Cb1–5 — ceratobranchials, numbered 1–5
Ch — ceratohyal
Cl — cleithrum
Cla — clavicle
Co.1–4 — coronoids, numbered 1–4
con.Part — contact surface with prearticular
c.p.l — cheek pit line
c.v.on — canal for orbitonasal vein

De — dentary
det — teeth fused to dentary
d.p — enlarged sensory pore within dentary
Dpl — dermopalatine
d.s — dorsum sellae
D1.b — basal plate supporting first dorsal fin
D2.b — basal plate supporting second dorsal fin

Eb1–4 — epibranchials, numbered 1–4
Ecl — extracleithrum
Ecpt — ectopterygoid
e.s.p — enlarged pores leading from jugal sensory canal

ET	ethmosphenoid portion of braincase	f.VII.m.int	foramen for internal mandibular ramus of VII
eth.com	ethmoid commissure		
exc.Po	excavation within postorbital	gr.a.stt.com	groove for anterior branch of supratemporal commissure
Exo	exoccipital		
Ext	extrascapular	gr.eth.com	groove for ethmoid commissure
Ext.l	lateral extrascapular		
Ext.m	median extrascapular	gr.j.v	groove for jugular vein
		gr.na.pap	groove for posterior nasal papilla
fa.Hy	facet for hyomandibular		
fa.i.j	facet for intracranial joint	gr.ot.s.c	groove for otic sensory canal
f.br.ot.VII	foramen for branches of the otic ramus of facial supplying otic sensory canal	gr.p.ros	groove for posterior openings from rostral organ
		gr.stt.X	groove for supratemporal branch of vagus
f.br.s.opth	foramen for branch of the superficial ophthalmic	gr.VII.m.ext	groove for external mandibular ramus of VII
f.bucc	foramen for buccal nerves	Gu	gular plate
fen	fenestra endonarina communis	gu.p.l	gular pit line
		Hy	hyomandibular
f.hy.VII	foramen for hyomandibular trunk of facial	hy.VII	hyomandibular trunk of the facial
f.i.c.a	foramen for internal carotid		
f.j.v	foramen for jugular vein		
f.m	foramen magnum		
f.m.s.c	foramen leading to mandibular sensory canal	i.ar–hy.lig	insertion point for articular–hyomandibular ligament
f.o.a	foramen for orbitonasal artery	Ib2	second infrapharyngobranchial
f.ot.VII	foramen for otic ramus of facial		
		i.c.a	internal carotid artery
f.pal.VII	foramen for palatine branch of facial	Icla	interclavicle
		Ih	interhyal
f.pop.s.c	opening for preopercular sensory canal	i.j	intracranial joint
		Ina	internasal
f.r	fin ray	ino.oss	interorbital ossification
f.s.opth	foramen for superficial ophthalmic	io.s.c	infraorbital sensory canal
		Int	intertemporal
f.stt.IX	foramen for supratemporal branch of the glossopharyngeal	j.c	jugular canal
		j.s.c	jugal sensory canal
f.stt.X	foramen for supratemporal branch of the vagus supplying otic sensory canal and supratemporal commissure.	j.v	jugular vein
		L.e	lateral ethmoid
		lev.pal	palatal levator muscle
f.V.m	foramen for branch of V mandibular	L.j	lachrymojugal
		l.l	lateral line
f.VII.m.ext	foramen for external mandibular ramus of VII	l.l.s	lateral line scale
		L.r	lateral rostral

M	mandible	P	palate
mand.V	mandibular branch of trigeminal	Pa	parietal
		pal.VII	palatine branch of facial
max+bucc.V	maxillary and buccal branches of trigeminal	pa.pr	parampullary process
		Par	parasphenoid
max.V	maxillary branch of trigeminal	Part	prearticular
me	mesomere	P.b	pelvic bone
Mm	mentomeckelian	Pb1	first pharyngobranchial
m.p.l	middle pit line	p.Cat	posterior catazygal
Mpt	metapterygoid	p.Co	principal coronoid
m.s.c	mandibular sensory canal	p.eth.com	pore leading to ethmoid commissure
my.ant	anterior myodome		
		PG	pectoral girdle
Na	nasal	pit.fos	pituitary fossa
n.a	neural arch	pit.v	pituitary vein
n.c	notochordal canal	p.l.l	pores opening from lateral line
nos.a	anterior nostril		
nos.p	posterior nostril	Pmx	premaxilla
n.p	notochordal pit	pmx.t	premaxillary teeth
		Po	postorbital
o.De	overlap surface for dentary	Pop	preoperculum
o.Gu	overlap surface for gular plate	pop.s.c	preopercular sensory canal
o.n.c	orbitonasal canal	post.pr	postaxial process
on.v	orbitonasal vein	Pp	postparietal
Op	operculum	p.p.l	posterior pit line
o.p.l	oral pit line	pr.con	processus connectens
op.lig	insertion point for opercular ligament	preax.rad	preaxial radials
		Preo	preorbital
o.Po	overlap surface for postorbital	pr.lig	process for prefacial ligament
o.Pop	overlap surface for preoperculum	Pro	prootic
		p.ros	posterior opening(s) from the rostral organ
o.Preo	overlap surface for preorbital		
OT	otico-occipital portion of braincase	psmax	pseudomaxillary fold
		p.so.s.c	matrix infilling of pores leading from supraorbital sensory canal
ot.s.c	otic sensory canal		
ot.s.c.m	medial branch of otic sensory canal		
		p.s.s.c	posterior semicircular canal
ot.sh	otic shelf	Pt	pterygoid
ot.VII	otic ramus of facial (now called middle lateral line nerve)	p.w.Pro	posterior wing of prootic
		Q	quadrate
o.v.pr.Pa	overlap surface for descending process of parietal	Rart	retroarticular
o.v.pr.Pp	overlap surface for descending process of postparietal	rart/hm.lig	retroarticular/hyomandibular ligament
o.v.pr.Stt	overlap surface for descending process of supratemporal	ros.m	median rostral
		ros.oss	rostral ossicles

Ros.Pmx	rostropremaxilla	t.p.Bb	basibranchial tooth plate(s)
		t.p.d	dentary tooth plates
sac.ch	saccular chamber		
Sb2	suprapharyngobranchial	Uhy	urohyal
Scc	scapulocoracoid		
So	supraorbital	v.f	vestibular fontanelle
S.o	sclerotic ossicles	v.l.fo	ventrolateral fossa
Soc	supraoccipital	Vo	vomer
Sop	suboperculum	v.pit	foramen for pituitary vein
sop.br	subopercular branch of the	v.pr.L.r	ventral process of lateral rostral
	preopercular canal	v.pr.Pa	ventral (descending) process
sop.fl	subopercular flap		of the parietal
s.opth	superficial ophthalmic	v.pr.Pp	ventral (descending) process
	(anterodorsal lateral line		of postparietal
	nerve)	v.pr.Stt	ventral (descending) process
so.s.c	supraorbital sensory canal		of supratemporal
Sp	spiracular		
sph.c	sphenoid condyle	v.X	visceral branch of vagus
Spl	splenial		
spt.fos	suprapterygoid fossa	y.s	yolk sac
spt.pr	suprapterygoid process		
Sq	squamosal	II	optic foramen
Stt	supratemporal	III	oculomotor foramen
stt.com	supratemporal commissure	IV	trochlear foramen
stt.X	supratemporal branch of	V.m	mandibular ramus of v
	vagus	V1	profundus foramen
sut.p.Co	sutural contact surface with	VI	abducens
	principal coronoid	VII	facial foramen
Sy	symplectic	VII.m.ext	external mandibular ramus of
			the facial
Te	tectal	VII.m.int	internal mandibular ramus of
t.fos	temporal fossa		the facial
t.p	tooth plate	IX	glossopharyngeal foramen
t.par	toothed area of parasphenoid	X	vagus foramen

REFERENCES

Abel, O. (1931) Schwimmfährten von Fischen und Schildkröten im lithographischen Schiefer Bayerns. *Natur und Museum, Frankfurt*, **61**: 97–106.

Adamicka, P. and Ahnelt, H. (1976) Beiträge zur funktionellen Analys und zur Morphologie des Kaptes von *Latimeria chalumnae* Smith. *Annalen des (K.K.) Naturhistorischen (Hof) Museums, Wien*, **80**: 251–271.

Agassiz, L. (1835) *Recherches sur les poissons fossiles.* Vol. II. 4th Livraison. Feuilleton, pp. 35–64. Neuchâtel.

Agassiz, L. (1839) *Recherches sur les poissons fossiles.* Vol. 2. Neuchâtel.

Agassiz, L. (1844) *Recherches sur les Poissons fossiles.* Neuchâtel **2**(2): 1–336.

Ahlberg, P.E. (1989a) Paired fin skeletons and the relationships of the fossil group Porolepiformes (Osteichthyes: Sarcopterygii). *Zoological Journal of the Linnean Society of London*, **96**: 119–166.

Ahlberg, P.E. (1989b) The anatomy and phylogeny of porolepiform fishes, with special reference to *Glyptolepis*, PhD thesis, University of Cambridge, 359 pp.

Ahlberg, P.E. (1991) A re-examination of sarcopterygian interrelationships, with special reference to the Porolepiformes. *Zoological Journal of the Linnean Society of London*, **103**: 241–287.

Ahlberg, P.E. (1992) Coelacanth fins and evolution. *Nature, London*, **373**: 459.

Ahlberg, P.E. (1995) *Elginerpeton pancheni* and the earliest tetrapod clade. *Nature, London*, **373**: 420–425.

Ahlberg, P.E. and Trewin, N. (1994) The post-cranial skeleton of the Middle Devonian lungfish *Dipterus valenciennesi*. *Transactions of the Royal Society of Edinburgh: Earth Sciences*, **85**: 159–175.

Ahlberg, P.E., Lukevĭcs, E. and Lebedev, O. (1994) The first tetrapod finds from the Devonian (Upper Fammenian) of Latvia. *Philosophical Transactions of the Royal; Society of London*, **343B**: 303–-328.

Ahlberg, P., Clack, J.A. and Lukevĭcs, E. (1996) Rapid braincase evolution between *Panderichthys* and the earliest tetrapods. *Nature, London*, **381**: 61–64.

Aldinger, H. (1930) Über das Kopfskelett von *Undina acutidens* Reis und den kinetischen Schädel der Coelacanthiden. *Zentralblatt für Mineralogie, Geologie und Paläontologie, Stuttgart, B*, **1930**: 22–48.

Aldinger, H. (1931) Über karbonische Fische aus Westfalen. *Palaeontologische Zeitschrift, Berlin*, **13**: 186–201.

Alexander, R. McN. (1973) Jaw mechanisms of the Coelacanth *Latimeria chalumnae. Copeia*, **1973**: 156–158.

Allis, E.P. (1889) The anatomy and development of the lateral line sytem in *Amia calva. Journal of Anatomy, Boston*, **2**: 463–540.

Allis, E.P. (1897) The cranial muscles and cranial and first spinal nerves in *Amia calva. Journal of Morphology, Boston*, **12**: 487–808.

Allis, E.P. (1899) On certain homologies of the squamosal, intercalar, exoccipital, and extrascapular bones of *Amia calva. Anatomischer Anzeiger, Jena*, **16**: 49–72.

Allis, E.P. (1900) The lateral sensory canals of *Polypterus bichir. Anatomischer Anzeiger, Jena*, **17**: 433–451.

Allis, E.P. (1922) The cranial anatomy of *Polypterus*, with special reference to *Polypterus bichir. Journal of Anatomy, London*, **56**: 189–294.

Allis, E.P. (1923) The cranial anatomy of *Chlamydoselachus anguineus. Acta zoologica, Stockholm*, **4**: 123–221.

Allis, E.P. (1934) Concerning the course of the latero-sensory canals in Recent fishes, prefishes and *Necturus. Journal of Anatomy, London*, **68**: 361–415.

Allis, E.P. (1935) On a general pattern of arrangement of the cranial roofing bones in fishes. *Journal of Anatomy, Cambridge*, **69**: 233–291.

Anderson, M.E., Hiller, N. and Gess, R.W. (1994)

The first *Bothriolepis*-associated Devonian fish fauna from Africa. *South African Journal of Science*, **90**: 397–403.

Andrews, S.M. (1971) The shoulder girdle of 'Eogyrinus'. In Joysey, K.A. and Kemp, T.S. (eds), *Studies in vertebrate evolution*, pp. 35–50, Edinburgh: Oliver & Boyd.

Andrews, S.M. (1973) Interrelationships of crossopterygians. In Greenwood, P.H., Miles, R.S. and Patterson, C. (eds), *Interrelationships of Fishes*: pp. 137–177, London: Academic Press.

Andrews, S.M. (1977) The axial skeleton of the coelacanth, *Latimeria*. In Andrews, S.M., Miles, R.S. and Walker, A.D. (eds), *Problems in Vertebrate Evolution*, pp. 271–288, London: Academic Press.

Andrews, S.M. (1985) Rhizodont crossopterygian fish from the Dinantian of Foulden, Berwickshire, Scotland, with a re-evaluation of this group. *Transactions of the Royal Society of Edinburgh: Earth Sciences*, **76**: 67–95.

Andrews, S.M. and Westoll, T.S. (1970a) The postcranial skeleton of *Eusthenopteron foordi* Whiteaves. *Transactions of the Royal Society of Edinburgh*, **68**: 207–329.

Andrews, S.M. and Westoll, T.S. (1970b) The postcranial skeleton of rhipidistian fishes excluding *Eusthenopteron*. *Transactions of the Royal Society of Edinburgh*, **68**: 391–489.

Anthony, J. (1976) *Operation Coelacanth*. 197 pp. Paris: Arthaud.

Anthony, J. and Robineau, D. (1976) Sur quelques charactères juvéniles de *Latimeria chalumnae* Smith (Pisces, Crossopterygii Coelacanthidae). *Compte rendu de l'Academie des Sciences, Paris*, Series D, **283**: 1739–1742.

Atz, J.W. (1976) *Latimeria* babies are born, not hatched. *Underwater Naturalist*, New York, **9**: 4–7.

Balon, E.K., Bruton, M.N. and Fricke, H. (1988) A fiftieth anniversary reflection on the living coelacanth, *Latimeria chalumnae*: some new interpretations of its natural history and conservation status. *Environmental Biology of Fishes*, **32**: 241–280.

Bang, B.S. (1976) *Avanceret kemisk praepartionmetodik egnet for palaeontologiske og kulturhistoriske objekter en material-og metodelaere.* 59 pp., København: Geologisk Museum.

Bassani, F. (1896) La Ittiofauna della Dolomia Principale di Giffoni (Provincia di Salerno). *Palaeontographia Italica* , Pisa, **1**: 169–210.

de Beer, G.R. (1926) Studies on the vertebrate head. II. The orbito-temporal region of the skull. *Quarterly Journal of Microscopical Science*, London, **70**: 263–370.

de Beer, G.R. (1937) *The Development of the Vertebrate Skull.* 552 pp. Oxford: University Press.

Belloti, C. (1857) Descrizione di alcune nuove specie di pesci fossili di Perledo e di altre località Lombarde. In Stoppani, A. (ed.). *Studii Geologici e Paleontologici sulla Lombardia*, pp. 419–438. Milano: Presso Carlo Turati Tipografo–Editore.

Beltan, L. (1968) *La faune ichthylogique de l'Eotrias du N.W. de Madagascar: le neurocrane.* 135 pp. Paris: C.N.R.S.

Beltan, L. (1972) La faune ichthyologique du Muschelkalk de la Catalogne. *Memorias de la Real Academia de Ciencias y Artes de Barcelona*, **41**: 280–325, 12 figs.

Beltan, L. (1979) Eotrias du Nord-Ouest de Madagascar: étude de quelques poissons dont un est en parturition. *Annales de la Societe Geologique du Nord, Lille*, **99**: 453–464.

Beltan, L. (1984) Quelques poissons du Muschelkalk superieur d'Espagne. *Acta Geologica Hispanica*, Barcelona, **19**: 117–127.

Bemis, W.E. and Hetherington, T.E. (1982) The rostral organ of *Latimeria chalumnae*: morphological evidence of an electroreceptive function. *Copeia*, **1982**: 467–471.

Bemis, W.E. and Northcutt, T.E. (1991) Innervation of the basicranial muscle of *Latimeria chalumnae*. *Environmental Biology of Fishes*, **32**: 147–157.

Bendix-Almgreen, S.E. (1976) Palaeovertebrate faunas of Greenland. In Escher, A. and Watt, W.S. (eds), *Geology of Greenland*, pp. 536–573, Copenhagen: The Geological Survey of Greenland.

Berg, L.S. (1940) Classification of fishes, both recent and fossil. *Trudy Zoologicheskogo Instituta, Akademiya Nauk SSSR*, Leningrad, **5**: 1–431. (In Russian with English summary)

Berg, L.S. (1955) *Classification of fishes and fish-like vertebrates* (Second Edition). *Trudy Zoologicheskogo Instituta, Leningrad*, **20**: 1–286.

Berger, H.A.C. (1832) *Die Versteinerungen der Fische und Pflanzen im Sandsteine der Coburger Gegend.* 29 pp., Coburg.

Bjerring, H.C. (1967) Does a homology exist between the basicranial muscle and the polar cartilage? *Colloques internationaux du Centre National de la Recherche Scientifique*, **163**, 223–268.

Bjerring, H.C. (1971) The nerve supply to the second metamere basicranial muscle in osteolepiform vertebrates, with some remarks on the basic composition of the endocranium. *Acta Zoologica*, Stockholm, **52**: 189–225.

Bjerring, H.C. (1972) The *nervus rarus* in coelacanthiform phylogeny. *Zoologica Scripta*, Stockholm, **1**: 57–68.

Bjerring, H.C. (1973) Relationships of coelacanthiforms. In Greenwood, P.H., Miles, R.S. and Patterson, C. (eds), *Interrelationships of Fishes*, pp. 179–205, London: Academic Press.

Bjerring, H.C. (1977) A contribution to the structural analysis of the head of craniate animals. *Zoologica Scripta*, Stockholm, **6**: 127–183.

Bjerring, H.C. (1978) The 'intracranial joint' versus the 'ventral otic fissure'. *Acta Zoologica*, Stockholm, **59**: 203–214.

Bjerring, H.C. (1985) Facts and thoughts on piscine phylogeny. In *Evolutionary Biology of Primitive Fishes*. NATO ASI series. Series A, Life Sciences 103, pp. 31–57. New York: Plenum Press.

Bjerring, H.C. (1986) Tofsstjärtfiskarnas elsinnorgan. *Fauna och Flora*, Uppsala, **81**: 215–222.

Bock, W. (1959) New eastern American Triassic fishes and Triassic correlations. *National Academy of Sciences, Geological Center Research Series*, Philadelphia, **1**: 1–184,.

Borgen, U.J. (1983) Homologizations of skull roofing bones between tetrapods and osteolepiform fishes. *Palaeontology*, Cambridge, **26**: 735–753.

Braus, H. (1901) Die Muskeln und Nerven der Ceratodusflosse. In Semon, R., *Zoologische Forsschungsreisen in Australien und dem Malayischen Archipelago. I. Ceratodus*. Part 3, pp. 139–300, Jena: Gustav Fischer Verlag.

Broili, F. (1933) Der Obere Jura von Montsech (Provinz Lérida) im Vergleich mit den Ob. Jura-Vorkommen von Cerin (Dept. Ain) und von Franken. *Congrés Geologique Internationale*. 14th Session, Barcelona, *Géologie de la Méditeranée Occidentale*, **2**, no. 16: 1–11.

Broom, R. (1905) On a species of *Coelacanthus* from the Upper Beaufort Beds of Aliwal North. *Records of the Albany Museum*, Grahamstown, **1**: 338–339.

Broom, R. (1909) The fossil fishes of the Upper Karoo Beds of South Africa. *Annals of the South African Museum*, East London, **7**: 251–269.

Brough, J. (1931) On fossil fishes from the Karroo System and some general considerations of the bony fishes of the Triassic period. *Proceedings of the Zoological Society of London*, **1931**: 235–269.

Brough, M.C. and Brough, J. (1967) Studies on early tetrapods. I. The Lower Carboniferous

microsaurs. II *Microbrachius*, the type microsaur. III The genus *Gephyrostegus*. *Philosophical Transactions of the Royal Society of London*, **252B** : 107–165.

Browne, M.W. (1995) Fish that dates back to age of dinosaurs is verging on extinction. *The New York Times*, Tuesday, April 18: C4.

Bruton, M.N. (1988) The Coelacanth Conservation Council. *Environmental Biology of Fishes*, **23**: 315–319.

Bruton, M.N. (1989) The coelacanth – can we save it from extinction? *WWF Reports*, October/November 1989: 10–12.

Bruton, M.N. and Armstrong, M.J. (1991) The demography of the coelacanth *Latimeria chalumnae*. *Environmental Biology of Fishes*, **32**: 301–311.

Bruton, M.N. and Coutouvidis, S.E. (1991) An inventory of all known specimens of the coelacanth *Latimeria chalumnae*, with comments on trends in catches. *Environmental Biology of Fishes*, **32**: 371–379.

Bruton, M.N. and Stobbs, R.E. (1991) The ecology and conservation of the coelacanth *Latimeria chalumnae*. *Environmental Biology of Fishes*, **32**: 313–339.

Bruton, M.N., Coutouvidis, S.E. and Pote, J. (1991) Bibliography of the living coelacanth *Latimeria chalumnae* with comments on publication trends. *Environmental Biology of Fishes*, **32**: 403–433.

Bruton, M.N., Cabral, A.J.P. and Fricke, H. (1992) First capture of a coelacanth, *Latimeria chalumnae* (Pisces, Latimeriidae), off Mozambique. *South African Journal of Science*, **88**: 225–227.

Bryant, W.L. (1934) New fishes from the Triassic of Pennsylvania. *Proceedings of the American Philosophical Society*, Philadelphia, **73**: 319–326.

Burger, J.W. and Hess, W.N. (1960) Function of the rectal gland in the spiny dogfish. *Science*, Washington, **131**: 670–671.

Burton, M. (1954) *Living Fossils*. 282 pp., London: Thames & Hudson.

Campbell, K.S.W. and Barwick, R.E. (1982) A new species of the lungfish *Dipnorhynchus* from New South Wales. *Palaeontology*, London, **25**: 509–527.

Campbell, K.S.W. and Barwick, R.E. (1983) Early evolution of dipnoan dentitions and a new genus *Speonesydrion*. *Memoirs of the Association of Australian Palaeontologists*, Sydney, **5**: 17–49.

Campbell, K.S.W. and Barwick, R.E. (1984a) *Speonesydrion*, an early Devonian dipnoan with primitive toothplates. *Palaeo Ichthyologica*, Munich, **2**: 1–48.

Campbell, K.S.W. and Barwick, R.E. (1984b) The choana, maxillae, premaxillae and anterior palatal bones of early dipnoans. *Proceedings of the Linnean Society of New South Wales*, Sydney, **107**: 147–170.

Campbell, K.S.W. and Barwick, R.E. (1985) An advanced massive dipnorhynchid lungfish from the Early Devonian of New South Wales. *Records of the Australian Museum*, Sydney, **37**: 301–316.

Campbell, K.S.W. and Barwick, R.E. (1987) Paleozoic lungfishes – a review. In Bemis, W.E., Burggren, W.W. and Kemp, N. (eds). *The Biology and Evolution of Lungfishes*, pp. 93–131, New York: Alan R. Liss.

Campbell, K.S.W. and Barwick, R.E. (1988) *Uranolophus*: a reappraisal of a primitive dipnoan. *Memoirs of the Association of Australian Palaeontologists*, Sydney, **7**: 87–144.

Campbell, K.S.W. and Barwick, R.E. (1990) Paleozoic dipnoan phylogeny: functional complexes and evolution without parsimony. *Paleobiology*, Chicago, **16**: 143–169.

Campos, A.D. and Wenz, S. (1982) Première découverte de Coelacanthes dans le Crétacé inférieur de la Chapada do Araripe (Brésil). *Compte Rendu de l'Academie des Sciences, Paris*, **294**: 1151–1153.

Carvalho, M.S.S. de (1982) O Gênero *Mawsonia* na ictiofáunula do Cretáceo do Estado da Bahia. *Anais da Academia Brasileira de Ciencias*, **54**: 519–539.

Casier, E. (1961) Materiaux pour la faune ichthyologique Eocretacique du Congo. *Annales Musee Royal de l'Afrique Centrale, Tervuren, Sciences Geologiques*, **39**: 1–96.

Casier, E. (1965) Poissons fossiles de la serie du Kwango (Congo). *Annales du Musee Royal de L'Afrique Centrale, Tervuren, Sciences Geologiques*, **50**: 1–64.

Casier, E. (1969) Addenda aux connaissances sur la faune ichthyologique de la série du Bokunga (Congo). *Annales du Musee Royal de L'Afrique Centrale, Tervuren, Sciences Geologiques*, **62**: 1–20.

Chabakov, A.V. (1927) Sur les restes des Crossoptérygiens du Carbonifère russe. *Izvestiya Geoloicheskago Komiteta*, Sanktpeterburg, **46**: 299–309, 1 pl. [French résumé.]

Chang, M.-M. (1982) *The braincase of* Youngolepis, *a Lower Devonian Crossopterygian from Yunnan, South-western China*. 113 pp., 21 figs, 13 pls. Stockholm: University of Stockholm.

Chang, M.-M. (1991a) 'Rhipidistians', dipnoans and tetrapods. In Schultze, H.-P. and Trueb, L. (eds), *Origins of Higher Groups of Tetrapods*, pp. 3–28, 37 figs, Ithaca, NY: Comstock Publishing Associates.

Chang, M.-M. (1991b) Head exoskeleton and shoulder girdle of *Youngolepis*. In Chang, M., Liu, Y. and Zhang, G. (eds), *Early Vertebrates and Related Problems of Evolutionary Biology*, pp. 355–378, Beijing: Science Press.

Chang, M.-M. (1995) *Diabolepis* and its bearing on the relationships between porolepiforms and dipnoans. *Bulletin du Muséum National d'Histoire Naturelle*, Paris, Section C, 8 Series, **17**: 235–268.

Chang, M.-M. and Yu, X. (1984) Structure and phylogenetic significance of *Diabolichthys speratus* gen. et sp. nov., a new dipnoan-like form from the Lower Devonian of Eastern Yunnan, China. *Proceedings of the Linnean Society of New South Wales*, Sydney, **107**: 171–184.

Cione, A.L. and Pereira, S.M. (1990) Los peces del Jurásico posterior a los movimientos intermálmicos y del Cretácico inferior de Argentina. In *Bioestratigrafia de los Sistemas Regionales del Jurásico y Cretácico en América del Sur. Vol. 1: Jurásico Anterior a los Movimientos Intermálmicos* (eds Volkheimer, W. and Musacchio, E.A.), pp. 385–395, Mendoza, Argentina.

Clack, J.A. (1988) New material of the early tetrapod *Acanthostega* from the Upper Devonian of East Greenland. *Palaeontology*, London, **31**: 699–724.

Clack, J.A. (1996) Otoliths in fossil coelacanths. *Journal of Vertebrate Paleontology*, Chicago, **16**: 168–171.

Cloutier, R. (1991a) Interrelationships of Palaeozoic actinistians: patterns and trends. In Chang, M.-M., Liu, Y.-L. and Zhang, G.-N. (eds) *Early Vertebrates and Related Problems of Evolutionary Biology*. pp. 379–428, Beijing: Science Press.

Cloutier, R. (1991b) Patterns, trends, and rates of evolution within the Actinistia. *Environmental Biology of Fishes*, **32**: 23–58.

Cloutier, R. (1996) The primitive actinistian *Miguashaia bureaui* Schultze (Sarcopterygii). In *Devonian Fishes and Plants of Miguasha, Quebec, Canada* (eds Schultze, H.-P. and Cloutier, R.), pp. 227–247, MünchenL Verlag Dr Friedrich Pfeil.

Cloutier, R. and Ahlberg, P.E. (1995) Sarcopterygian interrelationships: how far are we from a phyloegentic consensus? *Geobios*, Lyon, Special Memoir no. 19: 241–248.

Cloutier, R. and Ahlberg, P.E. (1996) Morphology, characters, and interrelationships of basal sarcopterygians. In *Interrelationships of fishes* (eds Stiassny, M.L.J., Parenti, L.R. and Johnson, G.D.), pp. 445–479, San Diego: Academic Press.

Cloutier, R. and Forey, P.L. (1991) Diversity of extinct and living actinistian fishes (Sarcopterygii). *Environmental Biology of Fishes*, 32: 59–74.

Coates, M. and Clack, J.A. (1990) Polydactyly in the earliest known tetrapod limbs. *Nature*, London, 347: 66–69.

Coates, M. and Clack, J.A. (1991) Fish-like gills and breathing in the earliest known tetrapod. *Nature*, London, 352: 234–235.

Cope, E.D. (1871) Contribution to the ichthyology of the Lesser Antilles. *Transactions of the American Philisophical Society*, Philadelphia, 14: 445–483.

Cope, E.D. (1873) On some new Batrachia and fishes from the coal measures of Linton, Ohio. *Proceedings of the Academy of Natural Sciences of Philadelphia*, 1873: 340–343.

Cope, E.D. (1894) New and little known Paleozoic and Mesozoic fishes. *Journal of the Academy of Natural Sciences of Philadelphia*, Second Series, 9: 427–448.

Costa, O.G. (1862) Studii sopra i terreni ad ittiolitti del Regno di Napoli diretti a stabilire l'età geologica dei medesimi. *Appendice Agli Atti Real Accadeimia delle Scienze*, Napoli, 12: 1–44.

Courtenay-Latimer, M. (1979) My story of the first coelacanth. *Occasional Papers of the California Academy of Sciences*, San Francisco, number 134: 6–10.

Courtenay-Latimer, M. (1989) Reminiscences of the discovery of the coelacanth, *Latimeria chalumnae*. *Cryptozoology*, Tucson, 8: 1–11.

Cracraft, J. (1968) Functional morphology and adaptive significance of cranial kinesis in *Latimeria chalumnae* (Coelacanthini). *American Zoologist*, 8: 354.

Darwin, C. 1859) *On the Origin of Species*. (1960 reprint), London: John Murray.

Davis, J.W. (1876) On a bone bed in the Lower Coal-Measures, with an enumeration of the fish remains of which it is principally composed. *Quarterly Journal of the Geological Society of London*, 32: 332–340.

Davis, J.W. (1883) On the fossil fishes of the Carboniferous Limestone Series of Great Britain. *Scientific Transactions of the Royal Dublin Society*, 1: 327–600.

Davis, J.W. (1884) On a new species of *Coelacanthus (C. tingleyensis)* from the Yorkshire Cannel coal. *Transactions of the Linnean Society of London*, 2: 427–433.

Davis, J.W. (1891) On *Coelacanthus phillipsii* Agassiz. *Geological Magazine*, London, [3], 7: 159–161.

de Saez, M. Dolgopol (1940) Noticias sobre peces fósiles Argentinos. *Notas de Museo de la Plata*, Buenos Aires, 5: 295–298.

De-Alessandri, G. (1910) Studii sui pesci Triasici della Lombardia. *Memorie della Società Italiana di Scienze Naturali e del Museo Civico di Storia Naturale di Milano*, 7: 1–145.

Dean, B. (1895) *Fishes Living and Fossil. An Outline of their Forms and Probable Relationships*. 300 pp. New York: Macmillan & Co.

Dehm, R. (1956) Ein Coelacanthide aus dem Mittleren Keuper Frankens. *Neues Jahrbuch für Geologie und Paläontologie, Monatsheft*. Stuttgart, 1956: 148–153.

Denison, R.H. (1968) The evolutionary significance of the earliest known lungfish, *Uranolophus*. In Ørvig, T. (ed.), *Current Problems of Lower Vertebrate Phylogeny. Nobel Symposium 4*, Stockholm, 4: 247–257.

Denison, R.H. (1978) Placodermi. In Schultze, H.P. (ed.), *Handbook of Paleoichthyology*, 2: 1–128, Stuttgart: Gustav Fischer Verlag.

Dixon, F. (1850) *The Geology and Fossils of the Tertiary and Cretaceous Formations of Sussex*. 422 pp., London: Richard & John Edward Taylor.

Dziewa, T.J. (1980) Early Triassic osteichthyans from the Knocklofty Formation of Tasmania. *Papers and Proceedings of the Royal Society of Tasmania*, Hobart, 114: 145–160.

Eastman, C.R. (1902) The Carboniferous fish-fauna of Mazon Creek, Illinois. *Journal of Geology*, Chicago, 10: 535–541.

Eastman, C.R. (1908) Notice of a new coelacanth fish from the Iowa Kinderhook. *Journal of Geology*, Chicago, 16: 357–362.

Eastman, C.R. (1914) Catalogue of the fossil fishes in the Carnegie Museum. Part III. Catalogue of the fossil fishes from the Lithographic Stone of Cerin, France. *Memoirs of the Carnegie Museum*, Pittsburg, 6: 349–388.

Echols, J. (1963) A new genus of Pennsylvanian fish (Crossopterygii, Coelacanthiformes) from Kansas. *University of Kansas Publications. Museum of Natural History*, Lawrence, 12: 475–501.

Edgeworth, F.H. (1935) *The Cranial Muscles of Vertebrates*. 493 pp. Cambridge: The University Press.

Egerton, T. (1850) Family Coelacanthidae, Agassiz. pp. 325–326. In King, W. (ed.) *A Monograph of the Permian Fossils of England.* pp. 235–236, London: Palaeontographical Society.

Egerton, P. (1861) *Holophagus gulo.* In Huxley, T.H., Preliminary essay upon the systematic arrangement of the fishes of the Devonian epoch. *Memoirs of the Geological survey of the United Kingdom,* London, Dec. **10**: 1–40.

Egerton, P. (1868) On the characters of some new fossil fish from the Lias of Lyme Regis. *Quarterly Journal of the Geological Society of London,* **24**: 499–505.

Eldredge, N. and Stanley, S.M. (eds) (1984) *Living Fossils,* New York: Springer Verlag.

Elliott, D.K. (1987) A new specimen of *Chinlea sorenseni* from the Chinle Formation, Dolores River, Colorado. *Journal of the Arizona–Nevada Academy of Science,* Tucson, **22**: 47–52.

Emeric, C.M. and Duncan, R.A. (1982) Age progressive volcanism in the Comores Archipelago, Western Indian Ocean and implications for Somali plate tectonics. *Earth and Planatary Science Letters,* **60**: 415–428.

Emmons, E. (1857). *American Geology, Containing a Statement of the Principles of the Science, with Full Illustrations of the Characteristic American Fossils.* Vol. I, Part 6, 158 pp. 13 pls. Albany, NY: Sprague & Co.

Escher, K. (1925) Das Verhalten der Seitenorgane der Wirbeltiere und ihrer Nerven beim Übergang zum Landleben. *Acta zoologica,* Stockholm, **6**: 307–414.

Fabre, J., Broin, F. de, Ginsburg, L. and Wenz, S. (1982) Les vertébrés du Berriasien de Canjuers (Var, France, et leur environnement). *Geobios,* Lyon, **15**: 891–923.

Fletcher, H. (1884) Report on the geology of Northern Cape Breton. *Geological Survey of Canada, Progress Report,* **1882–4 H**: 1–98.

Forey, P.L. (1980) *Latimeria*: a paradoxical fish. *Proceedings of the Royal Society of London,* **208B**: 369–384.

Forey, P.L. (1981) The coelacanth *Rhabdoderma* in the Carboniferous of the British Isles. *Palaeontology,* London, **24**: 203–229.

Forey, P.L. (1984) The coelacanth as a living fossil. In Eldredge, N. and Stanley, S.M. (eds), *Living Fossils.* pp. 166–169, New York: Springer Verlag.

Forey, P.L. (1987) Relationships of Lungfishes. In Bemis, W.E., Burggren, W.W. and Kemp, N. (eds). *The Biology and Evolution of lungfishes,* pp. 39–74, New York: Alan R. Liss.

Forey, P.L. (1988) Golden jubilee for the coelacanth *Latimeria chalumnae. Nature,* London, **336**: 727–732.

Forey, P.L. (1991) *Latimeria chalumnae* and its pedigree. *Environmental Biology of Fishes,* **32**: 75–97.

Forey, P.L. and Young, V.T. (1985) Acanthodian and coelacanthfishes from the Dinantian of Foulden, Berwickshire, Scotland. *Transactions of the Royal Society of Edinburgh: Earth Sciences,* **76**: 53–59.

Forey, P.L., Monod, O. and Patterson, C. (1985) Fishes from the Akkuyu Formation (Tithonian), Western Taurus, Turkey. *Geobios,* Lyon, **18**: 195–201.

Forey, P.L., Gardiner, B.G. and Patterson, C. (1992) The coelacanth, lungfish and cow revisited. In Schultze, H.-P. and Trueb, L. (eds), *Origins of the Higher Groups of Tetrapods,* pp. 145–172, Ithaca, NY: Comstock Publishing Associates.

Forster, G.R. (1974) The ecology of *Latimeria chalumnae* Smith: results of field studies from Grande Comore. *Proceedings of the Royal Society of London,* **186B**: 291–296.

Forster, G.R., Badcock, J.R., Longbottom, M.R., Merrett, N.R. and Thomson, K.S. (1970) Results of the Royal Society Indian Ocean deep slope fishing expedition. *Proceedings of the Royal Society of London,* **175B**: 367–404.

Forster-Coooper, C. (1937) The Middle Devonian fish fauna of Achanarras. *Transactions of the Royal Society of Edinburgh,* **59**: 223–239.

Fox, R.C., Campbell, K.S.W., Barwick, R.E. and Long, J.A. (1995) A new osteolepiform from the Lower Carboniferous Raymond Formation, Drummond Basin, Queensland. *Memoirs of the Queensland Museum,* Brisbane, **38**: 97–221.

Fricke, H. (1988) Coelacanths: the fish that time forgot. *National Geographic Magazine,* Washington, **173**: 824–838.

Fricke, H. and Frahm, J. (1992) Evidence for lecithotrophic viviparity in the living coelacanth. *Naturwissenschaften,* **79**: 476–479.

Fricke, H. and Hissman, K. (1990) Natural habitat of coelacanths. *Nature,* London, **346**: 323–324.

Fricke, H. and Hissmann, K. (1992) Locomotion, fin coordination and body form of the living coelacanth *Latimeria chalumnae. Environmental Biology of Fishes,* **34**: 329–356.

Fricke, H. and Hissmann, K. (1994) Home range and migrations of the living coelacanth *Latimeria chalumnae. Marine Biology,* Berlin, **120**: 171–180.

Fricke, H. and Plante, R. (1988) Habitat requirements of the living coelacanth *Latimeria chalum-*

nae at Grande Comore, Indian Ocean. *Naturwissenschaften*, **75**: 149–151.

Fricke, H. and Schauer, J. (1987) Im reich der lebenden Fossilien. *Geo*, Berlin, **10**: 14–34.

Fricke, H., Reinicke, O., Hofer, H. and Nachtigall, W. (1987) Locomotion of the coelacanth *Latimeria chalumnae* in its natural environment. *Nature*, London, **329**: 331–333.

Fricke, H., Schauer, J., Hissmann, K. Kasang, L. and Plante, R. (1991) Coelacanth *Latimeria chalumnae* aggregates in caves: first observations on their resting habitat and social behaviour. *Environmental Biology of Fishes*, **30**: 281–285.

Fricke, H., Hissmann, K., Schauer, J. and Plante, R. (1995) Yet more danger for coelacanths. *Nature*, London, **374**: 314.

Fritzsch, B. (1987) Inner ear of the coelacanth fish *Latimeria* has tetrapod affinities. *Nature*, London, **327**: 153–154.

Fritzsch, B. (1989) The origins of hearing are to be found underwater. *Reports of the Deutsche Forschungsgemeinschaft*, Weinheim, **2**: 26–28.

Gall, J.C., Grauvogel, L. and Lehman, J.P. (1974) Faune du Buntsandstein. V. Les Poissons Fossiles de la collection Grauvogel-Gall. *Annales de Paléontologie* (Vertébrés), Paris, **60**: 129–145.

Gardiner, B.G. (1960) A revision of certain actinopterygian and coelacanth fishes, chiefly from the Lower Lias. *Bulletin of the British Museum (Natural History) (Geology)*, London, **4**: 239–384.

Gardiner, B.G. (1963) Certain palaeoniscoid fishes and the evolution of the actinopterygian snout. *Bulletin of the British Museum (Natural History) (Geology)*, London, **8**: 255–325.

Gardiner, B.G. (1966) Catalogue of Canadian fossil fishes. *Contribution to Life Sciences at the Royal Ontario Museum*, Toronto, **61**: 1–153.

Gardiner, B.G. (1973) New Palaeozoic fish remains from Southern Africa. *Palaeontographica Africana*, Cape Town, **15**: 33–35.

Gardiner, B.G. (1984a) The relationships of the palaeoniscid fishes, a review based on new specimens of *Mimia* and *Moythomasia* from the Upper Devonian of Western Australia. *Bulletin of the British Museum (Natural History) (Geology)*, **37**: 173–428.

Gardiner, B.G. (1984b) Sturgeons as living fossils. In Eldredge, N. and Stanley, S.M. (eds), *Living Fossils*, pp. 148–152, New York: Springer Verlag.

Gardiner, B.G. and Bartram, A.W.H. (1977) The homologies of the ventral cranial fissures in osteichthyans. In Andrews, S.M., Miles, R.S. amd Walker, A.D. (eds), *Problems in Vertebrate Evolution*, pp. 227–245, London: Academic Press.

Garman, S. (1888) On the lateral canal system of the Selachia and Holocephala. *Bulletin of the Museum of Comparative Zoology*, Harvard, **17**: 57–119.

Gaudant, M. (1975) Sur la découverte de deux nouveaux coelacanthes fossiles au Liban et la disparition apparente des actinistiens au Crétacé. *Compte rendu de l'Academie des Sciences, Paris*, Series D, **280**: 959–962.

Golenberg, E.M., Gianassi, D.E., Clegg, D.E., Smiley, C.J., Durbin, M., Henderson, D. and Zurawski, G. (1990) Chlorplast DNA sequence from a Miocene magnolia species. *Nature*, London, **344**: 656–658.

Goodrich, E.S. (1925) On the cranial roofing bones in the Dipnoi. *Zoological Journal of the Linnean Society of London*, **36**: 79–86.

Goodrich, E.S. (1930) *Studies on the Structure and Development of Vertebrates*. 837 pp., London: Macmillan & Co. Ltd.

Gorr, T. and Kleinschmidt, T. (1993) Evolutionary relationships of the coelacanth. *American Scientist*, **81**: 72–82.

Gorr, T., Kleinschmidt, T. and Fricke, H. (1991) Close relationships of the coelacanth *Latimeria* indicated by haemoglobin sequences. *Nature*, London, **351**: 394–397.

Goujet, D. (1973) *Sigaspis*, un nouvel arthrodire du Dévonien inférieur du Spitsberg. *Palaeontographica*, Stuttgart, (A) **143**: 73–88.

Goujet, D. (1975) *Dicksonosteus*, un nouvel arthrodire du Dévonien du Spitsberg. Remarques sur le squelette viscéral des Dolichothoraci. *Colloques Internationaux du Centre National de la Recherche Scientifique*, Paris, **218**: 81–100.

Goujet, D. (1984) *Les poissons placoderms du Spitsberg*. 284 pp. Paris: Centre National de la Recherche Scientifique.

Graham-Smith, W. and Westoll, T.S. (1937) On a new long-headed dipnoan fish from the Upper Devonian of Scaumenac Bay, P.Q., Canada. *Transactions of the Royal Society of Edinburgh*, **59**: 241–266, 12 figs, 2 pls.

Gregory, W.K. (1911) The limbs of *Eryops* and the origin of paired limbs from fins. *Science*, Washington, **33**: 508–509.

Griffith, R.W. (1973) A live coelacanth in the Comoro Islands. *Discovery*, New Haven, **9**: 27–33.

Griffith, R.W. (1980) Chemistry of the body fluids of the coelacanth, *Latimeria chalumnae*. *Proceed-*

ings of the Royal Society of London, **208B**: 329–347.

Griffith, R.W. (1991) Guppies, toadfish, lungfish, coelacanths and frogs: a scenario for the urea retention in fishes. *Environmental Biology of Fishes*, **32**: 199–218.

Griffith, R.W. and Burdick, C.J. (1976) Sodium-potassium activated adenosine triphosphate in coelacanth tissue: high activity in the rectal gland. *Comparative Biochemistry and Physiology*, **54B**: 557–559.

Griffith, R.W. and Thomson, K.S. (1973) *Latimeria chalumnae*: reproduction and conservation. *Nature*, London, **242**: 617–618.

Gross, W. (1935) Histologische studien am aussesskelett fossiler Agnathan und Fische. *Palaeontographica*, Stuttgart, **83A**: 1–60.

Hammerberg, F. (1937) Zur kenntnis der ontogenetischen entwicklung des Schädels von *Lepidosteus platysomus*. *Acta Zoologica*, Stockholm, **18**: 209–337.

Hancock, A. and Atthey, T. (1872a) Descriptive notes on a nearly entire specimen of *Pleuracanthus rankinii*, on two new species of *Platysomus* and a new *Amphicentrum*, with remarks on a few other fish remains found in the Coal-measures at Newsham. *Annals and Magazine of Natural History*, London, 4th series, **9**: 249–262.

Hancock, A. and Atthey, T. (1872b) Description of a considerable portion of a mandibular ramus of *Anthracosaurus russelli*; with notes on *Laxomma* and *Archichthys*. *Transactions of the natural history society of Northumberland and Durham*, **4**: 385–397.

Hay, O.P. (1900) Descriptions of some vertebrates of the Caboniferous age. *Proceedings of the American Philosophical Society*, New York, **39**: 96–123, 1 pl.

Hedges, S.B., Moberg, K.D. and Maxson, L.R. (1990) Tetrapod phylogeny inferred from 28S ribosomal RNA sequences and a review of the evidence for amniote relationships. *Molecular Biology and Evolution*, Chicago, **7**: 607–633.

Heemstra, P.C. and Compagno, L. (1989) Uterine cannibalism and placental viviparity in the coelacanth. *South African Journal of Science* **85**: 485–486.

Heemstra, P.C., Freeman, A.L.J., Wong, H.Y, Hensley., D.A and Rabesandratana, H.D. (1996) First authentic capture of a coelacanth *Latimeria chalumnae* (Pisces: Latimeriidae), off Madagascar. *South African Journal of Science*, **92**: 150–151.

Heemstra, P.C. and Greenwood, P.H. (1992) New observations on the visceral anatomy of late-term fetuses of the living coelacanth fish and the oophagy controversy. *Proceedings of the Royal Society of London*, **249B**: 49–55.

Heide, S. van der (1955) La faune du Carbonifère inférieur de l'Egypte. *Mededelingen van de Geologische Stichting*, Maastricht, **8**: 73–75.

Heineke, E. (1906) Die Ganoiden und Teleostier des Lithographischen Schiefers von Nusplingen. *Geologische und Palaeontologische Abhandlungen*, Jena, New Series **8**: 159–213, 21 figs, 8 pls.

Hennig, E. (1951) *Trachymetopon liassicum* Ald., ein Riesen- Crossopterygier aus Schwäbischem Ober-Lias. *Neues Jahrbuch für Geologie und Paläontologie, Abhandlungen*, Stuttgart, **94**: 67–79.

Hennig, W. (1966) *Phylogenetic Systematics*. Chicago: University of Illinois Press. 263 pp.

Hensel, K. (1986) Morphologie et interprétation des canaux et canalicules sensoriels cephalques de *Latimeria chalumnae* Smith, 1939 (Osteichthyes, Crossopterygii, Coelacanthiformes)/ *Bulletin Musée nationale d'Histoire naturelle, Paris*, 4th series, A **8**: 379–407.

Hibbard, C.W. (1933) Two species of *Coelacanthus* from the Middle Pennsylvanian of Anderson County, Kansas. *The University of Kansas Science Bulletin*, **21**: 279–283.

Hillis, D.M. (1987) Molecular versus morphological approaches to systematics. *Annual Reviews of Ecology and Systematics*, **18**: 23–42.

Hillis, D.M. and Dixon, M.T. (1989) Vertebrate phylogeny: evidence from 28S ribosomal DNA sequences. In Fernholm, B., Bremer, K. amd Jörnvall, H. (eds), *The Hierarchy of Life*, pp. 355–367, Amsterdam: Elsevier Science.

Hillis, D.M., Dixon, M.T. and Ammerman, L.K. (1991) The relationships of the coelacanth *Latimeria chalumnae*: evidence from sequences of vertebrate 28S ribosomal RNA genes. *Environmental Biology of Fishes*, **32**: 119–130.

Holmes, E.B. (1985) Are the lungfishes the sister group of tetrapods? *Biological Journal of the Linnean Society of London*, **25**: 379–397.

Holmgren, N. (1933) On the origin of the tetrapod limb. *Acta Zoologica*, Stockholm, **14**: 185–295.

Holmgren, N. (1949a) Contributions to the question of the origin of tetrapods. *Acta Zoologica*, Stockholm, **30**: 459–484.

Holmgren, N. (1949b) On the tetrapod limb problem – again. *Acta Zoologica*, Stockholm, **30**: 485–508.

Holmgren, N. and Stensiö, E. (1936) *Kranium und Visceralskelatt der Akranier, Cyclostomen und Fische.* In Bolk, L., Göppart, E., Kallins, E., Lubosch, W. (eds), *Handbuch der vergleichendan Antomie der Wirbeltire*, Vol. 4, pp. 233–500, Berlin: Urban & Schwarjenberg.

Hughes, G.M. (1976) On the respiration of *Latimeria chalumnae. Zoological Journal of the Linnean Society of London*, **59**: 195–208.

Hughes, G.M. (1980) Ultrastructure and morphometry of the gills of *Latimeria chalumnae*, and a comparison with the gills of associated fishes. *Proceedings of the Royal Society of London*, **208B**: 309–328.

Hughes, G.M. and Itasawa, Y. (1972) The effect of temperature on the respiratory function of coelacanth blood. *Experientia*, **28**: 1247.

Hughes, G.M. and Morgan, M. (1973) The structure of fish gills in relation to their respiratory function. *Biological Reviews*, Cambridge, **48**: 419–475.

Hureau, J.-C. and Ozouf, C. (1977) Détermination de l'âge et croissance du coelacanthe *Latimeria chalumnae* Smith, 1939 (poisson, crossoptérygien, coelacanthidé). *Cybium*, Paris, series 3, **2**: 129–137.

Hussakof, L. (1908) Catalogue of types and figured specimens in the American Museum of Natural History. Part I – Fishes. *Bulletin of the American Museum of Natural History*, **25**: 1–103.

Hussakof, L. (1911) The Permian fishes of North America. *Publications of the Carnegie Institute, Washington*, **146**: 153–178.

Huxley, T.H. (1861) Preliminary essay upon the systematic arrangement of the fishes of the Devonian epoch. *Memoirs of the Geological Survey of the United Kingdom*, London, Dec. 10: 1–40.

Huxley, T.H. (1866) Illustrations of the structure of the crossopterygian ganoids. Figures and descriptions illustrative of British organic remains. *Memoirs of the Geological Survey of the United Kingdom*, London, Dec. 12: 1–44.

Huxley, T.H. (1872) Illustrations of the structure of the crossopterygian ganoids. Figures and descriptions illustrative of British organic remains. *Memoirs of the Geological Survey of the United Kingdom*, London, Dec. 13: 1.

Huxley, T.H. (1908) *Discourses: Biological and Geological.* London: Macmillan.

Jaekel, O. (1927) Der Kopf der Wirbeltiere. *Ergebnisse der Anatomie und Entwicklungsgeschichte*, Berlin, **27**: 815–897.

Jain, S.L. (1974) *Indocoelacanthus robustus* n. gen., n. sp. (Coelacanthidae, Lower Jurassic), the first fossil coelacanth from India. *Journal of Palaeontology*, Chicago, **48**: 49–62.

Jain, S.L. (1980) The continental Lower Jurassic fauna from the Kota Formation, India. In Jacobs, L.L. (ed.), *Aspects of Vertebrate History*, pp. 99–123, Flagstaff: Museum of Arizona Press.

Janvier, P. (1974) Preliminary report on late Devonian fishes from Central and East Iran. *Geological Survey of Iran Report*, Téhéran, **31**: 5–48.

Janvier, P. (1977) Les poissons dévoniens de l'Iran central et de l'Afganistan. *Mémoires de la Societé géologique de France*, Paris, **8**: 277–289.

Janvier, P. and Martin, M. (1979) Les vertébrés Dévoniens de L'Iran Central. II–Coelacanthiformes, Struniiformes, Osteolepiformes. *Geobios*, Lyon, **12**: 497–511.

Janvier, P., Lethiers, F., Monod, O. and Balkas, O. (1984) Discovery of a vertebrate fauna at the Devonian–Carboniferous boundary in SE Turkey (Hakkari Province). *Journal of Petroleum Geology*, **7**: 147–168.

Jardine, N. (1970) The observational and theoretical components of homology: a study based on the morphology of the dermal skull-roofs of rhipidistian fishes. *Biological Journal of the Linnean Society of London*, **1**: 327–361.

Jarvik, E. (1942) On the structure of the snout of crossopterygians and lower gnathostomes in general. *Zoologiska Bidrag från Uppsala*, **21**: 235–675.

Jarvik, E. (1944) On the dermal bones, sensory canals and pit-lines of the skull in *Eusthenopteron foordi* Whiteaves, with some remarks on *E. säve-söderberghi* Jarvik. *Kungliga Svenska Vetenskapsakademiens Handlingar*, Stockholm, **21**: 1–48.

Jarvik, E. (1947) Notes on the pit-lines and dermal bones of the head in *Polypterus. Zoologiska Bidrag från Uppsala*, **25**: 60–78.

Jarvik, E. (1948) On the morphology and taxonomy of the Middle Devonian osteolepid fishes of Scotland. *Kungliga Svenska Vetenskapsakademiens Handlingar*, Stockholm, **25**: 1–301.

Jarvik, E. (1954) On the visceral skeleton in *Eusthenopteron* with a discussion of the parasphenoid and palatoquadrate in fishes. *Kungliga Svenska Vetenskapsakademiens Handlingar*, Stockholm, **5**: 1–104.

Jarvik, E. (1963) The composition of the intermandibular division of the head in fish and tetrapods and the diphyletic origin of the tetrapod tongue. *Kungliga Svenska Vetenskapsakademiens Handlingar*, Stockholm, **9**: 1–74.

Jarvik, E. (1965) On the origin of girdles and paired fins. *Israel Journal of Zoology*, **14**: 141–172.

Jarvik, E. (1967) On the structure of the lower jaw in dipnoans: with a description of an early dipnoan from Canada, *Melanognathus canadensis* gen. et sp. nov. In Patterson, C. and Greenwood, P.H. (eds), *Fossil vertebrates*, pp. 155–183, London: Academic Press.

Jarvik, E. (1968) The systematic position of the Dipnoi. In Orvig, T. (ed.), *Current Problems of Lower Vertebrate Phylogeny*, pp. 223–245, Stockholm: Almqvist & Wiksell.

Jarvik, E. (1972) Middle and Upper Devonian Porolepiformes from East Greenland with special reference to *Glyptolepis groenlandica* n. sp. *Meddelelser om Grønland*, København, **187**(2): 1–307.

Jarvik, E. (1980) *Basic Structure and evolution of Vertebrates.* Volume I. 575 pp., London: Academic Press.

Jauregui-Adell, J. and Pechere J.-F. (1978) Parvalbumins from coelacanth muscle. III. Amino acid sequence of the major component. *Biochimica et Biophysica Acta*, **536**: 275–282.

Jessen, H.L. (1966) Die Crossopterygier des Oberen Plattenkalkes (Devon) der Bergisch– Gladbach–Paffrather Mulde (Rheinisches Schiefergebirge) unter Berücksichtung von amerikanischem und europäischem *Onychodus*–Material. *Arkiv för Zoologi*, Stockholm, **18**: 305–389.

Jessen, H.L. (1972) Schultergurtel und Pectoralflosse bei Actinopterygien. *Fossils and Strata*, **1**: 1–101.

Jessen, H.L. (1973) Weitere Fischreste aus dem Oberen Plattenkalk der Bergische-Gladbach-Paffrather Mulde (Oberdevon, Rheinisches Schieferbebirge). *Palaeontolgraphica*, Stuttgart, **143A**: 159–187.

Jessen, H.L. (1980) Lower Devonian Porolepiformes from the Canadian Arctic with special reference to *Powichthys thorsteinssoni* Jessen. *Palaeontographica*, Stuttgart **167A**: 180–214.

Johansen, Z. and Ahlberg, P.E. (1997) New tristichopterid (Osteolepiformes; Sarcopterygii) from the Mandagery Sandstone (Famennian) near Canowindra, N.S.W. *Transactions of the Royal Society of Edinburgh: Earth Sciences* (in press).

Jollie, M. (1975) Development of the head skeleton and pectoral girdle in *Esox*. *Journal of Morphology*, Boston, **147**: 61–88.

Jollie, M. (1984) Development of cranial and pectoral girdle bones of *Lepisosteus* with a note on scales. *Copeia*, **1984**: 476–502.

Jollie, M. (1985) A primer of bone names for the understanding of the actinopterygian head and pectoral girdle skeletons. *Canadian Journal of Zoology*, **64**: 365–379.

Jonet, S. (1981) Contribution à l'étude des vertébrés du Crétacé portugais et spécialement du Cénomanien de l'Estremadure. *Comunicações dos serviços geológicos de Portugal*, Lisbon, **67**: 191–306.

Joss, J.M.P., Cramp, N., Baverstock, P.R. and Johnson, A.M. (1991) A phylogenetic comparison of 18S ribosomal RNA sequences of lungfish with those of other chordates. *Australian Journal of Zoology*, **39**: 509–518.

Jubb, R.A. and Gardiner, B.G. (1975) A preliminary catalogue of identifiable fossil fish material from southern Africa. *Annals of the South Africa Museum*, Cape Town, **67**: 381–440.

Kemp, A. (1995) Marginal tooth-bearing bones in the lower jaw of the Recent Australian lungfish, *Neoceratodus forsteri* (Osteichthyes, Dipnoi). *Journal of Morphology*, New York, **225**: 345–355.

Kner, R. (1866) Die Fische der bituminösen Schiefer von Raibl in Kärnthen. *Sitzungsberichte der Akademie der Wissenschaften. Mathematisch–Naturwissenschaftliche Classe.* Wien, **53**: 152–197.

Koenen, A. von (1895) Über einige Fischreste des norddeutschen und böhemischen Devons. *Abhandlungen der Königlichen Gesellschaft der Wissenschaften zu Göttingen*, **40**: 1–37.

Krupina, N. (1992) Some data on the ontogeny of Devonian dipnoans. In Mark-Kurik, E. (ed.), *Fossil Fishes as Living Animals*, pp. 215–222, Tallin: Estonian Academy of Sciences.

Lagios, M.D. (1975) The pituitary gland of the coelacanth *Latimeria chalumnae* Smith. *General amd Comparative Endocrinology*, **25**: 126–147

Lagios, M.D. (1979) The coelacanth and the Chondrichthyes as sister groups: a review of shared apomorph characters and a cladistic analysis and reinterpretation. *Occasional Papers of the California Academy of Sciences*, San Francisco, **134**: 25–44.

Lagios, M.D. (1982) *Latimeria* and the Chondrichthyes as sister taxa: a rebuttal to recent attempts at refutation. *Copeia*, **1982**: 942–948.

Lagios, M.D. and McCosker, J.E. (1977) A cloacal excretory gland in the lungfish *Protopterus*. *Copeia*, **1977**: 176–178.

Lambers, P. (1992) On the ichthyofauna of the Solnhofen Lithographic Limestone (Upper Jurassic, Germany). PhD thesis, Rijksuniversiteit Groningen, 336 pp.

Lauder, G.V. (1980) The role of the hyoid apparatus in the feeding mechanism of the coelacanth *Latimeria chalumnae*. *Copeia*, **1980**: 1–9.

Le, H.L., Lecointre, G. and Perasso, R. (1993) A 28S rRNA-based phylogeny of the gnathostomes: first steps in the analysis of conflict and congruence with morphologically based cladograms. *Molecular Phylogenetics and Evolution*, New York, **2**: 31–51.

Lebedev, O. (1995) Morphology of a new osteolepidid fish from Russia. *Bulletin du Muséum National d'Histoire Naturelle*, Paris, Section C, 48 Series, **17**: 287–341.

Lebedev, O. and Clack, J.A. (1993) Upper Devonian tetrapods from Andreyevka, Tula region, Russia. *Palaeontology*, London, **36**: 721–734.

Lehman, J.-P. (1952) Etude complémentaire des poissons de l'Eotrias de Madagascar. *Kungliga Svenska Vetenskapsatademiens Handlingar*, **2**: 1–201.

Lehman, J.-P. (1966) Crossopterygii. In Piveteau, J. (ed.), *Traité de Paléontologie*, part 4, vol. 3, pp. 301–412, Paris: Masson.

Lehmann, W.M. and Westoll, T.S. (1952) A primitive dipnoan fish from the Lower Devonian of Germany. *Proceedings of the Royal Society of London*, **140B**: 403–421.

Lekander, B. (1949) The sensory line system and the canal bones in the head of some ostariophysi. *Acta Zoologica*, Stockholm, **30**: 1–131.

Lelièvre, H. and Janvier, P. (1988) Un actinistien (Sarcopterygii: Vertebrata) dans le Dévonien supérieur du Maroc. *Compte rendu de l'Academie des Sciences, Paris*, Series 2, **307**: 1425–1430.

Lemire, M. and Lagios, M. (1979) Ultrastructure du parenchyme sécréteur de la gland post-anale de coelacanthe *Latimeria chalumae* Smith. *Acta anatomica*, Basel, **104**: 1–15.

Lenoble, J. and Le Grand, Y. (1954) Le Tapis e l'oeil du coelacanthe (*Latimeria anjouae* (Smith)). *Bulletin du Muséum National d'Histoire Naturelle*, Paris, Section C, 28 Series, **26**: 460–463.

Ley, W. (1959) Exotic Zoology. 468 pp. New York: Viking Press.

Lindsey, C.C. (1978) Form, function and locomotory habits in fish. In Hoar, W.S. and Randall, D.J. (eds), *Fish Physiology*, Vol. 7. pp. 1–100, New York: Academic Press.

Liu, H.-T. (1964) A new coelacanth from the marine Lower Triassic of N.W. Kwangsi, China. *Vertebrata Palasiatica*, Peking, **8**: 211–214, 1 pl. [In Chinese with English summary.]

Locket, N.A. (1973) Retinal structure in *Latimeria chalumnae*. *Philosophical Transactions of the Royal Society of London*, Series B, **266**: 493–521.

Locket, N.A. (1980) Some advances in coelacanth biology. *Proceedings of the Royal Society of London*, **208B**: 265–307.

Lockett, N.A. and Griffith, R.W. (1972) Observations on a living coelacanth. *Nature*, London, **237**: 175.

Long, J.A. (1985) The structure and relationships of a new osteolepiform fish from the Late Devonian of Victoria, Australia. *Alcheringa*, Sydney, **9**: 1–22.

Long, J.A. (1987) An unusual osteolepiform fish from the Late Devonian of Victoria. *Palaeontology*, Cambridge, **30**: 839–852.

Long, J.A. (1989) A new rhizodontiform fish from the early Carboniferous of Victoria, Australia, with remarks on the phylogenetic position of the group. *Journal of Vertebrate Paleontology*, Chicago, **9**: 1–17.

Long, J.A., Campbell, K.S.W. and Barwick, R.E. (1994) A new dipnoan genus, *Ichnomylax*, from the Lower Devonain of Victoria, Australia. *Journal of Vertebrate Paleontology*, Chicago, **14**: 127–131.

Løvtrup, S. (1977) *The Phylogeny of the Vertebrata*. 330 pp. 101 figs, London: John Wiley & Sons.

Lund, R. and Lund, W. (1984) New genera and species of coelacanths from the Bear Gulch Limestone (Lower Carboniferous) of Montana (U.S.A.). *Geobios*, Lyon, **17**: 237–244.

Lund, R. and Lund, W.L. (1985) Coelacanths from the Bear Gulch Limestone (Namurian) of Montana and the evolution of the coelacanthiformes. *Bulletin of the Carnegie Museum of Natural History*, Pittsburg, **25**: 1–74.

Lutken, C.F. (1868) Om Ganoiderens Begraedsning og Inddeling. Kjøbenhaum: Bianco Lumos, pp. 1–82.

McAllister, D.E. (1971) Old fourlegs: a 'living fossil'. Toronto: National Museum of Natural Science, *Odyssey Seroes* 1: 1–25.

McAllister, D.E. and Smith, C.L. (1978) Mensurations morphologiques, dénombrements méristiques et taxonomie du coelacanthe, *Latimeria chalumnae*. *Naturaliste Canadien*, Montreal, **105**: 63–76.

McCosker, J.E. (1979) Inferred natural history of the living coelacanth, *Latimeria chalmunae*. *Occa-*

sional Papers of the California Academy of Sciences, San Francisco, **134**: 17–24.

McCosker, J.E. and Lagois, M.D. (eds) (1979) The biology and physiology of the living coelacanth. *Occasional Papers of the California Academy of sciences*, San Francisco, number 134: 1–175.

Maddison, W. (1994) Missing data versus mising characters in phylogenetic analysis. *Systematic Biology*, Washington, **42**: 576–581.

Maeda, N., Zhu, D. and Fitch, W.M. (1984) Amino acid sequences of lower vertebrate parvalbumins and thier evolution: parvalbumins of boa, turtle and salamander. *Molecular Biology and Evolution*, Chicago, **1**: 473–488, 4 figs.

Maisey, J.G. (1986a) Heads and tails: a chordate phylogeny. *Cladistics*, Westport, **2**: 201–256.

Maisey, J.G. (1986b) Coelacanths from the Lower Cretaceous of Brazil. *American Museum Novitates*, New York, **2866**: 1–30.

Maisey, J.G. (1991) *Santana fossils: an illustrated atlas*. 459 pp., New Jersey: T.F.H. Publications.

Mangum, C. (1991) Urea and chloride senstivities of coelacanth hemoglobin. *Environmental Biology of Fishes*, **32**: 219–222.

Mantell, G. (1822) *The Fossils of the South Downs; or Illustrations of the Geology of Sussex*. 327 pp., London: Lupton Relfe.

Mart, A. (1988) The tectonic setting of the Seychelles, Masarene and Amirante plateaus in the Western equatorial Indian Ocean. *Marine Geology*, Amsterdam, **79**: 261–274.

Martin, M. (1981) Les dipneustes et Actinistians du Trias supérieur continental marocain. *Stuttgarter Beiträge zur Naturkunde*, Serie B, Stuttgart, **69**: 1–29, 5 figs, 1 pl.

Martin, M. and Wenz, S. (1984) Découverte d'un nouveau Coelacanthidé, *Garnbergia ommata* n. gen., n. sp., dans le Muschelkalk supérieur du Baden-Württemberg. *Stuttgarter Beiträge zur Naturkunde, Serie B (Geologie und Paläontologie)*, **105**: 1–17.

Melton, W.G. (1969) A new dorypterid fish from Central Montana. *Northwest Science*, Cheney, **43**(4): 196–206.

Meyer, A. (1993) Coelacanth controversy. *American Scientist*, **81**: 209–210.

Meyer, A. and Dolven, S.I. (1992) Molecules, fossils and the origin of tetrapods. *Journal of Molecular Evolution*, New York, **35**: 102–113.

Meyer, A. and Wilson, A.C. (1990) Origin of tetrapods inferred from their mitochondiral affiliation to lungfishes. *Journal of Molecular Evolution*, New York, **31**: 359–364.

Meyer, A. and Wilson, A.C. (1991) Coelacanth's relationships. *Nature*, London, **353**: 219.

Miles, R.S. (1977) Dipnoan (lungfish) skulls and the relationships of the group: a study based on new species from the Devonian of Australia. *Zoological Journal of the Linnean Society*, London, **61**: 1–328.

Miles, R.S. and Westoll, T.S. (1963) Two new genera of coccosteid Arthrodira from the Middle Old Red Sandstone of Scotland, and their stratigraphical distribution. *Transactions of the Royal Society of Edinburgh*, **65**: 179–210.

Miles, R.S. and Westoll, T.S. (1968) The placoderm fish *Coccosteus cuspidatus* Miller ex Agassiz from the Middle Old Red Sandstone of Scotland. Pt. 1. Descriptive morphology. *Transactions of the Royal Society of Edinburgh*, **67**: 373–476.

Milinkovitch, M.C. (1993) Coelacanth controversy. *American Scientist*, **81**: 210.

Millott, J. (1955) First observation on a living coelacanth. *Nature*, London, **175**: 362–363.

Millot, J. and Anthony, T. (1956) L'organe rostral de *Latimeria* (Coelacanthidae). *Compte rendu de l'Academie des Sciences, Paris*, Series D, **239**: 1241–1243.

Millot, J. and Anthony, T. (1958) Anatomie de *Latimeria chalumnae*. 1. Squellette et muscles. 122 pp., Paris: C.N.R.S.

Millot, J. and Anthony, T. (1960a) Le cloaque chez les coelacanthes. *Bulletin du Muséum National d'Histoire Naturelle, Paris*, **32**: 287–289.

Millot, J. and Anthony, T. (1960b) Appareil génital et reproduction des coelacanthes. *Compte rendu de l'Academie des Sciences, Paris*, Series D, **251**: 442–443.

Millot, J. and Anthony, T. (1965) *Anatomie de Latimeria chalumnae. II Système Nerveux et Organes des Sens*. 131 pp., Paris: C.N.R.S.

Millot, J. and Anthony, T. (1972) Le glande postanale de *Latimeria*. *Annales des Sciences Naturelles, Zoologie, Paris*, Series 12, **14**: 305–317.

Millot, J. and Anthony, T. (1973) La position ventrale du rein de *Latimeria chalumnae* Smith (Poisson, Coelacanthidé). *Compte rendu hebdomadaire des Séances de l'Academie des Sciences, Paris*, Series D, **276**: 2171–2173.

Millot, J. and Anthony, T. (1974) Les oeufs du coelacanthe. *Science et Nature, Paris*, **121**: 3–4, 1 fig.

Millot, J. and Carasso, N. (1965) Note préliminaire sur l'oeil de *Latimeria chalumnae* (crossoptérygien, coelacanthidé). *Compte rendu hebdomadaire des Séances de l'Academie des Sciences, Paris*, Series D, **241**: 576–577.

Millot, J., Anthony, T. and Robineau, D. (1972) État commente des captures de *Latimeria chalumnae* Smith (poisson crossoptérygien, coelacanthidé) effectuées jusqu'au mois d'Octobre 1971. *Bulletin de Museum Nationale d'Histoire Naturelle*, Paris, Series 3, Zoologie, **39**: 533–548.

Millot, J., Anthony, J. and Robineau, D. (1978) Anatomie de Latimeria chalumnae. III. *Appareil digestif, appareil respiratoire, appareil urogénital, glands endocrines, appareil circulatoire, téguments, ecailles, conclusions générales.* 198 pp., Paris: C.N.R.S.

Moy-Thomas, J.A. (1935a) A synopsis of the coelacanth fishes of the Yorkshire Coal Measures. *Annals and Magazine of Natural History*, London, Ser. 10, **15**: 37–46.

Moy-Thomas, J.A. (1935b) The coelacanth fishes from Madagascar. *Geological Magazine*, London, **72**: 213–227.

Moy-Thomas, J.A. (1937) The Carboniferous coelacanth fishes of Great Britain and Ireland. *Proceedings of the Zoological Society of London*, B, **3**: 383–415.

Moy-Thomas, J.A. and Miles, R.S. (1971) *Palaeozoic Fishes*. 259 pp., London: Chapman & Hall.

Moy-Thomas, J.A. and Westoll, T.S. (1935) On the Permian coelacanth, *Coelacanthus granulatus*, Ag. *Geological Magazine*, London, **72**: 446–457.

Münster, G. von (1834) Mittheilungen an Professor Bronn gerichtet. *Neues Jahrbuch für Mineralogie Geognosie, Geologie und Petrefakten-Kunde*, Stuttgart, **1834**: 538–542.

Münster, G. von (1842a) Beitrag zur Kenntniss einiger neuen seltenen Versteinerungen aus den lithographichen Schiefern in Baiern. *Neues Jahrbuch für Mineralogie Geognosie, Geologie und Petrefakten-Kunde*, Stuttgart, **1842**: 35–46.

Münster, G. von. (1842b) Beitrage zur Petrefacten-Kunde. 5th Part. 131 pp. 15 pls. Bayreuth: Buchner'schen Buchhandlung.

Musick, J.A., Bruton, M.N. and Balon, E. (eds) (1991) *The Biology of* Latimeria chalmunae *and Evolution of Coelacanths*. 446 pp., Amsterdam: Kluwer Academic Publishers.

Nelson, G.J. (1969) Gill arches and the phylogeny of fishes, with notes on the classification of vertebrates. *Bulletin of the American Museum of Natural History*, **141**: 475–552.

Nelson, G.J. (1973) Relationships of clupeomorphs, with remarks on the structure of the lower jaw in fishes. In Greenwood, P.H., Miles, R.S. and Patterson, C. (eds), *Interrelationships of Fishes*: pp. 333–349, London: Academic Press.

Nelson, G.J. (1978) Ontogeny, phylogeny, paleontology, and the biogenetic law. *Systematic Biology*, Washington, **27**: 324–345.

Newberry, J.S. (1856) Description of several new genera and species of fossil fishes, from the Carboniferous strata of Ohio. *Proceedings of the Academy of Natural Sciences of Philadelphia*, **8**: 96–100.

Newberry, J.S. (1873a) Descriptions of fossil fishes. *Report of the Geological Survey of Ohio*, **7**: 245–355.

Newberry, J.S. (1873b) Notes on the genus *Conchiopsis*. *Proceedings of the Academy of Natural Sciences of Philadelphia*, **1873**: 425–426.

Newberry, J.S. (1878) Descriptions of new fossil fishes from the Trias. *Annals of the New York Academy of Sciences*, **1**: 127–128.

Newberry, J.S. (1889) The Palaeozoic fishes of North America. *Monographs of the U.S. Geological Survey*, Washington, **16**: 1–340.

Nielsen, E. (1936) Some few preliminary remarks on Triassic fishes from East Greenland. *Meddelelser om Grønland*, København, **112**: 1–55.

Nielsen, E. (1942) Studies on Triassic fishes from East Greenland. I. *Glaucolepis* and *Boreosomus*. *Palaeozoologica Groenlandica*, København, **1**: 1–403.

Nieuwenhuys, R., Kremers, J.-P.M. and van Huijzen, C. 1977. The brain of the crossopterygian fish *Latimeria chalumnae*: a survey of its gross structure. *Anatomica embryologica*, **151**: 157–169.

Nilsson, T. (1943) On the morphology of the lower jaw of stegocephalia with special reference to Eotriassic stegocephalians from Spitsbergen. *Kungliga Svenska Vetenskapsakademiens Handlingar*, Stockholm, **20**: 1–46.

Nixon, K.C. and Wheeler, Q.D. (1992) Extinction and the origin of species. In Novacek, M.J. amd Wheeler, Q.D. (eds), *Extinction and Phylogeny*, pp. 119–143, New York: Columbia University Press.

Nolf, D. (1985) *Otolithi Piscium*.Vol. **10**. In H.-P. Schultze (ed.), *Handbook of Paleoichthyology*. Stuttgart: Gustav Fischer Verlag. 145 pp.

Normark, B.B., McCune, A.R. and Harrison R.G. (1991) Phylogenetic relationships of neopterygian fishes, inferred from mitochondiral DNA sequences. *Molecular Biology and Evolution*, Chicago, **8**: 819–834.

Northcutt, R.G. (1987) Lungfish neural characters and their bearing on Sarcopterygian phylogeny. In Bemis, W.E., Burggren, W.W. and Kemp, N.

(eds). *The Biology and Evolution of Lungfishes*, pp. 277–297, New York: Alan R. Liss.

Northcutt, R.G. and Bemis, W.E. (1993) Cranial nerves of the coelacanth, *Latimeria chalumnae* [Osteichthyes: Sarcopterygii: Actinistia], and comparisons with other Craniata. *Brain and Evolution*, Basel, **42**, Supplement 1: 1–76.

Northcutt, R.G., Neary, T.J. and Senn, D.G. (1978) Observations on the brain of the coelacanth *Latimeria chalumnae*: external anatomy and quantitative analysis. *Journal of Morphology*, Philadelphia, **155**: 181–192.

Obrucheva, O.P. (1955) Upper Devonian fishes of Central Kazakstan. *Soviet Geology*, Moscow, **45**: 84–99. [In Russian.]

Olsen, P.E., McCune, A.R. and Thomson, K.S. (1982) Correlation of the early Mesozoic Newark Supergroup by vertebrates, principally fishes. *American Journal of Science*, New Haven, **282**: 1–44.

Ørvig, T. (1986) A vertebrate bone from the Swedish Paleocene. *Geologiska Föreningens i Stockholm Förhandlingar*, **108**: 139–141.

Osawa, S., Jukes, T.H., Watanabe, K. and Muto, A. (1992) Recent evidence for the evolution of the genetic code. *Journal of Microbiological Reviews*, **56**: 229–264.

Panchen, A.L. (1967) The nostrils of choanate fishes and early tetrapods. *Biological Reviews*, **42**: 374–420.

Panchen, A.L. and Smithson, T.R. (1987) Character diagnosis, fossils and the origin of tetrapods. *Biological Reviews*, **62**: 341–438.

Pang, P.K.T., Griffith, R.W. and Atz, J.W. (1977) Osmoregulation in elasmobranchs. *American Zoologist*, 17: 365–377.

Patterson, C. (1970) A clupeomorph fish from the Gault (Lower Cretaceous). *Zoological Journal of the Linnean Society*, London, **49**: 161–182.

Patterson, C. (1973) Interrelationships of holosteans. In Greenwood, P.H., Miles, R.S. and Patterson. C. (eds), *Interrelationships of Fishes*, pp. 233–305,27 figs, London: Academic Press.

Patterson, C. (1975) The braincase of pholidophorid and leptolepid fishes, with a review of the actinopterygian braincase. *Philosophical Transactions of the Royal Society of London*, **269B**: 275–579.

Patterson, C. (1977a) The contribution of paleontology to teleostean phylogeny. In Hecht, M.K. and Goody, P.C. (eds), *Major Patterns in Vertebrate Evolution*, pp. 579–643, New York: Plenum Press.

Patterson, C. (1977b) Cartilage bones, dermal bones and membrane bones, or the exoskeleton versus the endoskeleton. In Andrews, S.M., Miles, R.S. and Walker, A.D. (eds), *Problems in Vertebrate Evolution*, pp. 77–121, London: Academic Press.

Patterson, C. (1980) Origin of tetrapods: historical introduction to the problem. In Panchen A.L. (ed.), *The Terrestrial Environment and the Origin of Land Vertebrates*, pp. 159–175, Academic Press.

Patterson, C. (1982) Morphology and interrelationships of primitive actinopterygian fishes. *American Zoologist*, **22**: 241–259.

Patterson, C. (ed.) (1987) *Molecules and Morphology in Evolution: Conflict or Compromise*. 229 pp., Cambridge: Cambridge University Press.

Patterson, C. and Rosen, D.E. (1977) Review of ichthyodectiform and other Mesozoic teleost fishes and the theory and practice of classifying fossils. *Bulletin of the American Museum of Natural History*, **158**: 81–172.

Patton, W.W. and Tailleur, I.L. (1964) Geology of the Killik-Itkillik Region, Alaska. *Geological Survey of the United States, Professional Papers*, Washington, **303**G: 409–500.

Pearson, M.D. and Westoll, T.S. (1979) The Devonian actinopterygian *Cheirolepis* Agassiz. *Transactions of the Royal Society of Edinburgh*, **70**: 337–399.

Pechere J.-F., Rochat, H. and Ferraz, C. (1978) Parvalbumins from coelacanth muscle. II. Amino acid sequence of the two less acidic components. *Biochimica et Biophysica Acta*, **536**: 269–274.

Pehrson, T. (1944) Some observations on the development and morphology of the dermal bones in the skull of *Acipenser* and *Polyodon*. *Acta Zoologica*, Stockholm, **25**: 27–48.

Pimentel, R.A. and Riggins, R. (1987) The nature of cladistic data. *Cladistics*, London, **3**: 201–209.

Platnick, N.I., Griswold, C.E. and Coddington, J.A. (1991a) On missing entries in cladistic analysis. *Cladistics*, London, **7**: 337-343.

Platnick, N.I., Coddington, J.A. Forster, R.R. and Griswold, C.E. (1991b) Spinneret morphology and the phylogeny of Haplogyne spiders (Araneae, Araneomorphae). *American Museum Novitates*, New York, **3016**: 1–73.

Platt, J. (1896) Ontogenetic differentiation of the ectoderm in *Necturus*. *Quarterly Journal of Microscopical Science*, London, **38**: 485–547.

Priem, F. (1924) Paléontologie de Madagascar. X11. Les poissons fossiles. *Annales de Paléontologie*, Paris, **1924**: 105–132.

Pruvost, P. (1914) La faune continentale du terrain houiller du Nord de la France: son utilisation stratigraphique. *Compte-Rendu de la 12ième Session Congrès Géologique International*, Ottawa, **1913**: 925–937.

Quenstedt, F.A. (1858) *Der Jura.* 823 pp. 100 pls. Tubingen.

Rackoff, J.S. (1980) The origin of the tetrapod limb and the ancestry of tetrapods. In Panchen, A.L. (ed.), *The Terrestrial Environment and the Origin of Land Vertebates*, pp. 255–292, London: Academic Press.

Reif, W.-E. (1982) Evolution of the dermal skeleton and dentition in vertebrates. *Evolutionary Biology*, New York, **15**: 287–368.

Reis, O.M. (1888) Die Coelacanthinen, mit besonderer Berücksichtigung der im Weissen Jura Bayerns vorkommenden Gattungen. *Palaeontographica*, Stuttgart, **35**(1): 1–96.

Reis, O.M. (1892) *Zur Osteologie der coelacanthinen. I, Theil. Rumpskelet, Knochen des Schädels und der Wangen, Kiemenbogenskelet, Schultergürtel, Becken, Integument und innere organ.* 39 pp., München: K. Ludwigs-Maximilians–Universität zu München.

Reis, O.M. (1900) *Coelacanthus lunzensis* Teller. *Jahrbuch der Kaiserlich–Königlichen Geologischen Reichsanstalt*, Wien, **50**: 187–192.

Reuss, A.E. (1857) Neue Fischreste aus dem Böhmischen Pläner. *Denkschriften der Akademie der Wissenschaften, Wien. Mathematisch-Naturwissenschaftliche Klasse.* **13**: 1–10.

Richter, M. (1985) Situaçao pequisa paleoictiológica no Paleozóico Brasiliero. *Trabalhos Appresentados no. VII Congresso Brasiliero de Paleontologia 1983*, Brasilia: 105–110.

Rieppel, O. (1976) Die orbitotemporale Region im Schädel von *Chleydra serpentina* Linnaeus (Chelonia) und *Lacerta sicula* Rafinesque (Lacertilia). *Acta anatomica*, Basel, **96**: 309–320.

Rieppel, O. (1980) A new coelacanth from the Middle Triassic of Monte San Giorgio, Switzerland. *Eclogae Geologicae Helvetiae*, Basel, **73**: 921–939.

Robineau, D. and Anthony, J. (1973) Biomechanique du crane de *Latimeria chalumnae*. *Compte Rendu Hebdomadaire des Séances de l'Académie des Sciences*, Paris, **276D**: 1305–1308.

Romer, A.S. (1924) Pectoral limb and shoulder girdle in fish and tetrapods. *Anatomical Record*, **27**: 119–143.

Romer, A.S. and Byrne, F. (1931_ The pes of *Diadectes*: notes on the primitive tetrapod limb. *Palaeobiology*, Chicago, **4**: 25–48.

Rosen, D.E., Forey, P.L., Gardiner, B.G. and Patterson, C. (1981) Lungfishes, tetrapods, paleontology and plesiomorphy. *Bulletin of the American Museum of Natural History*, New York, **167**: 159–276.

Roux, G.H. (1942) The microscopic anatomy of the *Latimeria* scale. *South African Journal of Medical Science*, **7**: 1–18.

Saint-Seine, P. de (1949) Les poissons des calcaires Lithographiques de Cerin (Ain). *Nouvelles Archives du Musêm d'Histoire Naturelle de Lyon*, **2**: 1–357.

Saint-Seine, P. de (1950) Contribution à l'étude des vertebres fossiles du Congo Belge. *Annales du Musée Royal du Congo Belge, Tervuren Sciences Geologigues*, **5**: 1–32.

Saint-Seine, P. de (1955) Poissons fossiles de l'étage de Stanleyville. I. La faune des argilites et schistes bitumineux. *Annales du Musée Royal du Congo Belge Tervueen, Sciences Geologicas*, **14**: 1–126.

Sanz, Y.L., Wenz, S., Yebenes, A., Estes, R., Martinez-Delclos, X., Jimenez-Fuentes, E., Diéguez, C., Buscalioni, A.D., Barbadillo, L.J. and Via, L. (1988) An early Cretaceous faunal and floral continental assemblage: Las Hoyas fossil site (Cuenca, Spain). *Geobios*, Lyon, **21**: 611–635.

Sauvage, H.E. (1903) Noticia sobre los peces de la Caliza Litográfica de la Provincia de Lérida (Cataluña). *Memorias de la Real Academia de Ciencias y Artes de Barcelona*, **4**(35): 1–32.

Säve-Söderbergh, G. (1933) The dermal bones of the head and the lateral line system in *Osteolepis macrolepidotus* Ag. *Nova Acta Regiae Societatis Scientiarum Upsaliensis*, Series 4, **9**(2): 1–130.

Säve-Söderbergh, G. (1935) On the dermal bones of the head in labyrinthodont stegocephalians and primitive Reptilia. *Meddelelser om Grønland*, Copenhagen, **98**: 1–211.

Schaeffer, B. (1941) A revision of *Coelacanthus newarki* and notes on the evolution of the girdles and basal plates of the median fins in the coelacanthini. *American Museum Novitates*, **1110**: 1–17.

Schaeffer, B. (1948) A study of *Diplurus longicaudatus* with notes on the body form and locomotion of the coelacanthini. *American Museum Novitates*, **1378**: 1–32.

Schaeffer, B. (1952a) The Triassic coelacanth *Diplurus*, with observations on the evolution of the coelacanthini. *Bulletin of the American Museum of Natural History*, **99**(2): 25–78.

Schaeffer, B. (1952b) Rates of evolution in the coelacanth and dipnoan fishes. *Evolution*, **6**: 101–111.

Schaeffer, B. (1954) *Pariostegus*, a Triassic coelacanth. *Notulae Naturae*, Academy of Natural Sciences of Philadelphia, **261**: 1–6.

Schaeffer, B. (1962) A coelacanth fish from the Upper Devonian of Ohio. *Scientific Publications of the Cleveland Museum of Natural History*, New Series, **1**: 1–13.

Schaeffer, B. (1967) Late Triassic fishes from the western United States. *Bulletin of the American Museum of Natural History*, **135**: 285–342.

Schaeffer, B. and Gregory, J.T. (1961) Coelacanth fishes from the continental Triassic of the western United States. *American Museum Novitates*, **2036**: 1–18.

Schaeffer, B. and Mangus, M. (1976) An early Triassic fish assemblage from British Columbia. *Bulletin of the American Museum of Natural History*, **156**: 515–564.

Schaumberg, G. (1978) Neubeschreiburg von *Coelacanthus granulatus* Agassiz (Actinistia, Pisces) aus dem Kupferschiefer van Richelsdorf (Perm, W.-Deutschland). *Paläontologische Zeitschrift*, Stuttgart, **52**: 169–197.

Schliewen, U., Fricke, H., Schartl, M., Epplen, J.T. and Paabo, S. (1993) Which home for the coelacanth? *Nature*, London, **363**: 405.

Schultze, H.-P. (1972) Early growth stages in coelacanth fishes. *Nature New Biology*, London, **236**: 90–91.

Shultze, H.-P. (1973) Crossopterygier mit heterozerker Schwanzflose aus dem Oberdevon Kanadas, Nebst einer Beschreibung von Onychodontida-Resten aus dem Mittledevon Spaniens und dem Karbon der USA. *Palaeontographica*, Stuttgart, **143A**: 188–208.

Schultze, H.-P. (1975) Die Lungenfisch–Gattung *Conchopoma* (Pisces, Dipnoi). *Senckenbergiana Lethaea*, Frankfurt am Main **56**: 191–231.

Schultze, H.-P. (1980) Eier legende und lebend gebärende Quastenflosser. *Natur und Museum*, Frankfurt, **110**: 101–108.

Schultze, H.-P. (1981) Das Schädeldach eines ceratodontiden Lungen fisches aus der Trias Süddeutschlands (Dipnoi, Pisces). *Stuttgarter Beiträge zur Naturkunde*, Serie B (Geologie und Paläontologie), **70**: 1–31.

Schultze, H.-P. (1987) Dipnoans as sarcopterygians. In Bemis, W.E., Burggren, W.W. and Kemp, N. (eds). *The Biology and Evolution of Lungfishes*, pp. 39–74, New York: Alan R. Liss.

Schultze, H.-P. (1992) Coelacanth fish (Actinistia, Sarcopterygii) from the Late Pennsylvanian of Kinney Brick Company Quarry, New Mexcico. *Bulletin of the New Mexico Bureau of Mines and Mineral Resources*, Socorro, **138**: 205–209.

Schultze, H.-P. (1993) Osteichthyes: Sarcopterygii. In Benton, M.J. (ed.), *The Fossil Record 2*, pp. 657–663, London: Chapman & Hall.

Schultze, H.-P. (1994) Comparison of hypotheses on the relationships of sarcopterygians. *Systematic Biology*, **43**: 155–173.

Schultze, H.P. and Arsenault, M. (1987) *Quebecius quebecensis* (Whiteaves), a porolepiform crossopterygian (Pisces) from the late Devonian of Quebec, Canada. *Canadian Journal of Earth Sciences*, **24**: 2351–2361.

Schultze, H.P. and Campbell, K.S.W. (1986) Characterization of the Dipnoi, a monophyletic group. In Bemis, W.E., Burggren, W.W. and Kemp, N. (eds), *The Biology and Evolution of Lungfishes*, pp. 25–38, New York: Alan R. Liss.

Schultze, H.-P. and Chorn, J. (1988) The Upper Pennsylvanian vertebrate fauna of Hamilton, Kansas. In Mapes, G. and Mapes, R.H. (eds), *Regional Geology and Paleontology of Upper Paleozoic Hamilton Quarry Area in Southeastern Kansas*, pp. 147–154, Kansas: Kansas Geological Guidebook Series 6.

Schultze, H.-P. and Cloutier, R. (1991) Computed tomography and magnetic resonance imaging studies of *Latimeria chalumnae*. *Environmental Biology of Fishes*, **32**: 159–182.

Schultze, H.-P. and Cloutier, R. (1996) Comparison of the Escuminac Formation Ichthyofauna with other late Givetian/early Frasmian ichthyofaunas. In Schultze, H.-P. and Cloutier, R. (eds), In *Devonian Fishes and Plants of Miguasha, Quebec, Canada*, pp. 348–366, Munchen: Pfiel.

Schultze, H.-P. and Marshall, C.R. (1993) Contrasting the use of functional complexes and isolated characters in lungfish evolution. *Memoirs of the Association of Australian Palaeontologists*, Sydney, **15**: 211–244.

Schultze, H.-P. and Möller, H. (1973) Wirbeltierreste aus dem Mittleren Muschelkalk (Trias) von Göttingen, West Deutschland. *Palaeontologische Zeitschrift*, **60**: 109–129.

Schweizer, R. (1966) Ein Coelacanthide aus dem Oberen Muschelkalk Göttingens. *Neues Jahrbuch für Geologie und Paläontologie*, Abhandlungen, Stuttgart, **125**: 216–226.

Schwimmer, D.R., Stewart, J.D. and Dent Williams, G. (1994) Giant fossil coelacanths of the eastern United States. *Geology*, Boulder, **22**: 503–506.

Sedgwick, A. (1835) On the geological relations and internal structure of the Magnesian Limestone, and the lower portions of the New Red Sandstone Series in their range through Nottinghamshire, Derbyshire, Yorkshire and Durham, to the southern extremity of Northumberland. *Transactions of the Geological Society of London*, (2nd series): 37–124.

Semon, R. (1898) Die Entwickelung der paarigen Flossen des *Ceratodus forsteri*. In Semon, R., *Zoologische Forsschungsreisen in Australien und dem Malayischen Archipelago. I. Ceratodus*. Part 2, pp. 59–111, Jena: Gustav Fischer Verlag.

Setter, A.L. and Brown, G.W. (1991) Enzymes of the coelacanth *Latimeria chalumnae* evidenced by starch gel electrophoresis. *Environmental Biology of Fishes*, **32**: 193–198.

Sewertzoff, A.N. (1926) Studies on the bony skull of fishes. *Quarterly Journal of Microscopical Science*, **70**: 451–540.

Shainin, V.E. (1943) New coelacanth fishes from the Triassic of New Jersey. *Journal of Paleontology*, Chicago, **17**: 271– 275.

Sharpe, P.M., Lloyd, A.T. and Higgins, D.G. (1991) Coelacanth's relationships. *Nature*, London, **353**: 218–219.

Shubin, N. and Alberch, P. (1986) A morphogenetic approach to the origin and basic organisation of the tetrapod limb. *Evolutionary Biology*, New York, **20**: 319–387.

Simpson, G.G. (1944) *Tempo and mode in evolution*. 237 pp., New York: Columbia University Press.

Simpson, G.G. (1953) *The Major Features of Evolution*. 434 pp., New York: Columbia University Press.

Smith, A.B. (1989) RNA sequence data in phylogenetic reconstruction: testing the limits of its resolution. *Cladistics*, New York, **5**: 321–344.

Smith, A.B. (1994) *Systematics and the Fossil Record*. 223 pp., Oxford: Blackwell Scientific Publications.

Smith, C.L., Rand, C.S., Schaeffer, B. and Atz, J.W. (1975) *Latimeria*, the living coelacanth, is ovoviparous. *Science*, Washington, **190**: 1105–1106.

Smith, J.L.B. (1939a) A living fish of Mesozoic type. *Nature*, London, **143**: 455–456.

Smith, J.L.B. (1939b) The living coelacanthid fish from South Africa. *Nature*, London, **143**: 748–750.

Smith, J.L.B. (1939c) A living coelacanthid fish from South Africa. *Transactions of the Royal Society of South Africa*, Cape Town, **28**(1): 1–106.

Smith, J.L.B. (1953) The second coelacanth. *Nature*, London, **171**: 99–101.

Smith, J.L.B. (1956) *Old Fourlegs: the Story of the Coelacanth*. 260 pp, London: Longman Green.

Smith, M.M. (1979) Scanning electron microscopy of odontodes in the scales of a coelacanth embryo, *Latimeria chalumnae* Smith. *Archives of Oral Biology*, Oxford, **24**: 179–183.

Smith, M.M. and Chang, M.-M. (1990) The dentition of *Diabolepis speratus* Chang and Yu, with further consideration of its relationships and the primitive dipnoan dentition. *Journal of Verterbate Paleontology*, Washington, **10**: 420–433.

Smith, M.M., Hobdell, M.H. and Miller, W.A. (1972) The structure of the scales of *Latimeria chalumnae*. *Journal of Zoology, London*, **167**: 501–509.

Stensiö, E.A. (1918) Notes on a crossopterygian fish from the Upper Devonian of Spitzbergen. *Bulletin of the Geological Institution of the University of Uppsala*, **16**: 115–124.

Stensiö, E.A. (1921) *Triassic Fishes from Spitzbergen, Part 1*. xxviii + 307 pp., Vienna: Adolf Holzhausen.

Stensiö, E.A. (1922a) Über zwei Coelacanthiden aus dem Oberdevon van Wildungen. *Palaeontologischen Zeitschrift*, Berlin, **4**: 167–210.

Stensiö, E.A. (1922b) Notes on certain crossopterygians. *Proceedings of the Zoological Society of London*, **1922**: 1241–1271.

Stensiö, E.A. (1925) Triassic fishes from Spitzbergen. Pt. II. *Kungliga Svenska Vetenskapsakademiens Handlingar*, Stockholm, Series 3, **2**: 1–261.

Stensiö, E.A. (1932) Triassic fishes from East Greenland. *Meddelelser om Grønland*, Copenhagen, **83**: 1–305.

Stensiö, E.A. (1937) On the Devonian coelacanthids of Germany with special reference to the dermal skeleton. *Kungliga Svenska Vetenskapsakademiens Handlingar*, Stockholm, Series 3, **16**: 1–56.

Stensiö, E.A. (1947) The sensory lines and dermal bones of the cheek in fishes and amphibians. *Kungliga Svenska Vetenskapsakademiens Handlingar*, Stockhom, Series 3, **22**: 1–70.

Stensiö, E.A. (1959) On the pectoral fin and shoulder girdle of the arthrodires. *Kungliga Svenska Vetenskapsakademiens Handlingar*, Stockhom, Series 4, **8**: 1–229.

Stensiö, E.A. (1963) Anatomical studies on the

arthrodiran head. Part 1. *Kungliga Svenska Vetenskapsakademiens Handlingar*, Stockholm, **9**: 1–419.

Stobbs, R.E. (1989) The coelacanth enigma. *Phoenix*, Grahamstown, **2**: 8–15.

Stobbs, R.E. and Bruton, M.C. (1991) The fishery of the Comoros, with comments on its possible impact on coelacanth survival. *Environmental Biology of Fishes*, **32**: 341–359.

Stock, D.W. and Swofford, D.W. (1991) Coelacanth's relationships. *Nature*, London, **353**: 217–218.

Stock, D.W. and Whitt, G.S. (1992) Evidence for 18S ribosomal RNA sequences that lampreys and hagfishes form a natural group. *Science*, Washington, **257**: 787–789.

Stock, D.W., Gibbons, J.K. and Whitt, G.S. (1991a) Strengths and limitations of molecular sequence comparisons for inferring the phylogeny of the major groups of fishes. *Journal of Fish Biology*, **39** (Supplement A): 225–236.

Stock, D.W., Moberg, K.D., Maxson, L.R. and Whitt, G.S. (1991b) A phylogenetic analysis of the 18S ribosomal RNA sequence of the coelacanth *Latimeria chalumnae*. *Environmental Biology of Fishes*, **32**: 99–117.

Suyehiro, Y., Uyeno, T. and Suzuki, N. (1982) Coelacanth: dissecting a living fossil. *Newton Graphic Science Magazine*, Tokyo, **2**: 82–93. [In Japanese.]

Swofford, D.L. (1993) *Phylogenetic Analysis Using Parsimony. Version 3.1*. User Manual. Champaign, IL: Ilinois Natural History Survey.

Sylva, D.P. de (1966) Mystery of the the silver coelacanth. *Sea Frontiers*, **12**: 172–175.

Tabaste, N. (1963) Étude de restes de poissons du Crétacé Saharien. *Mémoires de l'Institut Français d'Afrique Noire*, Ifan-Dakar, **68**: 437–485.

Tabaste, N. (1973) Géologie et paléontologie du gisement de Gadoufaoua (Aptien du Niger). Thèse, Université Paris.

Taverne, L. (1973) Sur la présence d'un dermosupraoccipital chez les Mastacembelidae (Téléostéens Perciformes). *Revue de Zoologie et de Botanique Africaines*, Bruxelles, **87**: 825–828.

Teller, F. (1891) Über den Schädel eines fossilen Dipnoërs *Ceratodus sturii* nov. spec. aus den Schichten der oberen Trias der Nordalpen. *Abhandlungen der Kaiserlich–Königlichen Geologischen Reichsanstalt*. Vienna, **15**(3): 1–39.

Thiollière, V. (1854) *Description des Poissons fossiles provenant des gisements coralliens du Jura dans le Bugey*. 27 pp., Paris: J.-B. Baillière.

Thiollière, V. (1858) Note sur les poissons fossiles du Bugey, et sur l'application de la méthode de Cuvier à leur classement. *Bulletin de la Société Géologique de France*, Paris, **15**: 782–793.

Thomson, K.S. (1966) Intercranial mobility in the coelacanth. *Science*, Washington, **153**: 999–1000.

Thomson, K.S. (1967) Mechanisms of intracranial kinesis in fossil rhipidistian fishes (Crossopterygii) and their relatives. *Zoological Journal of the Linnean Society of London*, **46**: 223–253.

Thomson, K.S.W. (1968) A critical review of the diphyletic theory of rhipidistian–amphibian relationships. In Ørvig, T. (ed.) *Current Problems of Lower Vertebrate Phylogeny*, pp. 285–305, Stockholm: Almqvist & Wiksell.

Thomson, K.S. (1970) Intracranial movement in the coelacanth *Latimeria chalumnae* Smith (Osteichthyes, Crossopterygii). *Postilla*, New Haven, **149**: 1–12.

Thomson, K.S. (1972) New evidence of the evolution of the paired fins of Rhipidistia and the origin of the tetrapod limb, with description of a new genus of Osteolepidae. *Postilla*, New Haven, **157**: 1–7.

Thomson, K.S. (1973) New observations on the coelacanth fish. *Copeia*, **1973**: 813–814.

Thomson, K.S. (1991) *Living Fossil. The Story of the Coelacanth*. 252 pp., London: Hutchinson Radius.

Thomson, K.S. and Campbell, K.S.W. (1971) The structure and relationships of the primitive lungfish–*Dipnorhynchus sussmilchi* (Etheridge). *Bulletin of the Peabody Museum of Natural History*, **38**: 1–109.

Thys van den Audernaerde, D.F.E. (1984) Le coelacanthe des Comores, *Latimeria chalumnae*, curiosité zoologique, fossile vivant ou animal aberrant? *Africa-Tervuren*, **30**: 90–103.

Tima, V. (1986) Revision of *Macropoma speciosum* Reuss, 1857. (Crossopterygii, Coelacanthiformes). *Věstník Ústrdniho Ustavu Geologického*, Praha, **61**: 209–216.

Traquair, R.H. (1881) Notice of new fish remains from the Blackband Ironstone of Borough Lee, near Edinburgh. *Geological Magazine*, London, **1881**: 491–494.

Traquair, R.H. (1901) Notes on the Lower Carboniferous fishes of eastern Fifeshire. *Geological Magazine*, London, **4**: 110–114.

Traquair, R.H. (1903) On the distribution of fossil fish remains in the Carbonifereous rocks of the Edinburgh district. *Transactions of the Royal Society of Edinburgh*, **40**: 687–707.

Traquair, R.H. (1905) Notes on the Lower Carboniferous fishes of Eastern Fifeshire. *Proceedings of the Royal Physical Society of Edinburgh,* **16**: 80–86, 1 pl.

Travers, R.A. (1984) A review of the Mastacembeloidei, a suborder of synbranchiform teleost fishes. Part 1: Anatomical descriptions. *Bulletin of the British Museum (Natural History) (Zoology),* London. **46**: 1–133.

Uyeno, T. (1984) Age estimation of the coelacanth by scale and otolith. *Proceedings of the First Symposium on Coelacanth Studies,* Tokyo, **1**: 28–29.

Uyeno, T. (1991) Observations on locomotion and feeding of released coelacanths, *Latimeria chalumnae. Environmental Biology of Fishes,* **32**: 267–273.

Uyeno, T. and Tsutsumi, T. 1991. Stomach contents of *Latimeria chalumnae* and its feeding habits. *Environmental Biology of Fishes,* **32**: 275–280.

Vetter, B. (1881) Die Fische aus dem Lithographischen Schiefer in Dresdener Museum. *Mitteilungen aus dem Königlichen Mineralogisch-Geologischen Museum und der Prähistorischen Sammlung zu Dresden,* Cassel, **4**: 1–118.

Vorobjeva, E.I. (1977) Morfologiya i osobennosti wolyutsii kisteperykh ryb. *Trudy Paleontologicheskogo Instituta, Akademiya Nauk SSSR,* Moscow, **163**: 1–239.

Vorobjeva, E.I. and Obruchev, D.V. (1967) *Subclass Sarcoptrygii,* pp. 480–498. In Obruchev, D.V. (ed.). *Fundamentals of Palaeontology,* **11**, Jerusalem: Israel Program for Scientific Translations.

Vorobjeva, E.I. and Schultze, H.-P. (1991) Description and systematics of panderichthyid fishes with comments on their relationships to tetrapods. In Schultze, H.-P. and Trueb, L. (eds), *Origins of the Higher Groups of Tetrapods,* pp. 68–109, Ithaca, NY: Comstock Publishing Associates.

Wagner, A. (1863) Monographie der fossilen Fische aus den lithographischen Schiefern Bayerns. *Abhandlungen der Bayerischen Akademie du Wisenschaften. Mathematisch-Naturwissenschaftliche Abteilung,* Munchen 9: 611–748.

Wahlert, G. von (1968) Latimeria *und die Geschichte der Wirbesthiere. Eine evolutionsbiologische Untersuchung.* 125 pp. Stuttgart: Gustav Fischer Verlag.

Wang, N. and Liu, H. (1981) Coelacanth fishes from the marine Permian of Zhejiang, South China. *Vertebrata Palasiatica,* Peking 19: 305–312, 4 figs. [In Chinese with English summary.]

Ward, J. (1890) The geological features of the North Staffordshire Coal-Fields, their organic remains, their range and distribution; with a catalogue of the fossils of the Carboniferous sysytem of North Staffordshire. *Transactions of the North Staffordshire Institute of Mining and Mechanical Engineers,* Newcastle, **10**: 1–189.

Warthin, A.S. (1928) Fossil fishes from the Triassic of Texas. *Contributions from the Museum of Paleontology, University of Michigan,* **3**: 15–18.

Watson, D.M.S. (1913) On the primitive tetrapod limb. *Anatomica Anzeiger,* **44**: 24–27.

Watson, D.M.S. (1921) On the coelacanth fish. *Annals and Magazine of Natural History,* London, **8** (ninth series): 320–337.

Watson, D.M.S. (1927) The reproduction of the coelacanth fish *Undina. Proceedings of the Zoological Society of London,* **1927**: 453–457.

Watson, D.M.S. (1937) The acanthodian fishes. *Philosophical Transactions of the Royal Society of London,* **228B**: 49–146, 25 figs, 10 pls.

Watson, D.M.S. and Day, H. (1916) Notes on some Palaeozoic fishes. *Proceedings of the Manchester Literary and Philosophical Society,* **60**: 1–52.

Watson, D.M.S. and Gill, E.L. (1923) The structure of certain Palaeozoic Dipnoi. *Zoological Journal of the Linnean Society of London,* **35**: 163–216.

Webb, P.W. (1978) Fast-start performance and body form in seven species of teleost fish. *Journal of Experimental Biology,* **74**: 211–226.

Wehrli, H. (1931) Die Fauna der Westfälischen Strefen A und B der Bochumer Mulde zwischen Dortmund und Kamen (Westfalen). *Palaeontographica,* Stuttgart, **74**: 93–134.

Weiler, W. (1935) Ergebnisse der Forschungsneisen Prof. E. Stromers in den Wüsten Äegyptens. II. Wirbeltierreste der Baharîje-Stufe (untersetes Cenoman). 16. Neue Untersuchungen an den Fischresten. *Abhandlungen der Bayerischen Akademie der Wissenschaften, Mathematisch–naturwissenschaftliche Abteilung,* **32**: 1–57.

Weitzmann, S.H. (1962) The osteology of *Brycon meeki,* a generalized characid fish, with an osteological definition of the family. *Stanford Ichthyological Bulletin,* **8**(1): 1–77.

Wellburn, E.D. (1898) Fish fauna of the Lower Coal Measures of the Halifax and Littleborough districts. *Proceedings of the Yorkshire Geological and Polytechnic Society,* Leeds, **13**: 419–432.

Wellburn, E.D. (1901) On the fish fauna of the Yorkshire Coal Measures. *Proceedings of the Yorkshire Geological and Polytechnic Society,* Leeds, **14**: 159–174.

Wellburn, E.D. (1902a) On the fish fauna of the Pendleside Limestones. *Proceedings of the Yorkshire Geological and Polytechnic Society*, Leeds, **14**: 465–473.

Wellburn, E.D. (1902b) On the genus *Coelacanthus* as found in the Yorkshire Coal Measures, with a restoration of the fish. *Proceedings of the Yorkshire Geological and Polytechnic Society*, Leeds, **14**: 474–482.

Wellburn, E.D. (1903) On some new species of fossil fish from the Millstone Grit rocks, with an amended list of genera. *Proceedings of the Yorkshire Geological and Polytechnic Society*, Leeds, **15**: 70–78.

Wenz, S. (1975) Un nouveau coelacanthidé du Crétacé infrieur du Niger, remarques sur la fusion des os dermiques. *Colloques Internationaux du Centre National de la Recherche Scientifique*, Paris, **218**: 175–190.

Wenz, S. (1979) Découverte d'un coelacanthe dans le Jurassique supérieur de Normandie. *Bulletin Trimestriel de la Société Géologique de Normandie et des Amis Muséum du l'Havre*, **66**: 91–92.

Wenz, S. (198) Un coelacanthe géant, *Mawsonia lavocati* Tabaste, de l'Albien-base du Cénomanien du Sud Marocain. *Annales de Paléontologie, Vertebres*, Paris, **67**: 1–17.

Westoll, T.S. (1936) On the structures of the dermal ethmoid shield of *Osteolepis*. *Geological Magazine*, London, **78**: 157–171.

Westoll, T.S. (1938) Ancestry of the tetrapods. *Nature*, London, **141**: 127–128.

Westoll, T.S. (1939) On *Spermatodus pustulosus* Cope, a coelacanth from the Permian of Texas. *American Museum Novitates*, **1017**: 1–23.

Westoll, T.S. (1943) The origin of tetrapods. *Biological reviews*, **18**: 78–98.

Westoll, T.S. (1949) On the evolution of the dipnoi. In Jepsen, G.L., Mayr, E. and Simpson, G.G. (eds), *Genetics Paleontology and Evolution*. pp. 121–184, Princeton, NJ: Princeton University Press.

Westoll, T.S. (1958) The lateral fin fold theory and the pectoral fins of ostracoderms and early fishes. In Westoll, T.S. (ed.), *Studies on Fossil Vertebrates*, pp. 180–211, London: University of London Press.

White, E.I. (1939) One of the most amazing events in the realm of natural history in the twentieth century: the discovery of a living fish of the coelacanth group, thought to have been extinct 50 million years, off South Africa. *The Illustrated London News*, Supplement, March 1939.

White, E.I. (1954a) More about the coelacanths. *Discovery*, London, **15**: 332–335.

White, E.I. (1954b) The coelacanth fishes. *Annual Report of the Smithsonian Insitution*, Washington, **1953**: 351–360.

White, E.I. (1965) The head of *Dipterus valenciennesi* Sedgwick and Murchison. *Bulletin of the British Museum (Natural History) (Geology)*, London, **11**(1): 1–44.

White, E.I. and Moy-Thomas, J.A. (1937) The coelacanth genus *Graphiurus* Kner. *Geological Magazine*, London, **74**: 286.

Wiley, E.O. (1979) Ventral gill arch muscles and the interrelationships of gnathostomes, with a new classification of the Vertebrata. *Zoological Journal of the Linnean Society of London*, **67**: 149–180.

Willemoes-Suhm, R. (1869) Ueber *Coelacanthus* und einige verwandte Gattungen. *Palaeontographica*, Cassel, **17**: 73–89.

Williamson, W.C. (1849) On the microscopic structure of scales and dermal teeth of some ganoid and placoid fish. *Philosophical Transactions of the Royal Society of London*, **139B**: 435–475.

Winkler, T.C. (1874) Mémoire sur le *Coelacanthus harlemensis*. *Archives du Musée Teyler*, Haarlem, **3**: 101–116, 1 pl.

Winkler, T.C. (1880) Description de quelques restes de poissons fossiles des terrains Triassiques des environs de Wurzbourg. *Archives de Musée Teyler*, Haarlem, **5**: 109–149.

Wood, A. (1936) Fish remains from the North Wales coalfield. *Geological Magazine*, London, **73**: 481–488, 2 figs, 1 table.

Woodward, A.S. (1888) Synopsis of the Vertebrate fossils of the English Chalk. *Proceedings of the Geologists Association*, London, **10**: 273–338.

Woodward, A.S. (1890) Notes on name ganoid fishes from the English Lower Lias. *Annals and Magazine of Natural History*, London, Series 6, **5**: 430–436.

Woodward, A.S. (1891) *Catalogue of Fossil Fishes in the British Museum (Natural History)*. Volume 2, xliv + 567 pp., London: British Museum (Natural History).

Woodward, A.S. (1895) The fossil fishes of the Talbragar Beds (Jurassic?). *Memoirs of the Geological Survey of New South Wales, Palaeontology*, **9**: 1–31.

Woodward, A.S. (1898a) *Outlines of Vertebrate Palaeontology*. 470 pp., Cambridge: Cambridge University Press.

Woodward, A.S. (1898b) Note on a Devonian

coelacanth fish. *Geological Magazine*, London, **5**: 529–531.

Woodward, A.S. (1907) On the Cretaceous formation of Bahia (Brazil), and on vertebrate fossils collected therein. II. The vertebrate fossils. *Quarterly Journal of the Geological Society of London*, **63**: 131–139.

Woodward, A.S. (1908) On some fossil fishes discovered by Professor Ennes de Souza in the Cretaceous formation at Ilhéos (State of Bahia), Brazil. *Quarterly Journal of the Geological Society of London*, **64**: 358–362.

Woodward, A.S. (1909) *The Fossil Fishes of the English Chalk. Part 5.* pp. 153–184, London: Palaeontographical Society.

Woodward, A.S. (1910) On some Permo-Carboniferous fishes from Madagascar. *Annals and Magazine of Natural History*, London, Series 8, **5**: 1–6.

Woodward, A.S. (1912) Notes on some fish-remains from the Lower Trias of Spitzbergen. *Bulletin of the Geological Institution of the University of Uppsala*, **11**: 291–297.

Woodward, A.S. (1916) *The Fossil Fishes of the English Wealden and Purbeck Formations. Part 1.* pp. 1–48, London: Palaeontographical Society.

Woodward, A.S. (1942) Some new and little-known Upper Cretaceous fishes from Mount Lebanon. *Annals and Magazine of Natural History*, London, Series 11, **9**: 537–568.

Wourms, J.P., Grove, B.D. and Lombardi, J. (1988) The maternal–embryonic relationship in viviparous fishes. In Hoar, W.S. and Randall, D.J. (eds), *Fish Physiology*, Volume 11B, pp. 1–134, San Diego: Academic Press.

Wourms, J.P., Atz, J.W. and Stribling, M.D. (1991) Viviparity and the maternal–embryonic relationship in the coelacanth *Latimeria chalumnae*. *Environmental Biology of Fishes*, **32**: 225–248.

Yokobori, S., Hasegawa, M., Ueda, T., Okada, N., Nishikawa, K. and Watanabe, K. (1994) Relationships among coelacanths, lungfishes and tetrapods: a phylogenetic analysis based on mitochondrial cytochrome oxidase I gene sequences. *Journal of Molecular Evolution*, **38**: 602–609.

Young, G.C. (1978) A new Early Devonian petalichthyid fish from the Taemas/Wee Jasper region of New South Wales. *Alcheringa*, Sydney, **2**: 103–116.

Young, G.C., Barwick, R.E. and Campbell, K.S.W. (1989) Pelvic girdles of lungfishes (Dipnoi). In LeMaitre, R.W. (ed.), *Pathways in Geology: Essays in Honour of Edwin Sherbon Hills*. Melbourne: Blackwell Scientific, pp. 59–75.

Young, G.C., Long, J.A. and Ritchie, A. (1992) Crossopterygian fishes from the Devonian of Antarctica: systematics, relationships and biogeographic significance. *Records of the Australian Museum*, Sydney, Supplement **14**: 1–77.

Zardoya, R. and Meyer, A. (1996) Evolutionary relationships of the coelacanth, lungfishes, and tetrapods based on the 28S ribosomal RNA gene. *Proceedings of the National Academy of Sciences*, Washington, **93**: 5449–5454.

Zidek, J. (1975) Some fishes of the Wild Cow Formation (Pennsylvanian), Manzanita Mountains, New Mexico. *New Mexico Bureau of Mines and Mineral Resources*, Sorocco, *Circular*, **135**: 1–22.

Zittel, K.A. (1887) *Handbuch der Palaeontologie*. III, Vertebrata. 900 pp., Munich: Druck und Verlag van Oldenbarg.

GENERA AND SPECIES INDEX

Page numbers in **bold** indicate figures, page numbers in *italic* indicate tables.

GENERAL INDEX

Page numbers in **bold** indicate figures, page numbers in *italic* indicate tables.